MAIN GROUP
CHEMISTRY

ELLIS HORWOOD SERIES IN INORGANIC CHEMISTRY

Series Editor: J. BURGESS, Department of Chemistry, University of Leicester

Inorganic chemistry is a flourishing discipline in its own right and also plays a key role in many areas of organometallic, physical, biological, and industrial chemistry. This series is developed to reflect these various aspects of the subject from all levels of undergraduate teaching into the upper bracket of research.

MAIN GROUP CHEMISTRY

A. G. MASSEY B.Sc., Ph.D., D.Sc., FRSC, C.Chem.
Reader in Inorganic Chemistry
Loughborough University of Technology

ELLIS HORWOOD
NEW YORK   LONDON   TORONTO   SYDNEY   TOKYO   SINGAPORE

First published in 1990 by
ELLIS HORWOOD LIMITED
Market Cross House, Cooper Street,
Chichester, West Sussex, PO19 1EB, England

A division of
Simon & Schuster International Group

Printed and bound in Great Britain
by Hartnolls Ltd Bodmin Cornwall

British Library Cataloguing in Publication Data

Massey, A. G. (Alan G.)
Main group chemistry.
1. Chemical elements & chemical compounds
I. Title
540
ISBN 0–13–547084–6 (Library Edn.)
ISBN 0–13–547076–5 (Student Pbk. Edn.)

Library of Congress Cataloging-in-Publication Data

Massey, A. G.
Main group chemistry/A. G. Massey.
p. cm. — (Ellis Horwood series in inorganic chemistry)
ISBN 0–13–547084–6 (Library Edn.)
ISBN 0–13–547076–5 (Student Pbk. Edn.)
1. Chemistry, Inorganic   I. Title.   II. Series.
QD151.2.M37 1989
546—dc20   89–15274
                                              CIP

Table of Contents

Contents

Contents

8   **Contents**

Contents

Preface

It would hardly be denied that many students of natural sciences complete their time at school and the university with the hardened conviction that the study of inorganic chemistry is dull, monotonous, and a source of burdensome drudgery. This regrettable conclusion cannot be explained by any inherent lack of variety or interest in the subject, for the chemical individualism, the 'infinite variety' of chemical 'personality' among the elements, remains the most remarkable phenomenon of the universe. At the risk of maligning the class to which I myself belong, I offer a candid opinion: when a subject so vitally fascinating in itself appears laboured and uninspiring to the student it can only mean that the methods of teaching are seriously at fault.

> W. G. Palmer in the preface to his
> classic book *Experimental Inorganic
> Chemistry* (Cambridge University
> Press, 1954)

The chemical bond is a highly complex phenomenon which eludes all attempts at a simple description.

> Werner Kutzelnigg [quoted from *Angewandte Chemie* (International
> Edition in English) vol. 23 (1984) p. 292]

One hopes few readers will have been deterred by poor tuition from enjoying inorganic chemistry as Dr Palmer's students apparently were in the 1950s. Nature herself certainly appreciates the chemical diversity offered by the 'inorganic elements' and uses many of them in a wide variety of life-processes.

For some years there has been a need for a book which presents not only factual data but also adequate explanations of common phenomena in inorganic chemistry. This text attempts to cover both these aspects, although it must be admitted that certain topics, particularly those involving bonding, often defy simple discussion: for example, we still do not fully understand all the intricacies of even the most common of molecules, water.

It is a pleasure to acknowledge the patience and skill of my typists Miss Debbie Farmer and Mrs Janice Davidson; the technical expertise of Mr Alan Stevens, who transformed my meagre sketches into the crystal structure kit described in the Appendix, is evident in the excellent photographs taken by Mrs Jane Owens. I thank them most warmly for all their help.

Loughborough, January 1989   A. G. Massey

Suggested Further Reading

Ball, M. C. & Norbury, A. H. (1974) *Physical Data for Inorganic Chemists.* Longman

Cotton, F. A. & Wilkinson, G. (1988) *Advanced Inorganic Chemistry.* 5th ed. Wiley–Interscience

Douglas, B., McDaniel, D. H. & Alexander, J. J. (1983) *Concepts and Models of Inorganic Chemistry.* 2nd ed. Wiley

Ebsworth, E. A. V., Rankin, D. W. H. & Cradock, S. (1987) *Structural Methods in Inorganic Chemistry.* Blackwell Scientific Publications

Greenwood, N. N. & Earnshaw, A. (1984) *The Chemistry of the Elements.* Pergamon Press

Huheey, J. E. (1983) *Inorganic Chemistry.* 3rd ed. Harper & Row

Porterfield, W. W. (1984) *Inorganic Chemistry: A Unified Approach.* Addison-Wesley

Purcell, K. F. & Kotz, J. C. (1977) *Inorganic Chemistry.* W. B. Saunders Company

Trotman-Dickenson, A. F. (ed.). (1973) *Comprehensive Inorganic Chemistry* (five volumes). Pergamon Press

Wells, A. F. (1984) *Structural Inorganic Chemistry.* 5th ed. Oxford University Press

Introduction

Although at first glance it would appear that the study of over 100 elements presents the inorganic chemist with an overwhelmingly difficult task, nature has 'provided' the periodic table which enables us to correlate an enormous amount of information. The most stimulating way to study chemistry is to search for a set of rules which one can use in conjunction with the periodic table to predict, rather than passively learn, the properties of an unfamiliar element or compound: this is, of course, the technique employed by research chemists.

To take a simple example, we can recognize that lithium and magnesium are chemically similar owing to their diagonal relationship in the periodic table. Thus using a knowledge of the more common magnesium gained from previous elementary chemistry courses it is relatively easy to remember the ways in which lithium differs from the other alkali metals. I hope that some of the generalizations outlined in this Introduction will be of similar predictive value to the reader. It must be stressed, however, that a familiarity with electron orbitals and the other basic principles discussed in Chapter 1 is essential in order to have an intuitive grasp of inorganic chemistry.

SOLUBILITY OF SALTS IN WATER

It is extremely useful to know which simple salts are soluble in water because so much chemistry is carried out in this ubiquitous solvent: the extent of solubility usually defines the method of preparation or purification of a compound. Although there is no way of predicting or even calculating solubilities, the following generalizations can be made:

(a) All nitrates are soluble.
(b) Virtually all alkali metal salts are soluble.

(c) Ammonium salts are usually soluble.
(d) Chlorides, bromides and iodides of Group I–III metals are soluble, with the exception of those of thallium(I).
(e) The only soluble carbonates are those of the alkali metals and thallium(I).
(f) The common *insoluble* salts of Ca, Sr and Ba are the fluorides, carbonates, sulphates and phosphates.

Using this information it is possible to guess that barium sulphate, for example, can be prepared via a double decomposition (or metathetic) reaction:

$$Ba(NO_3)_2 \quad + \quad Na_2SO_4 \quad \rightarrow \quad BaSO_4\downarrow \quad + \quad 2NaNO_3$$
$$\text{(rule (a))} \qquad \text{(rule (b))} \qquad \text{(rule (f))} \qquad \text{(rule (a))}$$

The precipitated barium sulphate is collected on a sintered glass filter, washed thoroughly with water to remove soluble impurities and dried in an oven.

PREPARATION OF SALTS

There are four basic ways of preparing *soluble* salts:

(1) Metal plus acid. (Whether we need an oxidizing acid, such as concentrated sulphuric acid or nitric acid, will depend on the position of the metal in the electrochemical series.) Remember that Be and Al (diagonally related) are rendered passive by concentrated nitric acid and do not dissolve.
(2) Hydroxide plus acid.
(3) Oxide plus acid.
(4) Carbonate plus acid.

If the hydroxide or carbonate is soluble it is possible to add a stoichiometric volume (found by previous titration) to the acid; the resulting solution is then evaporated to crystallize out the salt, which must be purified by recrystallization. For insoluble oxides, hydroxides and carbonates it is normal practice to warm an excess of solid with the acid, filter and crystallize.

In a more unusual synthesis it is arranged for the waste by-products to be precipitated, so leaving the desired salt to be crystallized out after filtration of the solution:

$$LiCl + AgNO_2 \rightarrow AgCl\downarrow + LiNO_2$$

$$BaI_2 + ZnSO_4 \rightarrow BaSO_4\downarrow + ZnI_2$$

CARBONATES

Stable carbonates are formed only by those metals which possess basic oxides, typically the alkali metals and the Group II elements except beryllium. The chemistry of thallium(I) is somewhat similar to that of the alkali metals, and TlOH, being a strong base like KOH, will absorb carbon dioxide from the air to give Tl_2CO_3.

Table 1—Standard electrode potentials (electrochemical series)

Metal	Potential in volts (25°C)
Li^+–Li	−3.02
K^+–K	−2.93
Ca^{2+}–Ca	−2.87
Na^+–Na	−2.71
Mg^{2+}–Mg	−2.37
Al^{3+}–Al	−1.66
Zn^{2+}–Zn	−0.76
Fe^{2+}–Fe	−0.44
Cd^{2+}–Cd	−0.40
Tl^+–Tl	−0.34
Sn^{2+}–Sn	−0.14
Pb^{2+}–Pb	−0.13
H^+–H_2	0.00
Cu^{2+}–Cu	+0.34
Po^{2+}–Po	+0.65
Hg_2^{2+}–Hg	+0.79
Ag^+–Ag	+0.80

HYDROLYSIS OF SALTS

Metal cations are hydrated in solution, usually as hexahydrates, $M(H_2O)_6^{n+}$. The positive charge on the metal ion draws electron density away from the region of the O—H bonds, resulting in a tendency for the hydrogens to be lost reversibly as solvated protons:

$$(H_2O)_5\ M^{n+}\!\!-\!\!O\!\!\begin{array}{c} H \\ \\ H \end{array} \rightleftharpoons H_{aq}^+ + [(H_2O)_5MOH]^{(n-1)+} \rightleftharpoons \text{etc}$$

Further loss of protons from the other coordinated water molecules may also occur, especially if the pH of the solution is raised. As might be expected, the smaller the cation or the higher its charge n, the more pronounced will be the hydrolysis.

Gradual addition of base disturbs the equilibrium by reacting with the protons until eventually the metal hydroxide $M(OH)_n$ precipitates out of solution. Some of the species present in partially hydrolyzed solutions are quite complex because the OH^- group is a very good bridging ligand and multicentred anions are often present: e.g.

$$Be(OH_2)_4^{2+} \rightarrow [BeOH]_{aq}^+ \rightarrow [Be_2OH]_{aq}^{3+} \rightarrow [Be_3(OH)_3]_{aq}^{3+} \rightarrow Be(OH)_2\!\downarrow$$

$$\xrightarrow{\text{increasing pH}}$$

Conversely, the phenomenon of hydrolysis can be depressed by addition of acid to the salt solution, thus forcing the equilibria to the left, but there may then be problems caused by the acid's anion complexing with the metal ion, in which case the solution would not contain pure tetra- or hexahydrated metal ions: e.g.

$$Al_{aq}^{3+} + HF \rightarrow [FAl(OH_2)_5]^{2+} + [F_2Al(OH_2)_4]^+ + [F_3Al(OH_2)_3] + AlF_4^- + AlF_6^{3-}$$

It is common practice to assume that because the *large, univalent* perchlorate ion ClO_4^- has little affinity for cations in solution, perchloric acid may safely be used to suppress hydrolysis.

Hydrolysis is unimportant for the alkali metal or larger Group II ions but is troublesome for Be, the Group III metals, Sn and Pb.

LIGANDS AND COMPLEXES

Species with lone pairs of electrons are capable of either ion–dipole interactions with metal ions or donating their lone pairs into suitable empty orbitals on another atom to give a dative covalent bond. Such species are called *ligands* and the resulting compounds are known as *complexes*. In his wider definition of acids and bases Lewis considered ligands as the base species and their counterpart as acids: e.g.

$$\begin{array}{cccc} BF_3 & + & :NH_3 & \rightarrow & [F_3B\leftarrow:NH_3] \\ \text{Lewis acid} & & \text{Lewis base} & & \text{complex} \end{array}$$

Hydrated metal ions are thus complexes in which the negative end of the dipolar water molecule is electrostatically interacting with the positive ion:

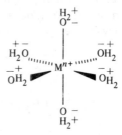

The laws of electrostatics state that a charged sphere (here a spherical ion) behaves as if the charge were located at its centre. Hence an ion–dipole interaction will be weakest when the ligand is binding to a large ion because the negative and positive charges are relatively far apart; conversely, the most stable ion–dipole complexes result from small, highly charged ions (e.g. $Na^+ < Mg^{2+} < Al^{3+}$ for an isoelectronic series and $Ca^{2+} > Sr^{2+} > Ba^{2+}$ for Group IIA).

Ligands which have more than one atom with lone pairs of electrons are called multidentate ligands (literally, many toothed) and can often wrap themselves around the Lewis acid to give complexes known as *chelates*. A multidentate ligand may also have the capacity to bind with two or more Lewis acids, in which case it acts as a *bridging* ligand. Some of the more common monodentate and multidentate ligands are shown in Table 2.

Table 2—Some typical ligands

Normally ligands contain at least one atom of a Group V, VI or VII element: *ions*—halide ions, NO_2^-, NO_3^-, SO_4^{2-}, CN^-; *neutral molecules*—NH_3, NR_3, PR_3, $RC\equiv N$, H_2O, R_3PO

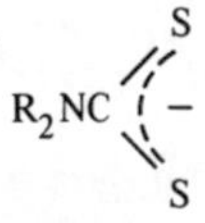

dithiocarbamates

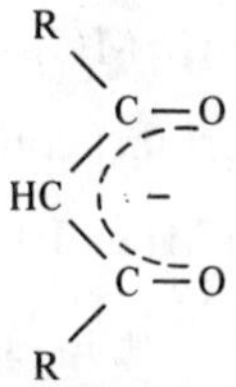

β-diketones
(R = Me: acetylacetonate)

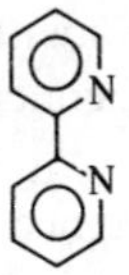

2,2'-bipyridine
(bipyridyl)

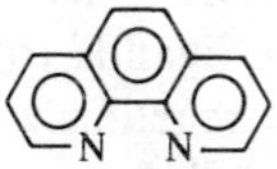

1,10-phenanthroline

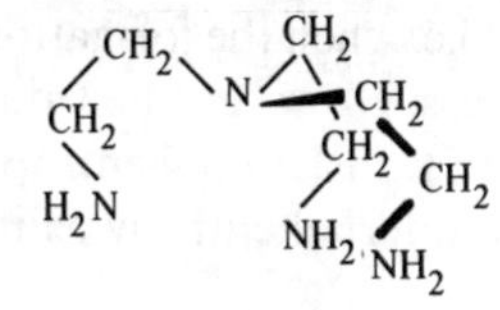

'tren'

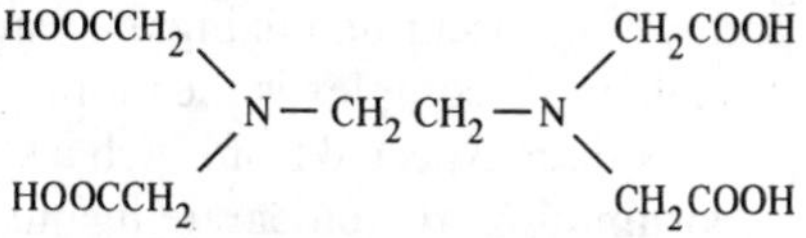

ethylenediamine tetraacetic acid
($EDTA^{4-}$: hexadentate ligand via two N and four acyl O atoms)

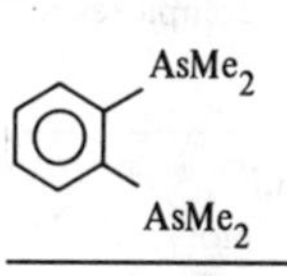

1,2-bis(dimethylarseno)benzene

THE CHELATE EFFECT

The name chelate is derived from a Greek word for a crab's claw because the metal is siezed in a vice-like grip by the ligand. Such complexes are often substantially more stable than those formed by comparable monodentate ligands, the increased stability being known as the chelate effect. A typical comparison would be between metal complexes of methylamine on the one hand and 1,2-diaminoethane on the other:

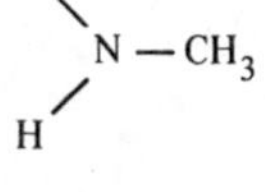

methylamine (L)

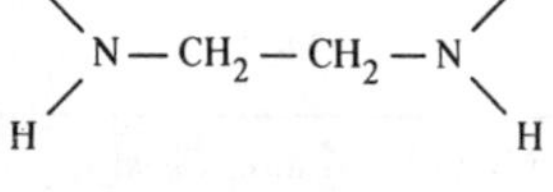

1,2-diaminoethane (L–L)

For the following reactions involving a metal cation M:

$$M + 2L \quad\rightleftharpoons ML_2 \qquad (1)$$

$$M + (L-L) \quad\rightleftharpoons M(L-L) \qquad (1')$$

$$ML_2 + (L-L) \rightleftharpoons M(L-L) + 2L \quad (2)$$

we can write their formation (equilibrium) constants β as

$$\beta_1 = \frac{[ML_2]}{[M][L]^2}, \quad \beta_{1'} = \frac{[M(L-L)]}{[M][L-L]} \quad \text{and hence} \quad K_2 = \frac{\beta_{1'}}{\beta_1}$$

or $\log K = \log \beta_{n'} - \log \beta_n$ in Table 3, with β_n and $\beta_{n'}$ being defined in the same way for the formation of ML_{2n} and $M(L-L)_n$. Since K, the equilibrium constant of a general competition reaction like equation (2), is one way of expressing the chelate effect thermodynamically, the data in Table 3 support the concept of more stable bidentate complexes. Table 3 also shows that $\Delta H^\ominus$ measured for the attachment of chemically similar monodentate and bidentate ligands to a metal ion are virtually equal and therefore $\Delta H^\ominus$ for the competition reaction will be approximately zero. Hence the equilibrium constant of the reaction, being related to the free energy change, must be mainly controlled by entropy effects because of the relationship $\Delta G^\ominus = \Delta H^\ominus - T\Delta S^\ominus$. Since there are more product molecules on the right-hand side of equation (2), an increase in entropy will occur on mixing the reactants if $K > 1$; i.e. when the formation of the bidentate ligand's complex is the more favourable process. However, the value of $\Delta S^\ominus$ will be less than expected from such a simple increase in translational entropy because, on coordination, the bidentate ligand will lose some internal entropy as its vibrational and rotational freedom becomes more restricted.

Table 3—Formation constants and thermodynamic parameters for some Cd^{2+} complexes at 25 °C[a]

Complex[b]	$\log \beta$ (mol l^{-1})^n	$\Delta H^\ominus$ (kJ mol^{-1})	$\Delta G^\ominus$ (kJ mol^{-1})	$\Delta S^\ominus$ (J mol^{-1} K^{-1})
$Cd(NH_2CH_3)_2^{2+}$	4.81	29.37	27.45	6.46
$Cd(en)^{2+}$	5.84	29.41	33.30	−13.05
Difference	1.03			−19.51
$Cd(NH_2CH_3)_4^{2+}$	6.55	57.32	37.41	66.94
$Cd(en)_2^{2+}$	10.62	56.48	60.67	−13.75
Difference	4.07			−80.69
$Cd(py)_2^{2+}$	2.2			
$Cd(bipy)^{2+}$	4.4			
Difference	2.2			
$Cd(py)_4^{2+}$	2.5			
$Cd(bipy)_2^{2+}$	7.3			
Difference	4.8			

[a] Data taken from D. Munro, *Chemistry in Britain* (1977) p. 100.
[b] en = 1,2-diaminoethane; py = pyridine; bipy = 2,2'-bipyridyl.

The chelate effect can easily be appreciated in terms of a simple physical picture: when one end of a bidentate ligand attaches to the metal ion, its other end is held in close proximity to the ion, thus creating a 'high local concentration' of the second ligand atom. Conversely, if the bidentate ligand tries to dissociate from an ion, the end to break free first cannot move very far away from its vacated site and thus has a good chance of reattachment before the second end is able to release itself from the metal. (On the other hand, a detached monodentate ligand molecule is able to diffuse rapidly away from the immediate vicinity of the ion and thus be lost in the solvent.)

In some cases adverse conformational effects may develop on formation of a chelate complex when the ligand is not flexible enough to satisfactorily encompass the cation, especially if the latter is either too small or too large; in biochemistry this may lead to specificity of a ligand for a particular cation owing to its size or preferred coordination stereochemistry. With simple polydentate ligands the formation of five- and six-membered rings normally results in the most stable complexes.

AMPHOTERISM

When a metal hydroxide dissolves in alkali, in addition to its normal reaction with acids, the hydroxide is said to be *amphoteric*. The dissolution of the hydroxide in alkali simply involves complex formation between hydroxyl ions OH^- and the metal ion to give soluble hydroxo-anions. Since small, highly charged ions form the strongest complexes with anions, amphoterism is typically shown by Be, Al, Ga and In, their hydroxo-anions $Be(OH)_4^{2-}$ and $M(OH)_4^-$ being no more unusual than, for example, BeF_4^{2-} or $AlCl_4^-$.

Oxo-anions, such as $[O-Al-O]^-$, result from amphoteric reactions carried out in the absence of water:

$$Al_2O_3 + SiO_2 \xrightarrow{\text{fuse}} Al_2(SiO_3)_3$$
$$\text{base} \quad \text{acid} \qquad\qquad \text{salt}$$

$$Al_2O_3 + CaO \xrightarrow{\text{fuse}} Ca(AlO_2)_2$$
$$\text{acid} \quad \text{base} \qquad\qquad \text{salt}$$

ISOELECTRONIC SPECIES

Species which have identical ligands and the same number of electrons on the central atom are said to be *isoelectronic* and almost invariably they have the same molecular structure. This useful fact helps in deducing the structure of molecules or complex ions which have not been encountered previously. For example, if one electron is added to boron or one electron removed from nitrogen then the resulting ions B^- and N^+ have the same number of electrons as C, making BF_4^-, CF_4 and NF_4^+ isoelectronic; four-coordinate carbon compounds such as CF_4 are well known to possess a tetrahedral conformation about the carbon atom, hence the two complex ions BF_4^- and NF_4^+ can confidently be predicted to adopt a tetrahedral conformation also.

More complicated examples might be the isoelectronic trio TeF_5^-, IF_5 and XeF_5^+ in which Te^-, I and Xe^+ have the same number of electrons. Some years ago the structure of IF_5 was shown by X-ray crystallography to be square pyramidal:

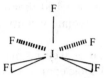

and hence it was predicted (and later verified) that the two ions had the same structure.

This principle can often be extended to include species which are not strictly isoelectronic but in which the central atoms have the same number of *outer* electrons rather than the same *total* number. The $BeCl_4^{2-}$ and BCl_4^- ions are pseudo-isoelectronic with tetrahedral $SiCl_4$ and can be expected to have the same structure. In the same way, since I and Cl are both halogens, BI_4^- and CCl_4 are pseudo-isoelectronic and hence isostructural.

Although the isoelectronic principle is an excellent aid for the memorizing of structures, one can also use it to predict new compounds (*and* their structures). Sulphur hexafluoride is a well-known and extremely stable compound. Isoelectronic with octahedral SF_6 are the ions AlF_6^{3-}, SiF_6^{2-}, PF_6^- and ClF_6^+, of which the latter looks highly unlikely. An obvious way to attempt the preparation of the anions would be to start with AlF_3, SiF_4 or PF_5 and treat them with a source of fluoride ions (F^-, having several lone pairs, is a good Lewis base):

$$Al_{aq}^{3+} + HF_{aq} \rightarrow AlF_6^{3-}$$

$$SiF_4 \text{ (or } SiO_2) + HF_{aq} \rightarrow SiF_6^{2-}$$

$$PF_5 + HF_{aq} \rightarrow PF_6^-$$

all of which are well-established reactions. The highest known fluoride of chlorine is ClF_5 and hence a very vigorous oxidizing agent, also capable of providing another fluorine ligand, is required for the synthesis of ClF_6^+:

$$KrF^+AsF_6^- + ClF_5 \rightarrow ClF_6^+AsF_6^- + Kr$$

The identity and structure of the ClF_6^+ ion originally made in this way were confirmed by X-ray crystallography; the demonstrated formation of this hexafluorochloronium cation naturally leads to the confident (and correct) expectation that the cations BrF_6^+ and IF_6^+ should also exist in stable salts.

Care must be taken not to extend such predictions too blindly. The halogens increase in size from F to I and there may be insufficient room to pack six of the heavier halogens around the central atom; also M—halogen bond energies decrease quite sharply as the atomic numbers of M and halogen increase, and hence high coordination numbers and high oxidation states of M may not be thermodynamically stable. Thus although salts containing PCl_6^- ions can be isolated, the isoelectronic SCl_6 is apparently too unstable to exist. It is possible that the unknown anion $SiCl_6^{2-}$

will ultimately be prepared because silicon, being larger than phosphorus, could easily accommodate six chlorine atoms; of course, we need to recognize that the effective positive charge attracting the chlorine ligands is in the order $P^V > Si^{IV} > Al^{III}$, which will thus be the order of stability of the hexachloro-anions.

Not unexpectedly, isoelectronic compounds have very similar bonding characteristics, best illustrated by the diatomic species N_2, CO, CN^-, C_2^{2-} and NO^+. Since these molecules and ions have the same number of electrons, a molecular orbital diagram similar to that shown on page 269 for N_2 adequately describes their bonding and demonstrates the presence of a triple bond in each case:

$$C\equiv O \qquad \text{carbon monoxide}$$

$$C\equiv N^- \qquad \text{cyanide ion}$$

$$C\equiv C^{2-} \qquad \text{carbide ion}$$

$$N\equiv O^+ \qquad \text{nitrosyl ion}$$

STABILIZATION OF LARGE IONS

Large anions are often found to be unstable in the solid state when accompanied by a small cation because the system would be thermodynamically more stable if the cation were to form a crystal with a smaller anion (owing to the increased lattice energy). This incipient decomposition is favoured by the presence of a small cation because of the relatively large difference in the lattice energies of the original salt and its decomposition product. Furthermore, a small cation will be strongly polarizing (i.e. will tend to deform the anion) which may *kinetically* assist the (*thermodynamically favoured*) decomposition. For this reason a large anion is often isolated in the presence of large cations such as Cs^+, NR_4^+ and PR_4^+. The relatively easy decomposition of lithium carbonate, nitrate, peroxide, and superoxide compared to the same derivatives of the larger alkali metals is a typical illustration of this phenomenon.

UNUSUAL PROPERTIES OF HEAD ELEMENTS IN GROUPS I–VII

Although the periodic table is extremely useful in correlating the chemistry of the non-transition elements into distinct groups, it is found that the first element in each group displays some anomalous properties relative to the later members. One manifestation of this is that the maximum coordination number for the head element is four while the maximum for the remainder of the group can be five, six or even seven (as in IF_7). Possibly this is due to the small size of the elements in the Li to Ne period—there may simply not be enough room round the atoms to accommodate more than four large ligand atoms.

It is useful to remember that the head elements of Groups I and II tend to show many properties similar to the elements diagonally to the right in the periodic table. Thus the pairs Li–Mg and Be–Al have several common aspects in their chemistry which result in Li and Be being somewhat atypical of their respective groups; comparisons within these pairs are given briefly on pages 136 and 157.

Another major difference between C, N and O and the other members of Groups IV–VI is that these three elements form particularly strong π bonds with themselves and with each other. This gives rise to allotropes (e.g. graphite, O_2, O_3) and compounds (e.g. alkenes, alkynes, aromatics, CO_2, NO_3^-, CN^-) denied to the heavier elements.

UNUSUAL PROPERTIES OF THE HEAVIER ELEMENTS OF GROUPS III–VI

There is an increasing tendency down Groups III–VI for the elements to exhibit stable oxidation states two units lower than the group oxidation states of $3 \rightarrow 6$. This is partly due to the progressive decrease in M—X covalent bond energies down these groups which thus become less able to compensate for the still relatively high promotion energies of the atoms from the ground state to the valence state; the promotion energy is obviously lowered if the outer ns^2 electrons are not involved. This preference for lower oxidation states, sometimes called the 'inert pair effect', is discussed further in Chapters 6 and 7.

Unlike the head elements, their congenors (i.e. members of the same group) have outer d orbitals available for bonding purposes. Although d orbitals are not extensively used in σ bonding, they appear to play an important role in d_π–p_π and d_π–d_π bonding (see page 62).

FORMATION OF HALIDES

Except for N, He, Ne and Ar all the main group *elements* will react *directly* with one or more of the halogens (although in some cases this may not be the best method for the preparation of a particular halide). A particular advantage of this simple synthesis is that the products are anhydrous; hydrolysis can sometimes be a problem when the formation of anhydrous metal halides is attempted by heating the corresponding hydrates. Unfortunately, for those elements with various oxidation states it may be difficult to remember which halide is formed; two generalizations may be of help:

(a) With few exceptions (e.g. N, Cl, Br, Xe) an excess of fluorine will oxidize an element to its highest known oxidation state. The three main reasons for this are (1) the low dissociation energy of F_2, (2) the exceptionally strong bonds fluorine forms with other elements and (3) the small atomic size of fluorine which allows high coordination numbers to be achieved if required. The tetrafluorides of S, Se and Te, rather than the hexafluorides, can be obtained by carrying out fluorination below $0°C$.

(b) Chlorine, bromine and iodine, in excess, give the maximum oxidation state with elements in Groups I–IV, with the exceptions that iodine will not oxidize thallium or lead to TlI_3 or PbI_4 respectively.

For the metallic elements another useful general synthesis of their anhydrous halides is to heat the metal with dry HF, HCl or HBr (HI is thermally unstable). Since these are non-oxidizing reagents, the dihalides of Sn and Pb are formed under these conditions. For those metals whose halides suffer ready hydrolysis the anhydrous halides can conveniently be made by refluxing the hydrated salts in either thionyl chloride (chlorides only) or 2,2-dimethoxypropane, when the solvents react readily with the coordinated water:

$$MCl_x.6H_2O + 6SOCl_2 \rightarrow MCl_x + 12HCl\uparrow + 6SO_2\uparrow$$

$$MX_x.6H_2O + 6CH_3C(OCH_3)_2CH_3 \rightarrow MX_x + 6(CH_3)_2C{=}O + 12CH_3OH$$

In the latter system the acetone, methanol and excess solvent are best removed by heating the anhydrous product gently under vacuum.

Halogen exchange reactions can be employed to good effect among the covalently bonded halides of the non-metallic elements. Since no valency change normally occurs during the exchange processes, they are particularly useful for making fluorides in lower oxidation states:

$$PCl_3 + ZnF_2 \text{ (or } CaF_2) \rightarrow PF_3$$

$$SCl_2 + NaF \rightarrow [SF_2] \xrightarrow{\ SCl_2\ } SF_4 + S_2Cl_2$$

$$SiCl_4 + HF \text{ (anhydrous)} \rightarrow SiF_4$$

$$PCl_5 + KI \xrightarrow[\text{solvent}]{\text{organic}} PI_5$$

$$MF_x + AlX_3 \rightarrow MX_x + AlF_3 \quad (M = \text{non-metal}; X = \text{Cl, Br or I})$$

SYNTHESIS OF ORGANO-ELEMENT DERIVATIVES

The simple method of 'direct synthesis', in which an alkyl or aryl halide is treated with a free main group element, is of wide applicability. The reaction conditions which have to be employed depend largely on the chosen element; for example, lithium and magnesium react smoothly with many organic halides at room temperature:

$$2Li + RX \xrightarrow[\text{hydrocarbon}]{\text{dry } N_2} LiR + LiX \quad (X = \text{Cl or Br}; R = \text{alkyl or aryl})$$

$$Mg + RX \xrightarrow[\text{ether}]{\text{dry } N_2} RMgX \quad (X = \text{Cl, Br or I}; R = \text{alkyl or aryl})$$

For the synthesis to work with less reactive elements a 'halogen getter' (e.g. sodium or copper) and/or elevated temperatures may be required; if an organic dihalide is used, heterocycles often result:

$$Si + CH_3Cl \xrightarrow[\text{heat}]{Cu} (CH_3)_2SiCl_2 + CH_3SiCl_3 + (CH_3)_3SiCl$$

The most widely used synthetic method involves the reaction of preformed LiR or RMgX with *covalent* halides of many main group elements carried out in ether solvents under dry nitrogen:

$$HgCl_2 + C_6H_5MgBr \rightarrow Hg(C_6H_5)_2$$

$$BF_3 + 3LiCH_3 \rightarrow B(CH_3)_3 \xrightarrow{LiCH_3} Li^+[B(CH_3)_4]^-$$

$$GeCl_4 + nLiR \rightarrow R_nGeCl_{4-n} \quad (R = \text{alkyl or aryl}; n = 1-4)$$

$$PCl_3 + CH_3MgI \rightarrow P(CH_3)_3 \xrightarrow{CH_3I} [P(CH_3)_4]^+I^-$$

$$SbCl_5 + 5LiCH_3 \rightarrow Sb(CH_3)_5 \xrightarrow{LiCH_3} Li^+[Sb(CH_3)_6]^-$$

Complications arise owing to the occurrence of Wurtz coupling when the heavier alkali metals are reacted with alkyl or aryl halides; their organo-derivatives are therefore usually made by treating diorganomercurials with the free metals in hydrocarbon solvents under a dry nitrogen atmosphere. Many other main group elements also react readily on heating with diorganomercurials to give a variety of products, including heterocycles:

$$HgR_2 + K \rightarrow 2KR + Hg \quad (R = \text{alkyl or aryl})$$

$$Hg(CH_3)_2 + Be \rightarrow Be(CH_3)_2$$

M= S:Se:Te.

NOMENCLATURE

Care should be taken not to confuse oxidation state and coordination number. The *oxidation state* of an element in a compound is given a Roman numeral and may be calculated as the theoretical charge it would carry if the electrons in *each bond* were assigned to the more electronegative atom:

$$SF_6 \rightarrow S^{6+} + 6F^-, \quad \text{i.e.} \quad S = +VI, \quad F = -I$$

In a complex ion such as a hydrate, $Mg^{2+}(H_2O)_6$, the electrons involved in the Mg—O interaction already 'belong' to the ligand water molecules and hence:

$$Mg^{2+}(H_2O)_6 \rightarrow Mg^{2+} + 6H_2O, \quad \text{i.e.} \quad Mg = +II$$

The *coordination number* of an element is defined as the number of other atoms interacting with it: hence in the examples used above both the coordination number and oxidation state of sulphur are six, whereas the coordination number of magnesium in the hexahydrate is six but its oxidation state is only two.

Although the vast array of known inorganic compounds makes systematic nomenclature difficult, particularly for the transition metals, the rules given below will help in naming most of the relatively simple main group derivatives discussed in this book.

Monatomic cations have the same name as the element, with the oxidation state, if ambiguous, added as a Roman numeral in brackets (without a space): e.g.

TlCl   thallium(I) chloride

Monatomic anions have -ide added to the name of the element, as in sodium chloride and lithium hydride.

Polyatomic cations of the main group elements have names based on the central atom and ending in -onium (except for the special case of NR_4^+ which is an ammonium ion): e.g. boronium BR_2^+, siliconium SiR_3^+, phosphonium PR_4^+, arsonium AsR_4^+, oxonium OR_3^+ and iodonium IR_2^+ as in:

$$PH_4^+ \quad \text{phosphonium}, \qquad OH_3^+ \quad \text{oxonium}$$

The names of *common polyatomic anions* end in -ide when free but -o when present as ligands in complexes:

OH^- hydroxide (hydroxo), O^{2-} oxide (oxo)

CN^- cyanide (cyano), I_3^- triiodide

NH_2^- amide (amido), S^{2-} sulphide

but more usually the names of such anions end in -ate (or -ato in complexes), although -ite can be used in a few specified cases to show a lower oxidation state of the central atom:

SO_4^{2-} sulphate (sulphato), ClO_3^- chlorate

SO_3^{2-} sulphite (sulphito), ClO_2^- chlorite

To number the *ligands* on the central atom of either a polyatomic cation or anion the prefixes mono, di, tri, tetra, penta, hexa, hepta, octa, ennea and deca are used:

Na_2SiF_6 sodium hexafluorosilicate(IV)

$TlBF_4$ thallium(I) tetrafluoroborate(III)

PCl_4PCl_6 tetrachlorophosphonium hexachlorophosphate(V)

For the nomenclature of mononuclear *oxo-acids* and their salts the following table proves very useful:

per-ic acid   → per-ate
     $\uparrow + [O]$
-ic acid   → -ate
     $\downarrow - [O]$
-ous acid   → -ite
     $\downarrow - [O]$
hypo-ous acid   → hypo-ite

and can be illustrated by perchloric, chloric, chlorous and hypochlorous acids which give rise to perchlorate, chlorate, chlorite and hypochlorite anions.

When an element can form a series of acids in which it has the same oxidation state, it is usual to call the mononuclear acid containing the maximum number of OH groups the ortho-acid:

$B(OH)_3$ $Si(OH)_4$ $O{=}P(OH)_3$
ortho-boric acid   ortho-silicic acid   ortho-phosphoric acid

Condensation via the loss of a molecule of water from two ortho-acid molecules gives the pyro-acid containing *one* M—O—M bridge:

$$2O{=}P(OH)_3 \xrightarrow{-H_2O} (HO)_2\underset{\overset{\|}{O}}{P}-O-\underset{\overset{\|}{O}}{P}(OH)_2$$

ortho-phosphoric acid

pyro-phosphoric acid
(or diphosphoric acid)

Further (theoretical) dehydration results in the formation of polymeric meta-acids which can have cyclic or linear structures:

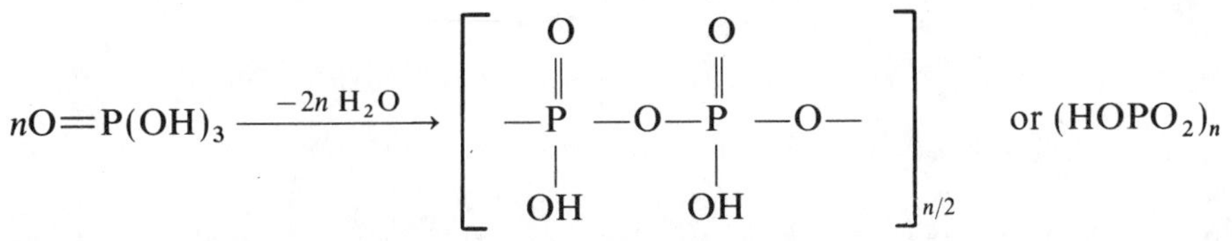

$$nO{=}P(OH)_3 \xrightarrow{\;-2n\,H_2O\;} \left[\begin{array}{cc} O & O \\ \parallel & \parallel \\ -P & -O-P & -O- \\ | & | \\ OH & OH \end{array} \right]_{n/2} \quad \text{or } (HOPO_2)_n$$

In many cases, especially among the silicates, the free acids are unknown even though their salts are often very stable.

In octahedral complexes containing two different ligands there are two possible isomers for the stoichiometries MA_4B_2 and MA_3B_3:

Owing to the high symmetry of an octahedron, these two complexes are identical and differ only in orientation. This is known as a *trans* complex.

These two identical *cis* complexes differ only in orientation.

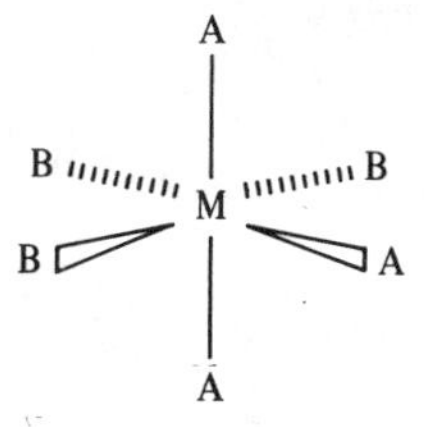

The three A and three B ligands lie on the 'meridian' of the octahedron and the compound is known as a *mer*-complex.

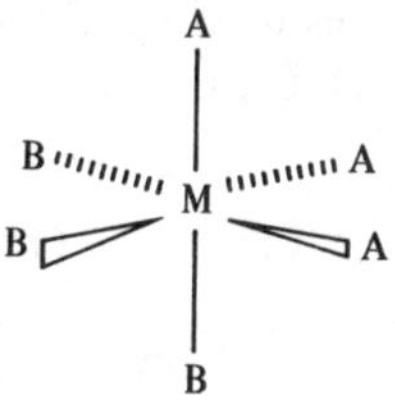

The three A and three B ligands lie on a (triangular) 'face' of the octahedron to give rise to a *fac*-complex.

STRUCTURAL POSSIBILITIES AMONG THE POLYMERIC OXO-ACIDS OF GROUPS IV–VII

The occurrence of far more polysilicates than polyanions of $P(V)$, $S(VI)$ and $Cl(VII)$ arises because the oxidation state of the central atom determines how many $-O^-$ groups from the 'parent' MO_4^{n-} ortho-anion can be changed into isoelectronic $-O-$ bridges. The perchlorate ion ClO_4^- cannot polymerize because uncharged dichlorine heptoxide results on oxygen bridging:

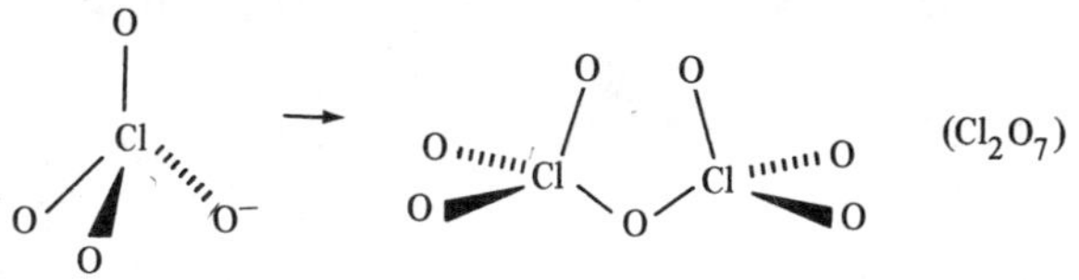

Similarly only a few polyanions can be derived from SO_4^{2-} because two terminal SO_3^- groups are always required to retain some anionic charge. This means that not all the sulphur atoms in a polysulphate ion can share two oxygens: for example, attempts to make a cyclic 'trisulphate' anion give only the γ form of sulphur trioxide:

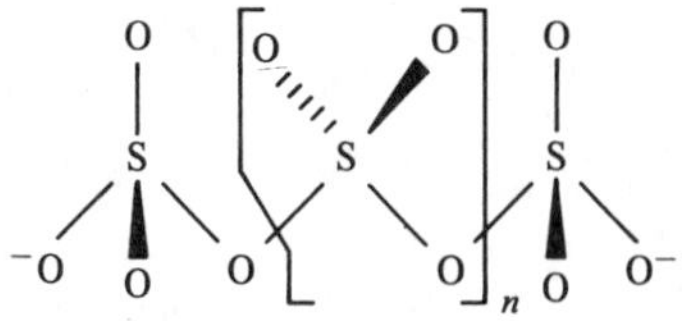

$n = 0, 1, 2, 3$ known 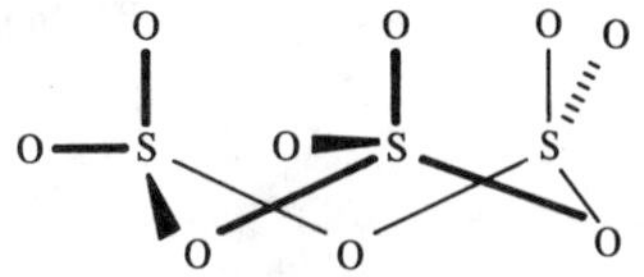 γ form of sulphur trioxide

However, many linear and cyclic polyanions can arise from PO_4^{3-} because two oxygens on each phosphorus atom can be shared:

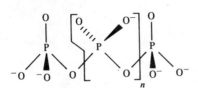

infinite polymers known;
smaller species with $n = 1$–15
isolated in a pure state

cyclic metaphosphates
isolated for $n = 3$–10

and in the so-called ultraphosphates some of the phosphorus atoms, but obviously not all, can even share three oxygens. The possibility of sharing one, two or three bridging oxygens whilst still retaining charged structures results in the immense structural variety found among the silicates. As the reader will have anticipated, it is also possible to have 'mixed' polyanions such as alumino-silicates, phosphate-sulphates and phosphate-arsenates.

Chapter 1

The Periodic Table

SCHRÖDINGER'S WAVE EQUATION AND QUANTUM NUMBERS

Classical laws which adequately describe movement of macroscopic bodies break down at the atomic level as a consequence of Heisenberg's uncertainty principle: it is fundamentally impossible to know precisely the values of *both* the energy and position of an electron in an atom. Hence, since the electron's energy can be measured very accurately using spectroscopy, we can only have a statistical, or probability, knowledge of its position. For submicroscopic species the wave characteristics described by de Broglie become important and a mathematical science developed by Schrödinger, called wave mechanics, must be used to describe their motion.

Although solutions can be found to wave mechanical equations describing one-electron systems such as H, He^+ and Li^{2+}, it has proved impossible to solve such equations for multi-electron atoms. The reader should realize that the orbitals described later in this chapter and used throughout chemistry are those calculated for the hydrogen atom: they are *assumed* to remain unchanged for all other elements in the periodic table.

When writing the wave equation for the hydrogen atom, Schrödinger considered the electron as a three-dimensional standing wave and connected the wavefunction ψ to the total electronic energy E by the expression

$$\frac{1}{r^2}\frac{\partial}{\partial r}\left(r^2\frac{\partial \psi}{\partial r}\right) + \frac{1}{r^2 \sin\theta}\frac{\partial}{\partial \theta}\left(\sin\theta\frac{\partial \psi}{\partial \theta}\right) + \frac{1}{r^2 \sin^2\theta}\frac{\partial^2 \psi}{\partial \phi^2} + \frac{8\pi^2 \mu}{h^2}\left(E + \frac{e^2}{r}\right)\psi = 0$$

where μ is the reduced mass of the electron and its polar coordinates are as shown in Fig. 1. Fortunately chemists do not need to understand this equation in order to use the results of its solution for E and ψ. The wavefunction, or *orbital*, ψ has no physical significance and hence we are unable to make a meaningful model of it. However, since ψ is an amplitude function, Schrödinger, and Born, suggested that ψ^2 measures the intensity of the electron wave, and from this idea it has become common practice to describe an orbital in terms of a smeared-out 'electron cloud'

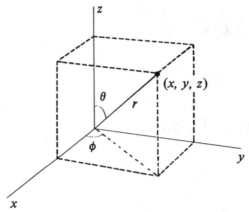

Fig. 1—The position of the electron in the hydrogen atom relative to the nucleus, in terms of Cartesian and polar coordinates. The nucleus is at the origin $(0, 0, 0)$. $z = r \cos \theta$, $y = r \sin \theta \sin \phi$. $x = r \sin \theta \cos \phi$.

having a probability, or *density*, equal to ψ^2 at any given point. It is found that the allowed values of the electronic energy E are governed by four *quantum numbers*. The principal quantum number n summarizes the predominant part of an orbital's energy whilst the secondary (or azimuthal) quantum number l and the magnetic quantum number m_l mainly describe the shape and orientation of a particular electron orbital in space; a fourth quantum number m_s describes the 'spin' state of the electron. These quantum numbers may take a range of values, *but once n has been chosen* the corresponding values of l and m_l are defined by the following relationships:

n can be any whole number from one to infinity: 1, 2, 3, ..., ∞.
l is then limited to 0, 1, 2, 3, ..., $n - 1$.
m_l, being governed by l as its subscript suggests, has the values 0, ± 1, ± 2, ..., $\pm l$.
m_s is limited to $+\frac{1}{2}$ and $-\frac{1}{2}$.

The quantum numbers* are widely used in shorthand notation to label orbitals: orbitals having $l = 0$ are called s orbitals whilst those with $l = 1$, 2 or 3 are p, d or f orbitals respectively. After $l = 3$ the letters follow in alphabetical sequence so that orbitals having $l = 4$ are g orbitals, those having $l = 5$ are h orbitals, etc. However, only s, p, d and f orbitals are of chemical significance and we may ignore the others. A numerical prefix to the orbital letter describes the principal quantum number to which the orbital belongs: a 2s orbital is one having $n = 2$ and $l = 0$, and a 3d orbital has $n = 3$, $l = 2$ for example. Although basically the periodic table is an array of the elements in order of their atomic number, it will be seen shortly that it is the Pauli exclusion principle which gives us an understanding as to why elements with similar chemical properties fall into such well-defined groups. One way of stating this principle is that no two electrons *within the same atom* may have an identical set of the four quantum numbers.

The sequence of orbital energies for the hydrogen atom is shown in Fig. 2. It will be noticed that the energies of the orbitals are negative. This implies that the energy

*The quantum numbers are, strictly, only legitimate for the single electron of a hydrogen atom. For polyelectron atoms in the rest of the periodic table they are better regarded as simply being convenient labels for the various electrons.

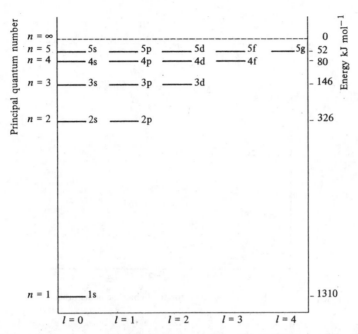

Fig. 2—Some electronic energy levels in the hydrogen atom. (Note: the energy of the nth level is equal to $-1310/n^2$ kJ mol^{-1}). As n increases the orbitals become larger and the electrons are further from the nucleus; by Coulomb's law this results in a lowering of the electron binding energy.

of the hydrogen atom, whether the atom is in the ground state ($n = 1$) or an excited state ($n \geqslant 2$), is lower than the energy of the isolated electron and the isolated nucleus ($E = 0$ kJ), a condition which is required for a stable atom.

The magnetic quantum number m_l may take values from $+l$ to $-l$, including zero. For an s orbital ($l = 0$) m_l can only be zero, but for a p orbital ($l = 1$) m_l has the values $+1$, 0, -1, which means that there are always three p orbitals for $n \geqslant 2$. In the absence of a magnetic field the three p orbitals are degenerate (i.e. they have identical energies), but when a magnetic field is applied to the hydrogen atom this degeneracy is removed. This occurs because the orientation in space of the three p orbitals is different and therefore each orbital interacts with an applied magnetic field in a different manner (Fig. 3). Similarly, d orbitals and f orbitals are fivefold ($m_l = 2, 1, 0, -1, -2$) and sevenfold ($m_l = 3, 2, 1, 0, -1, -2, -3$) degenerate respectively.

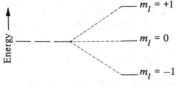

Fig. 3—The lifting of the degeneracy of p orbitals in a strong magnetic field.

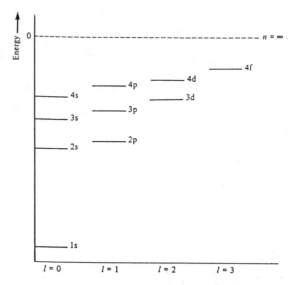

Fig. 4—The sequence of orbital energy levels for the lighter elements other than hydrogen.

From Fig. 2 it is obvious that, for the hydrogen atom, the s, p, d and f orbitals of a particular quantum shell are of equal energy—in other words, the energy of the orbitals is defined completely by the principal quantum number n. In all other atoms the energy of each orbital depends on both n and l; the effect of this on the sequence of orbital energies for the light elements is shown in Fig. 4. To understand this partial dependence of the energy on the secondary quantum number l requires a discussion of orbitals.

s ORBITALS

All hydrogen s orbitals are spherically symmetrical because their wavefunctions contain no expression involving the angular coordinates θ and ϕ of the electron, as shown in the exponential equations for 1s and 2s orbitals:

$$\psi_{1s} = \left(\frac{1}{\pi a^3}\right)^{1/2} \exp\left(-\frac{r}{a}\right)$$

$$\psi_{2s} = \frac{1}{4}\left(\frac{1}{\pi a^3}\right)^{1/2}\left(2 - \frac{r}{a}\right)\exp\left(-\frac{r}{2a}\right)$$

where the constant a is known as the Bohr radius and is equal to 0.53 Å. Various ways of representing these wavefunctions are demonstrated in Fig. 5 and show clearly that we have only a statistical knowledge of the electron's whereabouts, as demanded by Heisenberg's uncertainty principle (because the energy of each orbital can be measured very precisely).

Of particular interest to chemists when studying the periodic table is the *radial electron density function* $4\pi r^2\psi^2$, which measures the electron density within the

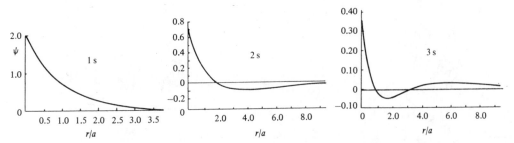

Fig. 5(a)—The wave function ψ plotted against distance from the nucleus (in units of the Bohr radius a) for 1s, 2s and 3s orbitals. Note that ψ changes sign at each node in the 2s and 3s orbitals.

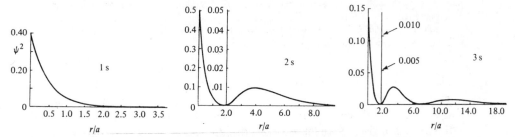

Fig. 5(b)—The variation of ψ^2 with distance from the nucleus for 1s, 2s and 3s orbitals. Note the change in units on the ψ^2 axis for the outermost parts of the 2s and 3s orbitals. Since ψ^2 represents the probability of finding an electron at a particular point in an orbital, there is obviously a high electron density at, and close to, the nucleus for s orbitals; this contrasts sharply with all other orbital types (including p, d and f orbitals) in which ψ^2 is zero at the nucleus; see Fig. 8. [Figs 5(a) and 5(b) reproduced with permission from *Atoms and Molecules* by M. Karplus and R. N. Porter. Copyright 1970 Benjamin/Cummings Publishing Co.]

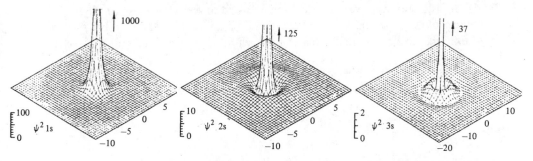

Fig. 5(c)—Computer simulations of the graphs shown in (b) rotated through 360° about the ψ^2 axis. [Reproduced with permission from *Journal of Chemical Education* vol. 47 (1970) p. 672.]

volume element $4\pi r^2\, dr$ between two spheres, one of radius r and the other of infinitesimally larger radius $r + dr$. Unlike ψ^2, which is the electron density per unit volume, this function is the probability of finding the electron at distance r Å from the nucleus summed over all directions (i.e. the electron density on the surface of a sphere of radius r). Plots of this radial function are given for the three orbitals in Fig. 6.

The graphs are unusual in that there are two variables in the function on the y axis; for example, in the radial function for the 1s orbital one variable is decreasing

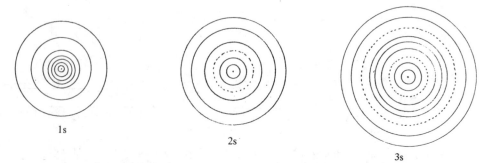

1s

2s

3s

Fig. 5(d)—Contour diagrams of the electron density 'mountains and valleys' shown in (c) demonstrate the circular (or, in three dimensions, spherical) nature of s orbitals. More usually only the outer contour is drawn as a circle (sphere) encompassing some percentage (say 95%) of the total electron density; it is obvious from (b), (c) and (d) that *the electron density within a spherical s orbital is not homogeneous.*

exponentially whilst the other increases quadratically. At $r = 0$, although ψ^2 is at a maximum, r^2 is zero and hence the radial function is also zero; at first the increase in r compensates for the decreasing electron density and the value of the radial distribution function rises. As r increases further the curve first reaches a maximum and then, as the (exponentially) decreasing ψ^2 becomes more and more dominant, it approaches zero asymptotically. Somewhat surprisingly the maximum occurs when r is 0.53 Å, which is identical to the radius calculated by Bohr for the first orbit of his planetary model of the hydrogen atom.

It should be noted from Fig. 6 that although the electron 'spends some of its time' further out from the nucleus the higher the value of n, there is still a finite chance of even a 3s electron being near to the nucleus as evidenced by the small 'humps' in the radial distribution curve. The electrons are said to *penetrate* close to the nucleus. The shapes of the ψ^2/r and $4\pi r^2\psi^2/r$ graphs are particularly important because they show that the electron density distribution within the spherical s orbitals is not homogeneous.

Wavefunctions of the spherical s orbitals having $n > 1$ are zero at certain values of r (for example, owing to the $2 - r/a$ term in the 2s wavefunction given on page 40).

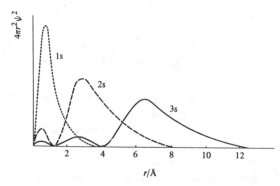

Fig. 6—Radial electron density functions for the 1s, 2s and 3s orbitals of a hydrogen atom.

This results in spherical nodal surfaces in these orbitals; at a node, besides the electron density being zero, the wavefunction ψ changes sign (Fig. 5). Although difficult to interpret in simple physical terms, the sign of ψ can be considered to describe the symmetry characteristics of the electron wave. Care should be taken to include symmetry signs on orbital diagrams because these properties of ψ are of primary importance when discussing bonding: only orbitals having wavefunctions with the same symmetry sign may 'overlap' to form a bond. In a rough analogy with light waves, two electron waves with the same ('vibrational') symmetry *reinforce* each other in the regions where they overlap, thus producing an increase in electron density in the region between the nuclei of the bonded atoms. When their symmetries are different the electron waves 'interfere', resulting in the formation of a node between the nuclei: Coulombic repulsion would then force the two nuclei apart because their positive charges are not shielded from each other by the electron cloud. Clearly this is an antibonding situation.

p and d ORBITALS

Unlike those of s orbitals, the wavefunctions for p and d orbitals depend on both r (the distance of the electron from the nucleus) and the angular coordinates of the electron, θ and ϕ, defined in Fig. 1. For example:

$$\psi_{2p_z} = \frac{1}{4}\left(\frac{1}{2\pi a^3}\right)^{1/2}\left[\frac{r}{a}\exp\left(-\frac{r}{2a}\right)\right](\cos\theta)$$

$$\psi_{3d_{xy}} = \frac{1}{81}\left(\frac{1}{2\pi a^3}\right)^{1/2}\left[\left(\frac{r}{a}\right)\exp\left(-\frac{r}{3a}\right)\right](\sin^2\theta\sin 2\phi)$$

i.e.

$$\text{wavefunction} = (\text{a constant}) \times (\text{a radial part}) \times (\text{an angular part})$$

As can be seen, it is possible to divide these wavefunctions into two parts, one being the *radial function* $R(r)$ which contains an expression described only in terms of r, and the other an *angular function* $A(\theta, \phi)$ which depends only on the direction defined by the angles θ and ϕ.

These two functions are normally interpreted in such a way that $R^2(r)$ describes the probability of finding the electron at a distance r from the nucleus but *without defining direction*, and $A^2(r)$ describes the probability of finding the electron in the direction θ, ϕ from the nucleus *regardless of the distance*.

A plot of $4\pi r^2 R^2(r)$ against r gives the variation of electron density with distance from the nucleus. From Fig. 7 it can be seen that, for a given principal quantum number, an s orbital spends more of its time close to the nucleus than does a p electron, and a p electron similarly spends more time near to the nucleus than does a d electron. Thus the penetration effect for electrons in the same quantum shell is in the order s > p > d > f.

It is the graphical plots of $A(\theta, \phi)$ or $A^2(\theta, \phi)$ which are usually shown when drawing diagrams of p and d orbitals. However, it should be realized that neither plot can represent the *whole* orbital because the radial part of the wavefunction is

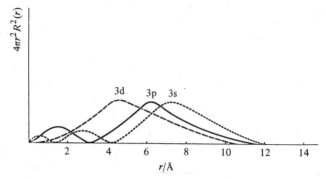

Fig. 7—Radial electron density function for 3s, 3p and 3d orbitals of the hydrogen atom. Note that there are $n-1$ nodes in an s orbital, $n-2$ nodes in a p orbital and $n-3$ nodes in a d orbital for a given principal quantum number n.

ignored; therefore, stating that the 'orbitals' as drawn in Fig. 8 contain 90%, 95% or 99% of the electron density is meaningless because *size* cannot be represented without the inclusion of the radial function (as is done in the contour diagrams of ψ^2 shown in Fig. 8). Since the orbital nodes arise only in the radial part of the wave functions, the $A(\theta, \phi)$ plots for *all* p and d orbitals are the same as those shown in Fig. 8 for 2p and 3d, irrespective of the principal quantum number.

The radial distribution functions calculated for the various orbitals of the hydrogen atom necessarily correspond to the situation in which all the orbitals except the one under study are empty. In a multi-electron atom such as sodium, many of the orbitals

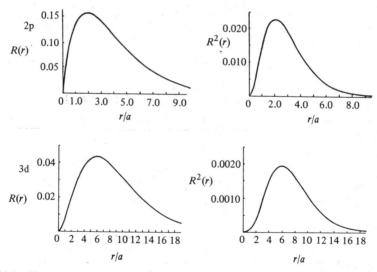

Fig. 8(a)—The variation of $R(r)$ and $R^2(r)$ with distance from the nucleus (in units of the Bohr radius) plotted for 2p and 3d orbitals. Both functions are zero at the nucleus because of the r/a and $(r/a)^2$ terms which occur in the equations for ψ given on page 43; p, d and f orbitals differ in this respect from s orbitals which, as shown in Fig. 5, have a high electron density at, and near to, the nucleus. [Reproduced with permission from *Atoms and Molecules* by M. Karplus and R. N. Porter. Copyright 1970 Benjamin/Cummings Publishing Co.]

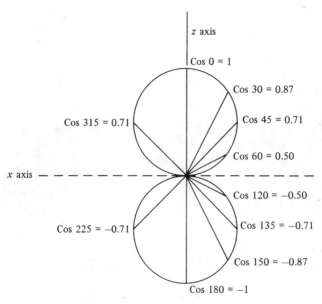

Fig. 8(b)—As stated in the text, the angular part of the wavefunction for a $2p_z$ orbital is $\cos\theta$. One way of plotting $\cos\theta$ is to draw, in the xz or yz plane, a series of lines from the origin (i.e. the nucleus) making an angle θ to the z axis and of length $\cos\theta$ (in centimetres or inches). Joining the extremities of an infinite number of such lines gives two tangential circles touching at the nucleus. When this is repeated for all possible planes containing the z axis (more easily done by rotating this diagram about the z axis), the resulting shape is two *spheres* in contact.

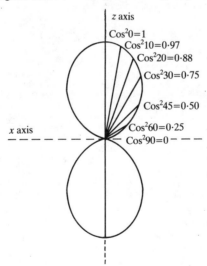

Fig. 8(c)—Some books show diagrams of 2p orbitals as 'figures-of-eight'. These represent $A^2(\theta,\phi)$, which for $2p_z$ is simply $\cos^2\theta$; this function can be drawn in the same way as was done for $A(\theta,\phi)$ but ignoring that $\cos^2 A$ is *positive* in all four quadrants.

Fig. 8(e)—Contour diagram of constant ψ^2 for 2p, 3p and 3d orbitals. It is impossible to plot in three dimensions the variation of the four variables $\psi(\theta,\phi,r)$, θ, ϕ and r. This difficulty can be overcome to a large degree by holding $\psi(\theta,\phi,r)$ constant and plotting the spatial variations of $\psi^2(\theta,\phi,r)$, which in two dimensions results in a contour map of the electron density within

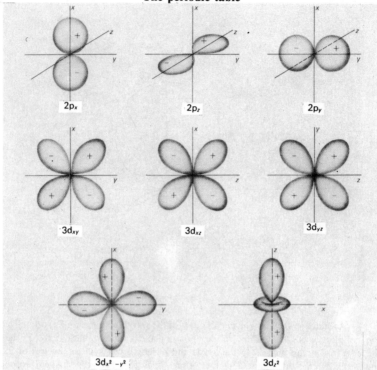

Fig. 8(d)—The angular part of the wavefunction, $A(\theta,\phi)$, for the 2p and 3d orbitals of the hydrogen atom. Note the convention for denoting the different p orbitals and d orbitals which differ in their orientation relative to the axes x, y and z.

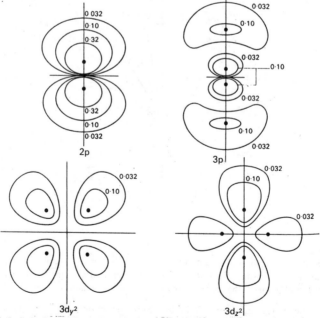

the orbital. It is to be realized that in three dimensions the contours shown above become contour surfaces. The figures represent the ratio ψ/ψ_{max}; the dots show the positions of maximum electron density in each part of the orbitals.

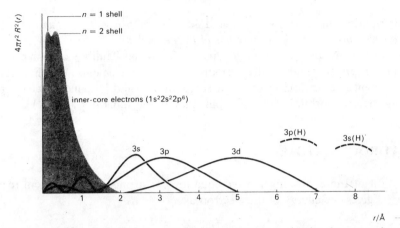

Fig. 9—Radial electron density functions for the sodium atom, showing the contraction effect of orbital penetration on the 3s, 3p and 3d orbitals. The position of the maximum for radial electron density functions of hydrogen 3s and 3p orbitals is indicated by the broken curves labelled 3s(H) and 3p(H). The hydrogen 3d orbital is similar in position to that shown for sodium.

are occupied; it is interesting to see the effect of orbital penetration on the outer 3s, 3p and 3d orbitals in such an atom (Fig. 9). Without the effect of orbital penetration it would be expected that the electron in sodium's $n = 3$ quantum shell would experience the attraction of the nuclear charge ($+11$) moderated by the ten electrons in the $n = 1$ and $n = 2$ shells, *which are between it and the nucleus* (remembering that the electrons having the highest value of the principal quantum number are farthest out from the nucleus). The ten inner-shell electrons will repel the outermost electron, which is equivalent to reducing the attractive force of the nucleus: the $n = 3$ electron is said to be *shielded* from the full nuclear charge by the inner electrons. The effective nuclear charge felt by the $n = 3$ electron might then be expected to equal $+1$ (i.e. $11-10$), in which case the radial distribution functions would appear as in Fig. 7, the 3d, 3p and 3s orbitals having their maximum 'humps' at about 4.5, 6 and 6.5 Å from the nucleus respectively. In sodium the 3d orbital certainly has its hump at 4.5 Å from the nucleus, but the 3s and 3p orbitals have their maxima very much closer than expected. This is because the 3s and 3p orbitals penetrate close in to the nucleus through the inner-shell electrons and therefore experience a nuclear charge considerably greater than $+1$; the effect is more marked for s orbitals, as mentioned above when discussing Fig. 7. Hence a 3s electron in the sodium atom would experience a higher effective charge than would a 3p electron. The 3d orbital shows no penetration effect, and an electron in such an orbital, well outside the inner shell of ten electrons, would experience an effective nuclear charge of about $+1$.

Because of this penetration effect*, a 3s electron in the sodium atom is more strongly bound than a 3p electron, which in turn is more strongly bound than a 3d electron. The overall result of penetration on the sequence of orbital energies for the first two dozen or so elements in the periodic table is shown in Fig. 4. As is obvious from the

---

* Note that although the penetration effect increases the attraction between an electron and the nucleus there will also be an increase in repulsion between this electron and the core electrons as they come closer together. It so happens that the attractive energy gained by penetration outweighs this repulsion.

radial distribution functions for 3s, 3p and 3d orbitals in Fig. 9, *the smallest orbitals are those which are the most strongly bound to the nucleus.*

Using the sequence of orbital energies given in Fig. 4 and Pauli's exclusion principle, we can now begin to build up the periodic table of the elements. We shall use the notation ↑ to mean a single electron (with, say, spin $+\frac{1}{2}$) and ↑↓ to mean two electrons occupying the same orbital with their spins paired (i.e. spins $+\frac{1}{2}$ and $-\frac{1}{2}$).

THE HYDROGEN ATOM

The solution of the Schrödinger wave equation for a one-electron system (e.g. H, He$^+$, Li^{2+}) gives the energy of the electron as

$$E = -\frac{2\pi^2 me^4 Z^2}{h^2} \times \frac{1}{n^2}$$

where e is the electronic charge, Z is the nuclear charge and n is the principal quantum number. For hydrogen, $Z = 1$ and hence

$$E = -1310 \times \frac{1}{n^2} \text{ kJ (mol H)}^{-1}$$

Fig. 10 illustrates the ground state configuration of the hydrogen atom. When two electrons are placed in the 1s orbital, as in the hydride ion H$^-$, they repel one another so strongly that the binding energy of the electrons falls from -1310 kJ to about -63 kJ. From this it can be recognized that the repulsion energy between the two 1s electrons in H$^-$ must be $1310 - 63 = 1247$ kJ. Since the electrons are much less strongly bound to the nucleus in H$^-$ than in H owing to their mutual repulsion, it follows that the hydride ion is considerably larger (radius $\simeq 1.5$ Å) than the hydrogen atom (radius 0.53 Å).

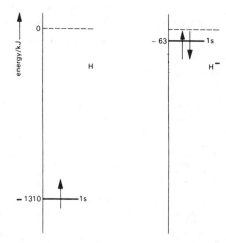

Fig. 10(a)—The electronic structures of the hydrogen atom and the gaseous hydride ion in their ground states (i.e. $n = 1$).

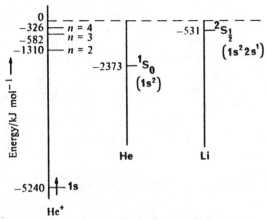

Fig. 10(b)—Energy level diagrams for He⁺, He and Li. Energy level diagrams showing *orbitals* are only strictly correct for one-electron systems like H and He⁺; in multi-electron species the *term symbol* corresponding to a given electronic configuration should be used. Thus the singlet ground state of He (symbol $^1S_0$) arises from the $1s^2$ configuration and lies 2373 kJ mol⁻¹ below the state corresponding to He⁺ and an isolated electron (i.e. 2373 kJ mol⁻¹ is the first ionization energy of helium). The energy of the helium atomic 1s orbital is presumably -5240 kJ mol⁻¹ but strong repulsion between the two 1s electrons markedly destabilizes the atom and makes it relatively easy to lose the first electron.

The 2s electron of Li is shielded (repelled) by the inner $1s^2$ pair from the full attraction of the nuclear charge; it is also further away from the nucleus, on average, than 1s electrons. Both these effects result in a low first ionization energy of 531 kJ mol⁻¹.

THE HELIUM ATOM

The binding energy of the electron in He⁺ is four times that in the hydrogen atom because of the increased nuclear charge ($Z = 2$): 1310×2^2 or 5240 kJ (mol He⁺)⁻¹. This results in a greatly reduced size of the 1s orbital in He⁺ compared to the hydrogen 1s orbital. The energy required to remove one electron from the helium atom (i.e. the first ionization energy of He) is 2373 kJ, which means that the repulsion energy between the two 1s electrons is $5240 - 2373$ or 2867 kJ mol⁻¹: an enormous amount of energy when we are used to thinking of electrons which are spin-paired in the same orbital as constituting a 'stable' system; note that the reduced orbital size increases the electron repulsion energy in He relative to H⁻. It is possible to calculate the *total* electron binding energy for the $1s^1 2s^1$ state of helium as $5240 + 1310$ or 6550 kJ mol⁻¹ *if electron repulsion effects are ignored* compared with the value $2373 + 5240$ or 7613 kJ mol⁻¹ *measured* for the $1s^2$ state. It is clearly much more favourable to pair the electrons in the 1s orbital and lose a considerable amount of repulsion energy than to place the second electron in the next available orbital. This calculation illustrates the *Aufbau* (or building-up) principle used to decide into which orbitals the available electrons are to be placed; the occupied orbitals are those which give rise to the most stable state (called the *ground state*) of the atom.

THE FIRST SHORT PERIOD: LITHIUM TO NEON

There is only one orbital in the $n = 1$ quantum shell and, from Pauli's exclusion principle, we know that a maximum of two electrons can be accommodated in it.

Hence this quantum shell is completed at helium. The next element, lithium, contains three electrons, and from Fig. 4 it is obvious that the third electron must occupy the 2s orbital since this is the most stable orbital (the one with the most negative energy) after the completed $n = 1$ shell. The electron configuration of lithium in its ground state is therefore $1s^2 2s^1$. The outer 2s electron is relatively easily lost as shown by the low ionization energy of 531 kJ mol^{-1}, which is less than half that of hydrogen. This occurs because although the nuclear charge is $+3$ for lithium, the two 1s electrons shield the 2s electron from the full attraction of the nucleus, resulting in an effective nuclear charge experienced by the 2s electron of only about $+1$; the fact that a 2s electron is further from the nucleus reduces its binding energy still more relative to one in a hydrogen 1s orbital. Lithium therefore readily forms salts containing the Li$^+$ ion.

Beryllium has the electronic structure $1s^2 2s^2$ because this is 264 kJ mol^{-1} more stable than the configuration $1s^2 2s^1 2p^1$, notwithstanding the considerable repulsion energy which arises by having two electrons in the 2s orbital. It is interesting to compare the $1s^2 2s^2 \rightarrow 1s^2 2s^1 2p^1$ promotion energy for beryllium (264 kJ) with the huge $1s^2 \rightarrow 1s^1 2s^1$ promotion energy for helium (1913 kJ) where the electron has to be promoted to another quantum shell (involving a change in the principal quantum number). Thus although helium cannot be expected to form divalent compounds, the s $\rightarrow$ p promotion for beryllium (involving only a change in l, the secondary quantum number) is relatively easy and beryllium forms many covalent compounds in which the beryllium atom has its orbitals sp, sp^2 or sp^3 hybridized. The first (900 kJ) and second (1756 kJ) ionization energies of beryllium are apparently much too high to allow any salts containing the bare Be^{2+} ion to form.

Boron has five electrons and in the ground state adopts the configuration $1s^2 2s^2 2p^1$ since the 2p orbital is lowest in energy after the 2s. At carbon, which has six electrons, there is something of a dilemma. From Fig. 4 it is obvious that the ground state must be $1s^2 2s^2 2p^2$, but this can represent the configuration of several different energy states, e.g. $1s^{\uparrow\downarrow} 2s^{\uparrow\downarrow} 2p_x^{\uparrow} 2p_y^{\uparrow}$, $1s^{\uparrow\downarrow} 2s^{\uparrow\downarrow} 2p_x^{\uparrow} 2p_y^{\downarrow}$ or $1s^{\uparrow\downarrow} 2s^{\uparrow\downarrow} 2p_x^{\uparrow\downarrow}$; guidance on deciding the ground state configuration in such cases is given by the following rules:

(a) *Electrons avoid being spin-paired in the same orbital as far as possible to reduce their Coulombic repulsion*; the extent of this repulsion varies with the size of the orbital. Two electrons confined to the small 1s orbital of Be^{2+} (isoelectronic with helium) repel each other to the extent of 6157 kJ mol^{-1}, but this falls to about 120 kJ mol^{-1} for two paired electrons in a larger carbon 2p orbital.

(b) *When unpaired electrons occupy degenerate orbitals, the energy of the system is lowest when their spins are parallel (Hund's rule of maximum multiplicity).* This is due to a quantum mechanical phenomenon, called the *exchange energy*, which has no classical counterpart in our macroscopic world. Tiny spinning particles such as electrons and neutrons tend to avoid occupying similar regions of space when their spins are parallel; for electrons this will obviously result in a lowering of electrostatic repulsion. The effect can be surprisingly large, as shown by the difference of 77 kJ mol^{-1} which exists between the $1s^{\uparrow} 2s^{\uparrow}$ and $1s^{\uparrow} 2s^{\downarrow}$ configurations of helium.

From these rules it can be deduced that a carbon atom in its ground state has the electron configuration $1s^{\uparrow\downarrow} 2s^{\uparrow\downarrow} 2p_x^{\uparrow} 2p_y^{\uparrow}$. The energy required to promote carbon to the configuration $1s^{\uparrow\downarrow}2s^{\uparrow}2p_x^{\uparrow}2p_y^{\uparrow}2p_z^{\uparrow}$ is reasonably low (403.7 kJ mol^{-1}) and results in the formation of CX_4 being energetically more favourable than the formation of CX_2; hence carbon is almost always four-covalent in its compounds (see page 208).

The electronic configurations of the remaining elements in the first short period are built up following the rules used above:

$$N \quad 1s^{\uparrow\downarrow}2s^{\uparrow\downarrow}2p_x^{\uparrow}2p_y^{\uparrow}2p_z^{\uparrow}$$

$$O \quad 1s^{\uparrow\downarrow}2s^{\uparrow\downarrow}2p_x^{\uparrow\downarrow}2p_y^{\uparrow}2p_z^{\uparrow}$$

$$F \quad 1s^{\uparrow\downarrow}2s^{\uparrow\downarrow}2p_x^{\uparrow\downarrow}2p_y^{\uparrow\downarrow}2p_z^{\uparrow}$$

$$Ne \quad 1s^{\uparrow\downarrow}2s^{\uparrow\downarrow}2p_x^{\uparrow\downarrow}2p_y^{\uparrow\downarrow}2p_z^{\uparrow\downarrow}$$

Although in theory these last four elements might increase their usual valencies via promotion of an electron to the empty 3s orbital, the energy required for this (between 800 and 1600 kJ, depending on the element) is prohibitively large and the covalency maxima remain 3, 2, 1 and 0 for nitrogen, oxygen, fluorine and neon respectively.

The apparently erratic variation in first ionization energies across the $n = 2$ quantum shell from lithium to neon (Fig. 11(b)) is explicable in terms of the electronic configurations given above for these elements. The value for beryllium is higher than that for lithium because the increase in nuclear charge more than compensates for

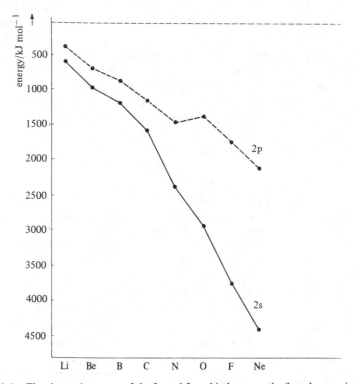

Fig. 11(a)—The change in energy of the 2s and 2p orbitals across the first short period.

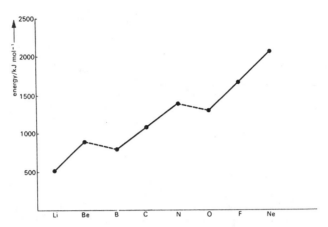

Fig. 11(b)—The variation in first ionization energies from lithium to neon.

the electronic repulsions between the two 2s electrons of beryllium. At boron, configuration $1s^2 2s^2 2p^1$, the least strongly bound electron is now in a 2p orbital, and since 2p orbitals are not stabilized by penetration effects to the same extent as 2s orbitals, the ionization energy of boron is lower than expected by extrapolation from lithium ($1s^2 2s^1$) and beryllium ($1s^2 2s^2$). The ionization energies rise smoothly between boron and nitrogen as one electron is placed successively in each of the three 2p orbitals. As shown in Fig. 8, the p orbitals are orientated at right angles to each other, and because of this p electrons do not shield each other very effectively from the Coulombic atraction of the nucleus; therefore the increase in nuclear charge from boron to nitrogen means that the 2p electrons become more and more strongly bound and this is reflected in the increasing ionization energies. The drop in ionization energy between nitrogen and oxygen occurs because from oxygen to neon the electrons in the 2p orbitals are being paired up, which results in a repulsion term not present for boron, carbon and nitrogen. The increased nuclear charge, penetration effects and shielding combine to make the first ionization energy of neon about four times larger than that of lithium, thus precluding the formation of ionic salts of neon containing the cation Ne^+. Similarly, derivatives of F^+ are highly unlikely. The above three effects fall off somewhat with increasing atomic number down a particular group; among the halogens, the ionization energy falls steadily until at iodine it is only 1009 kJ, and a few derivatives of I^+ have been isolated.

Figure 11(a) represents the change in 2s and 2p orbital energies from lithium to neon; note the similarities to Fig. 11(b). The difference between the 2s and 2p energies for a given atom gives the magnitude of the 2s orbital penetration effect, which even at lithium amounts to 175.7 kJ. As the 2s orbital becomes more strongly bound to the nucleus from lithium to neon, its size decreases and it therefore shields the 2p orbitals more effectively from the full attraction of the nucleus: hence the divergence of the 2s and 2p orbital energies, especially at oxygen, fluorine and neon, where electronic repulsions cause further destabilization of the 2p orbitals. The 2s–2p separation, being very sensitive to the increasing nuclear charge, rises to 3000 kJ at sodium and a staggering 16 000 kJ at copper.

THE SECOND SHORT PERIOD: SODIUM TO ARGON

At neon the $n = 2$ quantum shell is completed, so that the next element, sodium, has its last electron placed in the lowest-energy orbital of the $n = 3$ shell, the 3s, giving sodium the electronic configuration $1s^2 2s^2 2p^6 3s^1$. The next seven elements complete the 3s and 3p orbitals in a regular manner following the rules adopted for the $n = 2$ shell:

$$
\begin{array}{ll}
\text{Na} & 1s^2 2s^2 2p^6 3s^{\uparrow} \\
\text{Mg} & 1s^2 2s^2 2p^6 3s^{\uparrow\downarrow} \\
\text{Al} & 1s^2 2s^2 2p^6 3s^{\uparrow\downarrow} 3p_x^{\uparrow} \\
\text{Si} & 1s^2 2s^2 2p^6 3s^{\uparrow\downarrow} 3p_x^{\uparrow} 3p_y^{\uparrow} \\
\text{P} & 1s^2 2s^2 2p^6 3s^{\uparrow\downarrow} 3p_x^{\uparrow} 3p_y^{\uparrow} 3p_z^{\uparrow} \\
\text{S} & 1s^2 2s^2 2p^6 3s^{\uparrow\downarrow} 3p_x^{\uparrow\downarrow} 3p_y^{\uparrow} 3p_z^{\uparrow} \\
\text{Cl} & 1s^2 2s^2 2p^6 3s^{\uparrow\downarrow} 3p_x^{\uparrow\downarrow} 3p_y^{\uparrow\downarrow} 3p_z^{\uparrow} \\
\text{Ar} & 1s^2 2s^2 2p^6 3s^{\uparrow\downarrow} 3p_x^{\uparrow\downarrow} 3p_y^{\uparrow\downarrow} 3p_z^{\uparrow\downarrow} \\
\end{array}
$$

THE FIRST LONG PERIOD

Potassium has one more electron than argon and one might predict that, because the 4s orbital *is less stable* than the 3d set (Figure 4), this electron would be found in one of the five degenerate 3d orbitals giving potassium the configuration $[\text{Ar}]3d^1$. However, the effects of penetration and inter-electron repulsions *within the atom as a whole* are such that the ground state of potassium is that in which the 4s orbital is occupied: $1s^2 2s^2 2p^6 3s^2 3p^6 4s^1$. The ground state turns out to be about 255 kJ mol^{-1} more stable than the excited state with the $[\text{Ar}]3d^1$ configuration. A 4s and a 3d orbital are of approximately the same size at this position in the periodic table so that electrons in 4s orbitals do not shield 3d electrons very effectively from the attraction of the nucleus. At calcium, the increase in nuclear charge and the poor shielding effect of a single 4s electron make the 3d orbitals more stable than they were at potassium; however, the calcium atom is most stable when the twentieth electron is accommodated in the 4s orbital giving the electron configuration $1s^2 2s^2 2p^6 3s^2 3p^6 4s^2$.

It is only at the next element, scandium, that the 3d orbitals start to fill giving scandium the configuration $[\text{Ar}]3d^1 4s^2$. Although there is room for ten electrons in the 3d orbitals it will be noticed that two of scandium's electrons are placed in the less stable 4s orbital. This situation arises because the ground state of scandium hinges on a subtle balance of penetration, inter-electron repulsions and poor mutual 3d–4s shielding which can only be achieved *if the 4s orbital is fully occupied*. When the three outer electrons of scandium are all placed in 3d orbitals it is found that the resulting $[\text{Ar}]3d^3$ configuration represents an excited state some 400 kJ above the ground state. This illustrates an important point: the outer electrons in some atoms are not necessarily placed in orbitals having the lowest energy. The actual orbitals occupied are always those which result in the lowest energy for the atom as a whole and this depends on several factors only one of which is the orbital energy. However, such a situation can only occur when the two sets of orbitals in question (for example, the 3d and 4s orbitals) are of similar energy. On ionization of the first row transition metals the first two electrons come from the *least stable orbital* which is the 4s and hence their M^{2+} and M^{3+} ions always have $3d^n$ configurations.

The influence of an increased effective nuclear charge on the very delicate balance of factors affecting the 3d–4s occupancy can be judged from the ground state electron configurations of Sc^+ (isoelectronic with Ca) and Sc^{2+} (isoelectronic with K) which are $[Ar]3d^14s^1$ and $[Ar]3d^1$ respectively. In this region of the periodic table the difference in 4s and 3d orbital energies is approximately proportional to the square of the effective nuclear charge which, for the scandium species, is in the order $Sc^{2+} > Sc^+ > Sc$.

The nine elements after scandium are derived by filling up the five d orbitals. Slight variations from regularity occur at Cr ($3d^54s^1$) and Cu ($3d^{10}4s^1$) owing to the similarity of the 3d and 4s orbital energies which allows relatively small effects to dictate the final electron configuration. For example, at chromium where the difference between the $3d^54s^1$ and $3d^44s^2$ configurations amounts to only 90 kJ, it is slightly more favourable to have the electrons occupying orbitals with parallel spins to minimize the effect of both electronic repulsions and the exchange energy. For the second- and third-row transition metals the increased nuclear charge results in the d and s orbitals becoming even closer in energy, and it is then impossible to predict their ground states with certainty using our simple rules (see the periodic table printed on the foldout at the end of the book). These minor variations are of relatively little consequence; much more important to chemists are the metal ions M^{2+} and M^{3+} which *invariably* have the d^ns^0 outer electron configuration.

The $4s^1$ outer configuration of copper and potassium suggests that they might be chemically comparable. However, the poor mutual shielding effect of 3d and 4s electrons is sufficient to make the first ionization energy of copper nearly twice that of potassium. On the other hand, the second ionization energy of copper (where the electron is removed from a 3d orbital) is almost *half* that of potassium (where the second electron comes from a strongly bound 3p orbital. Copper is thus able to have variable valency, commonly occurring as Cu^+ and Cu^{2+} ions. Furthermore, although copper has ten more electrons than potassium, the effect of poor mutual shielding of the filled 3d orbitals in Cu^+ is to make the ion very much smaller than K^+ ($r_{Cu^+} = 0.93$ Å, $r_{K^+} = 1.33$ Å, $r_{Na^+} = 0.95$ Å). Hence there is little similarity in the chemistry of copper on the one hand and that of the alkali metals on the other.

When the 3d orbitals are filled at zinc, the next available orbitals are the three degenerate 4p orbitals, and gallium thus has the configuration $1s^22s^22p^63s^23p^63d^{10}4s^24p^1$ which places it in Group III below aluminium. The effect of poor shielding by the filled 3d orbitals on the outer 4s and 4p electrons is apparent in the ionization energies (see page 177) of gallium which are almost equal to those of aluminium, showing that the electrons in gallium are much more tightly bound than would be expected by simple extrapolation of the trends in ionization energies noted in Groups I and II (where these energies decrease smoothly down the groups with increasing atomic number of the elements). Continued filling of the 4p orbitals in the normal systematic manner gives the Group IV (germanium), V (arsenic), VI (selenium), VII (bromine) and VIII (krypton) elements, which thus completes the first long period.

THE SECOND LONG PERIOD

The sequence of orbitals in this period closely resembles that of the first long period: the 5s orbital is occupied at rubidium and strontium before the 4d orbitals begin to fill up giving rise to the ten elements of the second transition series. From In to Xe the outer electrons enter the 5p orbitals and build up the next set of main group elements.

THE HEAVY ELEMENTS

In line with all the preceding periods, the 6s orbital now begins to fill at what would have been expected to be the third long period. Thus caesium and barium have the $6s^1$ and $6s^2$ outer electron configurations respectively, and with the next element, lanthanum, the series of 5d transition elements apparently begins. However, at this point in the periodic table the combination of penetration effects, shielding effects and increasing nuclear charge significantly stabilizes the 4f orbitals. (At caesium the 4f orbital energy is virtually identical to that of the 4f orbitals in hydrogen, so efficiently are the f orbitals shielded by the inner electrons.) These effects drop the 4f orbital energy at cerium to a level below that of the 5d and 6p orbitals, and hence cerium has the outer configuration $4f^2 5s^2 5p^6 6s^2$; continued filling up of the f orbitals in a more or less straightforward fashion gives rise to the fourteen lanthanide elements. When the 4f orbitals are full, the third d transition series is completed by the filling of the 5d orbitals, which are still lower in energy than the 6p orbitals.

Therefore the next non-transition element after barium occurs twenty-five places further on in the periodic table as the 6p orbitals begin to be occupied at thallium. The ionization energies of thallium (see page 177) are higher than expected by extrapolation from either indium or strontium owing to the poor shielding properties of both the 5d and 4f electrons (which cause the outer $6s^2 6p^1$ electrons to be strongly bound to the nucleus). The combination of high ionization energies and low Tl—X bond energies (bond energies normally decrease down a group owing to the size and diffuseness of orbitals increasing with atomic number, which leads to poor orbital overlap) makes thallium reluctant to form either Tl^{3+} ions or TlX_3 covalent derivatives; more often it exists as the thallium(I) ion Tl^+ which has a $4f^{14} 5d^{10} 6s^2$ configuration. The chemistry of thallium in the monovalent state is similar to that of the heavier alkali metals (e.g. many of their salts are isomorphous), although solubilities (e.g. of the halides) closely resemble those of the corresponding silver salts. This tendency not to use the tightly bound ns^2 outer electrons is also marked with lead and bismuth, and was becoming apparent even at indium and tin in the previous period (where it was due to the poor shielding of the 4d electrons).

Predictably, after the 6p orbitals have been filled at radon the next two elements, francium and radium, have the outer configurations $7s^1$ and $7s^2$. However, the similar energies of 6d and 5f orbitals after actinium ($6d^1 7s^2$) result in several irregular electron arrangements among the remaining elements. Although the IUPAC ordering of the groups into A and B subsections is used in this book, there is a growing tendency to adopt a newly proposed system in which the periodic table is divided into eighteen groups starting with the alkali metals as Group 1 and ending with the rare gases in Group 18; Group IIIB (B, Al, Ga, In, Tl) would then become Group 13 because of the intervening transition metals in Groups 3–12.

SYNTHESIS OF ELEMENTS AFTER URANIUM AND POSSIBLE EXTENSION OF THE PERIODIC TABLE

Bismuth is the last element in the periodic table to have a stable isotope so that all the later elements are continuously undergoing radioactive decay. A few isotopes with very long half-lives, such as $^{235}_{92}U$ and $^{238}_{92}U$, have survived since the formation of the solar system, and their decay processes provide a source of short-lived 'daughter' elements spanning the gap in the periodic table between uranium and bismuth. However, chemists are faced with the task of devising syntheses for those elements out beyond uranium.

One of the simplest methods is to irradiate a 'target' element with neutrons in an atomic reactor when some nuclei will absorb a neutron, the excess energy being released as a γ photon. Such an (n, γ) reaction will only produce an isotope of the target because there has been no change in the number of protons in the nucleus during the reaction; however, the product nucleus often has a short half-life and decays via expulsion of a β^- particle (a very high-energy electron) as a neutron is converted into a proton:

$$n \rightarrow p^+ + e^- + neutrino$$

Thus there will be a gradual build-up of a new element containing one proton more in its nucleus than the target element. Since this element will also be undergoing radioactive decay, the amount of it which can be produced will depend on its half-life because a balance will be achieved between its production rate and its rate of decay. Plutonium-239, which has a half-life of 24 000 years, can be made in kilogram quantities by this method, whereas the yield of $^{257}_{100}Fm$ ($t_{1/2} = 100$ days) is measured in micrograms. Some typical syntheses which have been achieved are shown below:

$$^{238}_{92}U \xrightarrow{(n, \gamma)} {}^{239}_{92}U \xrightarrow[t_{1/2} = 23.5 \text{ min}]{\beta^-} {}^{239}_{93}Np \xrightarrow[t_{1/2} = 2.35 \text{ days}]{\beta^-} {}^{239}_{94}Pu \ (t_{1/2} = 24\,000 \text{ yr})$$

$$^{239}_{94}Pu \xrightarrow{(n, \gamma)} {}^{240}_{94}Pu \xrightarrow{(n, \gamma)} {}^{241}_{94}Pu \xrightarrow{(n, \gamma)} {}^{242}_{94}Pu \xrightarrow{(n, \gamma)} {}^{243}_{94}Pu \xrightarrow[t_{1/2} = 5 \text{ h}]{\beta^-} {}^{243}_{95}Am$$

$$(t_{1/2} = 7370 \text{ yr})$$

$$^{243}_{95}Am \xrightarrow{(n, \gamma)} {}^{244}_{95}Am \xrightarrow[t_{1/2} = 10 \text{ h}]{\beta^-} {}^{244}_{96}Cm \quad (t_{1/2} = 17.6 \text{ yr})$$

These (n, γ) reactions, followed by β^- emissions, allow us to move only one place to the right of the target element in terms of the periodic table. However, as shown above it is possible to have a series of consecutive $(n, \gamma)/\beta^-$ reactions, because as the new elements build in quantity over a period of time, they too will provide sufficient numbers of target nuclei. The whole process starts in an atomic reactor with the last element in the periodic table which is available in quantity: uranium. It would appear at first sight that to leave a reactor working for a long time would, in the end, result in the build-up of many actinide elements. Unfortunately, this situation does not occur beyond $^{257}_{100}Fm$ because addition of a neutron to this isotope produces $^{258}_{100}Fm$ which undergoes fission within seconds; no significant quantities of fermium-258 can thus build up and the production of further new elements ceases.

A much less convenient way to synthesize new elements, but one which allows us to jump past fermium-258, is to bombard a target with a beam of fully ionized atoms such as protons ($^1_1H^+$), alpha particles ($^4_2He^{2+}$) or $^{22}_{10}Ne^{10+}$ ions:

$$^{253}_{99}Es + ^4_2He^{2+} \rightarrow ^{256}_{101}Md + n$$

$$^{238}_{92}U + ^{22}_{10}Ne^{10+} \rightarrow ^{255}_{102}No + 5n$$

The 'stripped atom' projectiles have to be accelerated to high velocities (for example, in a cyclotron) in order to overcome the huge repulsion forces experienced as a target nucleus is approached; relative to the number of neutrons in a reactor, the intensity of these ion beams is very low, often resulting in the production of only a few new atoms per experiment. The yields are also adversely affected by the tiny amounts of target isotopes which are available. Care must be taken that the projectile has enough kinetic energy to overcome the nuclear repulsion, but not too much in excess otherwise the product nucleus will disintegrate before it has time to 'cool down' from its excited state by the expulsion of one or more neutrons:

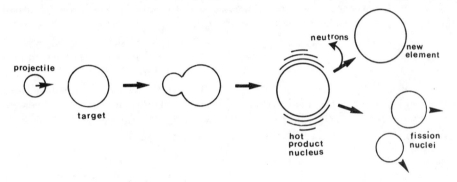

The ultimate in this type of process was achieved in 1982 when German scientists succeeded in synthesizing *one atom* of element [109] by hurling a fully ionized iron nucleus into a target bismuth atom:

$$_{26}Fe^{26+} + _{83}Bi \rightarrow [109] + n$$

Much unwanted debris was formed in side-reactions from which the single atom of [109] had to be isolated using a device similar to a mass spectrometer—but the separation had to be achieved quickly because [109] decayed away in only five-thousandths of a second:

$$[109] \xrightarrow{-\alpha} [107] \xrightarrow{-\alpha} [105] \xrightarrow{-\beta^+} [104] \rightarrow fission$$

By analysing this sequence of disintegration on the collector plate, the workers were able to deduce unambiguously the identity of [109].

As elements were made with atomic numbers above 100 it was found that the half-lives were becoming shorter, until it begain to appear that there would be a practical limit to extensions of the periodic table. However, over the years physicists have realized that certain numbers of protons and neutrons result in nuclei which are particularly stable towards disintegration; the 'magic numbers' are 2, 8, 20, 28, 50 and 82. Theoretical estimates suggest that the next combination of magic numbers will occur in a nucleus containing 114 protons and 184 neutrons, resulting in the element directly under lead in the periodic table. (It is interesting to note that the two previous magic proton numbers also correspond to members of Group IV: tin (50 protons) and lead (82 protons).) Original estimates suggested very long half-lives for [114] and its immediate neighbours, thus predicting that an *island of stability* should occur in this region of the periodic table. These ideas resulted in a careful, but fruitless, search for these 'super-heavy' elements both in nature and in meteorites. Revised calculations, although suggesting the half-lives had previously been over-estimated, still predict a modest stability towards both fission and α decay. However, many unsuccessful attempts have been made to 'jump the gap' from the present end of the periodic table to the island of stability by fusing two heavy nuclei together. With the targets and projectiles used up to now we appear to be in a Catch 22 situation: if the projectile energy is too low then the nuclei will not fuse together, whereas when the bombarding energy is high enough for fusion the product nuclei undergo immediate fission.

It will be unfortunate if we are denied access to the super-heavy elements, because predictions suggest their properties could be rather unusual. Direct comparisons with other members of the same group are less accurate in this part of the periodic table because of relativistic effects. Thus [112] and [114], which follow mercury and lead, could be very inert and exist as volatile liquids. The three 7p orbitals are expected to be non-degenerate, with one orbital being less stable than the other two. If this energy gap were large, an element like [115], which is below bismuth and possesses three outer 7p electrons, would behave as though its third p electron was outside a closed shell and hence have a very stable +1 oxidation state.

EFFECT OF SMALL ATOMIC SIZE ON BONDING

On page 52 it was shown how penetration and shielding effects resulted in steadily increasing ionization energies across the Li–Ne period. Since ionization energies reflect the nuclear attraction for the outer electrons it is not surprising to find that there is a corresponding decrease in atomic size across the period, and this has some interesting chemical consequences. An almost linear correlation occurs if dissociation energies are plotted against atomic number for the dihalogen molecules Cl_2, Br_2 and I_2. Extrapolation to find the expected dissociation energy of difluorine gives 266 kJ mol^{-1}, which is considerably greater than the measured value of 157.7 kJ mol^{-1}. At least three factors are probably responsible for this unexpectedly low dissociation energy (which makes fluorine the most reactive halogen):

(a) Supplementary p_π–d_π bonding can occur in Cl_2, Br_2 and I_2 because filled outer p orbitals on one atom have identical symmetry to empty d orbitals on the other; how much this contributes to the overall bonding is not known, but it will be favoured by the contraction in d orbital size caused by the high effective nuclear charge on the halogen atoms and has been shown by accurate calculations to be important in FCl.

(b) Electron pairs on one halogen will repel those on the other. This repulsion will be greatest for F_2, which has the shortest bond of all the halogen molecules, and will result in the lone pairs being brought very close together. It is presumably somewhat significant that the C–C dissociation energy in ethane, where there are no lone pairs on carbon, is more than twice the F_2 dissociation energy. However, although the bond in I_2 is longer than that in F_2, the non-bonding electrons of iodine are correspondingly further out from the nucleus and will also give rise to some inter-atom repulsion.

(c) Addition of an electron to an atom when forming either a covalent bond or an anion will naturally cause an increase in the inter-electron repulsion energy, which will be *largest* for the *smallest* halogen, fluorine. Therefore both the halogen electron affinity (E_X) and the dissociation energy of the dihalogen molecules (D_{X_2}) will be reduced by this interelectron repulsion.

Semiquantitative estimates of E_F and D_{F_2}, derived by extrapolation of trends shown by the other halogens, are *both* about 109 kJ mol^{-1} higher than those observed experimentally. Since these measurements correspond to processes involving an isolated F^- ion in one case and a difluorine molecule in the other, it has been suggested by Politzer that factors (a) and (b) above are of relatively minor importance in explaining the anomalously low dissociation energy of difluorine; rather it is the exceptionally small size of the fluorine atom which makes (c) the dominant factor. All bonds involving fluorine presumably suffer destabilization by about 109 kJ mol^{-1} because the bonding electrons will not be attracted by the fluorine nucleus quite as strongly as expected owing to the strong inter-electron repulsions (shielding); of course, in difluorine this will amount to 218 kJ mol^{-1} since two fluorine atoms are involved in sharing the bonding electrons. The high electronegativity of fluorine ensures that bonds between fluorine and all other elements in the periodic table will be polarized $X^{\delta +}$—$F^{\delta -}$, resulting in very substantial 'electrostatic contributions' to the bond energy; furthermore, the element bonded to fluorine normally has no lone pairs to cause strong repulsions. Hence, almost paradoxically it would seem after the discussion on F_2, fluorine forms some of the strongest known bonds to other elements.

Politzer has also drawn attention to the relative weakness of O—O and N—N single bonds and suggested that factors similar to (c) described above for fluorine are responsible for this; however, lone pair repulsions no doubt make a substantial contribution to the bond weakening as well. It is interesting to note that although atomic size is in the order N > O > F, the M—M single-bond lengths in N_2H_4 (1.47 Å), H_2O_2 (1.47 Å) and F_2 (1.43 Å) are remarkably similar. It has been pointed out that this could arise from the lone pair repulsions, since these molecules have one, two and three lone pairs on the central atom respectively and thus the repulsions could outweigh the normal atomic size contraction.

The difference in energy of the 2s and 2p orbitals is very considerable for oxygen and fluorine and as a result little s–p mixing (hybridization) occurs in their compounds. For example, although sp^3 hybridization of the oxygen orbitals is often used to describe the bonding in an isolated water molecule, there is now evidence from photoelectron spectroscopy measurements that the 2s orbital may not be involved in bonding to any significant extent. A more realistic, though still only approximate, view of the bonding assumes that only the hydrogen 1s and oxygen 2p orbitals interact. The resulting bond angle of 90° would cause strong steric repulsion between the two hydrogen atoms and hence the HOH angle opens up to be the observed value of 104.5°. In hydrogen sulphide the S—H bonds (1.35 Å) are longer than the O—H bonds of water (0.96 Å) and hence the HSH angle (92°) need not open too far from 90° in order to relieve the H–H strain.

The change in s–p separation across the Li–Ne period is also responsible for the observed ordering of molecular orbitals in homonuclear diatomic species formed by these elements. The expected energy level diagram for a typical diatomic molecule such as N_2 is shown in Fig. 76(a) where the orbital $\sigma(2p_z)$ lies below the degenerate pair of π bonding orbitals, $\pi(2p_x)$ and $\pi(2p_y)$. However, $\sigma(2s)$ and $\sigma(2p)$ have the same symmetry label σ_g and can mix, which results in the lower-energy orbital $\sigma(2s)$ becoming more stable; $\sigma(2p)$ is correspondingly destabilized. A similar situation arises with the antibonding orbitals of σ_u symmetry in that $\sigma^*(2s)$ falls in energy and $\sigma^*(2p)$ rises, but the mixing in this case is relatively small because the energy difference between the orbitals is large; orbitals can only mix significantly if their energy difference is small. When this mixing is taken into account the energy level diagram takes the form shown in Fig. 76(b) and the electron configurations of some known diatomic molecules become:

$B_2(1s)^2(1s)^2(\sigma 2s)^2(\sigma^*2s)^2(\pi 2p_x)^{\uparrow}(2p_y)^{\uparrow}$ bond order $= 1.0$; paramagnetic

$C_2(1s)^2(1s)^2(\sigma 2s)^2(\sigma^*2s)^2(\pi 2p_x)^2(\pi 2p_y)^2$ bond order $= 2.0$; diamagnetic

$N_2(1s)^2(1s)^2(\sigma 2s)^2(\sigma^*2s)^2(\pi 2p_x)^2(\pi 2p_y)^2(\sigma 2p_z)^2$ bond order $= 3.0$; diamagnetic

These configurations have been confirmed by both magnetic measurements and photoelectron spectroscopy, the latter being a technique which can measure the energy of each orbital.

However, when O_2 was studied by photoelectron spectroscopy the orbital order was found to be that predicted by simple theory as shown by the electron configuration in Fig. 85. This is due to the large 2s–2p energy difference in the oxygen atom which does not allow sufficient s–p mixing to force the $\sigma(2p_z)$ orbital above the π bonding orbitals.

THE LANTHANIDE CONTRACTION AND SIMILAR EFFECTS

There is a steady decrease in size as the 4f orbitals are being filled between lanthanum and lutetium which amounts to a total contraction across the lanthanide series of about 0.2 Å. This is sufficient to make the radii of zirconium and hafnium virtually identical and results in very similar properties for these two elements. Understandably, it was over 130 years after zirconium had been recognized that hafnium was discovered in many zirconium ores. The *lanthanide contraction* is responsible for many of the close similarities between the second- and third-row transition metals.

Lloyd has suggested that penetration effects are mainly responsible for the lanthanide contraction. The 4f orbitals are deeply 'buried' below the filled 5s and 5p orbitals which provide the outer electrons in the lanthanide tripositive ions; hence it is these orbitals which largely determine the ionic size. The four inner radial 'humps' of the 5s orbital and the three inner radial 'humps' of the 5p orbitals are all between the nucleus and the 4f orbitals' single radial maximum; thus the 5s and 5p orbitals penetrate the 4f orbitals very effectively. As the lanthanide series of elements is built up from La to Lu, protons are added to the nuclei, resulting in a steadily increasing nuclear attraction (owing to penetration) of the 5s and 5p electrons and hence a contraction in size.

The effects of relativity increase in importance as properties of the heavier elements in the periodic table are examined. When an electron is close to a heavy (and thus highly charged) nucleus its velocity increases and becomes comparable to the speed of light: a relativistic increase in mass occurs and with it a corresponding increase in binding energy E since E is proportional to the electron mass. This change in binding energy causes a contraction in size of the electron's orbital and the penetration properties of orbitals make these relativistic effects act in the order $s > p > d > f$. Even at the comparatively light lanthanides relativistic effects are thought to be responsible for up to 15 % of the lanthanide contraction and will become increasingly important further on in the periodic table.

Shielding and relativity effects result in the similar sizes of the second- and third-row transition metals and are also responsible for the fact that *all* the sixth-period elements have higher ionization energies than their congeners in period five. Among the heavier non-transition elements the effects cause a reluctance of the $6s^2$ electrons to take part in bonding (the 'inert pair effect') and hence a stabilization of Tl(I), Pb(II) and Bi(III) derivatives relative to compounds showing the normal group valencies.

The increasing effective nuclear charge, which occurs across the various periods owing to relativity and underlying d or f orbitals, makes the Sn^{2+}, Pb^{2+} and Bi^{3+} cations *very* susceptible to hydrolysis by markedly enhancing their polarizing effects. Thus the water ligands experience an abnormally high effective charge on ions that are also reduced in size relative to their positions in the periodic table.

A substantial size decrease also occurs between the elements Sc and Zn, sometimes referred to as the 'scandinide or scandide contraction'. At this point in the periodic table the 3d orbitals are being filled but, unlike the 4f orbitals in the lanthanide series, they are outermost orbitals and contribute substantially in determining atomic and ionic sizes. Owing to their spatial orientation the different d orbitals do not shield each other very effectively from the nuclear charge; Fig. 35. Thus as the scandinide elements are built up by addition of protons to the nucleus, the 3d electrons experience an increasing effective nuclear charge; furthermore, penetration close in to the nucleus by the 3p and 4s orbitals ensures that *all* the outer, size-determining orbitals experience an increasing contraction across the transition series.

Owing to the scandinide contraction there is not a smooth gradation in properties between respective elements in periods three and four, the latter being smaller and having higher ionization energies than predicted. When the lanthanide contraction is also taken into account we can expect that there will be considerable chemical differences to be found among the elements as we follow them down Groups III–VII, quite unlike the relatively smooth variation in properties which occurs in Groups I and II.

In summary, we have discussed size contractions involving s orbitals (H–He), p orbitals (C–Ne), d orbitals (Sc–Cu) and f orbitals (La–Lu); what should be noted is that there are similar contractions across *all* the periods so that, for example, the 'yttride' and 'actinide' contractions are just as important in determining the chemistry of the following elements as were the scandinide and lanthanide contractions.

π BONDING IN COMPOUNDS OF THE HEAVY ELEMENTS

The elements outside the Li–Ne period differ from their lighter congeners in two important ways: their p orbitals possess nodes and their outer d orbitals are, theoretically, available for bonding Hence there are several types of π bonding possible for the heavier members of Groups IV–VI:

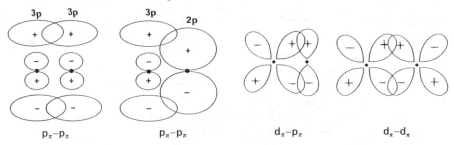

It is found that p_π–p_π bonds involving np orbitals with $n \geqslant 3$ are relatively weak, probably due to increased inner-shell repulsions and the relatively low electron density present in the outermost parts of such p orbitals giving rise to poor overlap. The nodes in the higher p orbitals also result in some electron density being 'lost' for bonding purposes because the inner-orbital lobes are too small to overlap (i.e. they are close in to the nucleus). For example, photoelectron spectra of the thermally fragile species HC≡P, FC≡P and C_6H_5C≡P show that their $2p_\pi$–$3p_\pi$ bonds are about 200 kJ mol^{-1} less stable than those in the corresponding RC≡N derivatives; chemically they display properties typical of unsaturated molecules: e.g.

Only recently has it proved possible to synthesize stable derivatives containing homonuclear double bonds between the heavier elements because attempts to make compounds such as R_2Si=SiR_2 and RAs=AsR usually result only in the formation of polymers. Hence one strategy used has been to build into the unsaturated molecules a *kinetic* resistance to polymerization by having bulky substituent groups present:

(stability sequence)

Cis and *trans* isomers of disilenes can be made if two different R groups are present, but, unlike alkenes, facile interconversion occurs at room temperature. The activity in this area of research has been so intense that hundreds of multiply bonded compounds of the heavier elements have now been synthesized.

The decrease in bond length between M—M and M=M is about 12% for carbon compared with 10% for silicon and only 2% for tin; in fact the M—M bond lengths exceed the sum of the Pauling double-bond radii for all the unsaturated derivatives of Si, Ge and Sn so far studied. Moreover, many M_2R_4 compounds are non-planar and some tin species even exist as MR_2 in the gas or liquid phases:

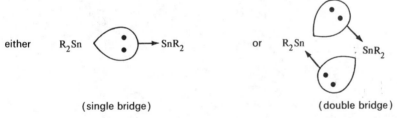

$\theta = 0°–18°$ for Si
$\theta = 15°–32°$ for Ge
$\theta \simeq 40°$ for Sn

Hence the question arises as to whether these compounds do contain substantial double bonds. Lappert, and Pauling, have suggested that the tin compounds are actually stannylenes which dimerize under some conditions via donation of a lone pair on one tin atom into a vacant orbital on the other:

either R_2Sn ⟶ SnR_2 or R_2Sn SnR_2

(single bridge)   (double bridge)

In a planar $R_2M=MR_2$ derivative the filled π molecular orbital and the empty σ* antibonding orbital have different symmetry, but when molecular bending occurs they are able to interact. At carbon these orbitals are of quite disparate energies and little or no stabilization is to be gained by having a non-planar structure. However, increasingly down Group IV the π and σ* orbitals become closer in energy and considerable stabilization relative to the planar state occurs as they mix, resulting in non-planar Si_2R_4 and Ge_2R_4 systems. The admixture of *antibonding* character lowers the electron energy in the new 'hybrid' orbital relative to a 'pure' π orbital but at the expense of weakening the M=M double bond.

On the other hand there is good agreement between the observed P—P and As—As bond lengths in a series of R_2M_2 derivatives and those calculated from the Pauling double-bond radii; from this it would appear that these double bonds have a comparatively strong $p_\pi–p_\pi$ contribution.

Until very recently phosphine complexes of transition metals such as $Ni(PF_3)_4$ were considered to represent good examples of simple $d_\pi–d_\pi$ bonding involving empty d orbitals on phosphorus. However, calculations show that two degenerate P—R antibonding σ* orbitals are the lowest-energy unoccupied orbitals on a phosphine PR_3. This pair, which possess mainly phosphorus 3p character, are of the correct symmetry to overlap with metal d orbitals in a 'back-bonding' π interaction:

initial σ donation
P-to-metal

back-bonding π donation
metal-to-P

Thus at the qualitative level there is no back-bonding requirement for the large, diffuse 3d orbitals on phosphorus but, quantitatively, the addition of a small amount of d character to the σ^* pair gives two hybrids which point *more effectively* towards the metal and so improve the back-bonding:

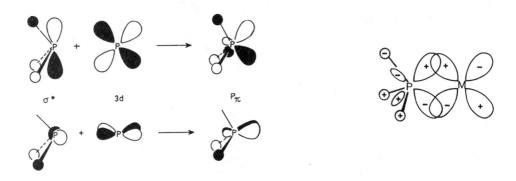

(For further discussion, see *J. Chem. Soc. Chemical Communications* (1985) p. 1310 from which this diagram was taken with permission.)

The use of PF_3 as a strong ligand towards transition metals might appear curious at first sight considering that the high electronegativity of fluorine makes it a *very ineffective* ligand towards main group elements. However, the weak P-to-metal σ bonding is more than offset by the influence of the fluorine atoms on the back-bonding interaction: (a) the P—F σ^* orbitals have a low energy relative to similar orbitals in other phosphines and are thus more compatible with the metal d orbitals; (b) in PF_3 the phosphorus 3d orbitals contract markedly and are able to hybridize with the σ^* orbitals more effectively than is possible in other phosphines. (Main group elements, having no available filled d orbitals, are unable to participate in similar back-bonding with ligands.)

The back-bonding described above places electron density in a P—R antibonding orbital; hence the stronger the back-bonding, the longer should be the P—R bond. This can be tested by choosing pairs of complexes such as $C_5H_5Co(PEt_3)_2$ and $[C_5H_5Co(PEt_3)_2]^+[BF_4]^-$ which differ only in the oxidation state of the metal. Loss of d electrons capable of back-bonding from the cobalt in the cation should result in a weaker (i.e. longer) Co—P and a stronger (i.e. shorter) P—C bond relative to the neutral complex. This is exactly what is found:

	Co—P distance (Å)	P—C distance (Å)
$C_5H_5Co(PEt_3)_2$	2.218 ± 0.001	1.846 ± 0.003
$[C_5H_5Co(PEt_3)_2][BF_4]$	2.230 ± 0.001	1.829 ± 0.003

Although the changes in bond length are quite small they are definitely real when compared with the error margin involved in their measurement; furthermore, other pairs of complexes show exactly the same behaviour.

The P—O bonds in the phosphine oxides R_3PO are considerably shorter than the single-bond value of about 1.63 Å. Calculations on the hypothetical H_3PO molecule reveal that the P—O bond length changes from 1.60 to 1.47 Å on the inclusion of d orbitals in the equations. The effective charges on P and O in H_3PO are $+1.33$ and -0.97 respectively, which suggests a theoretical model involving a semipolar bond $\overset{\delta+}{P}\!\!-\!\!\overset{\delta-}{O}$ stabilized by fairly substantial $d_\pi-p_\pi$ back-bonding, probably of the partial triple-bond type. (The *two* 2p lone pair orbitals on oxygen have the correct symmetry to interact with two d orbitals which are at right angles to each other; see the diagram for OSF_4 on page 66).

Photoelectron spectroscopy (PES) supports the occurrence of $d_\pi-p_\pi$ bonding in many phosphine oxides and, for example, is found to be in the order $OPF_3 > OPCl_3 > OP(CH_3)_3$, possibly owing to the decreasing charge on phosphorus (which expands the d orbitals making them less effective for bonding) as the electronegativity of the substituents decreases. Similarly, PES studies on phosphorus ylides show that the CR'_2 groups have considerable carbanion character, $\overset{\delta+}{R_3}\!\!=\!\!\overset{\delta-}{CH_2}$.

The structure of $F_4S{=}CH_2$ is particularly unusual because the hydrogen atoms lie in the same plane as the axial fluorines, which is the most unfavourable steric conformation:

There is no spectroscopic evidence in support of any hydrogen bonding between the CH_2 group and the two axial fluorines (which are actually bent *away* from the methylene hydrogens). The observed molecular configuration is thus presumably adopted because electron density in the $p_\pi-d_\pi$ bond can be most easily accommodated in the relatively spacious equatorial plane. Replacement of the methylene hydrogen atoms by a less sterically demanding oxygen p orbital in the related molecule $F_4S{=}O$ results in the $F_{ax}SO$ angle (91.4°) being smaller than angle $F_{ax}SC$ (94.8°) in $F_4S{=}CH_2$; this would appear to be further conformation of the lack of F—H hydrogen bonding in $F_4S{=}CH_2$. The fact that the $F_{ax}SO$ angle is close to the 'ideal' value of 90° might suggest a $p_\pi-d_\pi$ bond also lies in the equatorial plane of $F_4S{=}O$, but more likely two π bonds at right angles are formed to give a 'partial triple bond' (oxygen has two 2p lone pairs unlike a CH_2 group):

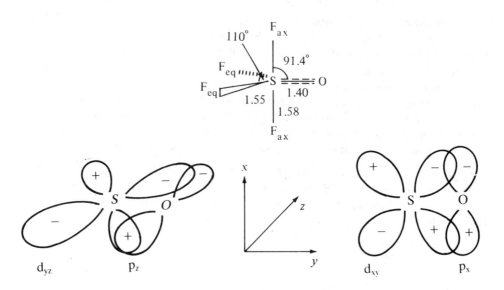

Nitrogen is also able to form $p_\pi-d_\pi$ bonds, particularly with phosphorus or sulphur. It is not too surprising to find that one of the shortest S—N bonds, for example, occurs in a highly fluorinated species, thiazyl trifluoride (NSF_3):

$$
\begin{array}{l}
\text{1.416 Å} \\
\text{N} \equiv \text{S}
\end{array}
\quad\quad \text{(cf. S—N single bond} \simeq 1.72\ \text{Å)}
$$

F
94°.
F

F

The planar Si_3N skeleton in trisilylamine, $N(SiH_3)_3$, has been attributed to Si—N π bonding in which the filled nitrogen 2p orbital interacts simultaneously with empty d orbitals on the silicon atoms:

$$
H_3Si — N\cdots \begin{array}{c} SiH_3 \\ SiH_3 \end{array} \quad\text{cf.}\quad H_3C\cdots \begin{array}{c} N \\ CH_3 \\ CH_3 \end{array}
$$

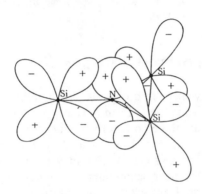

An alternative suggestion is that trisilylamine is forced into a planar configuration solely by the steric requirements of three bulky SiH_3 groups and repulsions arising from substantial polarity in the $\overset{\delta+}{Si} — \overset{\delta-}{N}$ bonds. Opponents of this 'no π bond approach' have pointed out that in crystalline trisilylamine the molecules are stacked with their nitrogen atoms one above the other, a situation implying an absence of significant Si—N polarity. The electronegativity difference between silicon and nitrogen would give rise to some $\overset{\delta+}{Si} — \overset{\delta-}{N}$ polarity in the σ bond, but this would be offset by movement of electronic charge from N to Si in a p_π–d_π interaction.

It should be realised that some chemists are extremely sceptical about the *substantial* involvement of d orbitals in π bonding, mainly on the grounds that such orbitals are of high energy and too diffuse to bond effectively. A short bond length in a molecule is often cited as evidence of π bonding; the problem arises when a short bond occurs in a molecule where π bonding cannot reasonably be expected. One such case is found in the tetrahedral ortho-nitrate ion NO_4^{3-} which has an N—O bond length of 1.39 Å, somewhat shorter than the presumed single bond found in hydroxylamine, NH_2OH (1.47 Å). In NO_4^{3-} it has to be assumed that bond polarity is mainly responsible for the N—O shortening, which suggests that a p_π–d_π interaction may not be the only important factor involved in the bonding of anions such as SO_4^{2-}, PO_4^{3-} and ClO_4^-; see Table 4 and page 313.

Table 4—Apparent shortening of M—O bonds in tetrahedral MO_4^{n-} ions

Ion	Experimental M—O distance (Å)	Calculated single M—O bond length (Å)[a]	Apparent shortening (Å)	Percentage shortening of M—O bond
SiO_4^{4-}	1.63	1.78	0.15	8.4
PO_4^{3-}	1.53	1.74	0.21	12.0
SO_4^{2-}	1.49	1.71	0.22	12.9
ClO_4^-	1.45	1.71	0.26	15.2

[a] Calculated using the Schomaker–Stevenson relationship

$$r_{AB} = r_A + r_B - 0.09(X_A - X_B)$$

where r_A and r_B are the single-bond radii of atoms A and B; X_A and X_B are the Pauling electronegativities of atoms A and B.

PARTICIPATION OF d ORBITALS IN σ BONDING

Chemists often cite involvement of d orbitals in σ bonding as the reason why only elements outside the first short period can have coordination numbers greater than four; however, there appears to be no experimental or theoretical justification for the very substantial d orbital contributions demanded by the suggested sp^3d and sp^3d^2 hybridization schemes. A simpler explanation may be that B, C, N and O are too small to link with five or six near neighbours in trigonal bipyramidal or octahedral fashion when forming normal, two-centre covalent bonds.

It has been calculated that, to reduce the steric repulsion to a reasonable level, the C—F bond lengths in CF_5^- would have to be at least 16% longer than those in CF_4. The resulting loss in C—F bond strength would be far too large to stabilize CF_5^- in relation to CF_4 and an uncoordinated fluoride ion. This inability of carbon to increase its coordination number even when bound to the smallest halogen explains why the carbon tetrahalides do not react with water although they are thermodynamically unstable with respect to hydrolysis. Nucleophilic attack by a water molecule or a hydroxyl group would require carbon to increase its coordination number to five:

$$CX_4 + H_2O \longrightarrow \left[X\substack{\\\cdots} \overset{\overset{X}{|}}{\underset{\underset{OH_2}{|}}{C}} - X \right] \longrightarrow CO_2 + 4HX$$

The ready hydrolysis of silicon tetrahalides, on the other hand, presumably occurs because the larger silicon is able to form the required five-coordinate transition state. Even outside the first short period unacceptable steric requirements can result in hydrolytic stability: although both SF_4 and SF_6 are thermodynamically unstable to attack by water, only the tetrafluoride hydrolyses under normal conditions. Even though the S—F bonds of SF_6 are relatively weaker, the sulphur atom is too small to form the seven-coordinate transition state required in the hydrolysis mechanism and hence the hexafluoride is hydrolytically stable. Note the difference here between thermodynamic and kinetic stability: although carbon tetrahalides and SF_6 are thermodynamically unstable towards water, they are able to resist its attack because of kinetic factors involved in the reaction mechanism. The ready hydrolysis of acyl chlorides, RCOCl, shows that there is nothing intrinsically 'stable' about the C—Cl bond: in this case the hydrolysis proceeds via a four-coordinate transition state which is well within the steric requirements of carbon:

$$CH_3C\overset{O}{\underset{Cl}{\big\backslash}} + H_2O \rightarrow \left[\begin{array}{c} H_3C \,\,_{\prime\prime\prime\prime} \\ H_2O \end{array} C \overset{O}{\underset{Cl}{\big\backslash}} \right] \rightarrow CH_3COOH + HCl$$

Owing to their relative lack of penetration through the inner-core electrons, the outer d orbitals of an atom are large and hence diffuse (i.e. being large, their electron density at any point is low). For example, van Wazer has estimated that a 3d orbital in the sp^3d configuration of phosphorus has a mean radius of about 3.8 Å compared with 1 Å for the 3s and 3p orbitals. Clearly such d orbitals are far too large to hybridize efficiently with the s and p orbitals. When highly electronegative ligands are attached to the phorphorus or other central atom, the outer d orbitals contract markedly because they experience an apparently higher effective nuclear charge as electron density is drawn towards the ligands; in such compounds the d orbitals are able to contribute more effectively to the bonding. In part, this may explain why PF_5 and SF_6 exist whereas PH_5 and SH_6 do not, even though the ligands are of very similar size.

The bonding in PF_5, SF_6 and similar 'hypervalent' compounds can be described mainly in terms of s and p orbitals with the large d orbitals having only a secondary, supporting role to play (but they still contribute about 170 kJ mol^{-1} to the stability of PH_3F_2, for example). This theory, first proposed by Rundle, suggests that in SF_6 the fluorine atoms are bound to the sulphur in pairs by linear three-centre molecular orbitals formed via the interaction of three p orbitals: one from the central atom and one from each of two fluorines lying diametrically opposite each other in the molecule. One of the resulting σ-type molecular orbitals will be bonding, one non-bonding and the third antibonding as shown in Fig. 12; four electrons (two from sulphur, one from each fluorine) fill the bonding and non-bonding orbitals giving rise to the name three-centre, four-electron (or 3c–4e) bonding.

The symmetry properties of the non-bonding orbital are identical to those of d_{z^2} and $d_{x^2-y^2}$ sulphur orbitals. Hence overlap of these two types of orbital will strengthen the S—F bonds because the originally non-bonding orbital is stabilized and thus becomes partially bonding; the total occupancy of the d orbitals is calculated to be only about 0.25 electrons in SF_6. Three-centre, four-electron bonding explains, in a natural fashion, the observed octahedral conformation of sulphur hexafluoride because the lobes of the sulphur p orbitals point towards the vertices of an octahedron. In PF_5, which has trigonal bipyramidal geometry, the three fluorines in the equatorial plane are held by normal covalent bonds involving sp^2 hybrid orbitals on phosphorus, while the two axial fluorine atoms participate in three-centre, four-electron bonding. The axial P—F bonds will be slightly weaker than the equatorial ones because

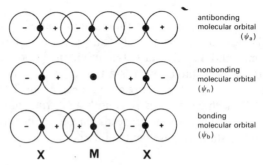

(a) Without d orbital participation.

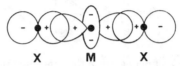

(b) Interaction between a d_{z^2} (or $d_{x^2-y^2}$) orbital and the non-bonding molecular oribtal; the latter becomes slightly bonding as a result. [After W. Kutzelnigg, *Angewandte Chemie* (1984) p. 272]

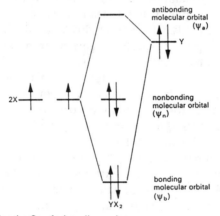

(c) Energy level diagram for the 3c–4e bonding scheme.

Fig. 12—Orbital interactions involved in three-centre, four-electron bonding.

essentially only two electrons hold three atoms together in the 3c–4e bonding scheme; this allows the prediction that the axial bonds will be longer than the equatorial bonds, a fact verified experimentally for PF_5 and many other trigonal bipyramidal molecules of main group elements:

REFERENCES

Atomic orbitals

Karplus, M. & Porter, R. N. (1970) *Atoms and Molecules: An Introduction for Students of Physical Chemistry*. Benjamin, Menlo Park

Kikuchi, Q. & Suzuki, K. (1985) Orbital shape representations. *J. Chem. Educ.* **62** 206

Massey, A. G. & Massey, S. (1976) Simulation of atomic and molecular orbitals using magnets. *Educ. Chem.* p. 111

Ogryzlo, E. A. & Porter, G. B. (1963) Contour surfaces of atomic and molecular orbitals. *J. Chem. Educ.* **40** 256

Periodic table

Huheey, J. E. & Huheey, C. L. (1972) Anomalous properties of elements that follow long periods of elements. *J. Chem. Educ.* **49** 227

Kratz, J. V. (1983) The search for superheavy elements. *Radiochim. Acta* **32** 25; see also Herrmann, G. (1988) *Angew. Chemie* (International Edition in English) **27** 1417

Lloyd, D. R. (1986) On the lanthanide and scandinide contractions. *J. Chem. Educ.* **63** 502

Aspects of bonding

Cowley, A. H. (1984) Stable compounds with double bonding between the heavier main-group elements. *Acc. Chem. Res.* **17** 386

Cruickshank, D. W. J. & Eisenstein, M. (1985) The role of d functions in ab-initio calculations. *J. Mol. Struct.* **130** 143

Fuess, H., Bats, J. W., Cruishank, D. W. J. & Eisenstein, M. (1985) Comparison of theoretical and experimental deformation densities in S—O bonds. *Angew. Chemie* (International Edition in English) **24** 509

Kutzelnigg, W. (1984) Chemical bonding in higher main group elements. *Angew. Chemie* (International Edition in English) **23** 272

Laing, M. (1984) No rabbit ears on water. *J. Chem. Educ.* **64** 124

Orpen, A. G. & Connelly, N. G. (1985) Structural evidence for the participation of P—X σ^* orbitals in metal—PX_3 bonding. *J. Chem. Soc. Chem. Commun.* p. 310

Politzer, P. (1969) Anomalous properties of fluorine. *J. Am. Chem. Soc.* **91** 6235; (1977) Some anomalous properties of oxygen and nitrogen. *Inorg. Chem.* **16** 3350

Reed, A. E. & Weinhold, F. (1986) On the role of d orbitals in SF_6. *J. Am. Chem. Soc.* **108** 3586

Chapter 2

Hydrogen

Notwithstanding its very simple atomic structure, that of a single electron and a unipositive nucleus, hydrogen is remarkable in forming more compounds than any other element in the periodic table. It is an essential constituent of animal and plant matter and has been detected spectroscopically in the Sun, many other stars and a variety of nebulae. Estimates have suggested that of all the matter in the universe 92% is hydrogen, 7% is helium and all the other elements together make up the remaining 1%.

ISOTOPES OF HYDROGEN

There are three isotopes of hydrogen:

(a) Protium (hydrogen-1) or, simply, hydrogen, ^{1}H: single proton in the nucleus; 99.04%.
(b) Deuterium (hydrogen-2), ^{2}H or D: one proton, one neutron in the nucleus; 0.016%.
(c) Tritium (hydrogen-3), ^{3}H or T: one proton, two neutrons in the nucleus.

Tritium is radioactive (half-life 12.26 years) and decays into helium-3 with the expulsion of a beta particle:

$$^3_1\text{H} \rightarrow {}^3_2\text{He} + \beta^-$$

It does not occur naturally in meaningful amounts (minute quantities are produced by the bombardment of nitrogen-14 by cosmic rays in the upper atmosphere) and is made artificially, for example, by bombarding lithium with low-energy neutrons in a nuclear reactor:

$$^6_3\text{Li} + {}^1_0\text{n} \rightarrow {}^3_1\text{H} + {}^4_2\text{He}$$

Both deuterium and tritium make useful tracers with which to follow reactions involving hydrogen. Although the hydrogen isotopes are chemically identical, quite wide variations in the rates of their reactions often occur because the zero-point vibration energy (which, like all vibration energies, is dependent on mass) of an X-to-hydrogen bond is lower for deuterium and tritium than for protium. This makes the bond dissociation energy of either an X—D or X—T bond slightly higher than that of the corresponding X—H bond, which in turn increases the activation energy of reactions involving X—H bond rupture when deuterium and tritium are substituted for the normal hydrogen isotope; see Figs 13 and 14. As an example, reactions involving carbon–hydrogen bonds are often found to proceed about seven times more slowly when protium is replaced by deuterium. In the biosphere, D_2O is found to be harmful to many living organisms because extensive exchange of deuterium occurs with N—H and O—H bonds, causing serious disturbances in the rates of delicately balanced biological reactions.

Deuterium is available commercially in large quantities, usually as D_2O, which is separated from normal water by a variety of methods, including electrolysis, fractional distillation, fractional diffusion and catalytic exchange of the deuterium between hydrogen gas and water. Several deuterium analogues of hydrogen compounds may be synthesized quite simply from D_2O using well-known reactions: e.g.

$$CaC_2 + D_2O \rightarrow DC{\equiv}CD \xrightarrow{\text{catalyst}} C_6D_6 \quad \text{(hexadeuterobenzene)}$$

$$Mg_3N_2 + D_2O \rightarrow ND_3$$

$$Na + D_2O \rightarrow D_2$$

$$LiC_6H_5 + D_2O \rightarrow C_6H_5D \quad \text{(singly labelled benzene)}$$

$$SO_3 + D_2O \rightarrow D_2SO_4$$

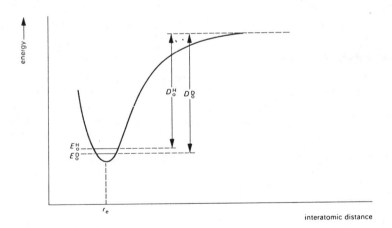

Fig. 13—Morse curves for X—H and X—D bonds relating potential energy and interatomic distance. The curves are essentially identical for hydrogen and deuterium, but the zero-point energies E_0^H and E_0^D differ owing to the difference in mass between the hydrogen and deuterium atoms. The chemical heats of discussion for the two bonds are given by D_0^H and D_0^D. The difference in the zero-point energies makes D_0^D larger than D_0^H by about 5 or 6 kJ mol^{-1}.

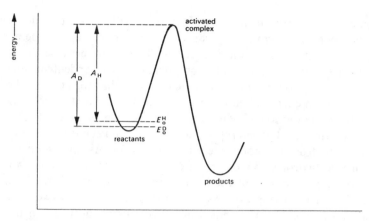

Fig. 14—Potential energy profile for a reaction involving the cleavage of a bond to hydrogen. The effect of the differing zero-point energies is to make the activation energy A_D of the deuterium compound larger than A_H, resulting in a slower rate of cleavage for the bond to deuterium.

Table 5 compares the physical properties of H_2O and D_2O and illustrates the changes which occur on substitution of hydrogen by deuterium. The dielectric constant of D_2O is somewhat less than that of H_2O, which accounts, at least in part, for the lower solubility of salts in D_2O relative to H_2O; the dielectric constant and stronger O—D bonds are responsible for the lower self-ionization of D_2O.

Industrially, hydrogen gas is made by any one of several reactions: the electrolysis of water; the steam–water-gas reaction; the thermal cracking of hydrocarbons in the petroleum industry. When high-purity hydrogen ($\sim99.99+\%$) is required, it is produced by diffusing hydrogen through heated films of palladium, when the impurities remain behind. Large quantities of liquid hydrogen are at present being used for rocket propulsion because hydrogen represents the ultimate in high-energy fuels.

Table 5—Properties of normal and heavy water

	H_2O	D_2O
Melting point	0.00 °C	3.82 °C
Boiling point	100.00 °C	101.42 °C
Temperature of maximum density	4 °C	11.6 °C
Dielectric constant	82	80.5
Specific gravity at 20 °C	0.9982	1.1059
Ionic product K_w for self-dissociation at 25 °C ($2H_2O \rightleftharpoons H_3O^+ + OH^-$), $K_w = [H_3O^+][OH^-]$	1×10^{-14}	3×10^{-15}
Solubility of NaCl	35.9 g/100 g	30.5 g/100 g
Solubility of $BaCl_2$	35.7 g/100 g	28.9 g/100 g
Ionic mobilities at 18 °C:		
$\quad$ K$^+$	64.2	54.5
$\quad$ Cl$^-$	65.2	55.3
$\quad$ H$^+$ (or D$^+$)	315.2	213.7

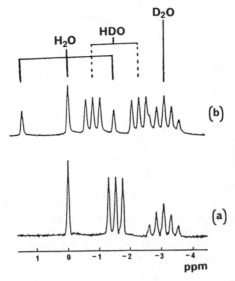

Fig. 15—^{17}O NMR spectrum (54 MHz) of a mixture of water isotopomers in $CDCl_3$: (a) proton-decoupled; (b) poroton-coupled. Isotopomers are species ('isomers') which differ only in their isotopic composition. The proton, nuclear spin $I = \frac{1}{2}$, couples to the ^{17}O nucleus to produce $(2nI + 1)$ lines, where n is the number of protons in a particular water isotopomer; thus H_2O gives a triplet which collapses to a singlet when the proton coupling is removed by a double-irradiation experiment. Coupling between the deuterium isotopes, $I = 1$, and ^{17}O is responsible for the extra peaks $(2nI + 1)$ in the spectrum as shown clearly in the proton-decoupled spectrum in (a). [The spectra were kindly provided by Professor G. D. Mateescu of Case Western Reserve University, Cleveland, OH.]

ORTHO AND *PARA* HYDROGEN

The proton, which forms the nucleus of protium, has spin $I = \frac{1}{2}$ (measured in units of $h/2\pi$). When two hydrogen atoms are coupled together in the H_2 molecule, the two nuclei have their spins either parallel or antiparallel to each other (Fig. 16).

Transitions between the *ortho* and *para* states are theoretically forbidden, but in practice they occur very slowly and as a result the pure gaseous species have half-lives of about 3 years at 20°C. At the temperature of liquid air (~ 80 K) the molecules are in their lowest rotational energy state and *para* hydrogen is the form thermodynamically favoured; at higher temperatures increasing amounts of *ortho* hydrogen are present in the equilibrium mixture to the limiting ratio of 3 *ortho*:1 *para* (which is essentially the composition of 20°C). To obtain pure *para* hydrogen the change

$$3o\text{-}H_2 + 1p\text{-}H_2 \rightarrow \text{pure } p\text{-}H_2$$

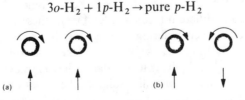

Fig. 16—(a) *Ortho* hydrogen; nuclear spins parallel; resultant spin equals 1. (b) *Para* hydrogen; nuclear spins antiparallel; resultant spin equals 0.

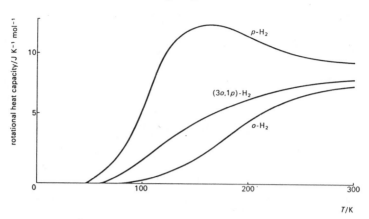

Fig. 17—The heat capacities of *ortho*, *para* and normal (3o, 1p) hydrogen. Mixtures of *ortho* and *para* hydrogen can be analysed by measuring their thermal conductivities, which are proportional to the heat capacities. The boiling points of *ortho* and *para* hydrogen are too similar to be of use in analysis.

is catalysed by adsorbing ordinary hydrogen on oxygenated charcoal at liquid air temperature and then pumping off the pure *para* hydrogen from the charcoal. It is also possible to separate the two forms of hydrogen by gas chromatography. The change from the *ortho* to the *para* form is exothermic to the extent of 703 J g^{-1}, the heat liberated being sufficient to evaporate over 60% of liquid hydrogen stored as a 3 *ortho*:1 *para* mixture; liquid hydrogen must therefore be catalytically converted into the pure *para* form before storage. The deuterium nucleus has a spin of 1 and hence a somewhat similar situation occurs in the D_2 molecule, of which the *ortho* and *para* forms have been isolated.

THE POSITION OF HYDROGEN IN THE PERIODIC TABLE

The electronic configuration of the hydrogen atom is 1s^1 and over the years various chemists have considered hydrogen as being the first member of no less than three groups in the periodic table. It has been classed with the alkali metals in Group I by virtue of the single s electron, with the halogens in Group VII because hydrogen requires only one electron to complete the quantum shell, and with carbon in Group IV because the outer quantum shell (1s) is, like that of carbon, only half full.

While it may appear extreme to consider hydrogen with *any* other group, it should be remembered that the elements within Groups III–VII have quite similar properties even though their electron 'cores', which govern such things as ionization energies and atomic size, are quite different (see Chapter 1). On these grounds alone a fairly strong case can be made for considering hydrogen with Group I, remembering that diverse ionization energies and (because of atomic and ionic sizes) hydration energies are to be expected. Although hydrogen is a diatomic gas under normal conditions and other Group I elements are metallic solids†, the situation is not too different

† Hydrogen is reported to become metallic when subjected to a pressure of two million atmospheres; the density under these conditions is about 1.06 g cm^{-3}.

Table 6—First ionization energies (kJ mol^{-1}) of several elements compared to hydrogen

$H(g) \rightarrow H^+(g) + e$	1310
$Li(g) \rightarrow Li^+(g) + e$	518.8
$Na(g) \rightarrow Na^+(g) + e$	493.8
$F(g) \rightarrow F^+(g) + e$	1682
$Cl(g) \rightarrow Cl^+(g) + e$	1255
$C(g) \rightarrow C^+(g) + e$	1088

from the cases of nitrogen and oxygen relative to the other members of Group V and Group VI respectively. We are at fault, of course, for expecting any of the elements to fall nicely into groups; a short study of the electronic configurations within the periodic table shows that it is indeed fortuitous that we are able to divide the table up into groups at all.

The ionization energy of hydrogen is very high compared to that of an alkali metal (Table 6). When hydrogen loses its single electron the exceedingly small proton H^+ (radius $\simeq 10^{-5}$ Å) is formed, which cannot exist except in the isolation of high vacuum; certainly it is not present in solid HCl, HF and H_2SO_4. On the other hand, the alkali metals exist as ions in the salts MCl, MF and M_2SO_4. The heat of hydration of the proton is exceptionally high, but because of the large ionization energy term contained in the process

$$\tfrac{1}{2}H_2(g) \xrightarrow{+217.5 \text{ kJ}} H(g) \xrightarrow{+1310 \text{ kJ}} H^+(g) \xrightarrow{-1172 \text{ kJ}} H^+(aq)$$

the enthalpy of the reaction $\tfrac{1}{2}H_2 \rightarrow H^+(aq)$ ($\Delta H = +355.5$ kJ (mol H)$^{-1}$) is more endothermic than that of the alkali metals:

$$Li(s) \xrightarrow{+159 \text{ kJ}} Li(g) \xrightarrow{+518.8 \text{ kJ}} Li^+(g) \xrightarrow{-514.6 \text{ kJ}} Li^+(aq)$$

$$Li(s) \rightarrow Li^+(aq), \quad \Delta H = +163.2 \text{ kJ mol}^{-1}$$

So again hydrated H^+ ions are less likely to occur than hydrated alkali metal cations.

By gaining an electron the hydrogen atom can form the hydride ion H^-, which might be considered analogous to halide ions formed by the halogens. One great difference is that the hydride ion is unstable in water:

$$H^- + H_2O \rightarrow H_2 + OH^-$$

Energetically, it is also found that the heat of formation of gaseous H^- is positive whilst that of the halide ions is highly negative: e.g.

$$\tfrac{1}{2}H_2 \rightarrow H^-(g), \quad +146.4 \text{ kJ}$$

$$\tfrac{1}{2}F_2 \rightarrow F^-(g), \quad -351.5 \text{ kJ}$$

$$\tfrac{1}{2}Cl_2 \rightarrow Cl^-(g), \quad -251.0 \text{ kJ}$$

Although the ionization energies of hydrogen and chlorine atoms are very similar (Table 6), the Cl^+ ion is very much larger than H^+ and hence the solvation energy of Cl^+ is much too small to stabilize it in any known solvent. It is of interest to note that the heat of formation of gaseous Na^- is actually lower than that of H^-:

$$Na(s) \rightarrow Na^-(g), \quad +38 \text{ kJ (mol Na)}^{-1}$$

and recently solutions and crystals have been prepared which contain sodide ions (see page 117 for a discussion of M^- ions of the alkali metals).

It is not very profitable to pursue the comparison of carbon with hydrogen except to mention that both elements form covalent compounds and have some thermodynamic properties (e.g. ionization energies and electron affinity) which are rather similar.

ATOMIC HYDROGEN

As expected from the high heat of dissociation of the H_2 molecule (431 kJ mol^{-1}), hydrogen atoms are formed only at high temperature: at one atmosphere pressure, the degree of dissociation is about 0.001 at 2000°C and 0.95 at 5000°C. Hydrogen atoms can be made more conveniently either by passing an electrical discharge through hydrogen gas at low pressure or by exposing mixtures of hydrogen and mercury to the ultraviolet radiation produced by a mercury arc. At low pressure the atoms recombine only relatively slowly with a half-life of about 1 s at 0.2 mm pressure because a collision between two hydrogen atoms cannot give a molecule unless a third body (e.g. a vessel wall or another hydrogen atom) is present to remove the energy of reaction.

When hydrogen molecules enter into a reaction, one of the steps is the breaking of the strong hydrogen–hydrogen bond. Because of the high energy required for this dissociation, the activation energy of molecular hydrogen reactions is also high and hence many take place only very slowly. Atomic hydrogen is much more reactive because dissociation has occurred prior to reaction, and it is found that hydrogen atoms attack a wide variety of non-transition elements such as germanium, tin, arsenic, antimony and tellurium: e.g.

$$As + 3H \rightarrow AsH_3$$

As might be expected, elements which form very unstable hydrides (e.g. lead and bismuth) do not appear to react with atomic hydrogen.

THE CHEMISTRY OF HYDROGEN

Basically, there are three main options open to the hydrogen atom when it is undergoing chemical reaction:

(a) It can lose an electron to form H^+.
(b) It can gain an electron to form H^-.
(c) It can form normal covalent bonds by sharing two electrons with another atom.

Although the vast majority of hydrogen compounds are covered by these options, hydrogen will participate in three-centre bonds, as in the hydrides of boron, aluminium and beryllium, form interstitial hydrides with transition metals, and when coupled to a highly electronegative atom (fluorine, oxygen or nitrogen) will also exhibit an unusual attractive force called 'hydrogen bonding'. Each of these various types of hydrogen derivative will now be discussed more fully.

Hydrogen present as H^+

A large amount of energy is required to form H^+ from the H_2 molecule:

$$\tfrac{1}{2}H_2 \xrightarrow{\text{217.5 kJ}} H(g) \xrightarrow{\text{1310 kJ}} H^+(g) + e$$

$$\tfrac{1}{2}H_2 \rightarrow H^+(g), \quad +1527 \text{ kJ (mol H)}^{-1}$$

The product of this ionization is the bare proton, which cannot exist except in the isolation of high vacuum because of its extremely small size (its radius is only about $1/50\,000$ that of the Li^+ ion). When generated in the presence of solvents having lone pair electrons, the proton can be stabilized by solvation as, for example, in water:

the oxonium ion H_3O^+ (see page 308)

In the oxonium ion the three hydrogen atoms are more positively charged than hydrogen atoms in the bulk of the surrounding water owing to partial electron migration towards the central (charged) oxygen atom. This will lead to the binding, via the phenomenon of hydrogen bonding (see page 98), of at least three more molecules of water to H_3O^+:

(i.e. $H_9O_4^+$)

In a classical proof of the existence of oxonium ions, Bagster and Cooling mixed H_2O and HBr in anhydrous liquid sulphur dioxide and found that the two compounds dissolved in equimolar proportions, thus indicating that a $1:1$ interaction had occurred. On electrolysis of the solution, water and hydrogen were formed at the cathode and bromine at the anode in proportions expected from Faraday's law if the $1:1$ product was $H_3O^+Br^-$.

Nuclear magnetic resonance and X-ray diffraction techniques have since verified the existence of H_3O^+ in aqueous solutions; also, solid hydrates of a variety of acids have been shown to be oxonium salts:

$$HClO_4.H_2O: \quad H_3O^+ClO_4^-$$

$$HNO_3 . H_2O \quad : \quad H_3O^+ NO_3^-$$

$$H_2SO_4 . 2H_2O \quad : \quad (H_3O^+)_2 SO_4^{2-}$$

$$H_2PtCl_6 2H_2O \quad : \quad (H_3O^+)_2 PtCl_6^{2-}$$

Higher proton hydrates such as $H_5O_2^+$, $H_7O_3^+$ and $H_9O_4^+$ are found in the crystalline polyhydrates of strong acids.

Owing to thermal agitation, the various water molecules in the vicinity of a dissolved proton are continually being changed as is found to be the case with metal cations (page 000). The ionic mobility of H^+ (and OH^-) ions in water is exceptionally high and this is assumed to be due to a 'proton-hopping' mechanism which removes the necessity for the (much slower) physical movement of solvated protons through the solution:

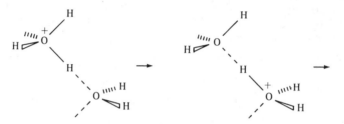

Pure water exhibits a minute electrical conductivity ($\sim 4 \times 10^{-6}\ \Omega^{-1}\ m^{-1}$) due to a slight self-ionization

$$2H_2O \rightleftharpoons H_3O^+ + OH^-, \quad K_w = [H_3O^+][OH^-] \simeq 1 \times 10^{-14}\ mol^2 l^{-2} \text{ at } 25°C$$

which forms the basis of the familiar pH scale of 'hydrogen ion' concentration. By definition, aqueous acids are those substances which dissolve in water to increase the concentration of the H_3O^+ ion and, conversely, bases are those substances which increase the concentration of hydroxyl ions. Solvated protons can exist in a variety of solvents and are not peculiar to aqueous media. For example, in liquid ammonia the proton is solvated to give the NH_4^+ ion, widely known in numerous ammonium salts. Liquid ammonia, like water, possesses a minute electrical conductivity due to the self-ionization

$$2NH_3 \rightleftharpoons NH_4^+ + NH_2^-$$

It is therefore possible to define acids in liquid ammonia as those solutes which increase the concentration of solvated protons:

$$HCl(g) + NH_3(l) \rightarrow NH_4^+ + Cl^- \quad \text{(by reaction with solvent)}$$

$$NH_4Cl(s) \xrightarrow{NH_3(l)} NH_4^+ + Cl^- \quad \text{(by direct ionization)}$$

Bases in this solvent are those substances which dissolve to increase the concentration of amide ions (NH_2^-) in the solvent (see Chapter 12 for a discussion of acids and bases in non-aqueous solvents).

Hydrogen present as H^-

The alkali and alkaline earth metals form colourless, salt-like hydrides on being heated in a hydrogen atmosphere. Only lithium hydride is sufficiently stable to be fused (melting point 688 °C), and electrolysis of the melt produces hydrogen at the *anode* in amounts required by Faraday's law if H^- ions are present. Similar electrolytic experiments have been carried out on solutions of the other hydrides in molten salts or salt eutectics (e.g. the LiCl–KCl eutectic melting at about 360 °C). Electrolysis cannot be carried out in aqueous media because the hydrides are hydrolytically unstable.

In alkali metal hydrides, which have the sodium chloride structure, the ionic radius of H^- is found to be somewhat dependent on the alkali metal, varying from 1.44 Å in lithium hydride to 1.54 Å in caesium hydride (cf. the radius of F^-, which is 1.36 Å). Thus the hydride ion is larger than the isoelectronic helium atom, radius 1.2 Å; this is due to hydrogen having the smaller nuclear charge, resulting in a weaker nuclear attraction of the two electrons in H^- compared to those in the helium atom. On the other hand, Pauling calculated that the ionic radius of the free hydride ion should be 2.08 Å, and the reason that the crystallographic radius is smaller than this is probably due to the high compressibility of the rather tenuous electron cloud in the hydride ion. The densities of saline hydrides are *higher* than those of the free metals (the increase in density is almost 50% for the alkali metal hydrides) owing to the strong Coulombic attraction between the metal and hydride ions which draws them close together and results in a relatively dense crystal.

Covalent hydrides

The derivatives discussed in this section are those in which hydrogen is coupled to another element by normal (i.e. two-electron, two-centre) covalent bonds; the elements forming such hydrides are to be found in Groups IV–VII. Hydrogen atoms are small and their bonds tend to become weaker with the increasing atomic number of the element because the disparity in orbital size makes overlap poor for the heavier elements in each group: for example, methane is thermally very stable whereas plumbane, PbH_4, has only a very fleeting existence. This decreasing stability of the M—H bond with increasing atomic number of the element M means that catenation in the hydrides also decreases down a given group of the periodic table. This is particularly marked in Group IV, where carbon, and to lesser extents silicon and germanium, form long-chain hydrides, but the maximum chain length for tin is two (and even Sn_2H_6 is highly unstable) and for lead is one.

Group IV hydrides

Although a serious discussion of the hydrocarbons will not be attempted here, it is to be remembered that the saturated hydrocarbons C_nH_{2n+2} can exist in a number of isomeric forms if n is greater than 3; thus when $n = 20$ there are over 360 000 structural isomers possible. Not unexpectedly it has been shown that a variety of structural isomers exist for the higher hydrides of silicon and germanium; for example, three pentagermanes Ge_5H_{12} are known:

n-pentagermane   isopentagermane

neopentagermane

The heats of formation of the simple Group IV hydrides are

$$CH_4, \quad -75 \text{ kJ mol}^{-1}$$

$$SiH_4, \quad +33.5 \text{ kJ mol}^{-1}$$

$$GeH_4, \quad +92 \text{ kJ mol}^{-1}$$

$$SnH_4, \quad +163 \text{ kJ mol}^{-1}$$

from which it is apparent that silane, germane and stannane are metastable compounds at room temperature; however, silane only shows noticeable signs of decomposition at temperatures around $300\,°C$. In the complete absence of oxygen, stannane deposits tin very readily at room temperature:

$$SnH_4 \rightarrow Sn + 2H_2$$

but even a trace of oxygen markedly retards the decomposition rate, probably by oxidizing the metallic tin surface upon which the stannane decomposition occurs.

Carbon forms a number of unsaturated hydrocarbons (e.g. olefins and acetylenes) in which some of the carbon atoms are linked together by $p_\pi-p_\pi$ multiple bonds. The other elements in Group IV do not form strong $p_\pi-p_\pi$ bonds, both because there are nodes present in the np orbitals and also inner-shell electron repulsions are thought to be large. Owing to this weak π bonding, no silicon or germanium analogues of ethylene or acetylene (or hydrides containing $Si=C$ and $Ge=C$ bonds) have yet been isolated, although some organo-substituted derivatives are now known (see page 62).

A convenient synthesis for the hydrides of silicon, germanium and tin involves the reduction of the corresponding chlorides with lithium tetrahydroaluminate:

$$SiCl_4 + LiAlH_4 \rightarrow SiH_4 \text{(monosilane; 100\% yield)}$$

$$Si_2Cl_6 + LiAlH_4 \rightarrow Si_2H_6 \quad \text{(disilane)}$$

$$Si_3Cl_8 + LiAlH_4 \rightarrow Si_3H_8 \quad \text{(trisilane)}$$

Acid hydrolysis of either magnesium silicide or magnesium germanide yields a complex mixture of hydrides which can be separated by vapour phase chromatography; in the case of silicon, 21 silanes have been detected, the heaviest being n-octasilane, Si_8H_{18}. When treated with phosphoric acid, a finely ground *mixture* of magnesium silicide and magnesium germanide generates 'mixed' hydrides (e.g. $H_3Si-GeH_3$) as well as silanes and germanes. The highly unstable plumbane PbH_4 can only be prepared in trace quantities via the acid hydrolysis of magnesium–lead alloy. It has been characterized by mass spectrometry but has proved to be far too unstable for its chemistry to be studied.

An interesting property of the higher silanes is that, like the hydrocarbons, they can be 'cracked' into other silicon hydrides on heating: e.g.

$$Si_5H_{12} \rightarrow 2(SiH)_x + Si_2H_6 + SiH_4$$

Conversely, monosilane and monogermane can be 'polymerized' into higher species by subjecting them to electrical discharges, the products again being separated by vapour phase chromatography.

Alkali metal derivatives of the MH_4 hydrides are known for silicon, germanium and tin, and these are the analogues of the metal methyls; for example, potassium germyl is formed when potassium metal reacts with monogermane in liquid ammonia:

$$K + GeH_4 \rightarrow KGeH_3 + \tfrac{1}{2}H_2$$

It reacts in a typical manner with methyl chloride to give methylgermane:

$$KGeH_3 + CH_3Cl \rightarrow CH_3GeH_3 + KCl$$

but the products of reaction between $KGeH_3$ and metal halides are almost always found to be unstable. A rare exception is germylmanganese pentacarbonyl:

$$KGeH_3 + BrMn(CO)_5 \rightarrow H_3GeMn(CO)_5 + KBr$$

Group V hydrides

As found for the Group IV hydrides, the stability of the hydrides in this group decreases with increasing atomic weight of the element, so that bismuthine BiH_3 decomposes rapidly at room temperature. Except for that of bismuth, it is possible to synthesize all the H_2M-MH_2 hydrides and these probably adopt the *gauche* configuration in the liquid and gaseous phases:

hydrazine   end-on view of N_2H_4

Although there is good evidence for the fleeting existence of diazine, N_2H_2, the other elements in the group are not expected to form such a hydride, because the nitrogen atoms in diazine are held by $p_\pi-p_\pi$ multiple bonds:

trans-diazine (di-imide)   *cis*-diazine (di-imide)

The HMH bond angles in phosphine, arsine and stibine suggest that the central atoms are using their p orbitals to bond to hydrogen (see Table 7). The variations from the expected 90° angle can be explained by assuming that hydrogen–hydrogen steric repulsions are partially relieved by a slight opening up of the HMH angle, less opening being required as the M—H bond length increases. The lone pair electrons on the central atoms are in a spherical s orbital, which will make phosphine, arsine and stibine less potentially useful as donor molecules than ammonia, which has the nitrogen lone pair electrons in a highly direction sp^3 hybrid orbital. The steric requirements of the lone pair electrons in ammonia are considered responsible for the slight closing of the HNH angle from the expected tetrahedral (sp^3) value of 109°28′.

Table 7—Bond angles and bond lengths in the MH_3 hydrides

	HMH bond angle (deg)	M—H distance (Å)
NH_3	107	1.015
PH_3	93.5	1.42
AsH_3	92	1.52
SbH_3	91.5	1.71

Group VI hydrides

A normal hydride MH_2 is formed by all the Group VI elements, although H_2Po was only detected by tracer experiments using about 10^{-10} g of the very radioactive polonium. The Group VI atoms are thought to use their outer p orbitals to bind to hydrogen, the deviation in HMH bond angles from the expected value of 90° presumably being due to H---H steric repulsions (see Table 8). H_2S, H_2Se and

Table 8—Physical data on the Group VI hydrides MH_2

	HMH bond angle (deg)	M—H bond length (Å)	$K_1{}^a$	$K_2{}^b$
H_2O	104.5	0.96	—	—
H_2S	92	1.32	10^{-7}	10^{-15}
H_2Se	91	1.47	1.3×10^{-4}	10^{-11}
H_2Te	89.5	1.69	2.3×10^{-3}	2×10^{-11}

$^a K_1$ corresponds to the reaction $MH_2 \xrightarrow{aq} MH^- + H^+(aq)$.
$^b K_2$ corresponds to the reaction $MH^- \xrightarrow{aq} M^{2-} + H^+(aq)$.

H_2Te are usually prepared by the action of acids on metallic sulphides, selenides, and tellurides; they are all very poisonous gases which have offensive smells. Although the heats of formation of H_2Se and H_2Te are positive, the compounds decompose only slowly at room temperature. Oxygen, sulphur and (probably) selenium form M_2H_2 hydrides which are non-planar molecules:

end-on view showing lone pairs

Sulphur readily forms chains with itself (cf. the behaviour of elemental sulphur, page 317) and is the only member of Group VI to form hydrides containing chains of more than two M atoms. These higher sulphanes are all thermodynamically unstable with respect to H_2S and sulphur, and owe their metastable existence at room temperature to a rather high activation energy for the decomposition (the activation energy in the case of H_2S_2 is about 105 kJ mol^{-1}). The components of a sulphane mixture have to be separated by distillation, which causes difficulties due to simultaneous decomposition, but it has proved possible to isolate sulphanes from H_2S_3 to H_2S_8 in a pure state, whilst others up to H_2S_{17} or H_2S_{18} have been obtained in an impure form. Kilogram quantities of the sulphanes may be made either via the acid hydrolysis of sodium polysulphides or from chlorosulphanes:

$$Na + S \rightarrow Na_2S_x$$

$$Na_2S_x \xrightarrow{\text{HCl}} H_2S_n \xrightarrow[\text{thermal cracking}]{\text{distillation and}} H_2S_n$$

$$n = 4, 5, 6 \qquad\qquad n = 2, 3, 4, 5, 6$$

$$2H_2S_x + S_yCl_2 \rightarrow 2HCl + H_2S_{2x+y} \quad \text{(up to } H_2S_{18})$$

Group VII hydrides

The hydrides are known for all the members of this group. Only hydrogen fluoride is strongly associated in the solid state, a zigzag polymer being formed due to hydrogen bonding. The liquid is also associated and the vapour contains polymers up to 60 °C,

the main species present probably being a cyclic hexamer. Hydrogen iodide is rather unstable thermodynamically, the heat of formation being about $+25$ kJ mol^{-1}.

THE HYDRIDES OF BORON

Diborane, B_2H_6

The reduction of a boron halide with lithium tetrahydroaluminate results in the formation of B_2H_6 (diborane) and not the expected monomer BH_3 (borane):

$$BF_3 + LiAlH_4 \xrightarrow{\text{ether}} B_2H_6;$$

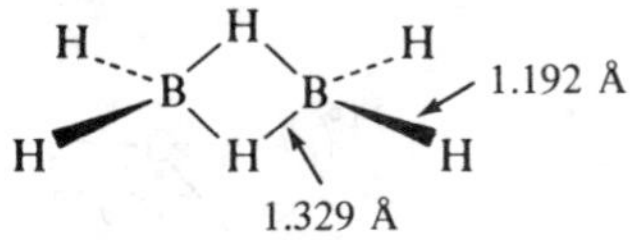

The eight atoms in diborane are held together by only twelve electrons, so clearly there are insufficient electrons to form normal two-centre, two-electron bonds between the atoms; for this reason diborane is often said to be an 'electron-deficient molecule'. An approximate description of the bonding in B_2H_6 can be obtained by assuming that the boron atoms use sp^3 hybrid orbitals to form normal σ bonds to the four terminal hydrogens, whereas in a B—H—B 'bridge' unit two sp^3 hybrid orbitals, one from each boron, simultaneously interact with a hydrogen 1s orbital to produce a bonding molecular orbital which encompasses all three atoms:

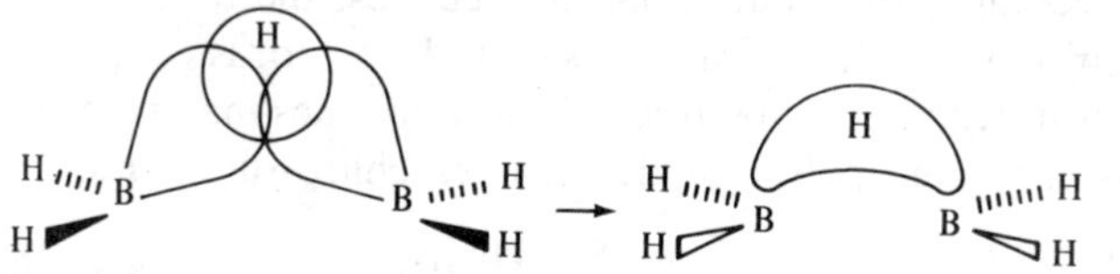

atomic orbital interaction   three-centre molecular orbital
in *each* B—H—B bridge   containing two electrons

Although both boron atoms are able to achieve an octet of electrons in this way, it is found that the B—H bonds within the bridges are about 10% longer than their terminal counterparts; this is not unexpected because only two electrons are available to bind all three bridge atoms together.

The electron deficiency of diborane makes it particularly vulnerable to attack by Lewis bases—molecules which have available lone pairs of electrons. Many bases cleave the diborane molecule *symmetrically* into two BH_3 fragments: e.g.

$$B_2H_6 + 2NMe_3 \rightarrow \quad 2$$

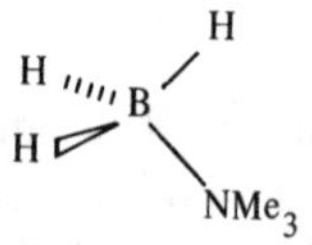

However, when ammonia and monomethylamine are used, a different reaction occurs to give an ionic boronium salt:

$$B_2H_6 + 2NH_3 \xrightarrow[\text{temperature}]{\text{low}}$$

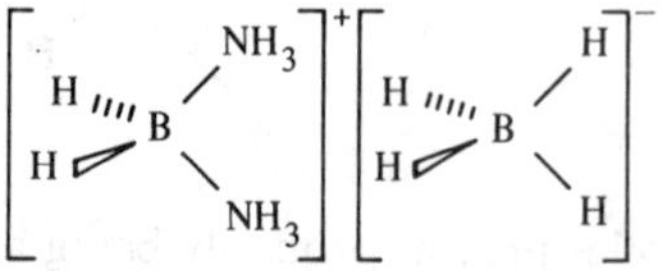

The two types of product probably arise from the differing steric requirements of trimethylamine and ammonia. The initial attack on the diborane is presumably identical for all bases:

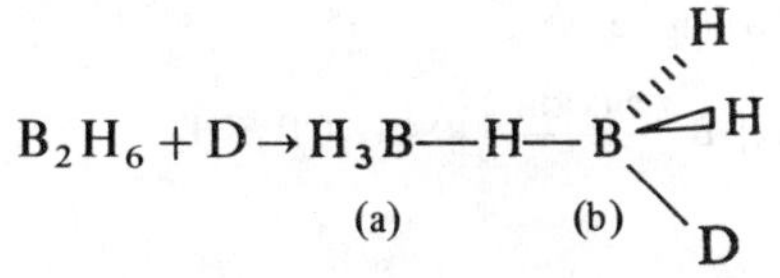

$$B_2H_6 + D \rightarrow \underset{(a)}{H_3B}-H-\underset{(b)}{B}\overset{H}{\underset{D}{\diagdown}}$$

(cf. $H_3BNMe_3 + \frac{1}{2}B_2H_6 \overset{-78}{\rightleftharpoons} H_3B-H-BH_2NMe_3$ (known reaction))

but whereas a small base like ammonia can add to the same boron atom (b) to give *unsymmetrical cleavage* products, a bulky base is forced to attack the less-crowded boron atom (a). The steric requirements must be closely balanced with dimethylamine since both symmetrical and unsymmetrical cleavage of the diborane molecule is observed. Among the more unusual bases which attack diborane are the hydride ion and carbon monoxide:

$$2LiH + B_2H_6 \xrightarrow{\text{ether}} 2LiBH_4 \quad \text{(lithium tetrahydroborate)}$$

$$2CO + B_2H_6 \rightleftharpoons 2H_3BCO \quad \text{(borane carbonyl)}$$

the latter unstable carbonyl being one of the few examples in which carbon monoxide binds to a main group element.

Without doubt the most important property of diborane is its ability to add, in anti-Markovnikov fashion, to alkenes and alkynes, forming boron alkyls in very high yield:

$$6RCH{=}CH_2 + B_2H_6 \rightarrow 2B(CH_2CH_2R)_3$$

The particular significance of this *hydroboration* reaction lies in the fact that the B—C bonds of the boron alkyl are cleaved by a wide variety of reagents to produce many synthetically useful products (Table 9). The diborane may be prepared *in situ* or a commercially available borane complex, such as $H_3B.THF$, can be used directly:

$$12RCH{=}CH_2 + 3NaBH_4 + 4BF_3.OEt_2 \xrightarrow{\text{tetrahydrofuran}} 4(RCH_2CH_2)_3B + 3NaBF_4 + 4Et_2O$$

$$3RCH{=}CH_2 + H_3B.THF \xrightarrow{\text{THF solution}} (RCH_2CH_2)_3B$$

Many functional groups, including esters, nitriles and sulphides, can tolerate hydroboration, and in some cases the scope of the subsequent cleavage reactions can be extended by thermal isomerization of the intermediate boron alkyl:

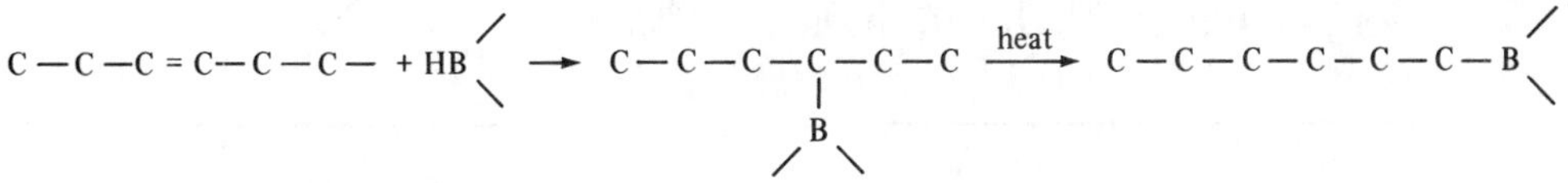

Higher boron hydrides

When diborane is heated it decomposes into a number of other boron hydrides. By subtle variation of the conditions employed during heating, the thermal cracking

Table 9—Reagents used to cleave BR_3 derivatives formed in hydroboration reactions

1 Carboxylic acids $\rightarrow$ RH (hydrogenation of original olefin), e.g.

$$RSCH_2CH{=}CH_2 \xrightarrow{BH_3} (RSCH_2CH_2CH_2)_3B \xrightarrow{R'COOH} RSCH_2CH_2CH_3$$

(note sulphide group)

2 Alkaline peroxide $(OH^-/H_2O_2) \rightarrow$ ROH (anti-Markovnikov hydration of original olefin), e.g.

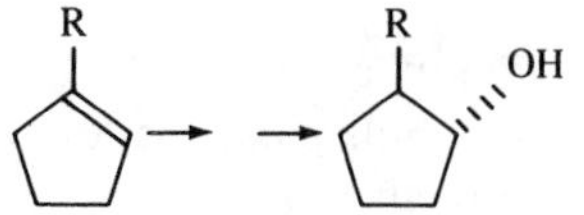

3 Chromic acid (oxidation) $\rightarrow$ ketones from

   acids from

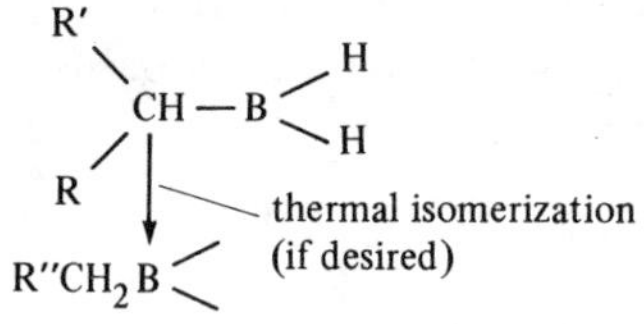

thermal isomerization (if desired)

4 Alkaline $AgNO_3 \rightarrow$ R—R (i.e. coupling; can also get 'mixed' coupling from BR_3 and BR'_3)

5 Carbon monoxide (1 atm); diglyme; 100–$125\,^{\circ}C \rightarrow (R_3CBO)_x \xrightarrow[H_2O_2]{NaOH} R_3COH$ (tertiary alcohol)

6 Carbon monoxide (1 atm); H_2O; diglyme; 100–$125\,^{\circ}C \rightarrow$

$\xrightarrow[H_2O_2]{OH^-} R_2C{=}O + ROH$

7 $I_2/NaOH \rightarrow$ RI, e.g.

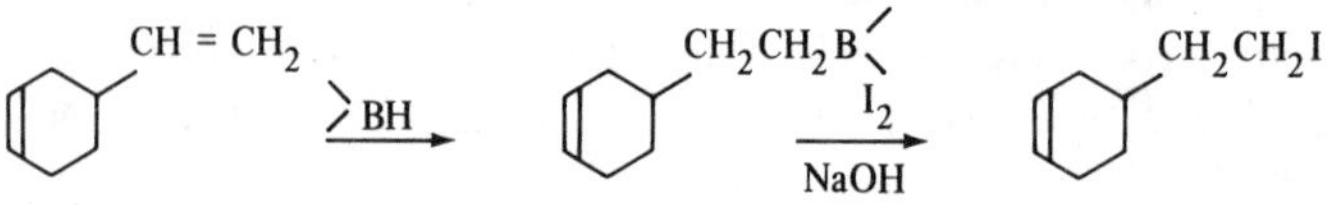

(note that conditions can be manipulated to leave one C=C bond unattacked)

8 N-chlorodialkylamine $(R'_2NCl) \rightarrow$ RCl (anti-Markovnikov addition of HI and HCl to initial alkenes in 7 and 8)

9 Alkali (on hydroboration production from alkene having chlorine on a β carbon) $\rightarrow$ cyclopropane, e.g.

$$CH_2{=}CHCH_2Cl \xrightarrow{BH_3} (ClCH_2CH_2CH_2)_3B \xrightarrow{OH^-} Cl^- +$$

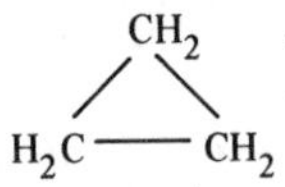

process can be used as a preparative technique which is essentially unique for a particular hydride, although sometimes a small amount of another hydride may be formed simultaneously; fractional distillation under conditions of high vacuum usually enables a ready separation of the two products. For example, when diborane is passed through a tube heated in a furnace to 180–$220\,^{\circ}C$, B_5H_9 (60% yield) and $B_{10}H_{14}$

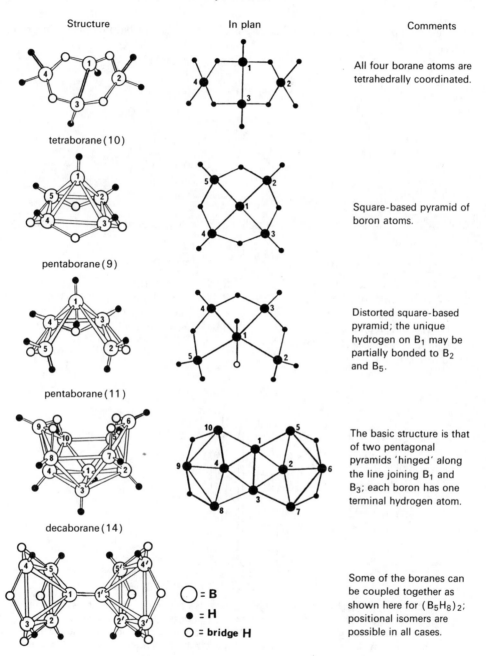

Structure	In plan	Comments
tetraborane(10)		All four borane atoms are tetrahedrally coordinated.
pentaborane(9)		Square-based pyramid of boron atoms.
pentaborane(11)		Distorted square-based pyramid; the unique hydrogen on B_1 may be partially bonded to B_2 and B_5.
decaborane(14)		The basic structure is that of two pentagonal pyramids 'hinged' along the line joining B_1 and B_3; each boron has one terminal hydrogen atom.
	○ = B ● = H O = bridge H	Some of the boranes can be coupled together as shown here for $(B_5H_8)_2$; positional isomers are possible in all cases.

Fig. 18—Structure of some higher boranes.

(6% yield) are formed; when the heating is carried out under static conditions, pure $B_{10}H_{14}$ results:

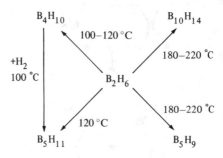

Almost all of the boron hydrides which have been prepared so far fall into two formula groups:

B_nH_{n+4} hydrides		B_nH_{n+6} hydrides	
B_2H_6	diborane; b.p. $-92.5\,°C$	B_4H_{10}	tetraborane; b.p. $18\,°C$
B_5H_9	pentaborane(9); b.p. $48\,°C$	B_5H_{11}	pentaborane(11); b.p. $63\,°C$
B_6H_{10}	hexaborane(10); m.p. $-62\,°C$	B_6H_{12}	hexaborane(12); m.p. $-82\,°C$
B_8H_{12}	octaborane(12);	B_8H_{14}	octaborane(14)
$B_{10}H_{14}$	decaborane(14); m.p. $99.5\,°C$		

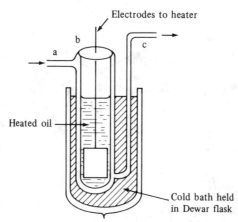

Fig. 19—Hot–cold reactor for pyrolysis of diborane. The pyrolysis apparatus (b) is essentially like an unsilvered Dewar vacuum flask provided with an entry side-arm (a) and an exit (c). The central tube contains electrically heated oil to provide a hot surface on which the diborane decomposes, any thermally unstable products being rapidly quenched out of the reaction zone by condensation onto the cold outer tube.

B_4H_{10} can be prepared by heating the oil to about $120\,°C$ and maintaining the cold bath at $-78\,°C$; when the bath is held at $-30\,°C$ tetraborane is not quenched out and reacts with more diborane to form B_5H_{11} (which freezes out at $-30\,°C$). When the hot and cold surfaces are maintained at 180 and $-78\,°C$ respectively under a static pressure of diborane for 2–3 days, B_5H_9 results. A convenient small-scale source of diborane for these syntheses is the reaction between commercially available sodium tetrahydroborate and iodine:

$$2NaBH_4 + I_2 \xrightarrow[\text{e.g. diglyme}]{\text{a polyether}} B_2H_6 + 2NaI + H_2 \quad (95\%\text{--}100\%)$$

Table 10—Some of the polyhedral cages present in $closo\text{-}B_nH_n^{2-}$ anions and $closo\text{-}B_{n-2}C_2H_n$ carboranes

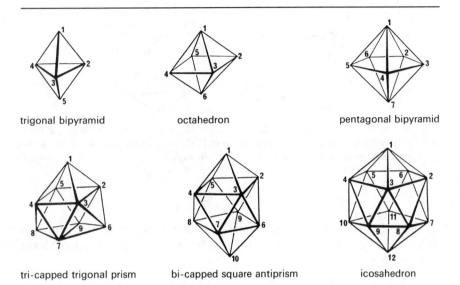

trigonal bipyramid	octahedron	pentagonal bipyramid
tri-capped trigonal prism	bi-capped square antiprism	icosahedron

At least two positional isomers are possible for the *closo*-carboranes; it has been found empirically that the most stable isomer is the one in which the carbon atoms are (a) as far from each other as possible and (b) bonded to the fewest neighbour atoms. Hence in a pyrolytic synthesis the most stable carborane isomer is normally the one isolated: e.g.

$$nido\text{-}2,3\text{-}C_2B_4H_8 \xrightarrow[\text{low pressure}]{450^\circ C} closo\text{-}1,5\text{-}C_2B_3H_5 + closo\text{-}1,6\text{-}C_2B_4H_6 + closo\text{-}2,4\text{-}C_2B_5H_7$$

$$40\% \qquad\qquad 20\% \qquad\qquad 40\%$$

whereas ultraviolet irradiation of $nido\text{-}2,3\text{-}C_2B_4H_8$ at room temperature results in the formation of $closo\text{-}1,2\text{-}C_2B_4H_6$.

(When the nomenclature is ambiguous, as for B_5H_9 and B_5H_{11}, the number of hydrogen atoms present is added to the name using parentheses: pentaborane(9) and pentaborane(11).)

The higher boranes have three-dimensional structures derived from clusters of boron atoms arranged in the form of incomplete 'cages' (Fig. 18); by using suitable reactions it is possible to provide the missing atoms and hence synthesize derivatives having closed cages of borons. As shown in Fig. 20, decaborane(14) is two atoms short of the 12 required to complete an icosahedral cage, and these can be added to it by treatment with Et_3NBH_3:

$$B_{10}H_{14} + 2Et_3NBH_3 \rightarrow [Et_3NH^+]_2[B_{12}H_{12}^{2-}]$$

All the closed-cage anions in the series $B_nH_n^{2-}$, where $n = 6\text{-}12$, have been made and some of their structures are shown in Table 10. The bonding in these ions is very complex, but it appears from theoretical calculations that a B_nH_n cage lacks the two electrons required to fill the last of $(n + 1)$ bonding molecular orbitals, which explains the formation of the *doubly charged* anion $B_{12}H_{12}^{2-}$ described above. A carbon atom is isoelectronic with B^-; thus a pair of carbon atoms substituted for two borons in

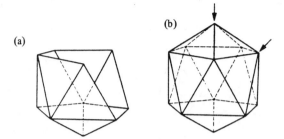

Fig. 20—(a) The boron skeleton of decaborane(14), $B_{10}H_{14}$. (b) The icosahedron, a figure having 20 equilateral triangular faces and a total of 12 vertices. Arrows indicate the two atoms added to the decaborane(14) skeleton to complete the cage.

a hypothetical B_nH_n cage molecule could provide the two extra electrons required for the cluster's stability and, in fact, such a *carborane* is formed when decaborane(14) reacts with ethyne:

$$B_{10}H_{14} \xrightarrow{Et_2S} \xrightarrow{C_2H_2} B_{10}C_2H_{12};$$

○ = BH
● = C

Owing to the high symmetry of an icosahedron, there are only three isomers possible for $B_{10}C_2H_{12}$, depending on the relative positions of the two carbon atoms in the cluster. The product of this particular synthesis is the 1,2-isomer from which the other two can be obtained by thermal isomerizations, the high temperatures involved emphasizing the amazing stability of these compounds:

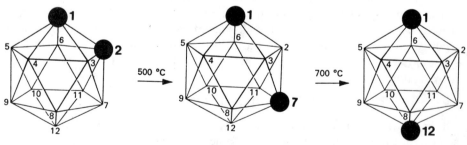

ortho-carborane or 1,2-carborane   *meta*-carborane or 1,7-carborane   *para*-carborane or 1,12-carborane
(most stable isomer)

Because many other carboranes are now known, the commonly used names for these icosahedral carborane isomers are rather ambiguous and the following nomenclature has been recommended:

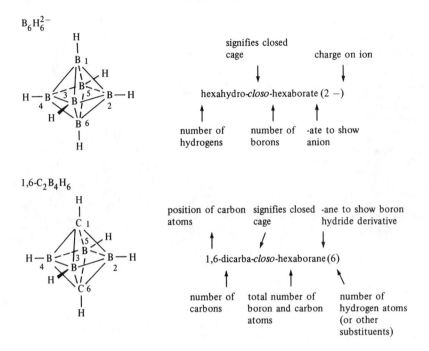

$B_6H_6^{2-}$

signifies closed cage → charge on ion →

hexahydro-*closo*-hexaborate (2 −)

↑ number of hydrogens   ↑ number of borons   ↑ -ate to show anion

$1,6\text{-}C_2B_4H_6$

position of carbon atoms   signifies closed cage   -ane to show boron hydride derivative

1,6-dicarba-*closo*-hexaborane (6)

↑ number of carbons   ↑ total number of boron and carbon atoms   ↘ number of hydrogen atoms (or other substituents)

A number of other species, including NH^+, N, P, As, S^+ and Se^+, are isoelectronic with BH^-, and closo-derivatives containing many of them have been synthesized in recent years; as with the carboranes, positional isomers are possible when two or more heteroatoms are present in the cage cluster:

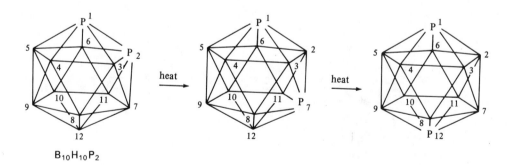

$B_{10}H_{10}P_2$

It is possible to extend these electronic arguments even further by realizing that a boron atom plus a bridge hydrogen, B—H—, is isoelectronic with a carbon atom; hence each BH_2 group (i.e. HB—H—) in a typical boron hydride such as hexaborane(10) can be replaced by CH, the resulting nest-shaped molecules being called *nido*-carboranes ('nido' is derived from the Greek work for a bird's nest):

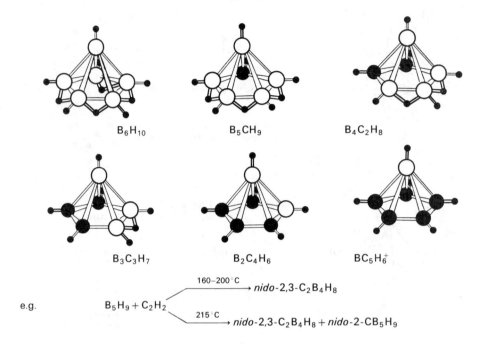

$$B_6H_{10} \qquad B_5CH_9 \qquad B_4C_2H_8$$

$$B_3C_3H_7 \qquad B_2C_4H_6 \qquad BC_5H_6^+$$

e.g. $\qquad B_5H_9 + C_2H_2 \xrightarrow{\text{160–200 °C}} nido\text{-}2,3\text{-}C_2B_4H_8$

$\qquad\qquad\qquad\qquad \xrightarrow{\text{215 °C}} nido\text{-}2,3\text{-}C_2B_4H_8 + nido\text{-}2\text{-}CB_5H_9$

Similarly, transition metal groups such as $Fe^0(CO)_3$ and $Co^{-1}C_5H_5$ are, like BH, four electrons short of a rare gas electronic configuration, and consequently it has proved possible to prepare both *nido*-metalloboranes and *closo*-metallocarboranes containing these fragments:

$$B_5H_9 + NaH \rightarrow H_2 + NaB_5H_8$$

$$NaB_5H_8 + NaC_5H_5 + CoCl_2 \xrightarrow{\text{tetrahydrofuran}} 2\text{-}(CoC_5H_5)B_5H_8$$

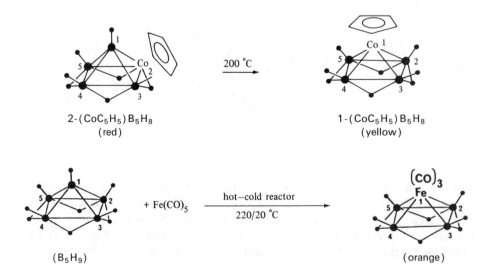

2-(CoC_5H_5)B_5H_8 $\xrightarrow{\text{200 °C}}$ 1-(CoC_5H_5)B_5H_8
(red)   (yellow)

(B_5H_9)   + Fe(CO)$_5$ $\xrightarrow[\text{220/20 °C}]{\text{hot–cold reactor}}$ (orange)

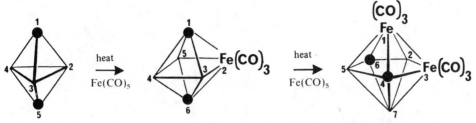

Historically, the first metallocarboranes to be isolated were synthesized by treating transition metal halides with the *nido*-carborane anion $B_9C_2H_{11}^{2-}$, itself derived via degradation of *closo*-1,2-$C_2B_{10}H_{12}$:

$$C_2B_{10}H_{12} \xrightarrow{\ OR^-\ } B(OR)_3 + C_2B_9H_{12}^- \xrightarrow{\ H^-\ } H_2 + C_2B_9H_{11}^{2-}$$

$$2C_2B_9H_{11}^{2-} + FeCl_2 \xrightarrow[\text{atmosphere}]{\ N_2\ } [Fe(B_9C_2H_{11})_2]^{2-}$$

$$\underset{\text{reduction}}{\overset{\text{oxidation by air}}{\rightleftharpoons}}$$

Fe^{2-} Fe^-

$$[Fe(B_9C_2H_{11})_2]^{2-} \qquad\qquad [Fe(B_9C_2H_{11})_2]^-$$

Here the 'ligand' $B_9C_2H_{11}^{2-}$ may be considered as arising from 1,2-$C_2B_{10}H_{12}$ by loss of BH^{2+}; *two* BH^{2+} units require, like *one* Fe^{II} atom, a total of 12 electrons to gain an inert gas electron structure and hence $(BH^{2+})_2$ and Fe^{II} are 'isoelectronic'. Put in terms of electron donation, each $B_9C_2H_{11}^{2-}$ ligand is a six-electron donor (like benzene and the cyclopentadienyl ion $C_5H_5^-$), and by interacting with two of them the iron is able to attain the effective atomic number of krypton.

METALLIC HYDRIDES

Several of the d and f transition elements absorb hydrogen to a widely variable degree, forming hydrides which exhibit many of the physical characteristics of metals: high thermal conductivity, high electrical conductivity, hardness and lustre. However, they are often more brittle than the parent metals. An unusual feature of these hydrides is that they are non-stoichiometric, the composition varying with temperature and external pressure of hydrogen gas (see Fig. 21). A considerable expansion of the metal crystal structure occurs during the process of hydrogen absorption, so that the hydrides are less dense than the parent metals; in many cases this lattice expansion is sufficient

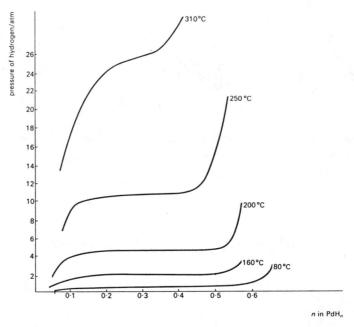

Fig. 21—Pressure–composition isotherms for the palladium—hydrogen system.

to cause a rearrangement of the original metal structure into a completely new one. Typically, the metals titanium, zirconium and hafnium form hydrides of composition between $MH_{1.6}$ and $MH_{1.8}$ which have the fluorite structure (Fig. 41, page 152) in which certain of the 'anion' lattice sites are vacant; with the absorption of more hydrogen the fluorite structure becomes unstable and a change to a face-centred tetragonal structure occurs having the composition $MH_{1.8}$–$MH_{1.98}$.

OTHER HYDRIDES OF THE TRANSITION METALS

In addition to the metallic hydrides described above, transition elements form many discrete species containing covalently bound hydrogen. Perhaps the most unusual of these compounds in terms of the sheer number of hydrogen atoms present is K_2ReH_9, formed by reducing potassium rhenate(VII), $KReO_4$, with potassium metal in damp 1,2-diaminoethane; in the ReH_9^{2-} ion, nine hydrogens surround the Re^{VII} atom in a tri-capped trigonal prismatic array:

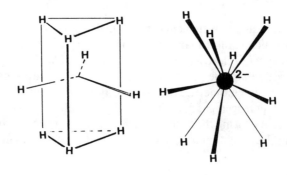

More commonly, however, the metal in these hydrides is in a much lower oxidation state and possesses a number of carbon monoxide, cyclopentadiene or organophosphine ligands. In relatively simple molecules the hydrogen is bound to only one metal:

$$Mn_2(CO)_{10} + Na/Hg \xrightarrow{\text{THF}} Na[Mn(CO)_5] \xrightarrow[\text{[H}^+\text{]}]{\text{acid}} HMn(CO)_5;$$
$$\text{octahedral Mn}$$

but in others, containing metal atom clusters, the hydrogen is often found to form *bridges* with two, or even three, metals:

$$Os_3(CO)_{12} \xrightarrow[\text{110°C; octane}]{H_2} H_2Os_3(CO)_{10};$$

$$Re_2(CO)_{10} \xrightarrow[\text{110°C}]{H_2} H_4Re_4(CO)_{12};$$

one hydrogen is above the centre of each face in the Re_4 tetrahedron and bridges three Re atoms

Very recently these studies have led to the isolation of complexes in which an *intact H_2 molecule* bonds sideways to a transition metal:

$$M(CO)_3(PPr_3^i)_2 \xrightarrow{H_2} M(CO)_3(PPr_3^i)_2(H_2) \quad (M = \text{Mo or W})$$

A whole range of compounds is known which have the stoichiometry metal:2H, but in most cases two metal–hydrogen bonds are present. However, neutron diffraction studies on single crystals of the above tungsten derivative showed that an H_2 ligand was present, the H—H distance being 0.82 Å (cf. 0.74 Å in H_2). Furthermore, the proton NMR spectrum of $W(CO)_3(PPr_3^i)_2(HD)$ showed a 1:1:1 triplet for the HD ligand (nuclear spin of D equals 1) with an H—D coupling constant of 33.5 Hz; the coupling constant for HD gas is 43.2 Hz and for $M{\overset{\text{D}}{\underset{\text{H}}{\diagdown}}}$ complexes is less than 2 Hz. In the iron and ruthenium phosphine complexes $MH_4(PR_3)_3$, neutron diffraction shows that one pair of hydrogen atoms is present as two normal M—H ligands whereas the other pair constitute an intact H_2 ligand, showing that both types can actually be present within the same molecule.

In these complexes an empty orbital on the metal probably accepts electron density from the σ bond of H_2 whilst a simultaneous 'back donation' of electrons occurs from a filled metal d orbital to the empty σ^* antibonding orbital in H_2:

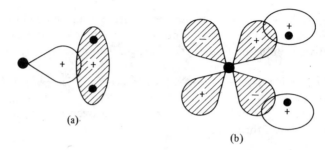

(a)

(b)

THREE-CENTRE, TWO-ELECTRON C—H—METAL BONDS

It is rapidly becoming apparent that C—H bonds of both alkyl and aryl groups are
sometimes able to interact with neighbouring metal atoms, presumably via 3c–2e
bonding. The phenomenon can usually be detected by the marked distortion of ligand
geometry required to place the hydrogen atom in its binding position relatively close
to an electropositive metal (which can be from either the main group or transition
elements). As an example, the methyl group in the titanium complex illustrated below
has one of its C—H—Ti angles reduced from the expected tetrahedral value to 94°
as the hydrogen 'tucks in' close to titanium:

THE HYDROGEN BOND

The rare gases helium, neon and argon do not form any chemical compounds and
yet at sufficiently low temperatures they first liquefy and then freeze to the solid state
(helium requires the application of about 25 atm pressure to enable it to solidify).
Hence weakly attractive forces (called van der Waals, or London, forces) must occur
between the atoms. This cohesion was first explained by London in 1930 by assuming
that although the time-averaged electron distribution in the rare gas atoms was
spherical, at any one instant in time it will be distorted and give rise to a dipole
owing to the vibration of the electron cloud with respect to the nucleus. This
instantaneous dipole will polarize the electron clouds of neighbouring atoms and
induce an appropriately orientated dipole in these neighbours; because these dipoles
are suitably orientated the net result is an attraction, the size of which is proportional
to α^2/r^6, where α is the polarizability of the rare gas atoms which are at distance r
apart. The polarizability increases with increasing size of the atoms so that the melting
and boiling points (which are a measure of the cohesive forces between the atoms)

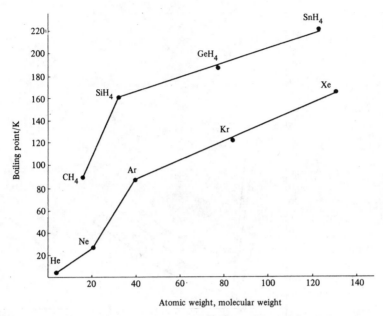

Fig. 22—Plot of boiling point against atomic weight or molecular weight for the rare gases and the hydrides of the Group IV elements.

will rise with increasing atomic size. As seen from Fig. 22, the boiling points of the rare gases increase in a fairly regular manner with increasing atomic weight of the elements. The same phenomenon is observed for the boiling points (and melting points) of the non-polar hydrides of carbon, silicon, germanium and tin.

However, when the boiling (or melting) points of the MH_3 hydrides of Group V, the MH_2 hydrides of Group VI or the MH hydrides of Group VII are plotted against their molecular weights, it is seen from Fig. 23 that NH_3, H_2O and HF have boiling (or melting) points which are very much higher than the values predicted from the trends within their group. In other words, the cohesive forces between the NH_3, H_2O or HF molecules are larger than those arising from just the normal van der Waals and dipole–dipole attractions. This extra cohesive force between these molecules is attributed to 'hydrogen bonding', and such bonding only occurs when a hydrogen atom is bonded to highly electronegative atoms such as nitrogen, oxygen and fluorine. The hydrogen atom taking part in hydrogen bonding is almost always found to lie on, or very near to a straight line joining the two other atoms involved in the interaction. This line usually points in the direction of regions of high electron density (e.g. lone pairs) on the atom to which the hydrogen is not covalently bonded; an example for oxygen is:

$$\overset{\delta-}{O^1}—\overset{\delta+}{H}--\left(\,\cdot\cdot\,\overset{..}{O}{}^2\right.$$

Owing to the electronegative nature of oxygen there will be an electron drift towards O^1 from the O^1—H covalent bond, which will leave the hydrogen atom positively charged; hence a crude description of hydrogen bonding could be given in terms of

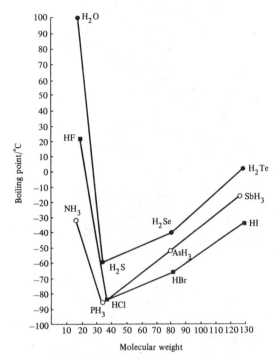

Fig. 23—Plot of boiling point against molecular weight for the hydrides of the elements of Groups V–VII.

a purely electrostatic attraction between the positive hydrogen atom and the electrons on O^2. However, the interaction is considerably more complex than this, so much so that a completely satisfactory interpretation of all the factors involved has yet to be given. Coulson suggests that four effects probably contribute to hydrogen bonding:

(a) Electrostatic attraction; $\overset{\delta-}{O^1}$—$\overset{\delta+}{H}$---$\overset{\delta-}{O^2}$.

(b) Delocalization (or covalent bonding) effect; the electrostatic attraction in (a) will polarize (i.e. pull) the electron clouds in O^2 towards the H—O^1 region of space.

(c) Electron cloud repulsion; the sum of the van der Waals radii for hydrogen and oxygen is 2.6 Å and yet in O---H hydrogen bonding the two atoms often approach to within 1.6 Å. Thus the normal electron–electron repulsive forces will occur when the charge clouds in the O^1—H bond and O^2 atom begin to overlap.

(d) London (or van der Waals) forces; as in all intermolecular interactions these forces contribute to the binding, but their effect is relatively small.

From Table 11 it is seen that the electron–electron repulsion term is numerically one of the largest terms in the hydrogen bond interaction and, if it became larger by only a factor of two, it would cancel out the other bonding contributions entirely. Other elements cannot bond in a similar manner to hydrogen because the presence of their inner electron shells would increase the repulsion term to such an extent that the overall interaction would be antibonding.

Table 11—Estimates of the energy contributions to each separate hydrogen bond in ice

Type of energy contribution	Energy $(kJ\ mol^{-1})$
(a) Electrostatic	25
(b) Delocalization	34
(c) Repulsion	-34
(d) London forces	13
Total	38

(cf. experimental value, 25.5 $kJ\ mol^{-1}$)

Detection of hydrogen bonding

(a) Abnormal melting and boiling points can be indicative of hydrogen bonding as shown in Fig. 23.

(b) Unexpectedly high molecular weights of hydrogen-containing species are often due to hydrogen bonding. For example, many carboxylic acids are dimeric in hydrocarbon solvents and in the vapour state: e.g.

$$R-C\begin{matrix} {\diagup}O{\cdots}H{-}O \\ {\diagdown}O{-}H{\cdots}O \end{matrix}C-R$$

These structures of gaseous formic and acetic acids ($R = H$, CH_3) have been confirmed by electron diffraction measurements.

(c) The infrared stretching frequency of an O—H or N—H group taking part in hydrogen bond formation will be lowered because the 'freedom' of the hydrogen atom is reduced by the hydrogen bond interaction. (In a simple analogy it is as if the weight of the hydrogen atom has been increased by attaching the second atom to it, thus making it vibrate more slowly.) For example, the O—H stretching frequency of H_2O is 3756 cm^{-1} in steam, 3453 cm^{-1} in liquid water and 3256 cm^{-1} in ice.

(d) X-ray and neutron diffraction studies on single crystals give the precise positions of the atoms within the crystal; the abnormally close approach of hydrogen atoms to electronegative groups can be interpreted in terms of hydrogen bonding. X-rays are diffracted by regions of high electron densities, and since hydrogen atoms possess relatively few electrons, X-ray diffraction techniques cannot always detect hydrogen positions accurately in crystals. Neutrons are diffracted by nuclei, and although the experimental difficulties are formidable, the positions of hydrogen atoms (or, more easily, deuterium atoms, since these have a larger neutron scattering radius) can be precisely determined.

Examples and importance of hydrogen bonding

Water

It is evident from Fig. 23 that without the effects of hydrogen bonding, water vapour would only begin to liquefy at about −70 to −80°C, or almost 200°C lower than the actual boiling point. Under such conditions life as we know it, with its absolute dependence on liquid water, could not have evolved. Also hydrogen bonding plays a major role in the chemistry of a wide variety of reactions basic to life, as discussed briefly in Chapter 13.

The isolated water molecule is often assumed to have the central oxygen atom sp^3-hybridized with two sets of lone electron pairs occupying two of the sp^3 orbitals. In ice the water molecules are found to be aligned so that hydrogen bonds are formed along the direction of the oxygen lone pairs, resulting in a tetrahedral coordination for each oxygen atom, the two covalently bonded hydrogens being closer to the oxygen than the hydrogen-bonded atoms:

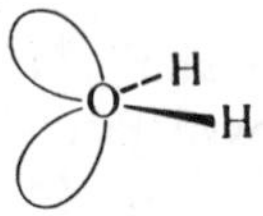

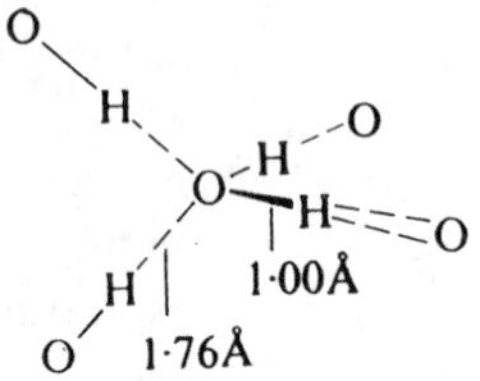

In a normal solid with no strong *directional* forces between the molecules the latter pack as close together as space and molecular shape allow. However, in ice the highly directional hydrogen bonds are stronger than other intermolecular forces and the water molecules have to adopt a much more open structure than would otherwise be the case. On warming, hydrogen bonds in ice begin to rupture, but a sufficient number still remain at 0°C for ice to retain an open structure and have a density lower than liquid water at the same temperature; about half the maximum number of hydrogen bonds are still present in water at 20°C.

Solid ammonium fluoride, NH₄F

Ammonium fluoride has, like water, a very open structure in the solid state. This is thought to be due to hydrogen bonding between the ammonium NH_4^+ and fluoride F^- ions, which results in tetrahedral coordination of both ions. In contrast ammonium chloride adopts the caesium chloride structure in which the ions have eight-coordination:

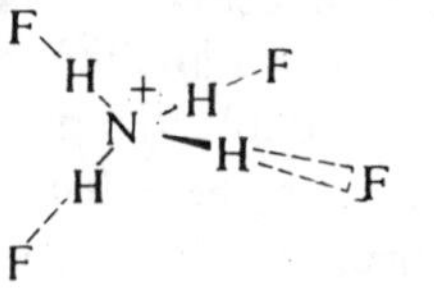

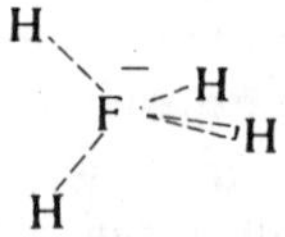

Hydrogen fluoride

The structure of solid hydrogen fluoride is that of a zigzag polymeric chain. Although the reason for the HFH bond angle of 120° is not clear, the angle is very different from that expected (180°) if the structure were dominated purely by electrostatic alignment of the highly polar HF molecules. Hydrogen bonds persist in liquid hydrogen fluoride and the vapour is even associated up to 60 °C, probably as $(HF)_4$ and $(HF)_6$ polymers.

'Hydrogen bifluoride' ion, HF_2^-

Fluoride ions readily attach themselves to free hydrogen fluoride molecules to give the 'hydrogen bifluoride' ion:

$$F^- + HF \rightarrow [F—H—F]^-$$

Unlike most other cases of hydrogen bonding, in which one bond to hydrogen is shorter than the other, the two H—F bond lengths in HF_2^- are identical. Careful X-ray measurements show that there is a substantial amount of electron density around the hydrogen nucleus so that the interaction cannot be a purely electrostatic one between two fluoride ions and a proton, i.e. $F^-H^+F^-$. The energy of the dissociation

$$[F—H—F]^- \rightarrow HF + F^-$$

is very much larger (~ 243 kJ mol^{-1}) than many other hydrogen bond energies; e.g. see those in water, Table 11. For this reason it has been suggested that the bonding in HF_2^- may be unique and perhaps involve three-centre bonding, the hydrogen 1s orbital interacting with two fluorine 2p orbitals.

Bicarbonates

In solid bicarbonates, such as $NaHCO_3$, hydrogen bonds link the planar CO_3^{2-} ions into polymeric chains, the crystal as a whole being bound together by electrostatic interaction between the metal cations and the negatively charged bicarbonate chains:

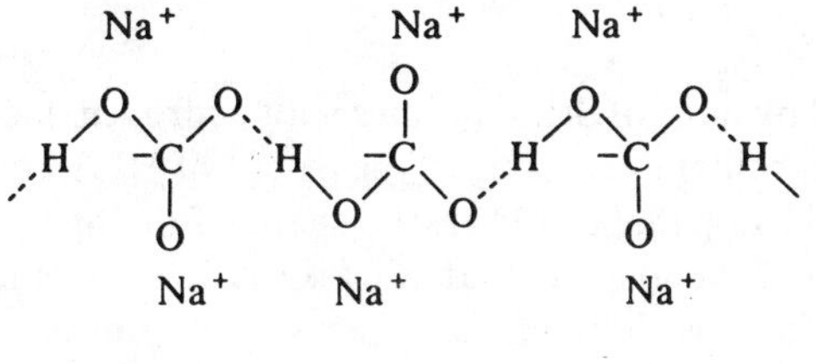

Nitrophenols

Of the three isomeric mononitrophenols, it is found that *ortho*-nitrophenol has the lowest melting point, the lowest boiling point and the highest solubility in organic solvents. This is considered to be due mainly to *intra*molecular hydrogen bonding in this isomer. The *meta* and *para* isomers form *inter*molecular hydrogen bonds, resulting in relatively high intermolecular forces in the solid and liquid states which affect physical properties such as melting and boiling points and solubility in non-polar solvents. Several other *ortho*-substituted phenols are also internally hydrogen-bonded:

salicyclic acid   salicylaldehyde

HYDROGEN BONDING PUT TO AN UNUSUAL USE

It is now possible to determine the gas phase structures of very weakly bonded molecules, in particular those which are held together by hydrogen bonds. Hydrogen fluoride is often a component of such systems and can thus be used to detect regions of high electron density in other molecules, as shown in the examples below:

(pyramidal O)   (pyramidal S)

The shapes adopted by most of the molecules with hydrogen fluoride are in agreement with simple theories of bonding, even to the extent of HF locating the region of the 'bent bonds' in cyclopropane; however, the shape of the hydrogen fluoride dimer is unexpected on steric grounds and that of $H_2O.HF$ requires further discussion. As discussed on page 59, it is thought that the oxygen atom does not use sp^3 hybrid orbitals when binding to hydrogen atoms in an isolated water molecule and hence there are no lone pair electrons residing in 'rabbit's ears' orbitals. However, such a bonding model would be expected to predict an $H_2O.HF$ molecule which was either planar or shaped like $H_2S.HF$. The actual configuration taken up by $H_2O.HF$ is only slightly more stable than the planar form, which shows that there is only a small, *but definite*, electron density increase in the region close to that usually associated with sp³ lone pairs. Hence our bonding theories seem somewhat inadequate in that sp³ hybridization of the oxygen atomic orbitals over-emphasizes the electron density from lone pairs whereas the pure p orbital description suggests there should be no electron density build-up at all in the region found to be occupied by the HF molecule.

This build-up of electron density localized in the 'lone pair region' of the water molecule can explain both the three-dimensional hydrogen-bonded structure of ice and the observation that in salt hydrates the coordinated water molecules take up a range of geometries. The orientation adopted by H_2O in many hydrates is apparently determined more by the possibility of the oxygen accepting hydrogen bonds from other water molecules in the crystal than by the characteristics of the M—O bond; hence the oxygen tends to assume a tetrahedral arrangement involving the metal atom, two covalently bound hydrogens and a hydrogen bond:

In solid hydrates θ varies between 0° and about 30°.

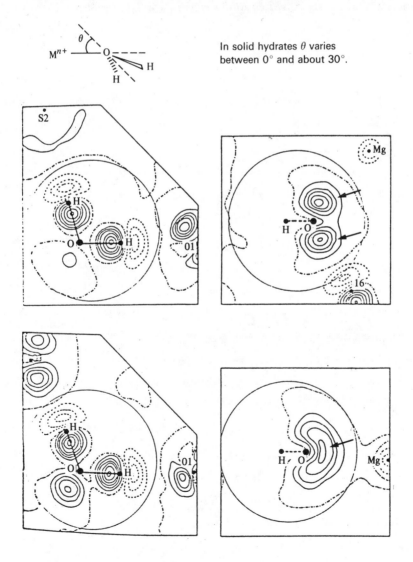

An illustration of how the electron density around the oxygen of a water molecule can apparently vary with the chemical environment is shown above in the electron density maps for two types of water molecule found in magnesium thiosulphate hexahydrate, $MgS_2O_3.6H_2O$. The left-hand diagrams show the density within the plane of the two water molecules and clearly indicate the build-up of electrons within the O—H bonding region; there is also a depletion of electron density (dotted lines) behind the hydrogen atoms which no doubt strongly aids hydrogen bonding.

The right-hand contour diagrams are taken at right angles to the water molecule plane. The H_2O molecule in the top diagram is involved in an Mg---O interaction *and* a hydrogen bond; the two lone pairs which are required for this are clearly visible (see arrows). The other water is making only one bond to the nearby magnesium and hence has trigonal geometry; here there is a single half-moon-shaped lone pair region with only *one* maximum in the electron density map (arrowed). [These diagrams are taken from *Acta Crystallographica* vol. B42 (1986) p. 26.]

Chapter 3

Group I: The Alkali Metals Lithium, Sodium, Potassium, Rubidium, Caesium and Francium

INTRODUCTION

Although the alkali metals all have the same outer electron configuration, ns^1, their ionization energies do not decrease smoothly from Li to Cs. In the following chapters this will be seen as a recurring feature of main group chemistry and it arises from the differing inner electron 'cores' which elements in the various periods possess. As the principal quantum number increases down Group I the ns electron becomes further from the nucleus and hence the first ionization energies will naturally decrease down the group; this general trend is shown by Li, Na and K, but the ionization energies of Rb and Cs are considerably higher than expected (Fig. 24).

Rubidium and caesium differ from the other alkali metals by having filled d orbitals within their electron cores and, due to the poor shielding characteristics of d orbitals discussed in Chapter 1, their outermost electrons experience an abnormally high effective nuclear charge. In the same way, filling of the 2p orbitals from B to Ne results in a steadily increasing effective nuclear charge across the period which has the effect of making the 3s electron of the next element, sodium, much more strongly held than would otherwise have been the case if all orbitals had identical shielding properties. (Of course, had there been no difference in orbital shielding characteristics the heavier atoms in the periodic table would be quite enormous; in fact there is only a modest increase in size from potassium to caesium; Fig. 25.)

The alkali metals are more chemically similar than any other group of elements within the periodic table, but lithium does sometimes show anomalous behaviour when properties depending markedly on size are considered. Often there is a remarkable and close resemblance in the chemical properties shown by lithium and magnesium which may be partially due to their very similar ionic radii; Fig. 25. It cannot yet be claimed that this *diagonal relationship* is fully understood but it does provide a useful way of remembering 'uncharacteristic' properties of lithium. If the

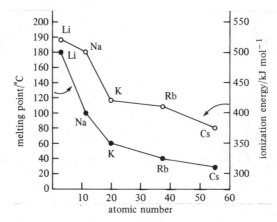

Fig. 24—Ionization energies (kJ mol^{-1}) and melting points (°C) of the alkali metals plotted against atomic number. The ionization energy of a hydrogen atom with its electron in an ns orbital is $1310/n^2$ kJ mol^{-1}; thus H($2s^1$) has an ionization energy of 327.5 kJ mol^{-1}. The comparative value for Li ($1s^2 2s^1$) is 520 kJ mol^{-1} and demonstrates the penetration effect of the 2s orbital (see page 42). This penetration results in the effective nuclear charge experienced by the 2s electron being about 1.28 rather than 1.00 as would be the case if the outer electron were perfectly shielded from the nuclear charge (3.00) by the two 1s electrons; penetration similarly increases the outer electron binding energy (and hence the ionization energy) of the other alkali metals. Poor mutual shielding of p and d orbitals during the build-up of elements preceding the alkali metals causes a gradual contraction of atomic size along that period and hence the outer s electrons of Na, K, Rb and Cs are able to approach closer to the nucleus, and thus be more strongly bound, than would otherwise be expected. It must be noted that *once the p and d orbitals are full they shield any outer electrons in subsequent elements very effectively.*

phenomenon were simply due to similar atomic or ionic sizes there would be so many other diagonal relationships as to make the periodic table almost redundant.

All the alkali metals adopt the body-centred cubic structure in the solid state, each atom providing only its outer s electron to hold the crystal lattice together; see page 125. Most other metals possess at least twice as many lattice-binding electrons and consequently the Group I elements are found to be comparatively soft and have

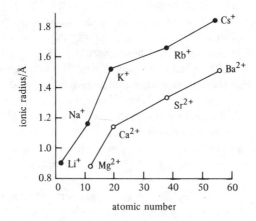

Fig. 25—Ionic radius plotted against atomic number for Group I and Group II ions.

low melting points. The steady decrease in melting point shown in Fig. 24 is due to two main factors: (a) the progressively inefficient overlap that occurs with the larger bonding s orbitals and (b) the increasing repulsive forces which arise from inner-shell electrons.

FORMATION OF M⁺ CATIONS

The electronic configuration of the alkali metals indicates that the simplest way they can achieve a rare gas electron structure is to form M^+ ions via loss of their outer s electron. Hence in the majority of their compounds the elements are found to exist as 'bare ions' in the solid state or as solvated ions in suitable solvents. However, even the formation of M^+ ions is an energy-consuming process and it is of some interest to inquire where the energy required for this ionization comes from.

Thermochemical measurements have shown that the reaction between sodium and chlorine takes place with the evolution of 410.9 kJ of heat for each mole of crystalline sodium chloride formed. Using the Born–Haber cycle, this reaction can be broken down, theoretically, into a series of more fundamental processes:

$$Na(s) \quad + \quad \tfrac{1}{2}Cl_2(g) \xrightarrow[\text{25°C, 1 atm}]{\Delta H_f^\ominus} NaCl(g)$$

$$\downarrow +\Delta H_{\text{sublimation}} \qquad \downarrow +\tfrac{1}{2}\Delta H_{\text{dissociation}} \qquad \uparrow -\text{lattice energy } L$$

$$Na(g) \quad + \quad Cl(g) \xrightarrow[-EA_{Cl}]{+IE_{Na}} Na^+(g)+Cl^-(g)$$

where EA_{Cl} is the electron affinity of a chlorine atom. By Hess's law, the standard heat of formation of sodium chloride, $\Delta H_f^\ominus$, is equal to the sum of all the terms in the cycle:

$$\Delta H_f^\ominus = \Delta H_{\text{sublimation}}^\ominus + \tfrac{1}{2}\Delta H_{\text{dissociation}}^\ominus + IE_{Na} - EA_{Cl} - L$$

$$= 108.7 + \tfrac{1}{2}(242.7) + 496.0 - 351 - 786$$

$$= -410.9 \text{ kJ mol}^{-1}$$

Thus the main source of energy which compensates for the formation of Na^+ ions is seen to be the lattice energy of the crystalline reaction product, sodium chloride; Fig. 26. When the ions are formed in aqueous solution the ionization energy of sodium is offset by the high heats of hydration of the cation and anion: e.g.

$$Na^+(g)+Cl^-(g) \xrightarrow[\text{water}]{\text{dissolve in}} Na^+(aq)+Cl^-(aq) \quad -759.8 \text{ kJ mol}^{-1}$$

The presence of M^+ ions in a number of crystalline salts of the alkali metals has been demonstrated by accurate X-ray diffraction measurements. Such experiments enable the number of electrons around the cation and anion nuclei to be calculated because the extent of diffraction is a function of the electron density at the ion sites in a crystal; Fig. 27.

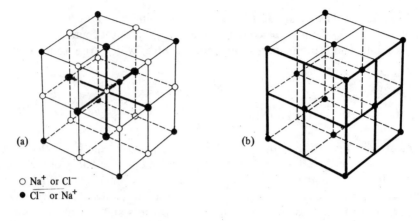

○ Na⁺ or Cl⁻
● Cl⁻ or Na⁺

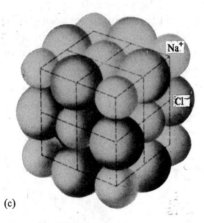

(c)

Fig. 26—(a) The sodium chloride structure, showing the octahedral coordination of the ions. (b) Because identical ions (either sodium or chloride) are located at the corners and the face centres of a cube, sodium chloride is said to have a 'face-centred cubic' structure. (c) the packing of the sodium and chloride ions in crystalline sodium chloride. For further details, see the Appendix.

POSSIBILITY OF DIVALENT SPECIES

Similar Born–Haber cycles to the one above can be used to investigate the feasibility of making stable salts of divalent alkali metals. Fluorine is the most reactive of all the elements which attack the alkali metals and hence would stand the best chance of oxidizing these metals to the M^{2+} state via a reaction such as

$$NaF(s) + \tfrac{1}{2}F_2(g) \xrightarrow[25°C,\ 1\ atm]{\Delta H} NaF_2(s)$$

$$Na^+(g) + F^-(g) + F(g) \xrightarrow[IE_{Na^+}]{-EA_F} Na^{2+}(g) + 2F^-(g)$$

L_{NaF} $\tfrac{1}{2}\Delta H^{\ominus}_{\text{dissociation}}$ $-L_{NaF_2}$

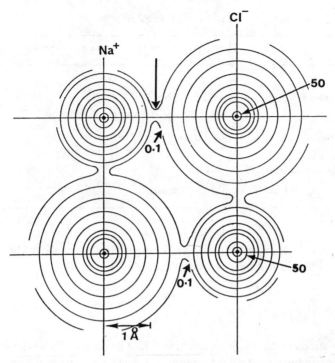

Fig. 27—Contour map of the electron density round the nuclei in crystalline sodium chloride. The figures give the electron densities in electrons per Å^3 round each nucleus. By integration, the number of electrons was calculated to be 10.05 round the sodium nucleus and 17.75 round the chlorine nucleus; the close agreement with the theoretical values of 10 electrons for Na$^+$ and 18 for Cl$^-$ demonstrates that the crystal is made up of ions. The position of the electron density minimum (arrowed) along a line joining the sodium and chlorine nuclei allows the radius of both ions to be calculated. [For further details see H. Witte and E. Wöolfel, *Zeitschrift für Physikalische Chemie*, vol. 3 (1955) p. 296, from which this diagram was copied.]

where IE_{Na^+} is the *second* ionization energy of sodium:

$$Na^+(g) \rightarrow Na^{2+}(g) + e^-$$

$$\therefore \Delta H = L_{NaF} + \tfrac{1}{2}\Delta H^{\ominus}_{dissociation} - EA_F + IE_{Na^+} - L_{NaF_2} \quad kJ \, mol^{-1}$$

If it is assumed that the lattice energy of NaF_2 would be approximately equal to that of MgF_2, then all the terms in this equation are known:

$$\Delta H = 899.4 + 157.7/2 - 349.5 + 4564 - 2908$$

$$= 2284.7 \, kJ \, mol^{-1}$$

Hence this reaction is thermodynamically unfavourable (as are all other routes to NaF_2) because, although the lattice energy of NaF_2 is very high due to a combination of small fluoride ions and doubly charged sodium cations, it is still insufficient to compensate for the (even higher) second ionization energy of sodium. This calculation demonstrates the well-known 'stability of the rare gas electron configuration': electrons in inner quantum shells, owing to a combination of their being closer to

Table 12—Some properties of the alkali metals and their ions

Property	Lithium	Sodium	Potassium	Rubidium	Caesium
Atomic number	3	11	19	37	55
Electron configuration	$[He]2s^1$	$[Ne]3s^1$	$[Ar]4s^1$	$[Kr]5s^1$	$[Xe]6s^1$
Melting point (°C)	180	98	64	39	29
Boiling point (°C)	1330	892	759	700	690
Ionic radius (Å) (from electron densities)	0.93	1.17	1.49	1.64	1.83
Ionization energy (kJ mol^{-1})	520.0	496.0	418.8	412.3	375.9

the nucleus and shielding effects, experience high effective nuclear charges and are consequently difficult to remove from an atom.

COVALENT DERIVATIVES

Not all compounds of the alkali metals are ionic. The vapour above the liquid metals, though predominantly monatomic, contains a small proportion of dimer molecules in which the bond is thought to arise by overlap of the atoms' outer s orbitals. Although the bonding is similar to that which occurs in H_2, the interaction in the metal dimers will be weaker because of (a) the strong repulsion forces arising between the inner electron shells of the two metal atoms and (b) the diffuse nature of, and the presence of nodes in, the outer ns orbitals which result in their poor overlap. The dissociation energy of the dimers can be expected to decrease from Li_2 to Cs_2 since both of these effects lead to a progressive weakening of the metal-to-metal bond; Table 13.

Furthermore, many organolithium derivatives appear to be highly covalent in that they dissolve in hydrocarbons and are often dimeric, tetrameric or hexameric when in the solid, gas or solution phases; the structures of these polymers are thought to involve multicentred bonds between lithium and carbon. A typical example is the methyllithium tetramer $(CH_3Li)_4$ in which the four lithium atoms are arranged in

Table 13—Bond lengths and dissociation energies of the alkali metal dimers at 25°C

Dimer	Bond length (Å)	Dissociation energy (kJ mol^{-1})
Li_2	2.673	113.8
Na_2	3.079	73.2
K_2	3.923	49.4
Rb_2		47.3
Cs_2		43.5
cf. H_2	0.74	435.9

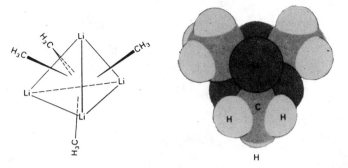

Fig. 28—The structure of the methyllithium tetramer.

the form of a tetrahedron with a methyl group placed above the midpoint of each tetrahedral face; Figs 28 and 29. If the metal orbitals are sp^3-hybridized then, by suitable orientation, three of the hybrid orbitals on each lithium will point towards the neighbouring methyl groups. Above each face of the tetrahedron, overlap of four sp^3 hybrid orbitals, one from each of the three lithium atoms and one from the methyl carbon atom, can then occur to form four molecular orbitals; there are just sufficient electrons available to fill the strongly bonding orbital, leaving the two non-bonding and the single antibonding orbitals empty (see Fig. 29). This description of the bonding leaves each lithium with one empty sp^3 orbital; in agreement with this, it is found that some Lewis bases will form complexes, $(MeLi)_4(base)_4$, with the methyllithium tetramer, presumably by donating electrons into these four empty orbitals.

GASEOUS HALIDES

On strong heating in a vacuum the alkali metal halides vaporize to give monomeric molecules MX which could conceivably be covalently bonded; however, their very high dipole moments show them to be simply ion pairs, M^+X^-. The alkali halide vapours also contain higher ion aggregates such as M_2X_2, M_3X_3 and, in some cases, even M_4X_4; see Table 14. Electron diffraction studies have shown that the dimers Li_2F_2 and Li_2Cl_2 are planar, diamond-shaped molecules and not tetrahedral clusters:

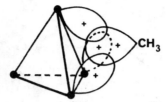

Fig. 29(a)—Overlap of four sp^3 hybrid orbitals in one face of the Li$_4$ tetrahedron in $(CH_3Li)_4$.

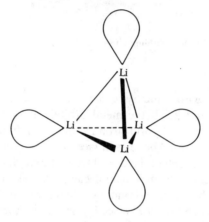

Fig. 29(b)—The four empty sp^3 orbitals which the methyllithium tetramer can use to form complexes.

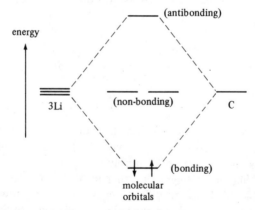

Fig. 29(c)—Energy level diagram for the four-centre bonding in *one* face of the Li$_4$Me$_4$ tetrahedron. (For simplicity, the carbon and lithium sp^3 hybrid atomic orbitals are shown as having identical energies.)

CATION HYDRATION

The unit charge on alkali metal cations results in their having a relatively weak ion–dipole interaction with water molecules when in aqueous solution; for example,

Table 14—Typical gaseous molecules of the alkali metals

Molecules[a]	Dipole moment of monomer (Debye)	Bond length of diatomic molecule (Å)
LiF, Li_2F_2, Li_3F_3	6.28	1.56
LiCl, Li_2Cl_2, Li_3Cl_3, Li_4Cl_4		
$LiNaF_2$ (i.e. 'mixed' dimer)		
NaF, Na_2F_2, Na_3F_3, Na_4F_4		
KCl, K_2Cl_2, K_3Cl_3	10.5	2.67
CsF, Cs_2F_2	7.9	2.35
CsCl, Cs_2Cl_2	10.4	2.91
CsI	12.1	3.315
NaOH, $Na_2(OH)_2$		
CsOH, $Cs_2(OH)_2$	7.1 (dimer is non-polar)	Cs—O (in monomer) = 2.40
$NaK(OH)_2$ (i.e. 'mixed' dimer)		
$NaNO_3$		
$NaNO_2$		
$RbNO_3$		
Cs_2O_2	0.0	
Cs_2SO_4	0.0	

[a] For a given halide, the polymers are less readily formed at a particular temperature the larger the alkali metal ion. For example, lithium chloride is predominantly a dimer in the gas phase at 820 °C whereas gaseous caesium chloride is monomeric at this temperature. This is mainly due to a decrease in the Coulombic attraction between the ions as the metal ion radius increases.

X-ray diffraction data on solutions of rubidium bromide suggest that only a single, *ill-defined* hydration shell exists around the cation within which the water molecules assume a wide range of coordination distances. On the other hand, the smallest cation in the group, Li^+, is octahedrally coordinated to six inner waters which then have further solvent molecules hydrogen-bonded to them; in very concentrated solutions of lithium chloride ($\sim 14.9M$) there is so little water available that up to three chloride ions are forced to enter the first coordination sphere of the cation in place of some of the solvent.

The relatively low affinity which the group I cations have for water results in many salts, particularly those of the heavier alkali metals K, Rb and Cs, separating from solution as anhydrous crystals. Within the group lithium forms by far the largest number of salt hydrates, many of which contain three water molecules: $LiX.3H_2O$, where $X = Cl^-$, Br^-, I^-, ClO_3^-, ClO_4^-, MnO_4^-, NO_3^- or BF_4^-. Lithium chloride, bromide and iodide are particularly unusual among the alkali halides in being *deliquescent* and in forming a range of hydrates:

$LiCl.H_2O$	$LiBr.H_2O$	$LiI.0.5H_2O$
$LiCl.3H_2O$	$LiBr.2H_2O$	$LiI.H_2O$
$LiCl.5H_2O$	$LiBr.3H_2O$	$LiI.2H_2O$
	$LiBr.5H_2O$	$LiI.3H_2O$

In $LiClO_4.3H_2O$, and probably the other trihydrates also, each lithium is surrounded by six water molecules which it shares with two neighbouring cations to form a linear polymer as the basic unit in the crystal:

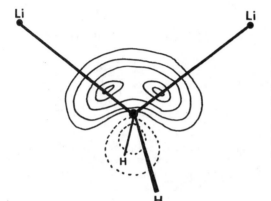

An electron density map, obtained from X-ray diffraction data on the crystalline hydrate of a lithium salt, showing the attraction between lithium ions and the two lone pairs of electrons on a water molecule.

The alums, double sulphates of univalent and trivalent metals having the general formula $M^I(H_2O)_6M^{III}(H_2O)_6(SO_4)_2$, represent the most common examples of crystalline salts which contain hexahydrated K^+, Rb^+ and Cs^+ ions; the six water molecules and a further six oxygen atoms from two sulphate groups complete 12-coordination around the heavier alkali metal cations. No alums have yet been isolated which contain lithium, presumably because this ion is too small to be accommodated satisfactorily into the alum crystal structure.

ALKALI METAL COMPLEXES

The single positive charge on the alkali metal cations is responsible for their having rather few well-characterized complexes because, as in the hydrates, the ion–dipole interactions between a potential ligand and the cations are weak. The stabilities of any complexes which do form are normally in the order $Li > Na \gg K > Rb > Cs$ expected from ionic size considerations; two typical examples formed by a simple ligand are the ammines $Li(NH_3)_4I$ and $Na(NH_3)_4I$ which contain tetrahedral cations. The fact that lithium chloride, bromide and iodide are somewhat soluble in polar organic solvents such as alcohols and ethers can be attributed to relatively strong coordination of solvent molecules around the small cation.

In recent years the number of known alkali metal complexes has been considerably increased by using multidentate ligands such as *crown polyethers* and *cryptands*:

A typical poly-ether ligand such as this is often known by its shorthand name dibenzo-18-crown-6 (there are *eighteen* atoms in the *crown*-shaped ring, *six* of which are oxygen atoms).

A cryptand; the metal ion is held in a cavity (i.e. a 'crypt') within the tricyclic ligand structure. When $x = y = z = 1$ the crypt ligand is usually referred to as [2,2,2]; similarly [2,1,1] would have $x = 1$, $y = z = 0$.

When dibenzo-18-crown-6 forms a complex, the metal ion is normally situated centrally in the ring with all the oxygen atoms planar. By making the polyether ring larger, as in dibenzo-30-crown-10, it is possible for the ligand to wrap completely around the metal so that all ten oxygen atoms can coordinate to the cation as shown in Fig. 30.

The cavity inside the cryptand [2,2,2] is about 2.8 Å across and it is easily able to accommodate either K^+ or Rb^+ giving a roughly spheroidal cation with an overall diameter of 10 Å. In K[2,2,2]I the potassium is eight-coordinated and lies in the centre of a trigonal antiprism with the nitrogen atoms capping both triangular faces; the iodide ions merely occupy holes in the lattice between these bulky, cryptated cations. By changing the values of x, y and z in the above formula it is possible to 'tailor' the size of the ligand cavity to fit a particular alkali metal cation; thus cryptands [2,1,1], [2,2,1] and [2,2,2] *preferentially* complex with Li^+, Na^+ and K^+ ions respectively.

The effect of ion charge is also important in determining ion selectivity as shown by the fact that cryptand [2,2,2] prefers Ba^{2+} relative to K^+ by a factor of 10^4. NMR studies of $Cs[2,2,2]^+$ in several solvents have demonstrated that the large Cs^+ ion can only just be held inside the ligand; in fact a rapid equilibrium is set up in solution between fully encapsulated and partially encapsulated caesium ions:

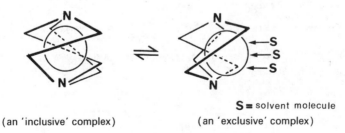

S = solvent molecule

(an 'inclusive' complex)   (an 'exclusive' complex)

METAL ANIONS

The crypt ligands have proved to be very useful aids during the synthesis of unusual anions by so increasing the size of neighbouring cations that essentially all their polarizing character is lost; for example, $Na[2,2,2]^+$ forms stable compounds with

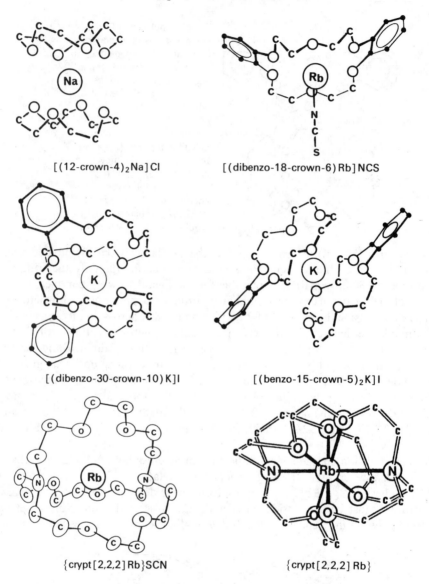

[(12-crown-4)₂Na]Cl

[(dibenzo-18-crown-6)Rb]NCS

[(dibenzo-30-crown-10)K]I

[(benzo-15-crown-5)₂K]I

{crypt[2,2,2]Rb}SCN

{crypt[2,2,2]Rb}

Fig. 30(a)—Some crown-ether and crypt complexes of alkali metal cations.

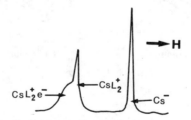

Fig. 30(b)—^{133}Cs NMR spectrum of caesium metal dissolved in a 5% solution of 12-crown-4 (L) in tetrahydrofuran. [For details of the experiment see *Journal of the American Chemical Society* (1988) p. 1618.]

Pb_5^{2-}, Sn_9^{4-} and Sb_7^{3-}, the ligand preventing a recombination of the sodium with tin, lead or antimony to reform the initial alloys used in the preparations.

In a logical, but rather extraordinary, extension to this latter series of anionic compounds it is found that golden crystals of $Na[2,2,2]^+Na^-$, isostructural with $Na[2,2,2]^+I^-$, form when sodium dissolves in ethylamine containing dissolved cryptand [2,2,2]. The sodide ion Na^- is spherically symmetrical with a diameter of ~ 4.6 Å and has two electrons in the outer 3s orbital. *Alkalide anions* are now known for all the alkali metals except lithium; it is found that lithium dissolves in a methylamine solution of cryptand [2,1,1] to give a dark blue lithium *electride* $Li[2,1,1]^+e^-$ rather than Li^-. Similar reactions occur in the presence of crown ethers. Thus when caesium dissolves in 12-crown-4 (L), NMR studies have confirmed the formation of the three species CsL_2^+, Cs^- and $CsL_2^+e^-$; mixed metal systems give CsL_2^+ and M^-, where $M = Na$, K or Rb. (Note the coordination of two molecules of the small 12-crown-4 ligand to Cs^+ in these derivatives.)

In the black electride $Cs(18\text{-crown-}6)_2^+e^-$ each electron is trapped and localized in an almost spherical, otherwise empty, cavity of radius 2.4 Å formed by the packing of eight complexed caesium ions.

SOLUBILITIES OF ALKALI METAL SALTS

The extent of a salt's solubility in water arises from a balance between the lattice energy of the solid (tending to hold the ions within the crystal) and the hydration energy of the constituent ions (trying to disperse the ions into solution). Therefore one must consider both of these factors when trying to rationalize solubilities. The fact that lithium fluoride is the least soluble of the alkali halides (0.13 g per 100 g H_2O at 20 °C) is probably due to its high lattice energy arising from the presence of the two smallest univalent ions in the crystal, but it should be realized that the hydration energy of these small ions will be high also; it just so happens that the lattice energy is the larger of the two terms. Virtually all alkali metal salts are soluble in water because usually their lattice energies, and the hydration energies of their ions, are approximately equal to each other. However, it is sometimes found that salts of K, Rb and Cs containing large anions are insoluble, presumably because the low hydration energies of these huge cations and anions are insufficient to overcome even the *correspondingly low* lattice energies:

$$NaB(C_6H_5)_4 + M^+ \rightarrow MB(C_6H_5)_4\downarrow \qquad \text{tetraphenylborate}$$

$$Na_3Co(NO_2)_6 + 3M^+ \rightarrow M_3Co(NO_2)_6\downarrow \qquad \text{hexanitrocobaltate(III)}$$

$$Na_2SiF_6 + 2M^+ \rightarrow M_2SiF_6\downarrow \qquad \text{hexafluorosilicate}$$

$$Na_2PtCl_6 + 2M^+ \rightarrow M_2PtCl_6\downarrow \qquad \text{hexachloroplatinate(IV)}$$

$$NaClO_4 + M^+ \rightarrow MClO_4\downarrow \qquad \text{perchlorate}$$

The solubilities of potassium, rubidium and caesium tetraphenylborates are sufficiently low to allow their use in the gravimetric estimation of these heavier alkali metals. Unfortunately, however, there are no salts of either lithium or sodium which are insoluble enough for accurate analytical work; probably the most insoluble sodium salt is the yellow zinc uranyl ethanoate $NaZn(UO_2)_3(CH_3COO)_9.9H_2O$ frequently used in qualitative tests for sodium.

FLAME COLORATION

Salts of the alkali metals give very characteristic colours when introduced into a bunsen flame: lithium, crimson; sodium, golden yellow; potassium, red-violet; rubidium, blue-violet; caesium, blue-violet. Except for lithium, where the crimson colour arises from the presence of excited LiOH molecules, these colorations are due to electronic transitions which occur in transient metal *atoms*. Flames are a rich source of electrons, some of which attach themselves momentarily to vaporized cations:

$$M^+ + e^- \rightleftharpoons M$$

Many of these atoms are in an excited state and will emit radiation as their outer electron undergoes transitions to orbitals of lower energy; some of these transitions, e.g. $ns \leftarrow np$ and $ns \leftarrow (n+1)p$, happen to be of the correct energy for their associated *emission* spectrum to fall within the visible region and impart a characteristic colour to the flame. In the *flame photometer* a given alkali metal salt solution is injected into an oxygen-gas flame and the intensity of the emission spectrum measured using filters to block out light at other wavelengths. By comparing this intensity to those of given standard solutions of the same metal, the concentration of the unknown sample can be determined.

SOLUTION CHEMISTRY

Alkali metal salts are completely ionized in aqueous solution and, because the univalent cations are relatively non-polarizing, hydrolytic reactions of the type encountered with small, highly charged ions (e.g. Be^{2+} and Al^{3+}):

$$(H_2O)_nM^{m+}-O\begin{array}{c}H\\\\H\end{array} \rightleftharpoons [(H_2O)_nMOH]^{(m-1)+} + H_{aq}^+$$

do not occur. The hydroxides are strong bases and hence salts of all but the very weakest acids may be made by simply mixing alkali with an equivalent amount of acid and crystallizing the solution: e.g.

$$NaOH + HA \rightarrow NaA + H_2O$$

The solution chemistry of the alkali metals is not very extensive because few, if any, complexes are stable in aqueous media. More commonly, their cations simply behave passively as 'bystanders' in other reactions such as neutralizations and metatheses:

$$(M^+) + OH^- + H_{aq}^+ + (Cl^-) \rightarrow (M^+) + (Cl^-) + H_2O$$

$$(M^+) + Cl^- + Ag^+ + (NO_3^-) \rightarrow (M^+) + (NO_3^-) + AgCl\downarrow$$

VARIATION IN PROPERTIES DOWN GROUP I

It is mainly in the solid state that differences in physical and chemical properties become apparent within the group. Some of the typical properties which change in the order $Li > Na > K > Rb > Cs$ are:

(a) hardness, melting points, sublimation energies of the metals;
(b) ease of thermal decomposition of compounds containing polyatomic anions (carbonates, nitrates, perchlorates, hydroxides, peroxides and polyhalides);
(c) lattice enthalpies of salts (although there are irregularities when the salts contain a small ion owing to inefficient packing in the lattice).

However, there are some properties for which the above order is not observed. A simple example is the reaction of the metals with water where the reactivity sequence is $Cs \sim Rb \sim K > Na > Li$. This may be a kinetic effect in that lithium is the only member of the group which does not melt during the reaction and so does not continually present a clean, reactive surface to the water. Ionic mobilities, which are a measure of the relative velocity of ions under a potential gradient, show exactly the opposite trend to that which might have been predicted from ionic sizes: small ions would be expected to experience less viscous drag when moving through a solution and hence have the higher velocities for a given potential gradient. However, Li^+ is heavily hydrated in solution and this coordinated water markedly increases the *effective size* of the lithium cation.

From Table 15 it is apparent that the electrode potential of lithium does not follow the trend for the rest of the group. Again it is possible to trace this back to the small size, and hence the high heat of hydration, of the lithium ion. The standard electrode potential of lithium, $E^\ominus$, is related to the free energy change $\Delta G^\ominus$ of the reaction

$$Li_{aq}^+(a = 1) + \tfrac{1}{2}H_2(1 \text{ atm}) \rightarrow Li^+(s) + H_{aq}^+(a = 1)$$

by the equation $-\Delta G^\ominus = nFE^\ominus$, where n is the number of electrons involved in the reaction ($n = 1$) and F is the Faraday constant. The half-reactions $Li_{aq}^+ \rightarrow Li(s)$ and $\tfrac{1}{2}H_2 \rightarrow H_{aq}^+$ involve the following (hypothetical) steps:

Table 15—Comparison of the terms involved in the calculation of electrode potentials for the alkali metals

Metal	Electrode potential, $M^+ + e^- \rightleftharpoons M$	Enthalpy of hydration, M^+ (kJ mol^{-1})	Ionization energy (kJ mol^{-1})	Heat of sublimation (kJ mol^{-1})	Calculated electrode potential (V)
Lithium	−3.03	−521	520	148	−3.0
Sodium	−2.71	−406	496	99	−2.6
Potassium	−2.93	−322	419	79	−2.7
Rubidium	−2.93	−301	403	76	−2.7
Caesium	−2.92	−277	376	67	−2.8

[Data taken from *Physical Data for Inorganic Chemists* by M. C. Ball and A. H. Norbury (Longmans, 1974); the enthalpies of hydration for the cations are based on the assumption that the hydration enthalpy of a proton is −1091 kJ mol^{-1}.]

$$Li_{aq}^+ \xrightarrow[\substack{(521)}]{-\Delta H_{hyd}} Li^+(g) \xrightarrow[\substack{(-520)}]{-IE_{Li}} Li(g) \xrightarrow[\substack{(-148)}]{-\Delta H_{sub}} Li(s) \quad \Sigma\Delta H_{Li} = -147 \text{ kJ mol}^{-1}$$

$$\tfrac{1}{2}H_2(g) \xrightarrow[\substack{(218)}]{\tfrac{1}{2}\Delta H_{diss}} H(g) \xrightarrow[\substack{(1310)}]{IE_H} H^+(g) \xrightarrow[\substack{(-1091)}]{\Delta H_{hyd}} H_{aq}^+, \quad \Sigma\Delta H_H = +437 \text{ kJ mol}^{-1}$$

If entropy changes in these two half-reactions are assumed to cancel, then $\Delta H^\ominus = \Delta G^\ominus$ for the whole reaction and hence $-(+290) = nFE_{Li}^\ominus$, giving $E_{Li}^\ominus = -3.01$ V. Table 15 shows that although the three terms making up $\Sigma\Delta H_m$ are numerically largest for lithium, the hydration enthalpy of Li^+ is mainly responsible for the unexpected sequence of electrode potentials. Although such calculations give the approximate order of potentials, the numerical values obtained for the heavier metals are not very accurate because the assumption that the entropy changes in the above reactions for H^+ and M^+ are equal is less valid for large cations.

COMPARISON OF ALKALI METAL CATIONS WITH AMMONIUM IONS

Salts of the ammonium ion may usually be prepared by treating the relevant acids with aqueous ammonia and crystallizing the resulting solutions. The unipositive NH_4^+ ion has a radius in between that of K^+ and that of Rb^+ which results in many ammonium salts being isomorphous to those of the heavy alkali cations. For example, NH_4Cl, NH_4Br and NH_4I adopt the caesium chloride structure in their low-temperature form but above a critical temperature (184, 138 and $-18\,°C$ respectively) undergo a transition to the sodium chloride structure. Ammonium fluoride is different in that it assumes an ice-like lattice due to $NH---F$ hydrogen bonding. Most ammonium salts are soluble in water, the few common exceptions being shown in Table 16 from which the striking similarity to K^+ and Rb^+ is obvious. Crypt complexes, in which the ligand fully encompasses the ammonium ion, are also known:

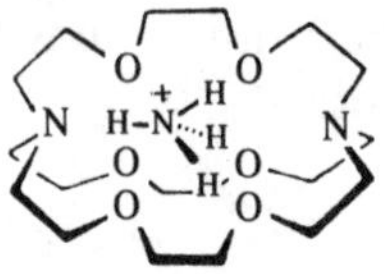

$X^- \quad (X^- = Cl^-, NO_3^-, ClO_4^- \text{ or } PF_6^-)$

The main difference between the chemistry of ammonium and alkali metal salts is the relative ease with which the former decompose to release ammonia. This can occur either by warming aqueous solutions with alkali:

$$NH_4Cl + NaOH \rightarrow NH_3\uparrow + H_2O + NaCl$$

or by gentle heating of solid salts:

$$NH_4Cl \rightarrow NH_3 + HCl$$

$$(NH_4)_2CO_3 \rightarrow 2NH_3 + H_2O + CO_2$$

$$NH_4NO_3 \rightarrow N_2O + 2H_2O$$

$$NH_4NO_2 \rightarrow N_2 + 2H_2O$$

$$3(NH_4)_2SO_4 \rightarrow 2SO_2 + 4NH_3 + 6H_2O + N_2$$

Table 16— Properties of aqueous alkali and ammonium cations

Reagent	Li^+	Na^+	K^+, Rb^+, Cs^+, NH_4^+
1 Sodium carbonate solution	White Li_2CO_3 ppted from conc solutions	No reaction (carbonate soluble)	No reaction (carbonate soluble)
2 Ammonium fluoride in the presence of aqueous NH_3	White gelatinous precipitate of LiF	No reaction (fluoride soluble)	No reaction (fluorides soluble)
3 Sodium hexanitrocobaltate(III) solution	Yellow precipitate of lithium hexanitrocobaltate(III) from extremely concentrated solutions	No reaction	Yellow precipitates of $M_2NaCo(NO_2)_6$ or $M_3Co(NO_2)_6$ depending on conditions
4 Tartaric acid solution	No reaction (tartarate soluble)	No reaction (tartarate soluble)	White precipites of hydrogen tartarates $MHC_4H_4O_6$ from concentrated solutions
5 Alkaline disodium hydrogen phosphate	White precipitate of Li_3PO_4	No reaction (phosphate soluble)	No reaction (phosphates soluble)
6 Sodium tetraphenylborate	No reaction (tetraphenylborate soluble)	No reaction (tetraphenylborate soluble)	White precipitates of $MB(C_6H_5)_4$
7 Perchloric acid solution	No reaction (perchlorate soluble)	No reaction (perchlorate soluble)	White precipitates of $MClO_4$ from concentrated solutions of K^+, Rb^+, Cs^+; no precipitate with NH_4^+ ions
8 Hexachloroplatinum(IV) acid	No reaction (Li_2PtCl_6 soluble)	No reaction (Na_2PtCl_6 soluble)	Yellow precipitates of M_2PtCl_6 from concentrated solutions
9 Zinc uranylethanoate solution	Yellow precipitate from concentrated solution	Yellow precipitate of $NaZn(UO_2)_3(CH_3COO)_9 \cdot 9H_2O$	Yellow precipitates from moderately concentrated solutions

A variety of alkyl-substituted ammonium halides may be prepared by the addition of an acid or an alkyl halide (usually the iodide) to amines:

$$EtNH_2 + HCl \rightarrow EtNH_3^+ Cl^-$$

$$Bu_2NH + HBr \rightarrow Bu_2NH_2^+ Br^-$$

$$NMe_3 + MeI \rightarrow NMe_4^+ I^-$$

$$NMe_3 + HF \rightarrow NMe_4^+ F^-$$

The halide may be converted into other salts by first passing it through an ion exchange column to convert it to the (strongly basic) hydroxide and then adding the acid of choice. These large, univalent cations, having a rather hydrophobic exterior, are not extensively hydrated in aqueous solution. In fact, to quote one author '... the hydration of NMe_4^+ is so weak that normal water persists right up to the surface of the ion'.

OCCURRENCE OF THE ALKALI METALS

Lithium

65 ppm total abundance; mainly in spodumene, $LiAl(SiO_3)_2$, and lepidolite, $Li_2(F,OH)_2Al_2(SiO_3)_3$; ^6Li 7.3%, ^7Li 92.7%.

Sodium

28 300 ppm total abundance; largest deposits are as rock salt, NaCl.

Potassium

25 900 ppm total abundance; mainly in natural brines (as KCl) and as carnallite, $KCl.MgCl_2.6H_2O$; of the three natural isotopes ^{39}K, ^{40}K and ^{41}K, one (^{40}K 0.01%) is radioactive (β^-) with a half-life of 1.5×10^{13} years.

Rubidium

310 ppm total abundance; no single source, occurs most often in the mica lepidolite (0%–3.5% Rb_2O); only about 50 kg prepared annually; ^{87}Rb ($\sim 28\%$ of natural rubidium) is radioactive (β^-) with a half-life of 6.3×10^{10} years.

Caesium

7 ppm total abundance; occurs in pollucite, $CsAl_2(SiO_3)_2.xH_2O$; main source, however, is lepidolite which is used for lithium extraction.

Francium

The longest-lived of the known isotopes is francium-223 which has a half-life of only

21 min; hence it does not occur in nature in meaningful quantities, although it is formed from actinium:

$$^{227}_{89}\text{Ac} \rightarrow {}^{223}_{87}\text{Fr} + \alpha$$

$$^{223}_{87}\text{Fr} \rightarrow {}^{223}_{88}\text{Ra} + \beta^-$$

EXTRACTION

All the metals are highly reactive towards water and cannot therefore be prepared by electrolytic reduction of salts in aqueous media. Industrially the electrolysis of the fused chloride or, sometimes, the hydroxide can be used. However, this method is normally used only for lithium and sodium; in the Downs process, for example, the electrolysis of the fused chloride is carried out. The operating temperature is kept at about 500 °C by adding calcium chloride (up to 67 %) to the sodium chloride so that the electrolysis is actually carried out on the eutectic, melting point 505 °C (the melting point of pure sodium chloride is 803 °C). The discharge potential for the sodium ion is lower than for calcium under these conditions and thus sodium is preferentially formed; the 1 %–2 % calcium liberated is insoluble in the (molten) sodium and precipitates back into the eutectic as it is formed. The low operating temperature does not cause difficulties due to the volatility of the sodium, boiling point 886.6°C. The older Castner process, using molten sodium hydroxide, has a much lower current efficiency and is not now widely used.

Electrolysis of fused potassium chloride presents problems owing to the extensive solubility of potassium in KCl and the high temperatures involved which cause vaporization of the isolated metal; attempts to reduce the operating temperature by adding other halides to the potassium chloride have not been very successful owing mainly to potassium ions having a higher discharge potential than the added metal. The modern method is to employ a large fractionating tower filled with stainless steel Raschig rings; molten potassium chloride is fed into the centre of the tower and meets a counter-current of sodium vapour introduced from the bottom:

$$\text{KCl} + \text{Na} \overset{850\,°\text{C}}{\rightleftharpoons} \text{K} \uparrow + \text{NaCl}$$

This equilibrium, which lies slightly to the right-hand side, is driven to completion by the high volatility of the potassium; purification of the metal is carried out by fractional distillation.

Caesium is prepared in a similar manner. Sources of rubidium almost always contain caesium as impurity and, because metallic rubidium and caesium are very difficult to separate, the rubidium salt is normally purified by repeated fractional crystallization prior to its reduction by sodium or calcium at 700–800 °C. If a mixture of rubidium and caesium halides should be reduced (as for potassium chloride), then the resultant metal must be subjected to careful fractional distillation, and even so the rubidium may contain up to 0.1 % caesium.

METALLIC BONDING

The solid alkali metals adopt the body-centred cubic structure described in the Appendix on page 521. In an alkali metal crystal, essentially a giant molecule consisting

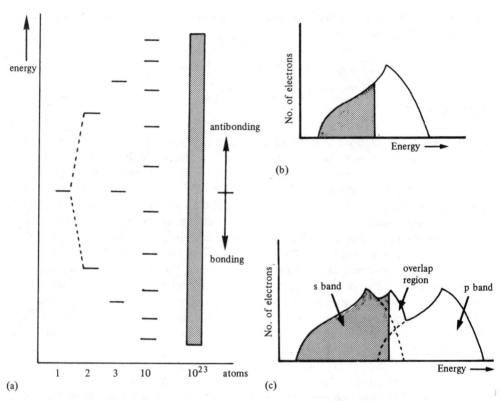

Fig. 31—(a) The development of an s energy band from the delocalized molecular orbitals in a one-dimensional metal containing n atoms. The band, shown shaded, contains 10^{23} orbitals of various energies. (b) An alternative way of describing an s band shows the variation in numbers of electrons having energies lying between the most bonding and most antibonding situations. For an alkali metal only the shaded portion of the band would contain electrons. (c) The s and p energy bands can overlap and here the shaded area represents the electrons of a typical Group II element populating these two bands.

of n bonded atoms, the outer s orbitals interact to give molecular orbitals delocalized over the whole crystal. As shown in the simplified diagram of Fig. 31 there will be a band of orbitals covering virtually the entire range of possible bonding, non-bonding and antibonding orbital energies. This *energy band* will contain n electrons, one from each atom, and hence about half the orbitals will contain two electrons. (The energy levels are so closely spaced that some of the last-occupied orbitals will contain only one electron in accord with Hund's rule of maximum multiplicity.)

The situation is slightly more complex than outlined above because the $ns–np$ separation for the Group I elements is quite small and hence considerable energy overlap occurs between the top end of the s band and the lower end of the p band as shown in Fig. 31; the latter band, of course, contains $3n$ orbitals arising from the three p atomic orbitals on each atom. The overlap fortunately persists in the Group II metals because otherwise the $2n$ electrons now available for bonding would fill *all* the orbitals in the s band, resulting in a non-bonding situation. However, some of the electrons are able to enter the lower bonding orbitals of the p band so leaving many of the antibonding s bond orbitals empty; Fig. 31. Since virtually two electrons per metal atom are involved in binding a Group II metal lattice together, these elements are tougher and have higher melting points than their alkali metal counterparts.

COMPOUNDS OF THE ALKALI METALS

Hydrides

All the metals react directly with hydrogen on heating to give colourless, ionic hydrides MH which have the sodium chloride structure. Only in the case of lithium hydride is it possible to accurately determine the position of the weakly scattering hydride ions by X-ray diffraction. Lithium hydride, unlike the others which decompose to metal and hydrogen, is stable in the molten state. The presence of hydride ions in the melt can be demonstrated by electrolysis when it is found that hydrogen is evolved at the *anode* in amounts required by Faraday's law. The hydrides react with water to give the hydroxide and hydrogen, and with the halogens or acids to give the corresponding salts:

$$M + \tfrac{1}{2}H_2 \xrightarrow{\ T\,°C\ } MH; \quad T(Li) = 700-800\,°C, \ T(Na-Cs) = 360-400\,°C$$

LiH: melting point $688\,°C$, stable at atmospheric pressure to $800-900\,°C$

NaH: decomposes at $350\,°C$

Sodium tetrahydroborate, $NaBH_4$, and lithium tetrahydroaluminate, $LiAlH_4$, are two complex, essentially ionic, hydrides which have assumed importance in recent years as exceedingly versatile reducing agents. They may be formally assumed to arise from the addition of borane (BH_3) or alane (AlH_3) to alkali metal hydrides, the hydride ion acting as an electron donor to the Group III acceptor:

$$:H^- + [BH_3] \longrightarrow \left[\begin{array}{c} H \\ H{\cdots}{\cdots}B{\nwarrow}^{H} \\ H \end{array} \right]^-$$

The tetrahedral ions BH_4^- and AlH_4^- are isoelectronic to methane and silane, SiH_4, respectively. Lithium tetrahydroaluminate is the more widely used of the two, its ether solubility being particularly useful to organic and inorganic chemists alike. Under the correct conditions a covalent halide of most elements in the period table will react to form the element hydride (in the transition metal series the metal usually has to be 'protected' by suitable organic ligands):

$$PCl_3 + LiAlH_4 \rightarrow PH_3$$

$$SiCl_4 + LiAlH_4 \rightarrow SiH_4$$

$$BrMn(CO)_5 + LiAlH_4 \rightarrow HMn(CO)_5$$

Typical of the reductions possible in organic chemistry are:

$$RC\overset{H}{\underset{\diagdown\!\!O}{\diagup}} \longrightarrow RCH_2OH$$

$$RC\overset{O}{\underset{OH}{\diagup}} \longrightarrow RCH_2OH$$

$$\begin{matrix}R\\\diagdown\\\diagup\\R'\end{matrix}C{=}O \longrightarrow \begin{matrix}R\\\diagdown\\\diagup\\R'\end{matrix}CH.OH$$

Oxides

The alkali metals form three main types of oxide: the monoxide M_2O, the peroxide M_2O_2 and the superoxide MO_2. Lithium, having a small, highly polarizing ion, does not form a superoxide, the anion being destabilized by the high electric field gradient produced by the metal ion; similarly, although lithium peroxide can be prepared by special means, it is unstable to heat and readily evolves oxygen to form the monoxide:

$$LiOH + H_2O_2 \xrightarrow[\text{solution}]{\text{alcohol}} Li_2O_2.H_2O_2 \xrightarrow[P_2O_5]{\text{desiccator}} Li_2O_2 \xrightarrow{300°C} Li_2O + \tfrac{1}{2}O_2$$

Monoxides

The metals will react with oxygen to form the monoxide but in all cases, except that of lithium, a deficiency of oxygen must be used otherwise higher oxides are also formed; the excess of metal remaining can be removed by heating the oxide in a vacuum, when the metal sublimes away. The antifluorite structure is adopted by most of the monoxides, sulphides, selenides and tellurides of this group; the metals occupy what would normally be the fluoride sites and the oxygen (or chalcogen) occupies the calcium positions in the fluorite crystal (see Fig. 41, page 152). Except for lithium, the monoxides are better made by reducing the nitrate or nitrite with the corresponding metal:

$$6M + 2MNO_2 \rightarrow 4M_2O + N_2$$

Peroxides

Lithium monoxide will not take up oxygen even under 12 atm pressure at 480°C, but the other monoxides readily absorb oxygen to give peroxides.

Superoxides

These are formed by heating the peroxides with oxygen (under pressure for sodium and potassium) or by treating solutions of the respective metals in liquid ammonia

Table 17

	Lithium	Sodium	Potassium	Rubidium	Caesium
M_2O	White	White	White, cold; yellow, hot	Pale yellow; golden yellow at 200°C	Orange-red; black at 250°C
M_2O_2	White	Pale yellow	Orange	Dark brown	Yellow; black when fused
MO_2	—	Yellow	Orange-yellow	Dark orange	Reddish yellow

with an excess of oxygen at $-78°C$; the crystal structures of the superoxides are similar to that of calcium carbide and contain O_2^- ions; see Fig. 42, page 154. The superoxide ion (see page 310) has one unpaired electron and consequently the alkali metal superoxides are found to be paramagnetic and have a magnetic moment close to the value of 1.73 Bohr magnetons calculated for a single electron.

Several of the alkali metal oxides are quite deeply coloured, partly owing to lattice defects in the crystals (see Table 17).

The three types of oxide, all of which react readily with water, can be distinguished by a study of their hydrolysis products:

$$M_2O + H_2O \rightarrow 2MOH \qquad \text{(no hydrogen peroxide, no oxygen)}$$

$$M_2O_2 + 2H_2O \rightarrow 2MOH + H_2O_2 \qquad \text{(hydrogen peroxide, no oxygen)}$$

$$MO_2 + 2H_2O \rightarrow MOH + H_2O_2 + \tfrac{1}{2}O_2 \qquad \text{(hydrogen peroxide \textit{and} oxygen)}$$

The action of ozone on solutions of alkali metal hydroxides in liquid ammonia is thought to give the paramagnetic ozonides MO_3, but nothing is known about their structure.

Sulphides

The metals react with sulphur to form compounds of formula M_2S_x, where $x = 1, 2, 3, 4, 5$ or 6; for the higher sulphides the reaction is best carried out in liquid ammonia. The polysulphide ions are present in the crystal lattice as zigzag chains; in Cs_2S_6 the S—S—S bond angle is 108.8° and there is an alternation in bond length in the chain with the terminal and central bonds being about 2.0 Å, the remainder 2.1 Å. Clearly the bonding in these sulphides is complex. Furthermore, there is a rather short van der Waals distance between the ends of the S_6 chains in Cs_2S_6, suggesting a weak interaction between the chains.

Nitrides

Lithium is unique among the alkali metals in reacting directly with nitrogen gas to give a ruby-red nitride:

$$6Li + N_2 \xrightarrow{\text{below 450°C}} 2Li_3N \xrightarrow{H_2O} LiOH + NH_3$$

This reaction presumably occurs because the product, made up of small ions (one of which is highly charged), has an extremely high lattice energy. Born–Haber calculations suggest that sodium nitride, Na_3N, should exist, but none can be detected when sodium is heated with nitrogen at temperatures up to $800\,°C$. An impure form of sodium nitride may be formed when atomic nitrogen is allowed to react with a sodium film.

The azides (salts of hydrazoic acid, HN_3) are known for all the metals and can be made by passing dinitrogen monoxide through the molten amide:

$$NaNH_2 + N_2O \rightarrow NaN_3 + H_2O$$

The N_3^- ion is linear, the two equal N—N bond lengths (1.15 Å) being indicative of multiple bonding; simple theory suggests a bond order of about two.

Carbonates

These salts are soluble in water but, as might be expected, lithium carbonate (small cation and doubly charged anion give a high lattice energy) is the least soluble and can be precipitated from moderately concentrated solutions of lithium salts. Although the solubility of lithium carbonate may be increased by saturating the solution with carbon dioxide, no bicarbonate can be isolated; the other alkali metals form solid bicarbonates which decompose ($Na > K > Rb > Cs$) on heating: e.g.

$$2NaHCO_3 \rightarrow Na_2CO_3 + H_2O + CO_2$$

On stronger heating the carbonates will also decompose, giving the metal monoxide and carbon dioxide. At $1000\,°C$ the dissociation pressures of carbon dioxide above the carbonates are: lithium carbonate, 90 mmHg; sodium carbonate, 19 mmHg; potassium carbonate, 8.3 mmHg. If the carbon dioxide is removed as it is formed by passing a stream of hydrogen over lithium carbonate, complete decomposition to lithium monoxide can be achieved at temperatures as low as $800\,°C$.

The decompositions of the bicarbonates and carbonates take place because the products are more stable, mainly owing to the large increases in lattice energy when forming M_2CO_3 and M_2O respectively. The relative ease of decomposition parallels the decrease in ionic radius of the cations from Cs^+ to Li^+, which in turn increases the lattice energy of the solid reaction products. A kinetic effect probably also operates in that the high polarizing power of the smaller metal ions (especially of Li^+) helps to deform the large anions, making them less stable.

Nitrates

Lithium nitrate and, perhaps unexpectedly, sodium nitrate are deliquescent. On heating, all the nitrates, except lithium nitrate, form the nitrite:

$$2MNO_3 \rightarrow O_2 + 2MNO_2$$

and it is only on very strong heating that the nitrite is further decomposed to the oxide. Lithium nitrate decomposes directly to the monoxide without the intermediate formation of nitrite.

Halides

Only the halides of lithium form hydrates (see page 115), and lithium fluoride, having the highest lattice energy of the series, is the most insoluble. In other solvents, where the solvent interacts more strongly with the metal ions, solvates of the other alkali metal halides can be isolated. The best-known example is perhaps $NaI.4NH_3$ formed by dissolving sodium iodide in anhydrous liquid ammonia and evaporating the solution; its melting point is $3\,°C$ so that this solvate is actually a liquid at room temperature and has an ammonia dissociation pressure of 420 mmHg at $25\,°C$. Like the alkali metal hydrates, the coordinated ammonia is held by ion–dipole forces only, although this gives rise to quite a rigid system, the tetrahedral $Na(NH_3)_4^+$ ion having an infrared spectrum similar to that of tetramethyllead, $Pb(CH_3)_4$. The fluorides crystallize from liquid hydrogen fluoride as solvates $MF.nHF$ in which the hydrogen fluoride is held to the anion by relatively strong hydrogen bonding:

$$[F—H—F]^-$$

All the halides except CsCl, CsBr and CsI adopt the face-centred cubic structure typical of sodium chloride (Fig. 26) in which each metal and halide ion is surrounded

Table 18—Solubilities of some alkali metal compounds in methanol

Compound	Temperature ($°C$)	Solubility (wt%)
NaCl	25	1.38
NaBr	25	14.82
NaI	25	43.35
KCl	25	0.52
KBr	25	2.05
KI	25	14.53
NaOH	28	23.5
KOH	28	35.5

Alkali metal halides are soluble in a number of coordinating solvents such as the lower alcohols, acetone and carboxylic acids.

In methanol the balance between decreasing lattice energy and decreasing solvation energy of the anion is obvious when the solubilities of either the sodium or the potassium salts are compared. However, the lowered lattice energies and poorer solvation of the cation result in potassium salts being less soluble than those of sodium, demonstrating the importance of considering all the energy terms when trying to rationalize solubilities. It is of interest to note that the solubility of the fluorides is in the order KF > NaF, which presumably shows that the high solvation energy of the (small) fluoride ion is able to compensate for the normally unfavourable difference in the lattice energies and cation solvation energies when sodium is replaced by potassium in a halide.

Hydrogen bonding of the solvent with OH^- ions probably accounts to a large degree for the high solubilities of the hydroxides in methanol and other alcohols.

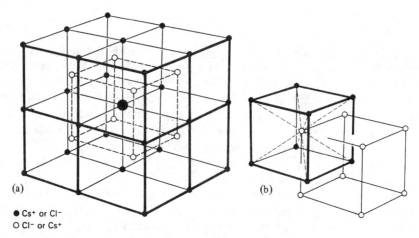

● Cs⁺ or Cl⁻
○ Cl⁻ or Cs⁺

Fig. 32—The caesium chloride structure. Identical ions are situated at the corners of a simple cube, at the centre of which is an ion of opposite charge (giving rise to the alternative, but incorrect, name 'body-centred cubic' structure). The structure as a whole may be regarded as arising from the interpenetration of two simple cubic lattices (b). Each ion, being at the centre of a cube, has a coordination number of eight (see Appendix).

octahedrally by six oppositely charged neighbours. With the bigger metal ions the most stable arrangement of the ions becomes the 8:8 coordination of the 'body-centred cubic' structure (see Appendix) assumed by caesium chloride, bromide and iodide under normal conditions; see Fig. 32. However, the lattice energies of the two types of cubic structure differ by only about 3% for the rubidium and caesium halides and hence some of the halides are dimorphic, adopting one structure or the other depending on the conditions. At 445°C caesium chloride changes to the sodium chloride structure and at low temperature rubidium chloride goes over to the body-centred cubic type. The change from face-centred to body-centred cubic results in a small reduction in volume and, as might be expected from Le Chatelier's principle, rubidium chloride, bromide and iodide change to the body-centred cubic crystal type on the application of about 5000 atm pressure.

The action of free halogen or an interhalogen on the alkali metal halides often results in the formation of polyhalide ions: e.g.

$$KI + I_2 \rightarrow KI_3$$

$$CsF + BrF_3 \rightarrow CsBrF_4$$

(for a more complete list see Table 43). The stability of these polyhalides depends on the polarizing power of the cation, and few, if any, have been prepared for lithium and sodium; this illustrates an often-used principle that to isolate salts of an unstable anion one has to use a large, virtually non-polarizing cation (e.g. Rb^+, Cs^+, NR_4^+). The products of decomposition of the polyhalides appear to be governed mainly by lattice energy considerations so that the smallest halogen remains with the metal as halide: e.g.

$$MICl_2 \rightarrow MCl + ICl$$

The alkali metal chlorates and perchlorates decompose on heating to give oxygen and the metal halide. The highly symmetrical perchlorate anion ClO_4^- is not strongly solvated in water, with the result that potassium, rubidium and caesium perchlorates are found to be insoluble.

Compounds with carbon

Only lithium reacts *directly* with carbon to give a carbide, Li_2C_2, which reacts with water to give ethyne and probably contains $[C\equiv C]^{2-}$ ions; electrolysis in molten lithium hydride produces carbon at the anode†. The other alkali metals give similar carbides when heated in ethyne. The heavier metals (potassium, rubidium and caesium) form interstitial carbides when heated with graphite, the graphite lattice expanding to accommodate metal atoms between the layers of carbon atoms (see page 226); the compounds formed are highly coloured and non-stoichiometric, having compositions approximating to the following:

$$C_{60}K \xrightarrow{K} C_{48}K \xrightarrow{K} C_{36}K \xrightarrow{K} C_{24}K + C_8K$$

dark grey   blue   bronze

The alkali metals normally undergo Wurtz coupling reactions when treated with organic halides:

$$2M + RX \rightarrow MX + [MR] \xrightarrow{RX} MX + R{-}R$$

However, with lithium this reaction stops at the intermediate stage to give an alkyllithium or aryllithium derivative in high yield:

$$2Li + RX \xrightarrow[\substack{or\ ether \\ under\ dry\ argon}]{hydrocarbon} LiX + LiR \quad (R = alkyl\ or\ aryl; X = Cl\ or\ Br)$$

These air- and water-sensitive *organometallic* compounds are typically covalent species which can be sublimed or distilled in a vacuum and which dissolve in many organic solvents, including simple hydrocarbons. In contrast, the highly reactive, pyrophoric alkyls and aryls of the other alkali metals are made by *transmetallation* reactions involving organomercury derivatives:

$$2M + HgR_2 \xrightarrow[hydrocarbon]{dry\ N_2} 2MR + Hg$$

A wide variety of organolithium reagents can be made via lithium–halogen or lithium–hydrogen exchange reactions using the commercially available *n*-butyllithium, the lithium being transferred to the alkyl or aryl group having the higher electronegativity:

$$LiBu + CH_2{=}CHBr \xrightarrow{dry\ N_2} BuBr + CH_2{=}CHLi \quad (vinyllithium)$$

$$LiBu + C_6F_5H \xrightarrow{dry\ N_2} BuH + C_6F_5Li \quad (pentafluorophenyllithium)$$

A COMPARISON OF COPPER WITH THE ALKALI METALS

Since both copper and the alkali metals have a single s electron in their outer quantum

---

† The C—C distance in the C_2^{2-} ion is 1.20 Å, identical to the $C\equiv C$ bond in ethyne (1.204 Å).

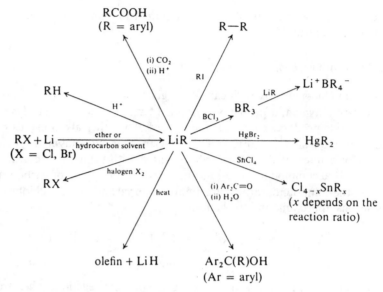

Fig. 33—Some typical reactions of organolithium compounds.

shell, one might expect them to have rather similar properties. However, the presence of the filled 3d shell (which is similar in energy to the 4s shell at copper) modifies the chemistry of copper to a remarkable degree, and perhaps the only resemblance of copper to the alkali metals is that the copper(I) ion is diamagnetic and colourless. The 3d electrons participate in metallic bonding with the result that copper is much harder and has a higher melting point than any of the alkali metals; the stronger metallic bonding of copper also makes it much less reactive than the alkali metals (Table 19).

The 4s electron in the copper atom is very poorly shielded from the nucleus by the filled 3d shell (see Figs 34 and 35) and is thus strongly attracted by the nuclear charge. The effect of this is to make the first ionization energy of copper greater than that of an alkali metal; it is, in fact, about twice the first ionization energy of potassium. The lack of penetration and poor mutual shielding (see Fig. 34) of the filled d orbitals also cause an increased nuclear attraction in all the inner electron shells, which results in the Cu^+ ion being much smaller than K^+ although ten elements occur between potassium and copper in the periodic table. (A contraction of atomic and ionic sizes due to penetration and shielding effects occurs across the first transition metal series and is of course, greatest at the last transition metal in the 3d group, copper.)

Table 19

		1st IE (kJ mol^{-1})	2nd IE (kJ mol^{-1})	3rd IE (kJ mol^{-1})	Radius M$^+$ (Å)	Melting point (°C)
Na	$1s^2 2s^2 2p^6 3s^1$	496	4564	6911	0.95	98
K	$1s^2 2s^2 2p^6 3s^2 3p^6 4s^1$	418.8	2932	4602	1.33	63.5
Cu	$1s^2 2s^2 2p^6 3s^2 3p^6 3d^{10} 4s^1$	745.0	1957	3554	0.93	1083

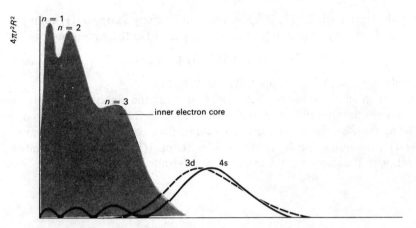

Fig. 34—Relative sizes of the 3d and 4s orbitals at potassium and copper. Because the 3d and 4s orbitals are of approximately the same size, 3d electrons do not shield a 4s electron from the attraction of the nuclear charge. As the 3d shell is filled, there is a corresponding increase in the nuclear charge, which results in the 4s electron being progressively more strongly held, until at copper (ten d electrons) the energy required to remove the 4s electron has increased to almost twice the ionization energy of potassium.

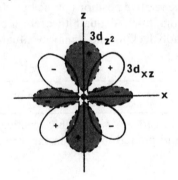

Fig. 35—Poor mutual shielding of the 3d orbitals of the copper atom. Owing to their different spatial orientations, the 3d orbitals of the copper atom do not sheild each other very effectively from the attraction of the nucleus. This is illustrated here for the $3d_{z^2}$ and $3d_{yz}$ orbitals. The lack of very extensive penetration of the d orbitals into the inner core results in the core electrons being strongly attracted to the nucleus (the charge of which increases by $+1$ for each d electron present). The combination of poor mutual shielding and lack of penetration of the 3d orbitals results in a contraction of all the electron shells at copper, thus making the copper(I) ion smaller than the potassium ion.

The second and third ionization processes at copper involve the loss of 3d electrons which, owing to the lack of penetration close in to the nucleus, are more readily lost than the $(n-1)$p electrons of the alkali metals. Hence the second and third ionization energies of copper are considerably lower than those of the alkali metals. The combination of high first ionization energy and relatively low second ionization energy makes the copper(I) state highly unstable in aqueous media towards the disproportionation

$$2Cu^+(aq) \rightarrow Cu + Cu^{2}(aq), \quad \frac{[Cu^{2+}(aq)]}{[Cu^+(aq)]^2} \simeq 10^6$$

The high hydration energy (2121 kJ) of the doubly charged copper(II) ion is probably the driving force of this reaction, since in the gas phase the equilibrium

$$2Cu^+(g) \rightleftharpoons Cu^{2+}(g) + Cu(cryst)$$

is strongly in favour of the copper(I) state.

Therefore the main difference between copper and the alkali metals is that copper is able to exist in more than one oxidation state (Cu^I, Cu^{II} and Cu^{III}), the copper(II) state being more stable under many conditions than copper(I). Furthermore, many copper(I) compounds, such as the halides (copper(I) fluoride is unknown), are covalent, with the copper atom usually being tetrahedrally coordinated.

SIMILARITIES IN THE CHEMISTRY OF LITHIUM AND MAGNESIUM

There is a surprising number of ways in which lithium behaves more like magnesium than an alkali metal, a fact which can prove useful when trying to memorize their chemical properties. This similarity was observed many years ago and called a diagonal relationship because of the relative positions of the two elements in the periodic table. Considering its early recognition, adequate theoretical explanation of the 'diagonal phenomenon' has been sadly lacking and it may eventually prove to be only a simple, but fortuitous, coincidence.

	Li and Mg	Na, K, Rb and Cs
1	Relatively inert; Li can even be melted and poured in the air without losing its brightness	Highly reactive; Rb and Cs catch fire in the air at room temperature
2	Relatively hard metals	Very soft metals
3	Form monoxides on burning in oxygen	Form peroxide (Na) or superoxide (K, Rb and Cs)
4	No superoxides exist under normal conditions	Superoxides MO_2 known for all four elements
5	Attack N_2 on heating to give the nitrides Li_3N and Mg_3N_2 which give ammonia on hydrolysis	No reaction with nitrogen
6	Hydrogen carbonates exist only in solution	Solid hydrogen carbonates known for all four elements
7	Carbonates, hydroxides and peroxides decompose on moderate heating	Similar decompositions only occur at much higher temperatures

	Li and Mg	Na, K, Rb and Cs
8	Nitrites and nitrates give the monoxides on moderate heating	Nitrates decompose to nitrites at about 500 °C; only at about 800 °C do the nitrites begin to form oxides
9	Fluorides, hydroxides, carbonates, oxalates and phosphates only slightly soluble in water	Corresponding salts soluble in water
10	Alkyls and aryls have low melting points; some liquid at room temperature	Alkyls and aryls have high melting points (or decompose without melting)
11	Alkyls and aryls soluble in a range of organic solvents, including hydrocarbons	Insoluble in all common solvents
12	Alkyls and aryls are covalent, often with polymeric structures	Alkyls and aryls are thought to be ionic

Chapter 4

Group IIA. Beryllium, Magnesium, Calcium, Strontium, Barium and Radium

INTRODUCTION

The divalent ions of the Group II elements are much smaller than their neighbours in Group I because the increased ionic charge (and hence effective nuclear charge) binds the remaining electrons very tightly. This will be expected to result in: (a) more hydrate formation by salts of the Group II elements due to stronger ion–dipole interactions; (b) much higher lattice energies for corresponding salts, often leading to lower solubilities especially for those containing multiply charged anions such as carbonate, sulphate and phosphate; (c) relatively low thermal stability of salts with polyatomic anions (e.g. carbonate, sulphate, nitrate).

AQUEOUS SOLUTIONS OF THE M^{2+} CATIONS

In solution, NMR and X-ray diffraction studies have shown that the beryllium ion is only large enough to accommodate four water molecules whereas Mg^{2+} is able to achieve octahedral coordination. The degree of hydration becomes less precise for the bigger ions; calcium, for example, is 10-coordinated in dilute solution but above about 1M the average number of associated water molecules begins to decrease, finally reaching ~ 5.5 at a concentration of 4.5M. As might be expected for small, doubly charged ions, the hydrated species $Be(H_2O)_4^{2+}$ and $Mg(H_2O)_6^{2+}$ are found to be surrounded by a further 12–15 water molecules in the second coordination sphere. The high affinity of the Mg^{2+} ion for water is presumably the main reason why magnesium sulphate is soluble in water whereas the sulphates of the larger calcium, strontium and barium are not.

When aqueous solutions of beryllium and magnesium salts are crystallized it is usual for hydrates such as $[Be(H_2O)_4]SO_4$ and $[Mg(H_2O)_6]Cl_2$ to separate out.

Indeed, so strongly polarizing is the bare Be^{2+} ion that the only ionic species of beryllium which can exist are those in which the cation is either hydrated or, less often, coordinated by other ligands; anhydrous compounds like $BeCl_2$ have all the properties associated with covalent species.

Solutions of beryllium salts have an acidic reaction owing to partial hydrolysis induced by the small Be^{2+} ion:

$$(H_2O)_3 Be^{2+} \leftarrow O\begin{subarray}{l} \nearrow H \\ \searrow H \end{subarray} \rightleftharpoons H_3O^+ + (H_2O)_3 Be^+ OH$$

Gradual addition of alkali forces the equilibrium to the right and eventually results in the precipitation of $Be(OH)_2$. A proton NMR study of the reaction has shown that the main species present in the initial stages are $Be(OH_2)_4^{2+} \rightarrow [Be_2OH]^{3+} \rightarrow [BeOH]_3^{3+}$, with the added possibility of higher polymers being formed near to the precipitation point. In both hydroxo species hydration allows each beryllium atom to achieve tetrahedral coordination; see Fig. 36.

The polarizing effects of the larger Mg^{2+} ion are insufficient to cause hydrolysis under ambient conditions, but at about $190\,^\circ C$ magnesium sulphate solutions begin to precipitate insoluble basic sulphates; under the same extreme conditions calcium sulphate and lithium sulphate are hydrolytically stable.

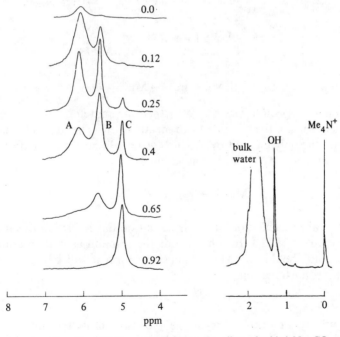

Fig. 36—A series of proton NMR spectra showing the effect of added Na_2CO_3 on $BeCl_2$ solutions. The spectra were recorded for solutions having Na/Be ratios ranging from 0.0 (pure $BeCl_2$) to 0.92; precipitation of $Be(OH)_2$ occurred at ratios slightly above 1.0. The three low-field peaks marked A, B and C are due to water of solvation in the species $Be(H_2O)_4^{2+}$, $[Be_2OH]_{aq}^{3+}$ and $[(BeOH)_3]_{aq}^{3+}$ respectively. Samples were held at $-50\,^\circ C$ to slow down the various water exchange processes occurring in solution. [Adapted with permission from J. W. Akitt and R. H. Duncan, *Journal of the Chemical Society Faraday I* vol. 76 (1980) p. 2212.]

POSSIBILITY OF MONOVALENT SPECIES

Although the total energy required to form the M^{2+} ions is very high, being of the order of 2100 kJ mol^{-1} for magnesium, Born–Haber calculations show that both the lattice energy and hydration energy of the alkaline earth salts are more than sufficient to overcome this large energy requirement. However, the high value of the second ionization energy does raise the question as to whether univalent salts of these metals could possibly be made. A suitable cycle to investigate this might be that involving the reaction of magnesium with chlorine gas:

$$
\begin{array}{ccc}
Mg(s) + \tfrac{1}{2}Cl_2(g) & \xrightarrow[\tfrac{1}{2}\Delta H_{diss}]{\Delta H_{sub}} & Mg(g) + Cl(g) \\
\Big\downarrow Q & & \Big\downarrow IE_{Mg} \quad \Big\downarrow {-EA_{Cl}} \\
MgCl(s) & \xleftarrow{\;-L_{MgCl}\;} & Mg^+(g) + Cl^-(g)
\end{array}
$$

Assuming the lattice energy of MgCl (L_{MgCl}) to be equal to that of sodium chloride, the theoretical heat of reaction Q can be calculated:

$$Q = \Delta H_{sub} + \tfrac{1}{2}\Delta H_{diss} + IE_{Mg} - EA_{Cl} - L_{NaCl}$$

$$= -125.5 \text{ kJ mol}^{-1}$$

Therefore the reaction

$$Mg + \tfrac{1}{2}Cl_2 \rightarrow MgCl$$

is thermodynamically quite feasible. Why then have attempts to isolate MgCl and other univalent salts of the alkaline earth metals failed? The reason is that the above reaction is not the only method by which MgCl might be synthesized; it should also be possible to reduce $MgCl_2$ with magnesium:

$$MgCl_2 + Mg \xrightarrow{Q'} 2MgCl$$

When Born–Haber calculations are carried out on this reduction it is found that Q' is equal to about $+389$ kJ. In other words the reaction as written above is highly endothermic and should any MgCl happen to form it will be very unstable with respect to disproportionation:

$$2MgCl \rightarrow MgCl_2 + Mg, \quad -389 \text{ kJ}$$

Thus when we say a compound is 'stable' we must clarify the statement by asking 'stable with respect to what?'. MgCl is stable towards decomposition into its elements but highly unstable towards disproportionation. There have been claims in the past to have made M^IX derivatives of Group II metals by heating the metal with its dihalide to $900°C$ in an atmosphere of hydrogen. Later work has shown that the compounds formed, which have strong reducing properties, are the hydride-halides $M^{II}HX$; they can be made more simply by heating together stoichiometric quantities of dihalide and dihydride:

$$MX_2 + MH_2 \rightarrow 2MHX \quad (M = Ca, Sr \text{ or } Ba; X = Cl, Br \text{ or } I)$$

The monohalides MX can be detected as short-lived ion pairs in the gas phase using spectroscopic techniques, a typical example being CaCl in which the Ca^+---Cl^- distance is 2.44 Å. These gaseous species, and the traces of M^+ ions which can be introduced as impurities in crystals of the alkali metal halides, are the only common univalent derivatives of the Group II elements. Such ions are paramagnetic and have characteristic visible spectra, both properties arising from the lone electron in the outer s orbital. Positively charged monohalides MCl^+, which contain divalent cations, have also recently been detected in the gas phase.

NO M^{3+} SPECIES POSSIBLE

The third ionization energy for these elements is much too high to allow the formation of M^{3+} salts. The large increase from the second to the third ionization energy is due to the fact that the third electron has to be removed from an inner quantum shell; electrons in such a shell are nearer to the nucleus and thus very strongly attracted to it.

COMPOSITION OF METAL VAPOUR

The vapour above the heated alkaline earth metals consists entirely of monatomic species because the outer electron configuration of the metals ns^2, does not allow the formation of dimers or polymers by any reasonable covalent bonding scheme. The situation is rather similar to that of helium, where the formation of a dimer He_2 would require the filling of an equal number of bonding and antibonding σ orbitals.

BERYLLIUM CHLORIDE

It has been mentioned already that the simple compounds of beryllium are covalent. For example, anhydrous beryllium dichloride is polymeric in the solid state with Be—Cl—Be bridging and in the gas phase forms a mixture of monomer and dimer, full dissociation to the monomer only occurring at about 1000°C:

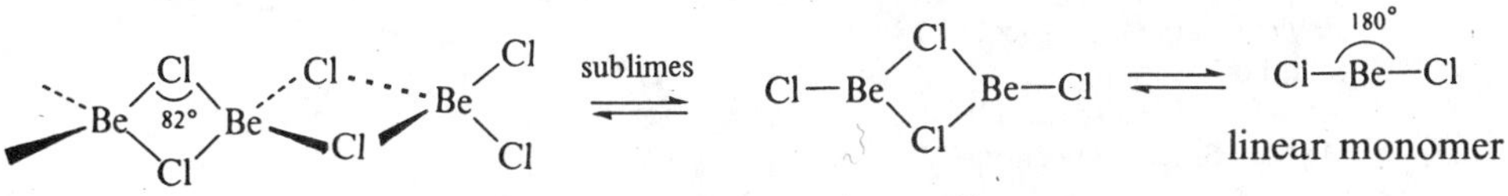

COMPLEXES OF THE GROUP II METALS

Complexes of the Group II elements are quite numerous, the tendency for their formation falling off with increasing size of the metal ion owing to the decrease in ion–dipole interactions: Be > Mg > Ca > Sr > Ba. Beryllium is almost unique among the Group II elements in forming halo-complexes, the stability of which are in the order $F \gg Cl > Br > I$. In aqueous solutions of BeF_2, for example, the species BeF^+, BeF_2, BeF_3^- and BeF_4^{2-} are present and many solid salts containing the latter ion are known. The tetrafluoroberyllate ion BeF_4^{2-} is similar in size to sulphate, which results in the two anions forming a number of isomorphous salts when paired with a common cation (e.g. $BaBeF_4$ and $BaSO_4$; $MBeF_4.6H_2O$ and $MSO_4.6H_2O$, where $M = Co^{II}$ or Ni^{II}).

In contrast, no tetrabromo- or tetraiodoberyllates have yet been isolated and tetrachloroberyllates have to be prepared under strictly anhydrous conditions. Thus although liquid $BeCl_2$ is a poor electrical conductor, the conductivity increases markedly on the addition of alkali metal chlorides, and crystalline tetrachloroberyllates can be obtained from the salt mixtures. Magnesium is the only other member of the group for which solid chloro-complexes have been described, $(NEt_4)_2MgCl_4$ being the best characterized.

The monodentate donor molecules which form complexes with the Group II metals include ammonia, ethers, alcohols, ketones, aldehydes and phenols. By far the widest range of complexes is formed by the beryllium halides and many have the formula $BeX_2.2D$ where D is a donor molecule; typical of these is the dietherate $BeCl_2.2Et_2O$ which is formed when anhydrous beryllium chloride dissolves in dry ether. It does not dissociate in benzene and has a dipole moment of 6.74 debye resulting from its tetrahedral shape:

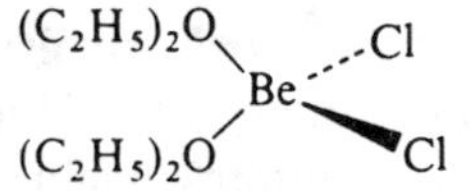

$$(C_2H_5)_2O \diagdown \qquad \cdots Cl$$
$$Be$$
$$(C_2H_5)_2O \diagup \qquad Cl$$

Some of the most stable Group II complexes are derived from multidentate ligands such as crown ethers, crypts and polycarboxylic acids, of which EDTA is the most common:

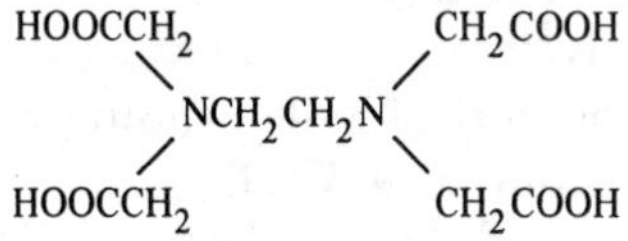

HOOCCH$_2$ CH$_2$COOH
$\diagdown$ $\diagup$
      NCH$_2$CH$_2$N
$\diagup$ $\diagdown$
HOOCCH$_2$ CH$_2$COOH

ethylenediamine tetraacetic
acid, EDTA; forms
octahedral complexes with
Mg, Ca, Sr and Ba cations

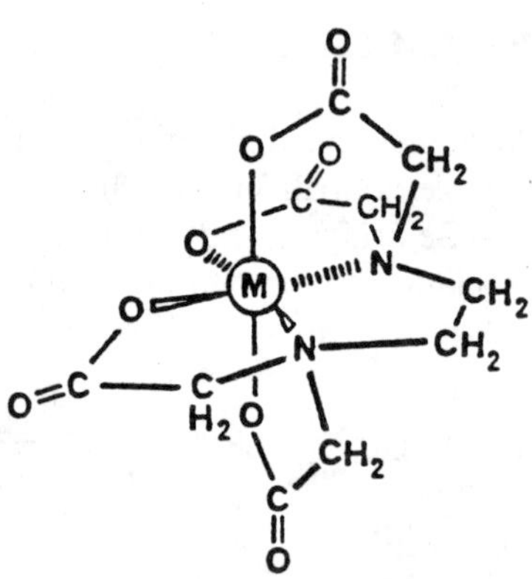

The heavier metals Ca, Sr and Ba form many strong complexes with suitably sized crown ethers, but it is unwise to assume that just because a crystal contains a crown ether molecule the ligand must necessarily encapsulate the metal ion. In $MgCl_2[12\text{-crown-}4]6H_2O$, for example, the cation is octahedrally coordinated by the six water

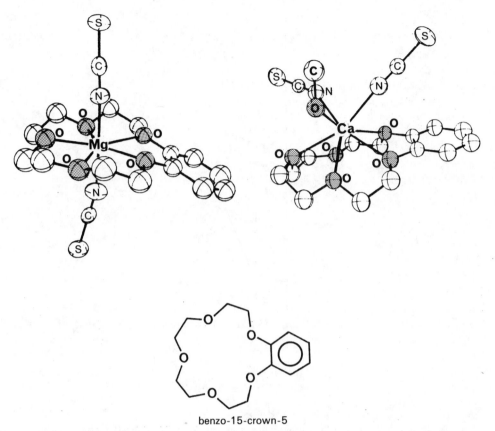

benzo-15-crown-5

Fig. 37—The structures of [Mg(benzo-15-crown-5)][NCS]$_2$ and [Ca(benzo-15-crown-5)] [NCS]$_2$.CH$_3$OH. [Reproduced with permission from *Journal of the Chemical Society, Dalton Transactions* (1978) p. 1418.]

molecules and the polyether merely resides passively in the crystal where it is hydrogen-bonded to the Mg(H$_2$O)$_6^{2+}$ units. Any possible interaction between the small crown ether and Mg^{2+} is clearly insufficient to overcome the strong hydration of the cation. (The reaction between a crown ether and a metal ion is thought to occur in two stages; first the ligand undergoes a rapid conformational change to the correct geometry and this is followed by a much slower process in which the ether begins to bind to the *hydrated* cation, sequentially stripping off the coordination water molecules as it sinuously envelops the metal.)

The 'double-crypt' ligand [1,1,1′,1′] is too small to hold a cation in each of its cavities but its 1:1 complex with Ca, Sr or Ba shows a particularly interesting property in that the encompassed cation oscillates between the two possible coordination positions. Presumably this ion-jumping process models very closely the movement of cations between the binding sites in membrane channels of cells (see page 451). When the ligand cavities are made larger by introducing further oxygen atoms, two cations can be accommodated within the same donor molecule as demonstrated

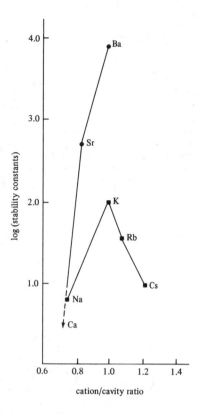

Fig. 38—Logarithm of stability (or formation) constants for complexes of 18-crown-6 plotted against cation radius/cavity radius ratio. A radius ratio close to unity results in a high stability constant because the greatest electrostatic interaction between cation and ligand is possible; the marked effect of ionic charge is apparent. The low, and uncertain, value of the stability constant for Ca^{2+} probably reflects the fact that water of solvation must be removed from the cation before complexation can occur. In *all* cases the complexes are partially destabilized owing to loss of the energy required to desolvate the cation, and this can be particularly serious for a strongly hydrated, divalent ion which does not closely match the size of the ligand's cavity. [Data taken from *Chemical Reviews* vol. 85 (1985) p. 277.]

crystallographically for the sodium derivative $(Na_2[2,2,2',2'])I_2$:

BASIC BERYLLIUM CARBOXYLATES

Remarkable among the many covalent compounds of beryllium are the basic carboxylates produced when $Be(OH)_2$ dissolves in a carboxylic acid. These compounds have the formula $Be_4O(OOCR)_6$, are volatile, soluble in organic solvents, possess sharp melting points and can be sublimed or distilled without decomposition; thus the basic ethanoate ($R = CH_3$) melts at 285–286°C, boils at 330°C and sublimes at its melting point under reduced pressure. Their common structure has been determined by X-ray diffraction techniques and shown to contain a central oxygen bonded to four beryllium atoms arranged tetrahedrally around it; the six carboxylate groups act as bidentate ligands and bridge the metal atoms. This gives rise to very compact, symmetrical and non-polar molecules in keeping with their volatility and solubility in organic solvents:

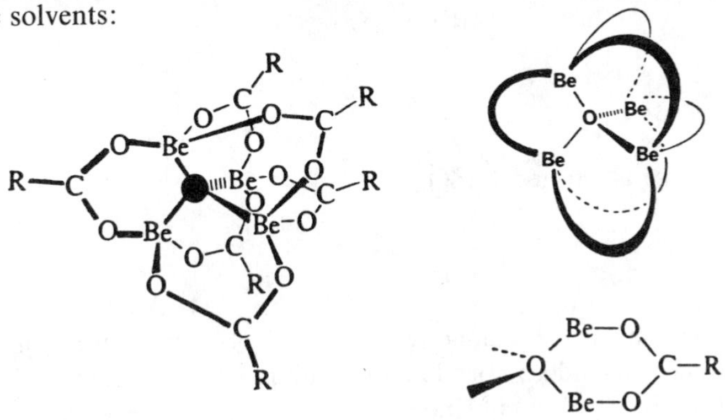

GROUP TRENDS

In summary, these elements tend to behave much less coherently as a group than was found for the alkali metals. Beryllium, as head element, shows anomalous behaviour in forming mainly covalent derivatives and thus there is an abrupt change in chemical properties between beryllium and magnesium. The members of the Ca, Sr and Ba triad, often called the alkaline earth metals, behave in a very similar manner and display only a gradual trend in properties due to their slightly different ionic sizes. Magnesium is considerably smaller than the alkaline earths and hence its salts tend (a) to be more soluble in water, mainly owing to the high hydration energy of Mg^{2+}, and (b) to show a lower thermal stability, particularly the carbonate, nitrate, hydroxide and peroxide.

OCCURRENCE OF THE GROUP II METALS

Beryllium

2–6 ppm of the Earth's crust; main minerals for the production of beryllium are beryl, $Be_3Al_2Si_6O_{18}$, and phenacite, Be_2SiO_4.

Magnesium

2.1% total abundance, of which about 0.13% occurs in sea-water (approximately 90% of all magnesium produced in the USA is now obtained from the sea).

Calcium

3.63 % abundance; widely distributed in nature, often occurring as limestone mountain ranges ($CaCO_3$).

Strontium

42 ppm of the Earth's crust; commercially important minerals are celestine, $SrSO_4$, and strontianite, $SrCrO_4$. The three main natural isotopes are ^{86}Sr 9.7%, ^{87}Sr 7.0% and ^{88}Sr 82.6%. Sometimes strontium-87 is found virtually pure in a mineral, having been formed by the radioactive decay of rubidium-87:

$$^{87}_{37}Rb \rightarrow {}^{87}_{38}Sr + \beta^-$$

Barium

39 ppm; main source is barytes, $BaSO_4$.

Radium

All isotopes of this element are radioactive so it does not occur naturally in significant amounts, however, being a decay product of uranium, it occurs in pitchblende to the extent of about 1 g in 7 t of pitchblende.

EXTRACTION

Beryllium

Beryl is heated with sodium fluorosilicate, Na_2SiF_6, at 700–750 °C and the product extracted with water, when the beryllium dissolves as either the fluoride or sodium fluoroberyllates. (The main impurity, aluminium, forms water-insoluble fluoride complexes and is removed at this stage by filtration.) Alkali is added to the solution to precipitate beryllium hydroxide which is then dissolved in ammonium hydrogen fluoride solution:

$$Be(OH)_2 + 2NH_4F.HF \rightarrow (NH_4)_2BeF_4 + H_2O$$

$$\Big\downarrow \text{evaporate and heat to 900 °C}$$

$$BeF_2$$

The metal is obtained by reducing beryllium fluoride with magnesium; the reaction proceeds rapidly at 900 °C but it is necessary to increase the temperature to 1300 °C (i.e. above the melting points of magnesium fluoride and beryllium) to aid separation of the reaction products.

Magnesium

Sea-water is agitated with calcined (heated) dolomite (which also contains magnesium) and the precipitated magnesium hydroxide filtered off; every tonne of magnesia requires the processing of about 300 t of sea-water. Treatment of the hydroxide with hydrochloric acid produces magnesium chloride which, after dehydration, is fused and electrolysed at temperatures above the melting point of magnesium ($651\,°C$).

Calcium

The carbonate is calcined to give calcium oxide which, after treatment with water and ammonium chloride, is converted to calcium chloride. The chloride is then fused and electrolysed.

Strontium and barium

Oxides are reduced with aluminium metal in a vacuum at such a temperature that the strontium and barium distil.

THE ELEMENTS

The two valence electrons which are available for binding the metallic lattice together make the Group II elements very much harder than their alkaline counterparts, the hardness decreasing down the group (calcium, for example, is slightly tougher than lead). The melting points vary in an apparently irregular manner: beryllium, $1283\,°C$; magnesium, $651\,°C$; calcium, $851\,°C$; strontium, $800\,°C$; barium, $850\,°C$. However, this variation is mainly due to differences in the crystal structures adopted by the elements. Beryllium and magnesium have a close-packed hexagonal structure, calcium and strontium a face-centred cubic structure and barium a body-centred cubic structure; when the melting points are compared for elements having the same structure, the expected decrease in melting point with increasing size does occur.

The alkaline earth metals dissolve in liquid ammonia to give blue solutions, but their solubilities are very much lower than those of the alkali metals. An interesting observation is that calcium, strontium and barium can be isolated from their ammonia solutions as solvates $M(NH_3)_x$, the value of x being somewhat variable but approximating to six. It is not clear what forces hold the ammonia molecules to the metal in these solvates, but their isolation does lend support to the idea that metals dissolve in liquid ammonia to give solvated species (see page 408). If the solvates are heated to drive off the coordinated ammonia, some amide formation occurs:

$$Ca(NH_3)_x \rightarrow Ca(NH_2)_2 + H_2 + (x-2)NH_3$$

COMPOUNDS OF THE GROUP II METALS

Hydrides

Unlike the other Group II metals, beryllium does not form a hydride on being

Table 20— Reactions of Group II elements

	Beryllium	Magnesium	Calcium, strontium, barium
Metal in air	Stable	Stable	Corrode in air
Salts *soluble* in water	F, Cl, Br, I, NO_3, SO_4, oxalate	Cl, Br, I, ethanoate, NO_3, SO_4, CrO_4, BrO_3, IO_3	Cl, Br, I, NO_3, ClO_4
Salts *insoluble* in water	OH, PO_4, AsO_4	OH, F, PO_4, CO_3, oxalate	SO_4, SO_3, IO_3, PO_4, oxalate, CO_3, F
Metal + NaOH	Dissolves with evolution of $H_2 \rightarrow Be(OH)_4^{2-}$	No reaction	No reaction

Reactions of salts with various aqueous reagents

	Beryllium	Magnesium	Calcium, strontium, barium
1 Aqueous NH_3	White precipitate of $Be(OH)_2$, insoluble in excess reagent	Partial precipitation of $Mg(OH)_2$	No precipitation (hydroxides soluble)
2 NaOH	White precipitate of $Be(OH)_2$, soluble in excess reagent to form beryllate ion $Be(OH)_4^{2-}$	White precipitate of $Mg(OH)_2$, insoluble in excess reagent	No precipitation
3 $(NH_4)_2CO_3$	White precipitate of basic carbonate, soluble in excess reagent owing to complex ions such as $Be_4O(CO_3)_6^{6-}$	White precipitate of basic carbonate on boiling solution, insoluble in excess reagent	White precipitates of carbonates, MCO_3
4 Na_3PO_4 in presence of NH_3 and NH_4Cl	White precipitate of NH_4BePO_4	White precipitate of NH_4MgPO_4	White precipitates of phosphates
5 Ammonium oxalate	No precipitation	No precipitation in presence of excess oxalate	White precipitates of oxalates
6 H_2SO_4	No precipitation	No precipitation	White precipitates of sulphates
7 NaF	No precipitation	White precipitate of MgF_2	White precipitates of fluorides

heated in hydrogen. However, BeH_2 can be prepared by the pyrolysis of di-*tert*-butylberyllium:

$$(CH_3)_3CMgBr + BeCl_2 \xrightarrow{\text{ether}} [(CH_3)_3C]_2Be$$

$$\Big\downarrow \text{heat}$$

$$2(CH_3)_2C=CH_2 + BeH_2$$

It is a thermally stable, polymeric solid in which the beryllium atoms are linked by hydrogen bridges.

The other hydrides of this group are salt-like and contain the hydride ion H^-. All react with water, liberating hydrogen.

Beryllium hydride, like aluminium hydride, reacts with diborane to form a volatile, covalent tetrahydroborate $Be(BH_4)_2$.

$$BeH_2 + B_2H_6 \longrightarrow$$

The Be—H—Be bridges in BeH_2 and the Be—H—B bridges in $Be(BH_4)_2$ are held together by two-electron, three-centre bonds very similar to those found in diborane (page 86).

Oxides

The divalent metal ions are sufficiently polarizing to make the peroxide, and especially the superoxide, ions unstable to heat, so that MO_2 and $M(O_2)_2$ do not form even when the metals are heated in pure oxygen; the monoxide MO is obtained under these conditions. Similarly this is the only oxide formed by the thermal decomposition of the carbonates, nitrates and hydroxides of these metals. However, strontium monoxide and barium monoxide absorb oxygen under pressure to give the peroxides. Peroxides of magnesium, calcium, strontium and barium can be also obtained when hydrogen peroxide is added to an aqueous solution of either their hydroxide or one of their salts:

$$Ca(OH)_2 + H_2O_2 \rightarrow CaO_2.8H_2O \xrightarrow{130°C} CaO_2 \xrightarrow{\text{heat}} CaO + \tfrac{1}{2}O_2$$

Beryllium monoxide assumes the covalent wurtzite (4:4) structure, unlike the other ionic monoxides which adopt the sodium chloride (6:6) lattice. The very high lattice energy of a sodium chloride structure containing *doubly* charged ions is reflected in the melting points of the monoxides: MgO, 2800°C; CaO, 1728°C; SrO, 1635°C; BaO, 1475°C; the fall in melting point along this series is due mainly to the increasing size of the cation from magnesium to barium which slightly reduces the lattice energy.

Hydroxides

Unlike the other Group II metals, beryllium has an amphoteric hydroxide, and as a consequence of this beryllium metal (cf. aluminium) dissolves in aqueous alkali, evolving hydrogen and forming soluble beryllates. All the hydroxides can be obtained either by adding water to the monoxides or by adding alkali to an aqueous solution of a suitable salt. The hydroxides of magnesium, calcium, strontium and barium follow the familiar pattern of thermal decomposition, the heavier metals forming the more stable hydroxides (decrease in polarizing power of M^{2+} and lower lattice energy of the MO product). The temperatures at which the pressure of water above the hydroxides is 10 mmHg are found to be approximately: $Mg(OH)_2$, 300 °C; $Ca(OH)_2$, 390 °C; $Sr(OH)_2$, 466 °C; $Ba(OH)_2$, 700 °C;

$$M(OH)_2 \rightarrow MO + H_2O$$

Nitrides

All the metals react on heating in nitrogen to give the nitrides M_3N_2, which evolve ammonia on treatment with water. Only lithium of the alkali metals reacts in this way, presumably because the formation of N^{3-} ions from the very stable N_2 molecule requires a large amount of energy, which can only be counterbalanced by the formation of nitrides having very high lattice energies (i.e. those of small monovalent ions like Li^+ or doubly charged ions like those of the alkaline earths and magnesium).

Nitrates

The hydrated nitrates can be obtained in the normal way by treating the oxides, hydroxides or carbonates, with nitric acid and crystallizing the solutions. Anhydrous beryllium nitrate results as a solvate, $Be(NO_3)_2.2N_2O_4$, when beryllium chloride is dissolved in dinitrogen tetroxide:

$$BeCl_2 + N_2O_4 \xrightarrow{\text{solvolysis}} Be(NO_3)_2.2N_2O_4 \xrightarrow[\text{in vacuo}]{50\,°C} Be(NO_3)_2$$

At about 125 °C in a vacuum $Be(NO_3)_2$ loses further quantities of nitrogen dioxide to give a basic nitrate, analysing as $Be_4O(NO_3)_6$, and which is thought to have a structure very similar to the basic beryllium carboxylates (page 145); however, it is now the nitrate entity which acts as the bidentate ligand:

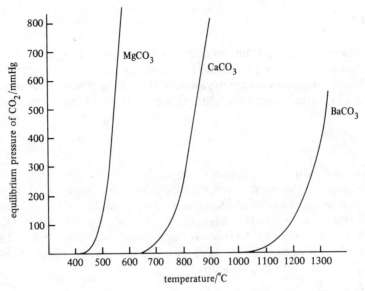

Fig. 39—Dissociation pressures of Group II carbonates at various temperatures. The approximate temperatures at which the dissociation pressures of carbon dioxide equal 0.1 mmHg are 400 °C (Mg), 500 °C (Ca) and 900 °C (Ba); the pressure above $CaCO_3$ at 1100 °C is 8740 mmHg compared with 18 mmHg for $BaCO_3$.

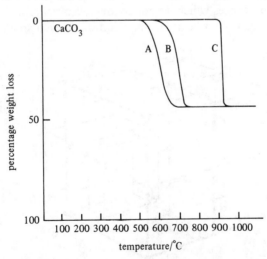

Fig. 40—Thermal decomposition of $CaCO_3$ under different conditions. The thermal decomposition of hydrates, peroxides, carbonates, sulphates, etc. can be monitored *thermogravimetrically* by continuously observing the weight of a pure specimen held on a sensitive balance, the sample pan of which is placed inside a furnace. The temperature is programmed to increase linearly and electronic components convert mass changes directly into percentage weight losses to be displayed on a chart by a pen recorder; only a few milligrams of sample are required. This illustration shows the effect of changing the composition of the atmosphere above three heated samples of calcium carbonate:

sample A   vacuum conditions; 10^{-4} mmHg
sample B   dry air at 760 mmHg; flow rate $5 \, l \, h^{-1}$
sample C   dry carbon dioxide at 760 mmHg; flow rate $5 \, l \, h^{-1}$.

Carbonates

These increase in stability from beryllium to barium, beryllium carbonate being so unstable that it can only be precipiated from solution under an atmosphere of carbon dioxide. Metal carbonates are always ionic, so that the only way beryllium carbonate can exist at all is to incorporate the hydrated ion $Be(H_2O)_4^{2+}$ into the crystal structure.

Sulphates

On heating, the anhydrous sulphates lose sulphur trioxide; by observing the initial change in slope of the weight–temperature curve for samples held on a thermo-gravimetric balance, the relative decomposition temperatures of the various sulphates can be obtained: $BeSO_4$, 580 °C; $MgSO_4$, 895 °C; $CaSO_4$, 1149 °C; $SrSO_4$, 1374 °C. The solubilities in water of the Group II sulphates are in the order $Be > Mg \gg Ca > Sr > Ba$, presumably reflecting the high solvation energies of the small and strongly polarizing beryllium and magnesium.

Halides

As already mentioned on page 141, the anhydrous beryllium halides are polymeric, covalently bonded substances which do not conduct electricity in the fused state and which sublime quite readily on heating. The addition of even small quantities of an

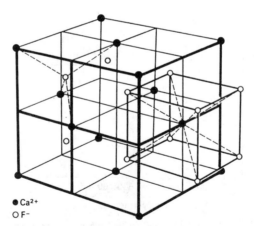

● Ca²⁺
○ F⁻

Fig. 41—The fluorite structure. The fluorite structure can be considered as derived from the caesium chloride lattice (Fig. 32) by removal of four cations from opposite corners of the faces in each cation cube; this gives rise to the 1:2 cation–anion ratio of the crystal as a whole. The anions are coordinated tetrahedrally to four calcium ions and the cations are surrounded by eight fluoride ions at the corners of a cube. The crystal is therefore said to show 4:8 coordination. Several compounds having the M_2X stoichiometry, for example, the alkali metal monoxides M_2O, adopt the 'antifluorite' structure in which the doubly charged anions replace the calcium cation and the singly charged cations replace the fluoride anion in the normal fluorite lattice. In the titanium–hydrogen, zirconium–hydrogen and hafnium–hydrogen systems, hydrogen occupies some of the fluoride ion sites in the fluorite structure, whilst the other fluoride ion sites remain vacant. The fluorite structure is not stable up to the composition MH_2 (see page 96).

alkali halide to the melts considerably increases their electrical conductivity because of the formation of complex ions such as $BeCl_4^{2-}$. With the exception of beryllium fluoride, the beryllium halides are soluble in polar organic media owing to a strong tendency towards complex formation with the solvent. For example, beryllium chloride is monomeric in pyridine because of the presence of the tetrahedral complex $Cl_2Be(NC_5H_5)_2$. The anhydrous halides readily absorb water to give the ionic tetrahydrates $Be(H_2O)_4X_2$ (except the iodide, which hydrolyses to hydrogen iodide) and, as expected of ionic compounds, these hydrated salts are insoluble in organic solvents.

Probably owing to their high lattice energies, the fluorides of magnesium, calcium, strontium and barium are only slightly soluble in water; increasing the size of the cation decreases the lattice energy and produces a slight increase in solubility along the series, so that barium fluoride is the most soluble (0.209 g dissolves in 100g of water at 25 °C). In contrast, the other halides of these elements are exceedingly soluble in water, anhydrous calcium chloride having a great enough affinity for water to be of use in the laboratory as a drying agent. Anhydrous magnesium halides are soluble in polar organic solvents (alcohols, ethers and ketones) and in some cases solid complexes such as $MgCl_2.6C_2H_5OH$ can be isolated by crystallizing the solutions. The halides probably enter the organic solvents as heavily solvated ions (solvation considerably lowers the lattice energy of the halide, making the process of solution energetically more feasible); this differs from the case of the beryllium halides which enter the organic solvents as molecular, as opposed to ionic, complexes.

Carbides

Only beryllium is small enough to be accommodated within the limited space available between the carbons of an M_2C carbide having the antifluorite structure. Beryllium carbide, Be_2C, prepared by direct reaction at about 1900°C, varies in colour from amber to dark brown and slowly liberates methane on hydrolysis.

The MC_2 carbides are formed by all the Group II elements though the preparative conditions differ slightly: beryllium and magnesium require heating in acetylene, whereas calcium, strontium and barium react directly with carbon at 2000 °C. Calcium, carbide, CaC_2, which can also be made by heating calcium carbonate with carbon, is a salt containing the $[C{\equiv}C]^{2-}$ ion; typically of a carbide containing the C_2^{2-} ion,

Table 21—Solubilities of Group II perchlorates[a] in water and alcohols (grams per 100 g solvent at 25°C)

Perchlorate	Water	Methanol	Ethanol
$Mg(ClO_4)_2$	100	52	24
$Ca(ClO_4)_2$	189	237	166
$Sr(ClO_4)_2$	310	212	181
$Ba(ClO_4)_2$	198	217	125

[a] The halides, particularly the bromides and iodides, are also soluble in alcohols.

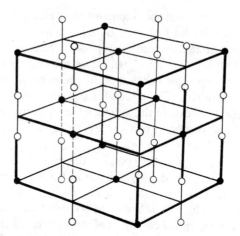

Fig. 42—The crystal structure of calcium carbide, CaC_2. The structure of calcium carbide is very similar to that of sodium chloride except that one axis of the crystal has lengthened to accommodate the non-spherical, but parallel-aligned, carbide C_2^{2-} ions, Crystalline potassium superoxide, KO_2, has an almost identical structure. See the Appendix for further discussion.

it liberates acetylene on treatment with water and is therefore an important commercial source of this hydrocarbon (Fig. 42). Calcium carbide absorbs nitrogen from the air at red heat giving calcium cyanamide $Ca(N{=}C{=}N)$, widely used as a nitrogenous fertilizer; the CN_2^{2-} ion is isoelectronic with CO_2.

Magnesium metal gives MgC_2 when heated in acetylene at around 400°C, but at 600°C a new carbide, Mg_2C_3, is obtained; this liberates propyne, $CH_3C{\equiv}CH$, on hydrolysis and is considered to contain the C_3^{4-} ion.

ORGANOMETALLIC COMPOUNDS

The most important of these derivatives are the Grignard reagents made by treating magnesium with either an alkyl or aryl halide (never the fluoride) in an anhydrous organic solvent, usually diethyl ether or tetrahydrofuran:

$$RX + Mg \xrightarrow[\substack{dry \\ N_2}]{ether} RMgX$$

They are normally given the simple formula RMgX, but this is an oversimplification. Solvation is important, as is shown by the crystallization from Grignard solutions of solid etherates such as

tetrahedrally
coordinated
magnesium

tetrahedrally
coordinated
magnesium

trigonal-bipyramidally
coordinated magnesium
(OC_4H_8 is tetrahydrofuran)

However, at concentrations above about 0.3 mol l^{-1}, diethyl ether solutions are found to contain polymeric species in which the magnesium atoms are probably linked by halogen bridges. The extent of polymerization depends not only on the concentration

$$\text{Et}_2\text{O} \diagdown \qquad \text{X} \qquad \text{X} \qquad \text{X}$$
$$\text{Mg} \qquad \text{Mg} \qquad \text{Mg}$$
$$\text{Et}_2\text{O} \diagup \qquad \text{R} \quad \text{O} \qquad \text{R} \quad \text{O} \qquad \text{R}$$
$$\text{Et}_2 \qquad \text{Et}_2$$

(bridging properties Cl > Br > I)

but also on the halide and solvent used. In strongly coordinating solvents such as triethylamine and tetrahydrofuran, Grignard reagents are monomeric over a wide range of concentrations, presumably because the Mg—N and Mg—O bonds in the solvated monomers are very much stronger than the magnesium–halogen bridges.

A further complicating factor is the possibility of a disproportionation

$$2\text{RMgX} \rightleftharpoons \text{MgR}_2 + \text{MgX}_2$$

but in simple ethers there is little or no evidence for the formation of significant amounts of either MgR$_2$ or MgX$_2$. On the other hand a solvent which complexes much more strongly with one member of the above equilibrium than the other two can cause the equilibrium, which is normally well over to the left, to shift to the right-hand side. Such a displacement occurs when dioxane is added to an ether solution of a Grignard reagent, a precipitate of MgX$_2$.2C$_4$H$_8$O$_2$ being formed:

$$2\text{RMgX} + 2\text{C}_4\text{H}_8\text{O}_2 \rightarrow \text{R}_2\text{Mg} + \text{MgX}_2.2\text{C}_4\text{H}_8\text{O}_2\downarrow$$

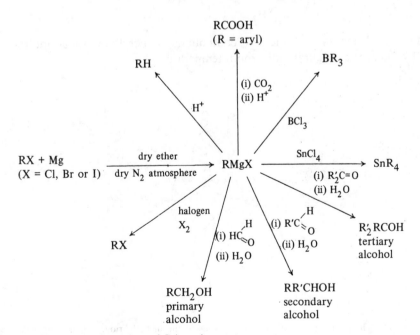

Fig. 43—Some general reactions of Grignard reagents.

Grignard reagents find wide application in organic chemistry, a few general reactions being illustrated in Fig. 43. In inorganic chemistry they are invaluable for the synthesis of a wide variety of organometallic compounds, normally accomplished by adding to the Grignard in ehter solution an anhydrous covalent metal halide: e.g.

$$BeCl_2 + 2RMgX \rightarrow BeR_2 + 2MgXCl$$

Dimethylberyllium and dimethylmagnesium are polymeric white solids linked by M—C—M bridges. There are clearly not enough electrons available to allow

the formation of normal two-centre molecular orbitals between the bonded atoms. It is thought that the bridges involve three-centre molecular orbitals formed by combination of sp^3 hybrid atomic orbitals from the two metals and the carbon atom in the bridge. The three-centre system with two electrons in the lowest-energy orbital constitutes a strongly bonding configuration (Fig. 44).

Other alkyls of beryllium are usually dimeric, unless steric interactions make the bridge system unstable and force the alkyl to be monomeric. An example of the latter effect is to be found in di-*tert*-butylberyllium which is a monomer under normal

conditions, being one of the few compounds of beryllium to adopt the linear (sp-hybridized) configuration at room temperature.

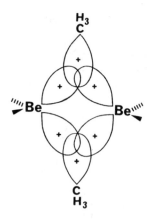

Fig. 44—The interaction of atomic sp^3 orbitals in the methyl bridges of dimethylberyllium polymer.

The difficulties previously experienced in making Grignard-like derivatives of Ca, Sr and Ba seem to have arisen from traces of alkali metals in the starting materials. Reactions of pure Group II elements with alkyl or aryl halides in ether have recently been shown to proceed smoothly and in high yield:

$$M + RX \rightarrow RMX$$

SIMILARITIES BETWEEN BERYLLIUM AND ALUMINIUM

Beryllium resembles aluminium in many respects, more so than it does magnesium. The main points of similarity arising from this diagonal relationship are summarized below:

(a) The metals are soluble in alkalis, evolving hydrogen.
(b) Both have amphoteric hydroxides which dissolve in alkalis to give beryllates and aluminates.
(c) The metals are rendered passive by concentrated nitric acid.
(d) The anhydrous halides have a strong tendency to form metal–halogen–metal bridges.
(e) The anhydrous halides are useful catalysts in Friedel–Craft reactions.
(f) Both elements form a carbide having what approximate to C^{4-} ions present in the lattice.
(g) Although the heats of formation of BeO and Al_2O_3 are high, the metals are protected from aerial oxidation by a thin film of oxide.
(h) Their alkyls contain M—C—M bridges held by three-centre bonds and the molecules are therefore 'electron-deficient'.
(i) Their anhydrous sulphates have approximately the same thermal stability; the vapour pressure of SO_3 at 750 °C is 365 mmHg over $BeSO_4$ and 900 mmHg above $Al_2(SO_4)_3$.
(j) Aqueous solutions of their salts suffer extensive hydrolysis.

Chapter 5

Group IIB. Zinc, Cadmium and Mercury

INTRODUCTION

Although zinc, cadmium and mercury have an ns^2 outer electron configuration, the presence of an underlying $(n-1)d^{10}$ shell makes their first and second ionization energies considerably larger than those of the corresponding alkaline earth metals calcium, strontium and barium. The filled 4f shell and relativity effects increase the binding energy of mercury's electrons still further, making the first ionization energy of mercury greater by about 130 kJ than that of *any* other metal and even higher than the *second* ionization energy of barium (see Fig. 45). Mercury is unique among

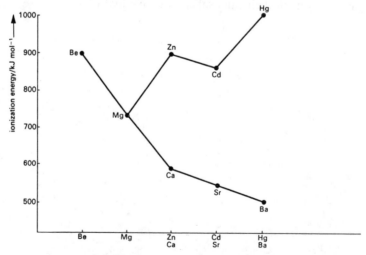

Fig. 45—The effect of poor d electron and f electron shielding on the first ionization energies of zinc, cadmium and mercury. The second ionization energies follow a similar trend.

Table 22—Zinc, cadmium and mercury

Element		1st IE (kJ mol⁻¹)	2nd IE (kJ mol⁻¹)	3rd IE (kJ mol⁻¹)	Radius of M²⁺ (Å)	$ns^2 \rightarrow ns^1np^1$ promotion energy (kJ mol⁻¹)
Zinc, Zn	$[Ar]3d^{10}4s^2$	906.4	1733	3831	0.88	431.7
Cadmium, Cd	$[Kr]4d^{10}5s^2$	867.7	1631	3616	1.09	407.6
Mercury, Hg	$[Xe]4f^{14}5d^{10}6s^2$	1006	1810	3302	1.10	524.2
cf.						
Calcium, Ca	$[Ar]4s^2$	589.4	1146	4942	1.14	
Strontium, Sr	$[Kr]5s^2$	549.4	1064	—	1.32	
Barium, Ba	$[Xe]6s^2$	502.9	965.2	—	1.49	

the metals in being a liquid at room temperature, a property again associated with the tightly bound nature of the 6s electrons which makes them relatively inaccessible for metallic bonding; in fact all three metals are unusually low-melting and volatile. The huge third ionization energies (Table 22) prevent the formation of any stable M^{3+} derivatives.

Owing to the high first and second ionization energies it is sometimes more advantageous for Zn, Cd and Hg to form two collinear bonds via $ns^2 \rightarrow ns^1 np^1$ promotion, a point particularly true in the case of mercury. This is to be contrasted with the Group IIA elements where only beryllium has any marked tendency towards linear two-coordination.

M_2^{2+} SPECIES

The values of the second ionization energies within the group suggest that the formation of M^+ derivatives might be possible. Although compounds of empirical formulae MX are known for the three elements, they are all diamagnetic (an M^+ ion would possess an unpaired electron in its outer s orbital and thus be paramagnetic) and contain $[M-M]^{2+}$ ions in which two M^+ species are covalently bonded together by overlap of their ns orbitals: the bonding situation is very similar to that occurring in the alkali metal dimers and the hydrogen molecule. Other evidence for the dimeric nature of the Hg_2^{2+} ion, for example, comes (a) from X-ray structural studies on mercury(I) salts, which show the close approach of the two mercury atoms to distances less than twice the van der Waals radius of mercury, and (b) from Raman spectra of aqueous solutions of mercurous nitrate, which contain a strong line at 172 cm^{-1} due to the vibrations of the Hg—Hg bond.

The mercury(I)–mercury(II) equilibrium

$$Hg^{2+} + Hg \underset{\text{soln}}{\overset{\text{aq}}{\rightleftharpoons}} Hg_2^{2+}, \quad \frac{[Hg_2^{2+}]}{[Hg^{2+}]} \simeq 116 \text{ at } 18°C$$

being only slightly in favour of the mercury(I) state even in the presence of free mercury, is easily disturbed by ligands which interact more strongly with mercury(II) ions than mercury(I) and hence decrease the relative concentration of free Hg^{2+}. For example, although mercury(II) cyanide is quite soluble in water it does not dissociate into Hg^{2+} and CN^- ions, and for this reason the addition of cyanide ions to a mercury(I) salt results in the immediate precipitation of mercury and the formation of $Hg(CN)_2$. Similarly, ammonia and amines, which form very stable complexes with the mercury(II) ion, cause the disproportionation of mercury(I) salts.

The vapour above heated mercury(I) chloride, although having a density corresponding to HgCl, is diamagnetic. This is because mercury(I) chloride disproportionates completely in the gas phase:

$$Hg_2Cl_2(g) \rightarrow HgCl_2(g) + Hg(g)$$

There is no evidence for Zn_2^{2+} or Cd_2^{2+} ions in aqueous solution, although the zinc(I) and cadmium(I) ions have been detected in $M-MCl_2$ melts using Raman spectroscopy; the Cd_2^{2+} ion can be isolated from such systems as the tetrachloroaluminate complex $Cd_2(AlCl_4)_2$ if the reaction is carried out in $CdCl_2/AlCl_3$ mixtures. When

Table 23—M—M stretching frequencies and bond lengths in some zinc(I), cadmium(I) and mercury(I) species

Compound	M—M frequency (cm⁻¹)	M—M bond length (Å)
Zn_2^{2+} dissolved in saturated $ZnCl_2$ soln	175	—
$Cd_2(AlCl_4)_2$	—	2.58
Hg_2F_2	186	2.51
Hg_2Cl_2	166	2.53
Hg_2Br_2	132	2.58
$Hg_2(NO_3)_2$ (aq soln)	172	—
$Hg_2(NO_3)_2.2H_2O$	180	2.54
$Hg_2(ClO_4)_2.4H_2O$	182	—

this green cadmium(I) derivative is added to water, an immediate disproportionation occurs with the production of cadmium metal:

$$Cd_2^{2+} \rightarrow Cd\downarrow + Cd^{2+}$$

Spectroscopic studies on M_2^{2+} ions suggest that the Zn—Zn and Cd—Cd bonds are considerably weaker than the Hg—Hg bond in the mercury(I) ion. This is, of course, against the normal trend within any group of the periodic table: in almost all other cases the bonds formed by the heaviest elements in a group are weaker than those involving the earlier members of the same group, and this is especially true when homonuclear (i.e. M—M) bonds are being considered. It has been argued that the strengths of the M—M bonds in these M_2^{2+} ions will be proportional to the electron affinities of the M^+ ions (the electron affinity of M^+ being the negative of the first ionization energy of M). The first ionization energy of mercury (and thus the electron affinity of Hg^+) is larger than that of zinc or cadmium and hence the M—M bond in the mercury(I) ion is expected, on these grounds, to be the strongest metal–metal bond (Table 23).

POLYATOMIC MERCURY CATIONS

Other homopolyatomic cations of mercury can be prepared by oxidation of Hg with either AsF_5 or SbF_5 in liquid sulphur dioxide, which gives successively Hg_2^{2+}, Hg_3^{2+} and Hg_4^{2+}. Salts of the last two ions, such as $Hg_3(AsF_6)_2$, $Hg_3(AlCl_4)_2$ and $Hg_4(AsF_6)_2$, can be isolated from the solutions and have been shown by X-ray crystallography to contain nearly linear Hg_n^{2+} ions; very weak Hg---Hg interactions *between* the ions result in the formation of infinite zigzag chains running through the crystals.

AQUEOUS M²⁺ IONS

Hexahydrated ions of the elements are produced when their perchlorates are dissolved in water; virtually no complexation occurs with the anion, but the solutions, particularly of the mercury derivative, are acidic owing to hydrolysis which gives rise to species such as $ZnOH^+$, $CdOH^+$, Cd_2OH^{3+} and $HgOH^+$. Dilute aqueous solutions of zinc chloride also contain $Zn(H_2O)_6^{2+}$ ions, but as the concentration is increased, chloro-complexes such as $ZnCl_{aq}^+$, $ZnCl_{2_{aq}}$ and particularly $ZnCl_{4_{aq}}^{2-}$ begin to form, some zinc moving towards the *anode* on electrolysis of solutions above 2M concentration; the position is further complicated by hydrolysis which lowers the pH to about 1.0 in 6M solution. Even at low concentrations cadmium chloride is strongly self-complexed in the form of $CdCl_x^{(2-x)+}$ ions, and mercury(II) chloride enters solution as virtually undissociated $HgCl_2$ molecules.

SYNTHESIS OF SIMPLE SALTS

Many oxo-salts separate from aqueous solution as hydrates containing $M(H_2O)_6^{2+}$ cations, some of the zinc salts being isostructural to their magnesium counterparts: $MgSO_4.7H_2O$ and $ZnSO_4.7H_2O$; $Mg(BrO_3)_2.6H_2O$ and $Zn(BrO_3)_2.6H_2O$; $Mg(ClO_4)_2.6H_2O$ and $Zn(ClO_4)_2.6H_2O$. A typically general preparation of many salts is to dissolve an excess of the oxide (or carbonate for Zn and Cd) in the corresponding acid and crystallize the filtered solutions; several examples for zinc salts are shown below:

$$ZnO + HCl \quad \rightarrow ZnCl_2 \quad \text{(several hydrates known)}$$

$$ZnO + H_2SeO_4 \rightarrow ZnSeO_4 \quad \text{(several hydrates known)}$$

$$ZnO + HClO_4 \quad \rightarrow Zn(ClO_4)_2.6H_2O$$

$$ZnO + HCOOH \rightarrow Zn(OOCH)_2.2H_2O \quad \text{(zinc methanoate)}$$

$$ZnO + H_2SiF_6 \quad \rightarrow ZnSiF_6.6H_2O$$

$$ZnO + HNO_3 \quad \rightarrow Zn(NO_3)_2 \quad \text{(several hydrates known)}$$

$$ZnO + H_3PO_4 \quad \rightarrow Zn_3(PO_4)_2$$

Metathetic reactions may also be used, and if the required salt is soluble, conditions are arranged for a highly insoluble by-product to be formed which can then be removed by filtration:

$$Hg(OOCCH_3)_2 + 2KSCN \rightarrow Hg(SCN)_2\downarrow + 2KOOCCH_3$$

$$Hg(NO_3)_2 + 2NaIO_3 \rightarrow Hg(IO_3)_2\downarrow + 2NaNO_3$$

$$ZnSO_4 + BaI_2 \rightarrow BaSO_4\downarrow + ZnI_2$$

$$CdSO_4 + Ba(SCN)_2 \rightarrow BaSO_4\downarrow + Cd(SCN)_2$$

COMPLEXES OF GROUP IIB METALS

All three metals form a wide variety of complexes, the stabilities of which are normally in the order Hg > Cd > Zn; the more common coordination numbers of the cations are two, four, five and six. Unusual 1:1 complexes have been prepared with cadmium

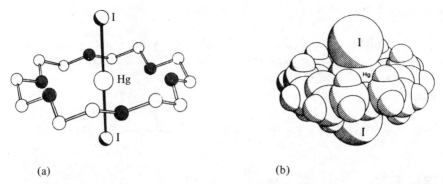

Fig. 46—Structure of HgI_2 (18-crown-6). (a) The linear HgI_2 molecule is threaded through the hole in the crown polyether, resulting in 8-coordination for mercury. The 18-crown-6 complexes of $CdCl_2$, $CdBr_2$, CdI_2 and $HgCl_2$ have the same structure (similar complexes have also been prepared for aluminium; see page 182). (b) Space-filling diagram of the complex. [Reproduced from *Acta Crystallographica* vol. C42 (1986) p. 51.]

or mercury dihalides and the polyether 18-crown-6, in which a linear dihalide molecule is 'threaded through' the crown ether cavity (Fig. 46). The coordination of the metal can be regarded as either two or eight depending on whether the six long M—O interactions are classed as 'bonds'; certainly the M---O distance is slightly less than the sum of the van der Waals radii of O and M. Since the ion $HgCl(18\text{-crown-6})^+$ can be detected in the mass spectrum of $HgCl_2(18\text{-crown-6})$, it can probably be assumed that the ether is actually bonded, albeit weakly, to the metal. Rebeck has suggested the following mechanism for the formation of this type of complex:

It is possible that zinc is too small to form even a weak interaction simultaneously with all six oxygens because initial attempts to make a 'threaded' complex of zinc iodide gave crystals of $ZnI_2(H_2O)_2(18\text{-crown-6}).H_2O$ in which the crown ether is only hydrogen-bonded to a tetrahedral $Zn(H_2O)_2I_2$ group and the lattice–water molecule.

OCCURRENCE OF ZINC, CADMIUM AND MERCURY

Zinc

130 ppm of the Earth's crust; major source is zinc blende, ZnS, which is often associated with sulphides of other metals, such as lead, copper, cadmium and iron.

Cadmium

0.15 ppm; cadmium minerals are rarely found alone but are more often associated with zinc ores (e.g. zinc sulphide often contains 0.1%–0.2% of cadmium).

Mercury

0.5 ppm; occurs native to a slight extent, but the main ore is cinnabar, HgS.

EXTRACTION

Zinc

Zinc blende is roasted to give zinc oxide and sulphur dioxide, the latter being oxidized to sulphur trioxide and converted into the valuable by-product sulphuric acid. The resulting zinc oxide is then reduced in a blast furnace using coke, the metallic zinc being condensed from the vapours issuing from the furnace.

Cadmium

This is obtained as a by-product of zinc smelting and, being considerably more volatile than zinc, can be separated from it by distillation.

Mercury

Cinnabar, HgS, is roasted in air, when sulphur dioxide and free mercury are formed (mercury(II) oxide is unstable above about 400°C); as the mercury distils away from the hot zone it is condensed in large, water-cooled metal condensers.

THE ELEMENTS

Both zinc and cadmium have a distorted hexagonal close-packed structure, the distortion being an elongation of the distance between the close-packed layers (see Fig. 47). Mercury is unique among the metals in being a liquid at room temperature; it freezes at −39°C to a rhombohedral structure in which all the metal atoms have a coordination number of six. The tightly bound nature of the two ns electrons makes them relatively unavailable for bonding in the metals and as a consequence zinc, cadmium and mercury are among the most volatile of the heavy metals. Their normal boiling points are: zinc, 906°C; cadmium, 765°C; mercury, 357°C. The vapours are essentially monatomic, as one would expect from the electron configuration of the metal atoms.

Notwithstanding their high ionization energies, the electrode potentials (Table 34) of zinc and cadmium have quite high negative values; these two metals therefore

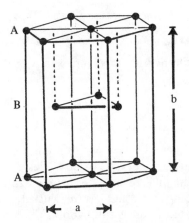

Fig. 47—The hexagonal close-packed structure adopted by metallic zinc and cadmium. The structure is slightly distorted, in that the axial ratio b/a is about 1.87 instead of the ideal value for this arrangement of 1.63. The black circles represent the positions taken up by the metal nuclei and hence a is equal to the length of *two radii*, i.e. the diameter of the metal atoms.

Table 24—Standard reduction potentials (V) at 25 °C

$Zn^{2+} + 2e \rightleftharpoons Zn$	-0.76
$Cd^{2+} + 2e \rightleftharpoons Cd$	-0.40
$2H^2 + 2e \rightleftharpoons H_2$	± 0.00
$Hg_2^{2+} + 2e \rightleftharpoons 2Hg$	$+0.79$
$Hg^{2+} + 2e \rightleftharpoons Hg$	$+0.85$
$2Hg^{2+} + 2e \rightleftharpoons Hg_2^{2+}$	$+0.91$
cf.	
$Sr^{2+} + 2e \rightleftharpoons Sr$	-2.89
$Ca^{2+} + 2e \rightleftharpoons Ca$	-2.87
$Mg^{2+} + 2e \rightleftharpoons Mg$	-2.37

dissolve in non-oxidising acids to give the corresponding salts and hydrogen. In apparent contradiction to this it is found that zinc can be electrically deposited from an aqueous solution of its salts; this is because the hydrogen overvoltage at a zinc surface amounts to 1 V (at a current of about 0.1 A). Mercury will dissolve only in oxidizing acids.

COMPOUNDS OF ZINC, CADMIUM AND MERCURY

Hydrides

Although transient diatomic species MH and MH$^+$ have been detected spectroscopically for all three elements, stable dihydrides are known only for zinc and cadmium; these are obtained by reduction of their alkyls with lithium tetrahydroaluminate in ether at low temperatures: e.g.

$$Zn(CH_3)_2 + 2LiAlH_4 \rightarrow ZnH_2 + 2LiAlH_3(CH_3)$$

Both colourless, involatile hydrides decompose at (ZnH_2) or below (CdH_2) room temperature.

Oxides and hydroxides

The oxides MO result either when the metals are heated in air or when the nitrates are decomposed thermally; mercury(II) oxide begins to dissociate into mercury and oxygen above 400 °C. The difference in size between zinc and cadmium is apparent when their oxides are compared: ZnO has the wurtzite structure in which the metal is tetrahedrally coordinated whereas CdO adopts the 6:6 lattice of sodium chloride.

$Zn(OH)_2$ and $Cd(OH)_2$ are precipitated from solutions of soluble salts by the addition of alkali; zinc hydroxide is soluble in excess of base owing to the formation of hydroxo-ions having the general decomposition $[Zn(OH)_m(H_2O)_n]^{2-m}$, and solid zincates typified by $NaZn(OH)_3$ and $Na_2Zn(OH)_4$ can be crystallized from the solutions. It is possible that cadmates may exist in solutions of very strong base because on addition of *solid* KOH to aqueous CdI_2 the initial precipitate of cadmium hydroxide dissolves when the mixture is heated; however, when the solutions are allowed to cool slowly, only crystals of $Cd(OH)_2$ separate out. (The solubility of cadmium hydroxide rises only from 0.0003 g/100 g in pure water to 0.13 g/100 g in 5M NaOH at room temperature.)

Zinc hydroxide also dissolves in aqueous ammonia owing to the formation of ammines and, since zinc migrates towards the cathode on electrolysis, it is assumed that cations such as $[Zn(NH_3)_4(H_2O)_2]^{2+}$ are present in solution. Cadmium hydroxide behaves in a similar fashion.

Mercury(II) hydroxide is unknown; addition of base to an aqueous solution of a mercury(II) compound results only in the precipitation of yellow mercury(II) oxide, which differs from the red form (prepared as above) only in particle size and is not a different crystal modification.

Sulphides, selenides and tellurides

The chalcogenides of the Group IIB elements may be made either by direct union on heating or by treatment of a soluble salt with H_2S, H_2Se or H_2Te:

$$Hg + S \rightarrow HgS$$

$$Cd(NO_3)_2 + H_2Te \rightarrow CdTe\downarrow + 2HNO_3$$

$$Hg(OOCCH_3)_2 + H_2S \rightarrow HgS\downarrow + 2CH_3COOH$$

The low-temperature zinc blende form of ZnS (Fig. 48) transforms into wurtzite at 1020 °C, although impurities can lower the transition temperature. Zinc sulphide is an industrially important phosphor used on television screens and in fluorescent or luminous paints; it emits radiation when excited by ultraviolet light, cathode rays, X-rays and γ rays.

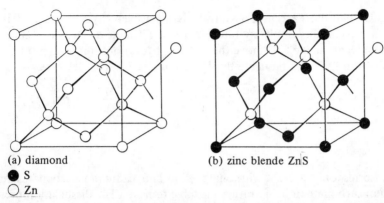

(a) diamond   (b) zinc blende ZnS

● S
○ Zn

Fig. 48—The structures of (a) diamond and (b) zinc blende. Zinc blende can be visualized as being derived from the diamond structure by replacing alternative carbon atoms with zinc and sulphur atoms. Each zinc atom and each sulphur atom is then tetrahedrally coordinated to its four neighbours. The zinc blende structure may also be considered to be a cubic close-packed arrangement of sulphur atoms with zinc atoms occupying half of the tetrahedral holes. The closely related wurtzite structure has the sulphur atoms in a hexagonal close-packed array with the zinc atoms again occupying half of the tetrahedral holes; see Appendix.

Zinc ethanoate

Dissolution of either zinc oxide or zinc carbonate in warm ethanoic acid gives zinc ethanoate, which separates from solutions as the crystalline dihydrate:

octahedrally coordinated zinc;
bidentate ethanoate groups

Dehydration of the crystals takes place at about $100\,^{\circ}C$ and, on further strong heating, partial decomposition occurs to give the volatile basic ethanoate $Zn_4O(CH_3COO)_6$, which is isostructural with its beryllium analogue (page 145). Unlike the latter, basic zinc ethanoate is rapidly hydrolysed by water, the difference in reactivity being due to the larger size of zinc which provides room for an attacking water molecule to coordinate initially.

Nitrates

All three metals dissolve in nitric acid to give the hydrated nitrates; in the case of mercury an excess of acid is required to ensure the formation of mercury(II) nitrate since any metallic mercury left in excess is capable of reducing the Hg^{2+} ion to Hg_2^{2+}

in aqueous solution. Mercury(II) nitrate is extensively hydrolysed and in dilute solution it breaks up completely into HgO and nitric acid. Decomposition occurs when attempts are made to remove the water of hydration from zinc nitrate, and the anhydrous nitrate has to be made by the reaction sequence

$$Zn + N_2O_4(l) \rightarrow NO + Zn(NO_3)_2.2N_2O_4 \xrightarrow[\text{vacuum}]{100\,°C} Zn(NO_3)_2$$

Carbonates

Mercury forms only a basic carbonate. Zinc and cadmium carbonates resemble magnesium carbonate in being rather unstable to heat. (The dissociation pressure of carbon dioxide is 1 atm at about 350 °C for both salts.) This is in line with the unexpectedly high polarizing power of the Zn^{2+} and Cd^{2+} ions; the pressure of carbon dioxide above calcium carbonate and barium carbonate only reaches 1 atm at 900 and 1330 °C respectively. Both carbonates dissolve in acids to give salts, and $ZnCO_3$ is also soluble in alkalis.

Halides

Zinc fluoride, ZnF_2 (rutile structure; Fig. 49), and cadmium fluoride, CdF_2 (calcium fluoride structure), are not very soluble in water and can be precipitated from solution using a soluble fluoride (cf. the low solubility of magnesium fluoride). Mercury(II) fluoride (calcium fluoride structure) is completely hydrolysed in aqueous solution.

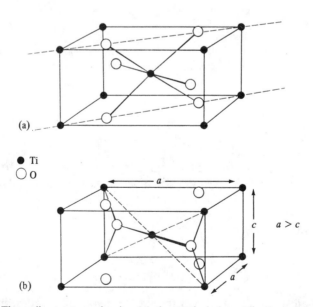

Fig. 49—The rutile structure, showing (a) the octahedral coordination of the cations and (b) the planar triangular coordination of the anions; see Appendix. Magnesium fluoride and zinc fluoride crystallize with this structure.

In sharp contrast, the other halides of zinc and cadmium are not only highly soluble in water but are also reasonably soluble in alcohols, ethers, amines, ketones, esters and nitriles (compare beryllium chloride, bromide and iodide). Aqueous solutions of zinc chloride have been shown by Raman spectroscopy to contain mainly $Zn(H_2O)_2Cl_4^{2-}$ ions together with $Zn(H_2O)_6^{2+}$, $ZnCl_2(aq)$ and $ZnCl^+(aq)$, but no evidence could be obtained for the presence of either $ZnCl_3^-$ or $ZnCl_4^{2-}$. Complexes between Cd^{2+} and halide ions are more stable than those of Zn^{2+} and as a result cadmium halides are found to dissolve in water giving $Cd^{2+}(aq)$, $CdX^+(aq)$, CdX_2, CdX_3^- and CdX_4^{2-}. The tetrahedral ZnX_4^{2-} and octahedral CdX_6^{4-} ions can be stabilized in the solid state, e.g. as in Cs_2ZnBr_4 and K_4CdCl_6.

Molten zinc halides are unusual in being highly viscous like BeF_2. An X-ray study of $ZnCl_2$ just above the melting point showed that no monomeric molecules were present, the system being highly polymeric and approximating to close-packed chlorine atoms with the zincs located in tetrahedral holes.

Mercury(II) chloride, bromide and iodide are only slightly soluble in water ($Cl > Br > I$); little or no dissociation to $Hg^{2+}(aq)$ occurs in such solutions. Halide complexes of the mercury(II) ion are relatively stable and in aqueous solution the common ones are HgX^+, HgX_2 and HgX_4^{2-}. The stability of the iodide complexes causes some interesting effects when a soluble iodide is added to a mercury(II) salt: although mercury(II) iodide is very insoluble (solubility $\simeq 10^{-4}$ mol l^{-1}) it does not precipitate from solution until about 0.2 mol of iodide ions have been added to the system because of the formation of the stable HgI^+ ion. In the presence of a large excess of iodide ion the red mercury(II) iodide dissolves to give the colourless HgI_4^{2-} ion.

In the vapour state the anhydrous chlorides, bromides and iodides of all three metals exist as the linear, covalent molecules X—M—X in which the metal atoms are presumably sp-hybridized. An interesting fact is that the zinc halides are appreciably more volatile in the presence of metallic zinc; it has been suggested that this may be due to the formation of gaseous ZnX or $(ZnX)_2$ molecules, but attempts to condense out any lower halides by quenching the vapour lead only to a mixture of zinc and ZnX_2.

Hydrogen reduces zinc bromide and zinc iodide to the metal on heating.

COMPLEXES OF ZINC, CADMIUM AND MERCURY

The presence of the filled $(n-1)d$, and to a lesser extent 4f, shells in the electronic configuration of these elements causes two effects which contribute to the higher stability of complexes formed by zinc, cadmium and mercury relative to their immediate counterparts, the alkaline earth metals calcium, strontium and barium. The first is that the poor shielding characteristics of the d and f orbitals reduce the size of Zn^{2+}, Cd^{2+} and Hg^{2+} ions relative to Ca^{2+}, Sr^{2+} and Ba^{2+} because the outer electrons in zinc, cadmium and mercury experience a relatively high effective nuclear charge (the radius of Hg^{2+} is smaller even than that of Ca^{2+}); the smaller the ion forming a complex, the higher will be the ion–dipole contribution to the bonding. The second effect is that filled $(n-1)d$ orbitals are more easily polarized ('pushed aside') by ligand electrons than are filled np orbitals; in this way a

Table 25—Coordination numbers in zinc complexes

Ligand atoms	Coordination numbers adopted
O only	4, 5, 6
O + halogen	4, 6
N only	4, 6
N + halogen	4, 5, 6
O + N	4, 5, 6
S only	4, 5

ligand on zinc, cadmium or mercury experiences a nuclear charge which is somewhat greater than the simple $+2$ charge on the ion.

Because the d orbitals of these elements are completely filled, there is no ligand field stabilization effect to contribute to the overall stability of their complexes.

The coordination numbers adopted by zinc, cadmium and mercury in their complexes can be two, four, five or six (Table 25). Although the stability sequence is often $Hg \gg Cd > Zn$ for tetrahedral complexes (e.g. the anionic halide derivatives MX_4^{2-}), mercury has a peculiar reluctance to form octahedral complexes.

Two-coordinate complexes

A number of covalent compounds of these metals ($HgCl_2$, $HgBr_2$, HgI_2, the gaseous halides of zinc and cadmium, alkyls or aryls of zinc, cadmium and mercury) all have a linear X—M—X structure.

Four-coordinate complexes

The metal atoms in these complexes have the four ligand groups arranged tetrahedrally around them:

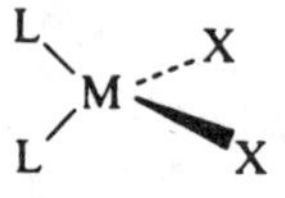

neutral

e.g. $Zn(py)_2Cl_2$
$Cd(py)_2Cl_2$
(py = pyridine)
$HgI_2[P(C_6H_5)_3]_2$
$HgCl_2[OAs(C_6H_5)_3]_2$
$Zn(C_6H_5)_2[P(C_6H_5)_3]_2$

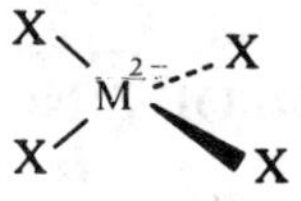

anionic

e.g. $[Cs]_2[ZnX_4]$ (X = Cl or Br)
$[Li]_2[Zn(CH_3)_4]$
$[K]_2[HgI_4]$
$[Cs]_2CdI_4$

Five-coordinate complexes

There are relatively few complexes of this type; three typical zinc complexes shown by X-ray crystallography to have a (distorted) trigonal bipyramidal shape are

$ZnCl_2$(terpyridyl), bis(acetylacetonato)zinc monohydrate and $NaZn(OH)_3$. The first complex, like its mercury analogue, has the tridentate ligand occupying one equatorial and the two axial positions of the trigonal bipyramid:

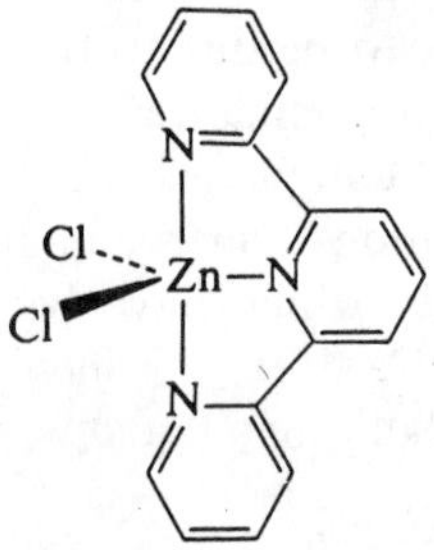

A number of $HgX_2.L$ complexes of monodentate ligands have trigonal planar $HgX_2.L$ groups, in which there is relatively strong bonding, linked by weaker axial halogen bridges from neighbouring molecules:

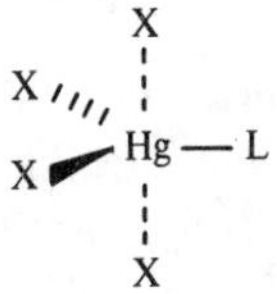

e.g. X = Cl, L = PEt₃
X = Cl, L = 2,4,6-Me₃ pyridine

Six-coordinate complexes

Many zinc and cadmium salts of oxy-acids contain the octahedral $Zn(H_2O)_6^{2+}$ and $Cd(H_2O)_6^{2+}$ ions: e.g. some common hydrates containing $Zn(H_2O)_6^{2+}$ ions are $ZnSO_4.7H_2O$, $Zn(BrO_3)_2.6H_2O$ and $Zn(ClO_4)_2.6H_2O$. Hexammines such as $M(NH_3)_6X_2$ and $M(en)_3X_2$, where en is ethylenediamine and X is Cl, Br or I, are also known for zinc and cadmium, but they are not particularly stable; the ammonia complexes exhibit a considerable dissociation pressure of ammonia at room temperature and must be stored in sealed tubes to prevent loss of the ligand.

Mercury forms very few octahedral complexes. Two recently reported examples are $[Hg(pyridine-N-oxide)_6][ClO_4]_2$ and $[Hg(dimethylsulphoxide)_6][ClO_4]_2$.

ORGANOMETALLIC COMPOUNDS

The monomeric dialkyls and diaryls of zinc, cadmium and mercury are readily obtained by the action of Grignard reagents or lithium reagents on ethereal solutions of the anhydrous metal halides: e.g.

$$CdCl_2 + 2RMgX \rightarrow CdR_2 + 2MgXCl$$

$$HgCl_2 + 2LiR \rightarrow HgR_2 + 2LiCl$$

Treatment of MR_2 with one mole of a dihalide MX_2 in an organic solvent results in the formation of the organometal halide RMX. Many alkylzinc halides, like the

Grignard reagents which they closely resemble, can be made directly from the corresponding alkyl halides and zinc dust.

Organozinc and organocadmium compounds are oxygen- and water-sensitive, the lower dialkyls of zinc being pyrophoric in air. The high stability of organomercurials towards air and water appears to be due to the fact that mercury has a very low affinity for oxygen and not to any enhanced strength of mercury–carbon σ bonds. Although the zinc alkyls do not conduct electricity, they dissolve alkyls of the alkali metals to give conducting solutions which contain complex anions such as ZnR_3^- and ZnR_4^{2-}; typical derivatives which have been isolated are $Na[Zn(C_2H_5)_3]$, $Li_2[Zn(CH_3)_4]$ and $K_2[Zn(C\equiv CH)_4]$. These complexes also result when an alkylzinc reacts with the free alkali metal:

$$2Rb + 3Zn(C_2H_5)_2 \rightarrow 2Rb[Zn(C_2H_5)_3] + Zn$$

When aromatic compounds are treated with mercury(II) acetate it is found that mercury will substitute one or more of the hydrogen atoms in the aromatic ring; the reactive species is thought to be $HgOAc^+$:

$$ArH + HgOAc^+ \rightarrow [ArHHgOAc]^+ \rightarrow ArHgOAc + H^+$$

Many RHgX derivatives have been widely used as pesticides and such a simple route to a potentially wide variety of products is of obvious industrial importance. However, the high toxicity of these mercurials is causing some concern at the present time: the high stability of the mercury–carbon bond towards air and water means that the rate of degradation of pesticides is so slow that 'pesticide residues' are beginning to find their way (still as RHgX) into the human food cycle.

MERCURY(I) DERIVATIVES

The chemistry of the mercury(I) state is not very extensive, mainly because the equilibrium

$$2Hg_2^{2+} \rightleftharpoons Hg + Hg^{2+}$$

is forced to the right under the conditions of many chemical reactions. Furthermore, the standard potentials at $25°C$ for the two reactions

$$Hg_2^{2+} + 2e \rightleftharpoons 2Hg, \quad +0.79v$$

$$Hg^{2+} + 2e \rightleftharpoons Hg, \quad +0.85 \text{ V}$$

are so close that when liquid mercury is treated with an excess of oxidizing agent the mercury(II) state always results because there are no known oxidizing agents having a potential lying between 0.79 and 0.85 V. In such reactions it is only by having metallic mercury in excess that the mercury(I) state can be obtained, by virtue of the reaction

$$Hg + Hg^{2+} \rightarrow Hg_2^{2+}$$

A typical example of this behaviour is the reaction of mercury with nitric acid.

Mercury(I) oxide and hydroxide are unknown; when alkali is added to a mercury(I) salt the black solid obtained has been shown to be an intimate mixture of mercury(II) oxide and mercury. All the halides exist, but the fluoride is completely hydrolysed by water to mercury and mercury(II) oxide. When mercury(I) chloride and bromide sublime, their vapours have densities corresponding to 'HgX' owing to the complete disproportionation

$$Hg_2Cl_2 \rightarrow HgCl_2 + Hg$$

The iodide is so unstable that it forms mercury and mercury(II) iodide simply on warming in water.

Mercury(I) nitrate is one of the most soluble salts, making it a convenient starting point for the preparation, by precipitation, of insoluble mercury(I) derivatives (e.g. Cl, Br, I, BrO_3, IO_3). It crystallizes from solution as the dihydrate $Hg_2(NO_3)_2.2H_2O$, which contains the linear ion $[H_2OHgHgOH_2]^{2+}$.

The number of known mercury(I) complexes is small because of the disproportionation of the mercury(I) ion in the presence of common ligands, including ammonia, many amines, CN^-, OH^-, SCN^-, thioethers and acetylacetone. The magnitude of those stability constants which have been measured for mercury(I) complexes suggests that the mercury(I) ion does not form weak complexes; the scarcity of mercury(I) complexes must be due to disproportionation because of the very strong complexing ability of mercury(II). It appears probable that the usual coordination numbers of Hg_2^{2+} are two and four. The two-coordinate complexes contain linear $[L—Hg—Hg—L]^{2+}$ ions as in the dihydrate mentioned above. Four-coordinate complex ions $Hg_2.4L^{2+}$ are formed when triphenylphosphine oxide or pyridine-N-oxide are added to the mercury(II) perchlorate solutions, but their stereochemistry is not known.

A COMPARISON OF ZINC WITH BERYLLIUM

The first and second ionization energies of beryllium and zinc are very similar and might be expected to lead to similarities in chemical properties, although it should be borne in mind that the zinc atom is bigger than that of beryllium, which may lead to differences in the coordination numbers of the two elements. As we have seen above, the high first and second ionization energies of zinc arise from the poor shielding properties of the filled 3d shell; those of beryllium are high because the electrons, coming from a small 2s orbital, experience a strong Coulombic attraction of the nucleus.

Some points of similarity between zinc and beryllium are:

(a) The free metals are isostructural and adopt a slightly distorted hexagonal close-packed arrangement (cadmium and magnesium have the same structure).

(b) Both metals dissolve in strong bases, evolving hydrogen owing to the formation of soluble beryllates or zincates; their oxides are thus amphoteric. (Cadmium oxide and magnesium oxide are not amphoteric and the metals do not dissolve in strong bases.)

(c) Although beryllium (unlike zinc) forms no anhydrous salts which contain the bare M^{2+} ion, both Be^{2+}(aq) and Zn^{2+}(aq) ions are known in aqueous solutions. Because of the disparity in size between Be^{2+} and Zn^{2+}, the number of water molecules in the inner hydration sphere is four for beryllium and six for zinc (Cd^{2+} and Mg^{2+} are similar to Zn^{2+}).

(d) Solutions of beryllium and zinc salts have an acidic reaction owing to hydrolyses of the type

$$M^{2+}(H_2O)_n \rightarrow HOM^+(H_2O)_{n-1} + H^+(aq) \rightarrow M(OH)_2 + H^+(aq)$$

This type of reaction is somewhat unexpected for an ion as large as zinc. It appears that the polarizing power of the Zn^{2+} ion (similarly the Cd^{2+} and Hg^{2+} ions) is increased above the expected value because the filled d shell is quite readily deformed by the presence of ligands and hence the ligands 'see' an effective ionic charge which is somewhat greater than $+2$.

(e) The $MX_2.2L$ complexes of both beryllium and zinc are tetrahedral and covalent; those of stoichiometry $BeX_2.4L$ are ionic and contain the tetrahedral cation BeL_4^{2+}. By virtue of its greater size, zinc is able to form complexes (e.g. hydrates and ammines) which contain the octahedral cation ZnL_6^{2+} (cadmium resembles zinc).

(f) Both beryllium and zinc form a basic ethanoate having the formulation $M_4O(O_2CCH_3)_6$; for the structure of these unusual basic acetates see page 145. (Cadmium and magnesium do not form this type of ethanoate.)

(g) Beryllium oxide and zinc oxide adopt the wurtzite structure in which the metal atoms have four-coordination (unlike magnesium oxide and cadmium oxide which have the sodium chloride structure and octahedral metal coordination). Four-coordination of beryllium and zinc also occurs in the ortho-silicates Be_2SiO_4 and Zn_2SiO_4, whereas magnesium is six-coordinate in Mg_2SiO_4.

Chapter 6

Group III. Boron, Aluminium, Gallium, Indium and Thallium

INTRODUCTION

Among the Group I and II elements the ionization energies were found to decrease with increasing atomic number; this is not the case in Group III, where the ionization energies vary down the group in an apparently erratic way (Table 26 and Fig. 50). On closer inspection, however, it is seen that the inner electronic configurations of the Group III elements are not identical; instead of a rare gas electron structure between them and the nucleus, the ns^2np^1 electrons of gallium, indium and thallium are outside a filled set of $(n-1)$d orbitals and have what might be called a 'd^{10} core'. The inner electron structure of thallium also contains fourteen electrons in the 4f orbitals filled during the building up of the preceding lanthanide series of elements.

The increasing effective nuclear charge and size contraction, which occur during the filling of d and f orbitals (see Chapter 1), combine to make the outer s and p electrons of Ga, In and Tl more strongly held than would be expected by simple extrapolation from boron and aluminium. The drop in ionization energies between Ga and In mainly reflects the fact that both elements have an identical inner electron core and hence there is an expected decrease in electron binding at indium because its outer electrons are further from the nucleus. There is also an increase in effective nuclear charge when progressing from left to right *across* any given period and this is responsible (together with the above 'core effects') for the observed variation in first ionization energies which are in the order Group I < Group II < Group III.

TRIVALENT CATIONS

The M^{3+} state for gallium, indium and thallium will be energetically less favourable than Al^{3+} because the high ionization energies of these three elements cannot always

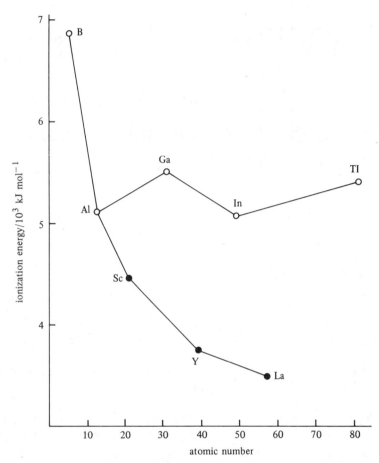

Fig. 50—The sum of the first three ionization energies plotted against atomic number for the Group IIIA (Sc, Y, La) and Group IIIB (B, Al, Ga, In, Tl) elements. The effect of the d^{10} and $f^{14}d^{10}$ inner electron cores on the ionization energies of the Group IIIB elements is strikingly apparent (the Group IIIA elements all have s^2p^6 cores).

be balanced by the lattice energies of possible reaction products (the lattice energies decrease down the aluminium, gallium, indium and thallium group owing to the increasing size of the M^{3+} cations) and, for the group as a whole, the M^{3+} state is the exception and not the rule. In fact, of the trihalides only the high-melting fluorides of Al, Ga, In and Tl are usually considered to be ionic. More commonly these elements achieve an oxidation state of three by forming covalent compounds via $ns \rightarrow np$ promotion.

Although simple M^{3+} cations are uncommon in anhydrous compounds of Group III elements, the hydrated ions of aluminium, gallium, indium and thallium are well known in aqueous solution. Nuclear magnetic resonance studies show that there are six water molecules held strongly by these ions in their primary coordination spheres and there is undoubtedly a layer of further solvent hydrogen-bonded to these hexahydrated species. At high salt concentration, ion association occurs and, for

Table 26—The Group III elements

Element	Atomic number		1st IE (kJ mol^{-1})	2nd IE (kJ mol^{-1})	3rd IE (kJ mol^{-1})	4th IE (kJ mol^{-1})	Radius of M$^+$ (Å)	Radius of M^{3+} (Å)
Boron, B	5	$1s^2 2s^2 2p^1$	800.3	2427	3658	25 030	—	—
Aluminium, Al	13	$[Ne]3s^2 3p^1$	564.2	1816	2744	11 580	—	0.68
Gallium, Ga	21	$[Ar]3d^{10}4s^2 4p^1$	564.2	1979	2962	6 193	—	0.76
Indium, In	49	$[Kr]4d^{10}5s^2 5p^1$	558.3	1820	2705	5 230	—	0.94
Thallium, Tl	81	$[Xe]4f^{14}5d^{10}6s^2 6p^1$	589.0	1970	2975	4 896	1.64	1.03

example, the complex ions $[InCl]^{2+}_{aq}$, $[InCl_2]^{+}_{aq}$ and $[InCl_4]^{-}_{aq}$ have been detected in aqueous indium chloride solution using Raman spectroscopy.

HYDROLYSIS OF M^{3+}_{aq} IONS

The high charge on the central metal of the hexahydrated cations induces hydrolysis via ionization of protons on the coordinated water molecules, an effect which will be greatest for Al^{3+}:

$$M(H_2O)_n^{3+} \rightleftharpoons HOM(H_2O)_{n-1}^{2+} + H^+(aq) \rightleftharpoons (HO)_2M(H_2O)_{n-2}^{+} + H^+(aq)$$

$$\rightleftharpoons M(OH)_3 + H^+(aq)$$

Such hydrolyses can be suppressed by the addition of acid but there is always the danger that complexes may be formed, particularly with the hydrogen halides; it is common practice to use perchloric acid for this purpose since the *large, univalent* ClO_4^- ion is the least likely of the common anions to coordinate with the cations.

Increasing the pH of $Al(H_2O)_6^{3+}$ solutions by gradual addition of base results in the hydrolytic equilibria being forced increasingly to the right-hand side, producing (solvated) species such as $Al(OH)^{2+}$, an oligomeric mixture with a stoichiometry close to $Al(OH)_{2.5}^{0.5+}$ and the tridecameric ion $[AlO_4Al_{12}(OH)_{24}(H_2O)_{12}]^{7+}$; this last ion is the principal species at low concentrations:

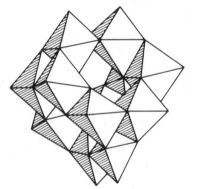

arrangement of twelve AlO_6 octahedra round the central AlO_4 tetrahedron in $[AlO_4Al_{12}(OH)_{24}(H_2O)_{12}]^{7+}$ [see *Journal of the Chemical Society, Dalton Transactions* (1988) p. 1347]

TRIVALENT BORON

The vast amount of energy required to remove three electrons from boron, and the high polarizing power which the resulting tiny ion would possess, combine to make the formation of salts containing bare B^{3+} cations unfavourable; even water of hydration is apparently too highly deformed by this cation and hence B^{3+}(aq) is also unknown in aqueous solution. Much less energy is required for $s \rightarrow p$ promotion and boron compounds are consequently covalent, the boron atomic orbitals being hybridized to either the sp^2 or sp^3 configuration. In a very few cases it has proved possible to increase the effective size of the B^{3+} ion, and at the same time to recoup some of the expended ionization energy in the form of strong ion–dipole forces, by complexing round it a number of suitable donor molecules; this has the effect of reducing the cation's polarizing power and so makes its salts more stable. Such a salt is formed when pyridine reacts with the addition compound trimethylamine-boron tribromide:

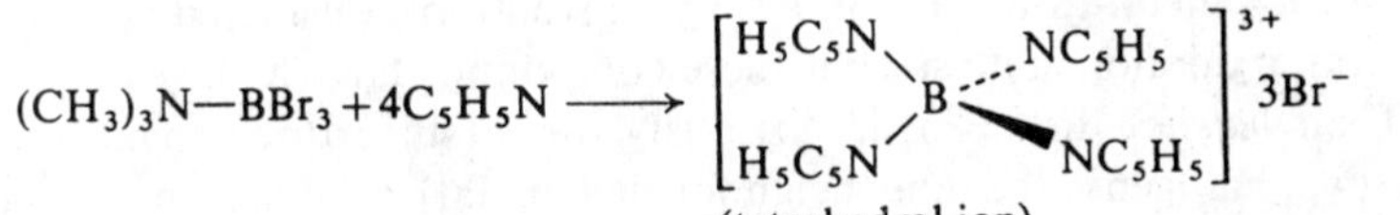

$$(CH_3)_3N-BBr_3 + 4C_5H_5N \longrightarrow \begin{bmatrix} H_5C_5N & & NC_5H_5 \\ & B & \\ H_5C_5N & & NC_5H_5 \end{bmatrix}^{3+} 3Br^-$$

(tetrahedral ion)

Solutions of the reaction product have an electrical conductivity expected for a salt containing four ions.

DIHALIDES OF GROUP III ELEMENTS

There have been many attempts to make paramagnetic, divalent derivatives of the Group III elements, usually via reduction of the trihalides:

$$Ga + GaCl_3 \rightarrow \text{'}GaCl_2\text{'}$$

$$In + InI_3 \rightarrow \text{'}InI_2\text{'}$$

However, the 'dihalides' so obtained are completely diagmagnetic and have been shown via a wide range of techniques to be mixed-valence salts $M^+[M^{III}X_4]^-$. Addition of ligands to the dihalides, or partial electrolytic oxidation of the metals in concentrated HCl, results in the formation of compounds containing M—M bonds:

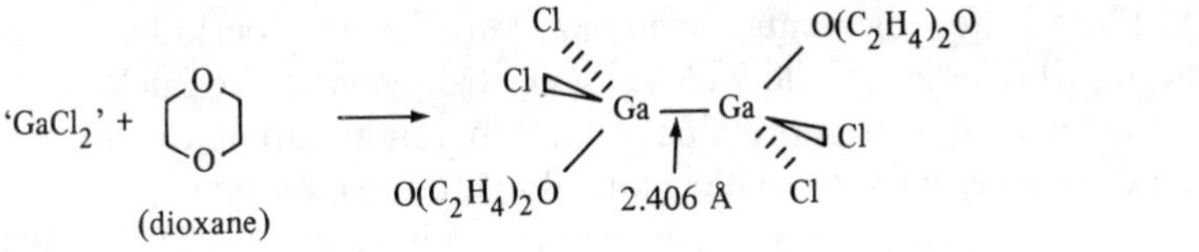

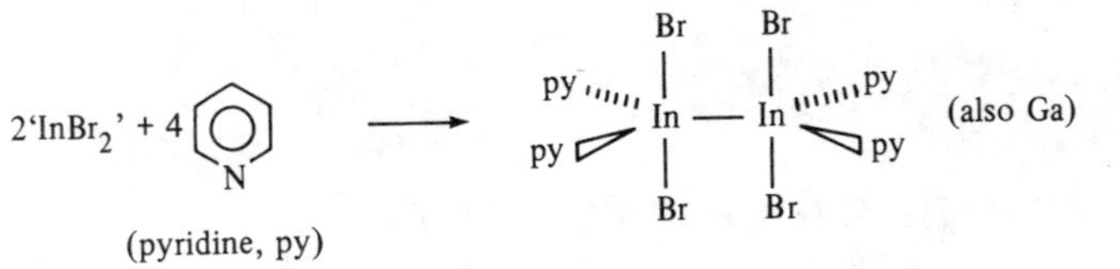

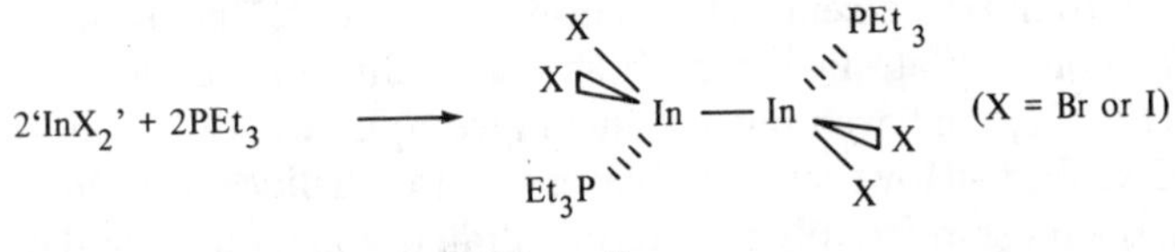

'GaCl$_2$' + (dioxane)

$$\begin{array}{c} Cl \\ Cl \diagdown Ga - Ga \diagup O(C_2H_4)_2O \\ O(C_2H_4)_2O \quad 2.406 \text{ Å} \quad Cl \end{array}$$

2'InBr$_2$' + 4 (pyridine, py)

$$\begin{array}{c} Br \quad Br \\ py \diagdown In - In \diagup py \\ py \diagup \quad \diagdown py \\ Br \quad Br \end{array}$$ (also Ga)

2'InX$_2$' + 2PEt$_3$

$$\begin{array}{c} X \quad PEt_3 \\ X \diagdown In - In \diagup X \\ Et_3P \diagup \quad X \end{array}$$ (X = Br or I)

$$Ga + \text{conc. HCl} \xrightarrow[\text{oxidation}]{\text{partial anodic}} [Cl_3Ga-GaCl_3]^{2-}$$

A notable feature of these compounds is the presence of a very strong band in their Raman spectra corresponding to the vibration frequency of the M—M bonds; for the $Ga_2X_6^{2-}$ ions dissolved in HX this band occurs at 233 cm^{-1} (Cl), 162 cm^{-1} (Br) and 122 cm^{-1} (I). When calculating the oxidation number of an element (see Introduction), a metal-to-metal bond is, by convention, considered not to contribute to the overall oxidation state so that these compounds have a *formal oxidation state* of two. Their *valency*, however, is obviously three (since three covalent bonds are made to two halogens and one neighbouring metal) and they have *coordination numbers* of either four or five depending on the number of added ligands (dioxane, pyridine, triethylphosphine or *halide*).

UNIVALENT DERIVATIVES

Increasingly down Group III there is a tendency to form univalent species, particularly at thallium where many Tl^I compounds are found to be more stable than the corresponding trivalent derivatives; the progressive weakening of M^{III}—X bonds from boron down to thallium is partly responsible for this. Furthermore, the outer ns^2 electrons remain tightly bound even for the heavier elements, one reason being the unusually high effective nuclear charges arising from filled d and f shells as discussed at the beginning of the chapter. Theoretical calculations show that *relativistic effects* also have to be considered: an electron close to a heavy (i.e. highly charged) nucleus reaches velocities comparable with that of light and as a result both its mass and binding energy increase whilst its orbital radius correspondingly decreases. Such effects probably contribute about 15% to the lathanide contraction and become even more important for elements occurring later in the periodic table. Owing to penetration, which governs how close an electron gets to the nucleus, the relativistic effects on orbitals of a given principal quantum number are in the order s > p > d > f. Thus there are several factors which contribute to the *inert pair effect*, a phrase coined many years ago to rationalize the fact that heavy elements (Tl, Pb, Bi) often show valencies two units less than their group valency.

COMPLEXES OF THE GROUP III ELEMENTS

A wide variation in stereochemistry and ligand type is to be found among the many complexes of Group III elements. Octahedral cations ML_6^{3+} are formed by the metals when L is a monodentate ligand such as water, dimethylsulphoxide or dimethyl-formamide (Fig. 51), and even one tetrahedral complex ion of boron, $B(pyridine)_4^{3+}$, has been described. However, complexes of the halides are by far the most numerous and almost invariably have tetrahedral, trigonal bipyramidal or octahedral coordination; they are easily prepared, usually by adding together stoichiometric quantities of trihalide and ligand (including halide ions) dissolved in an inert solvent.

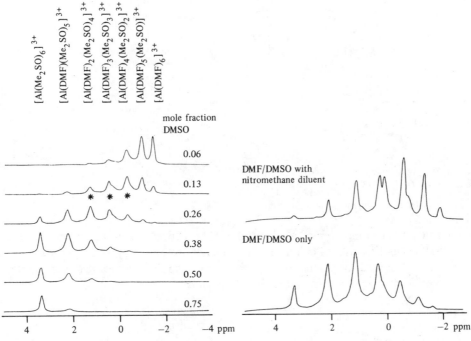

$[Al(Me_2SO)_6]^{3+}$

$[Al(DMF)(Me_2SO)_5]^{3+}$

$[Al(DMF)_2(Me_2SO)_4]^{3+}$

$[Al(DMF)_3(Me_2SO)_3]^{3+}$

$[Al(DMF)_4(Me_2SO)_2]^{3+}$

$[Al(DMF)_5(Me_2SO)]^{3+}$

$[Al(DMF)_6]^{3+}$

mole fraction
DMSO

0.06

0.13

0.26

0.38

0.50

0.75

4   2   0   −2   −4 ppm

DMF/DMSO with
nitromethane diluent

DMF/DMSO only

4   2   0   −2 ppm

^{27}Al NMR spectra at 20°C for several binary mixtures of DMF and Me$_2$SO. The mole fraction of Me$_2$SO is indicated.

^{27}Al spectra for a DMF–Me$_2$SO binary solvent mixture with Me$_2$SO of 0.260 and for the same solution diluted 1:3 v/v by nitromethane.

Fig. 51—^{27}Al NMR spectra of several DMF–Me$_2$SO complexes of Al^{3+}. Dimethylformamide (DMF) dimethylsulphoxide (Me$_2$SO) form strong complexes with the Al^{3+} ions and in a ligand mixture up to seven species are detectable. Three of the peaks, marked with an asterisk, show broadening due to the presence of isomers:

$$cis\text{-}Al(DMF)_4(Me_2SO)_2 \text{ and } trans\text{-}Al(DMF)_4(Me_2SO)_2$$

$$fac\text{-}Al(DMF)_3(Me_2SO)_3 \text{ and } mer\text{-}Al(DMF)_3(Me_2SO)_3$$

$$cis\text{-}Al(DMF)_2(Me_2SO)_4 \text{ and } trans\text{-}Al(DMF)_2(Me_2SO)_4$$

By using nitromethane as an inert diluent it is possible to partially resolve the peaks due to these isomers (right-hand spectra). (The shifts are relative to 0.1M Al(ClO$_4$)$_3$ in D$_2$O. [Adapted with permission from *Inorganic Chemistry* vol. 24 (1985) p. 883.]

Some typical examples are shown below:

X $''_{\prime\prime}$ B $\overset{X}{\underset{L}{\diagup}}$ (X = F, Cl, Br or I;
L = NMe$_3$ or pyridine)

(Similar complexes known for
other Group III elements)

$\left[X_{''\prime} M \overset{X}{\underset{X}{\diagup}} \right]^{-}$ (M = B, Al, Ga, In or Tl (not F);
X = F, Cl, Br or F)

Br $''_{\prime}$ $\overset{L}{\underset{L}{\mid}}$ Tl — Br (L = OP(C$_6$H$_5$)$_3$ or pyridine)
Br

Cl $''_{\prime}$ $\overset{L}{\underset{L}{\mid}}$ In $''''$ Cl (L = pyridine, NMe$_3$,
Cl   L   PMe$_3$ or OEt$_2$)

$$\left[\begin{array}{c} F \\ F_{\prime\prime\prime} \overset{|}{\underset{|}{Al}} \overset{\prime\prime\prime F}{\blacktriangleright} \\ F \end{array} \blacktriangleright OH_2 \right]^{2-} \qquad \left[\begin{array}{c} F \\ F_{\prime\prime\prime} \overset{|}{\underset{|}{Al}} \overset{\prime\prime\prime F}{\blacktriangleright} F \\ F \end{array}\right]^{3-} \qquad \left[\begin{array}{c} Cl \\ L_{\prime\prime\prime} \overset{|}{\underset{|}{Al}} \overset{\prime\prime\prime L}{\blacktriangleright} L \\ Cl \end{array}\right]^{+}$$

(L = tetrahydrofuran)

$$\left[\begin{array}{c} N \\ N_{\prime\prime\prime} \overset{|}{\underset{|}{In}} \overset{\prime\prime\prime N}{\blacktriangleright} N \\ N \end{array}\right]^{3+} \qquad \left[\begin{array}{c} X \\ X_{\prime\prime\prime} \overset{|}{\underset{|}{Tl}} \overset{\prime\prime\prime X}{\blacktriangleright} X \\ X \end{array}\right]^{3-} \quad (X = Cl \text{ or } Br)$$

(e.g. In (N⌒N)$_3$ X$_3$ when
X = Cl, Br or I)

(N⌒N = 1,2-diaminoethane or
2,2′-bipyridine)

Crown ether complexes of $AlCl_2^+$ and $TlMe_2^+$ cations have been made in which the MR_2 fragment either fits into the crown cavity to give bipyramidal coordination with axial R ligands or perches on top of the crown so that there is strong interaction between four oxygen atoms and a *cis* $AlCl_2^+$ group:

[TlMe$_2$(dibenzo-18-crown-6)]$^+$ isolated as the
2,4,6-trinitrophenolate

in [AlCl$_2$(benzo-15-crown-5)]$^+$[EtAlCl$_3$]$^-$

in [AlCl$_2$(12-crown-4)]$^+$[EtAlCl$_3$]$^-$

When the complex trimethylamine-boron trifluoride is treated with boron trichloride, ligand exchange occurs:

$$F_3B{-}NMe_3 + BCl_3 \rightarrow BF_3 + Cl_3B{-}NMe_3$$

A variety of similar 'competition' reactions involving other ligands show that the Lewis acid strength of the boron trihalides is in the order $BI_3 > BBr_3 > BCl_3 > BF_3$ and such a sequence has been confirmed by thermochemical measurements on the heats of formation of many boron trihalide complexes. This order of stabilities is the reverse of that predicted when the electronegativity and size of the halogens are considered: the high electronegativity of fluorine could be expected to make BF_3 more receptive to receiving a lone pair from a potential ligand, and the small size of fluorine would least inhibit the ligand's approach to boron during reaction.

π BONDING IN THE BORON TRIHALIDES

This apparent disagreement between the experimental and expected Lewis acidity sequences arises because of π bonding between halogen and boron in the trihalides:

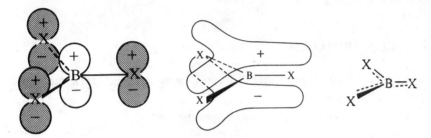

| the four interacting atomic p orbitals | the delocalized bonding molecular orbital formed | partial double-bond character introduced into the B—X bonds |

This has been calculated, and experimentally verified, to be in the order $BF_3 > BCl_3 > BBr_3 > BI_3$. During reaction with an incoming ligand there has to be a conversion of the BX_3 molecule from the planar to a pyramidal configuration:

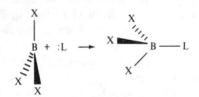

Such a stereochemical adjustment will naturally be resisted by the presence of any π bonding, and the complex formed will be destabilized by an amount corresponding to the loss of the π bonding energy; hence both kinetically and thermodynamically the formation of BF_3 complexes will be the least favoured among the trihalides.

Unlike the other halides of Group III, the boron trihalides are *monomeric* in all phases owing, it has been suggested, to the occurrence of boron–halogen π bonding which would stabilize the monomer relative to the dimer (π bonding could not occur in the dimer because no vacant boron orbitals would be available). It is also possible that boron is too small to form the stable ring system which would be required in a dimeric trihalide structure:

In Chapter 4 it was stated that beryllium dichloride forms Be—Cl—Be bridged molecules; however, the covalent radius of Be (1.06 Å) is considerably larger than that of B (0.88 Å), which presumably relieves steric congestion in bridged systems involving beryllium atoms.

The B—F distance in boron trifluoride (1.30 Å) is considerably shorter than that in the BF_4^- ion (1.43 Å) and many BF_3 complexes. This is partially attributable to boron–fluorine π bonding, but at least two other factors could contribute to the observed shortening: (a) steric crowding in BF_4^- or BF_3L will slightly increase the B—F distance; (b) there are only six electrons in the outer shell of boron in BF_3 and these will experience a greater attraction from the nucleus than when two further electrons are added in BF_4^- or BF_3L.

'MIXED' BORON TRIHALIDES

When two trihalides such as BCl_3 and BBr_3 are combined together, a rapid and mobile equilibrium is set up:

$$BCl_3 + BBr_3 \rightleftharpoons BCl_2Br + BBr_2Cl$$

Distillation and similar separation techniques give only pure trichloride and tribromide; thus the two 'mixed' halides can only be studied using spectroscopic techniques. The equilibrium mixture in the gas phase gives a mass spectrum in which all four components are identifiable by their mass (see Fig. 52).

Unlike the destructive action of mass spectrometry, Raman spectroscopy is non-invasive and can also be used to study the equilibrium in the liquid phase (Fig. 52). However, the technique suffers from the disadvantage that the peak at 408 cm^{-1} has to be *assumed* to arise from vibrations in $BBrCl_2$ since the peak is close to that of BCl_3 at 473 cm^{-1}; similarly, because the 346 cm^{-1} peak is nearer to BBr_3 at 278 cm^{-1} than BCl_3 at 473 cm^{-1} it is assigned to vibrations in BBr_2Cl. Calibration of peak intensities allows a *quantitative* estimate to be made of the concentrations of all four components in the equilibrium mixture to be made.

The isotope ^{11}B has a nuclear spin which enables it to be studied using nuclear magnetic resonance (NMR). Thus all ten components in a BCl_3–BBr_3–BI_3 ternary mixture have been detected using ^{11}B NMR spectroscopy. It is usually assumed that the mechanism for halogen exchange involves a dimeric species which can dissociate in two ways:

cleavage via (1) will give back the unchanged trihalides BX_3 and BY_3, whereas path (2) will give mixed halides BX_2Y and BXY_2. If this mechanism is correct then the formation of these mixed trihalides provides a most interesting way of detecting extremely minute amounts of (unstable) dimeric species in the boron halides (dimers must, of course, be presumed present at the same level of concentration in the pure trihalides if they exist in the mixed systems).

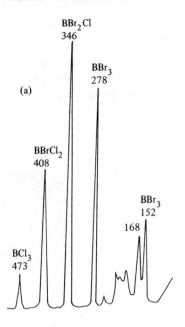

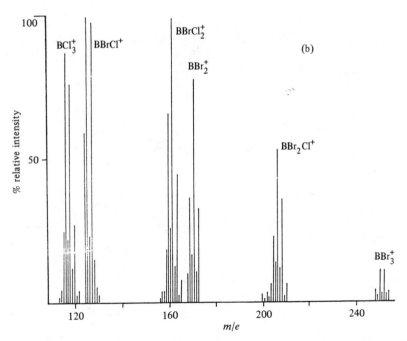

Fig. 52—A spectroscopic study of boron trihalide mixtures. (a) Raman spectrum of a BCl_3–BBr_3 mixture. Only the major peaks of each species are labelled; the peak positions are given in cm^{-1}. (b) Partial mass spectrum of a BCl_3–BBr_3 mixture. Boron, chlorine and bromine each possess two stable isotopes which give rise to peak clusters for each ion owing to the various possible isotopomers (isotopomers are isomers which differ in their isotopic make-up and hence mass: e.g. $^{11}B^{35}Cl_3$ and $^{10}B^{35}Cl_3$).

FRIEDEL–CRAFTS CATALYSIS

Aluminium halides, particularly $AlCl_3$, are widely used in organic chemistry as catalysts in the Friedel–Crafts alkylation or acylation of aromatic compounds. They behave as Lewis acids by strongly polarizing the attacking organic halide to form an $R^{\delta+}—X^{\delta-}$---AlX_3 intermediate in the reaction:

$$C_6H_6 + R—Cl \overset{\delta+ \; \delta-}{\text{---}} AlCl_3 \rightarrow \left[\text{arenium}^+ \; AlCl_4^- \right] \rightarrow C_6H_5R + HCl + AlCl_3$$

$$C_6H_6 + \underset{\underset{O}{\|}}{RC}—Cl \overset{\delta+ \; \delta-}{\text{---}} AlCl_3 \rightarrow \left[\text{arenium}^+ \; AlCl_4^- \right] \rightarrow C_6H_5COR + HCl + AlCl_3$$

$$\downarrow$$

$$\underset{R}{\overset{C_6H_5}{C}}=O—AlCl_3$$

With some tertiary alkyl halides and many acyl chlorides the polarization may be so extreme that the attacking species is essentially a 'free' carbonium ion, R^+. Beryllium, boron and gallium halides are also able to function in this way as Friedel–Crafts catalysts.

OCCURRENCE OF THE GROUP III ELEMENTS

Boron

About 3 ppm of the Earth's crust; main sources are the sodium borates $Na_2B_4O_7.4H_2O$ and $Na_2B_4O_7.10H_2O$.

Aluminium

7.5%; the most common of all metals; the most commercially important ore is the hydrated oxide, bauxite, $Al_2O_3.H_2O$.

Gallium

15 ppm; widely distributed in nature but only in minute concentrations; zinc blende ZnS often contains 0.1%–0.5% gallium.

Indium

0.1 ppm; no concentrated deposits have been found and, like gallium, it is obtained from the residues of zinc and lead smelting processes.

Thallium

0.3 ppm; obtained from the flue dust of sulphuric acid works.

EXTRACTION

Boron

Amorphous boron of about 95% purity is obtained by reducing boric oxide with either sodium or magnesium:

$$Na_2B_4O_7.10H_2O(borax) \xrightarrow{\text{acid}} B(OH)_3 \xrightarrow[\text{by heat}]{\text{dehydrate}} B_2O_3 \xrightarrow[\text{or Mg}]{\text{Na}} B$$

Pure, crystalline boron can be obtained in gram quantities by passing a hydrogen–boron trihalide mixture over a tantalum wire held at 1100–1300 °C.

Aluminium

Bauxite is purified by dissolving it in sodium hydroxide solution (to give sodium aluminate), when the oxides of iron present precipitate out. If the solution is held at 25–35 °C and a little $Al_2O_3.3H_2O$ added to it, most of the bauxite present crystallizes out as the trihydrate; this is heated to 1200 °C, dissolved in molten cryolite, Na_3AlF_6, and electrolysed between carbon electrodes at 960–980 °C.

Gallium, indium and thallium

As mentioned above, these metals are obtained as by-products in the production of zinc, lead and sulphuric acid. Gallium may be purified by electrolysis of an alkaline solution of one of its salts (gallium hydroxide is amphoteric and dissolves in alkali to give gallates).

THE ELEMENTS

The melting points within this group of elements vary widely: boron, ~ 2300 °C; aluminium, 639 °C; gallium, 29.6 °C; indium, 156 °C; thallium, 302.5 °C. Gallium, with a boiling point of about 2030 °C, has the longest liquid range of all the elements and has been used in high-temperature thermometers.

Boron is notable for the complexity of its crystalline forms, all of which are based

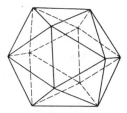

Fig. 53—An icosahedron has twenty equilateral triangular faces and a total of twelve vertices. The icosahedral B_{12} units is the basic 'building block' in the three known forms of crystalline boron.

on icosahedra of boron atoms; for example, the α rhombohedral modification consists of an approximately cubic close-packed arrangement of icosahedral B_{12} groups bound to each other by covalent bonds (Fig. 53). Aluminium has a face-centred cubic structure, but there is some evidence to suggest that all three outer electrons on the aluminium atoms are not used in bonding, which would account for the comparatively low melting point (only $8\,^\circ C$ above that of magnesium). Gallium has a complex structure in which the atoms have the following set of neighbours: one at 2.44 Å, two at 2.70 Å, two at 2.73 Å and two at 2.79 Å. It is thought that gallium may also contain some pairs of atoms in the liquid state because the X-ray diffraction pattern is different from that of a simple liquid like mercury. Indium has a slightly distorted face-centred cubic structure. Thallium is dimorphic, having hexagonal (α) and cubic (β) forms.

Elemental boron is highly inert and is not affected by acids, even boiling hydrofluoric; fused sodium hydroxide slowly attacks it but only above $500\,^\circ C$. By contrast, the other elements are quite reactive and dissolve readily in acids giving salts, and aluminium, gallium and indium, being amphoteric, dissolve in alkalis evolving hydrogen and forming aluminates, gallates and indates.

COMPOUNDS OF THE GROUP III ELEMENTS

Hydrides

(See Chapter 2 for a discussion of the hydrides of boron.) The expected monomeric hydrides of this group, MH_3, are unstable and can only be detected at low pressures where the polymerization rate is slow. For example, when aluminium is slowly vaporized off a tungsten filament in a current of hydrogen, the formation of AlH_3 is revealed by mass spectral analysis; if the amount of AlH_3 in the vapour phase is increased (by evaporating the aluminium metal more quickly) then the dimer Al_2H_6 can also be detected. Under similar conditions the much less thermally stable GaH_3 and InH_3 can be produced, but there was no indication of dimer formation. However, digallane has been isolated very recently as a volatile species which decomposes above

$-10\,^{\circ}\text{C}$ into gallium metal and hydrogen:

$$\text{Me}_3\text{SiH} + \text{Ga}_2\text{Cl}_6 \xrightarrow{-20\,^{\circ}\text{C}} (\text{HGaCl}_2)_2 \xrightarrow[-20\,^{\circ}\text{C}]{\text{Me}_3\text{SiH}}$$

$$(\text{H}_2\text{GaCl})_2 + \text{LiGaH}_4 \xrightarrow{-30\,^{\circ}\text{C}}$$

Although the free monomeric hydrides of Group III do not exist under normal conditions, their complexes with suitable donor molecules, such as amines, are well known: e.g.

$$\text{B}_2\text{H}_6 + 2\text{D} \rightarrow 2\text{H}_3\text{B}\!-\!\text{D}$$

$$\text{AlCl}_3 \xrightarrow[\text{ether}]{\text{LiAlH}_4} (\text{AlH}_3)_n + n\text{D} \xrightarrow{\text{ether}} n(\text{H}_3\text{Al}\!-\!\text{D})$$

$$\text{LiAlH}_4 + \text{Me}_3\text{N}.\text{HCl} \xrightarrow{\text{ether}} \text{LiCl} + \text{H}_2 + \text{Me}_3\text{N}.\text{AlH}_3$$

$$\text{Me}_3\text{N}.\text{AlH}_3 + \text{NMe}_3 \xrightleftharpoons{20\,^{\circ}\text{C}} (\text{Me}_3\text{N})_2\text{AlH}_3 \quad \text{(trigonal bipyramid with axial amines)}$$

The donor species D can also be the hydride ion H^-, in which case the reaction product is the tetrahedral anion MH_4^-:

$$\text{B(OMe)}_3 + 4\text{NaH} \xrightarrow{250\,^{\circ}\text{C}} \text{NaBH}_4 + 3\text{NaOMe}$$

$$\text{AlCl}_3 + \text{LiH} \xrightarrow{\text{ether}} (\text{AlH}_3)_n \xrightarrow{\text{LiH}} \text{LiAlH}_4$$

$$\text{NaH} + \text{Al} + \tfrac{3}{2}\text{H}_2 \xrightarrow{140\,^{\circ}\text{C}} \text{NaAlH}_4$$

$$\text{MCl}_3 + 4\text{LiH} \xrightarrow{\text{ether}} \text{LiMH}_4 \quad (\text{M} = \text{Ga or In})$$

Lithium tetrahydroaluminate is a useful reducing agent and, for example, reacts with many covalent metal and metalloidal halides to give the corresponding hydrides, often in high yield: e.g.

$$\text{SiCl}_4 + \text{LiAlH}_4 \rightarrow \text{SiH}_4 \quad (100\%)$$

Sodium tetrahydroborate, NaBH, which is soluble in water and a variety of organic solvents, readily reduces aldehydes to primary alcohols:

The specificity of sodium tetrahydroborate for the aldehydic carbonyl group is shown by the fact that aldehydes have been reduced in the presence of the following functional groups: olefin, ester, epoxy, nitro and carboxylic acid. Ketones are reduced less rapidly, although the reaction time is usually less than 90 min, the product being the corresponding secondary alcohol:

$$
\diagdown C = O \; + \; BH_4^- \quad \longrightarrow \quad \text{...}C \diagdown^{OH}_{}{}_{H}
$$

Oxides

The normal oxide M_2O_3 is formed by heating the elements in oxygen; on strong heating, $Tl_2^{III}O_3$ loses oxygen to form Tl_2^IO, the dissociation pressure of oxygen being about 115 mmHg at 700 °C. Gallium also gives a lower oxide when Ga_2O_3 is heated with the free metal at 700 °C:

$$
Ga_2O_3 + 4Ga \underset{800\,^\circ C}{\overset{700\,^\circ C}{\rightleftharpoons}} 3Ga_2O
$$

Hydroxides

Down any group of the periodic table the metallic character (and hence basicity) of the elements is found to increase. Thus in Group III: $B(OH)_3$ is acidic; $Al(OH)_3$, $Ga(OH)_3$ and $In(OH)_3$ are amphoteric; $Tl(OH)_3$ apparently does not exist, but the oxide Tl_2O_3 is basic and dissolves in acids to give thallium(III) salts.

Boric oxide is hygroscopic and readily takes up water to form boric acid, $B(OH)_3$. This is a very weak acid, but complex formation with polyhydroxyl compounds such as glycerol causes the release of *one* proton and it can then be titrated against sodium hydroxide using phenolphthalein as the indicator:

$$
\underset{\underset{H}{O}}{\underset{|}{\overset{HO}{\diagup}}} \overset{OH}{\underset{B}{\diagdown}} \;+\; \underset{-C-OH}{\overset{-C-OH}{||}} \rightleftharpoons \underset{-C}{\overset{-C}{\diagup^{O}\diagdown_{O}}}B-OH + H_2O \rightleftharpoons \text{...} B \text{...} + H^+(aq) + H_2O
$$

Addition of base to solutions of aluminium and gallium salts results in the precipitation of the white hydroxides which dissolve in excess base to give $M(OH)_4^-$ ions. Hot 15M alkali is required to dissolve indium hydroxide; cooling results in the formation of crystalline indates such as $K_3[In(OH)_6].2H_2O$ and $Rb_2[In(OH)_5.H_2O]$.

Borates

A wide range of metal borates is known, the anions of which are based on triangular BO_3 units, tetrahedral BO_4 units or a combination of both. Some of the common borate structures are shown in Fig. 54; even within simple species such as the borates containing triangular BO_3 units

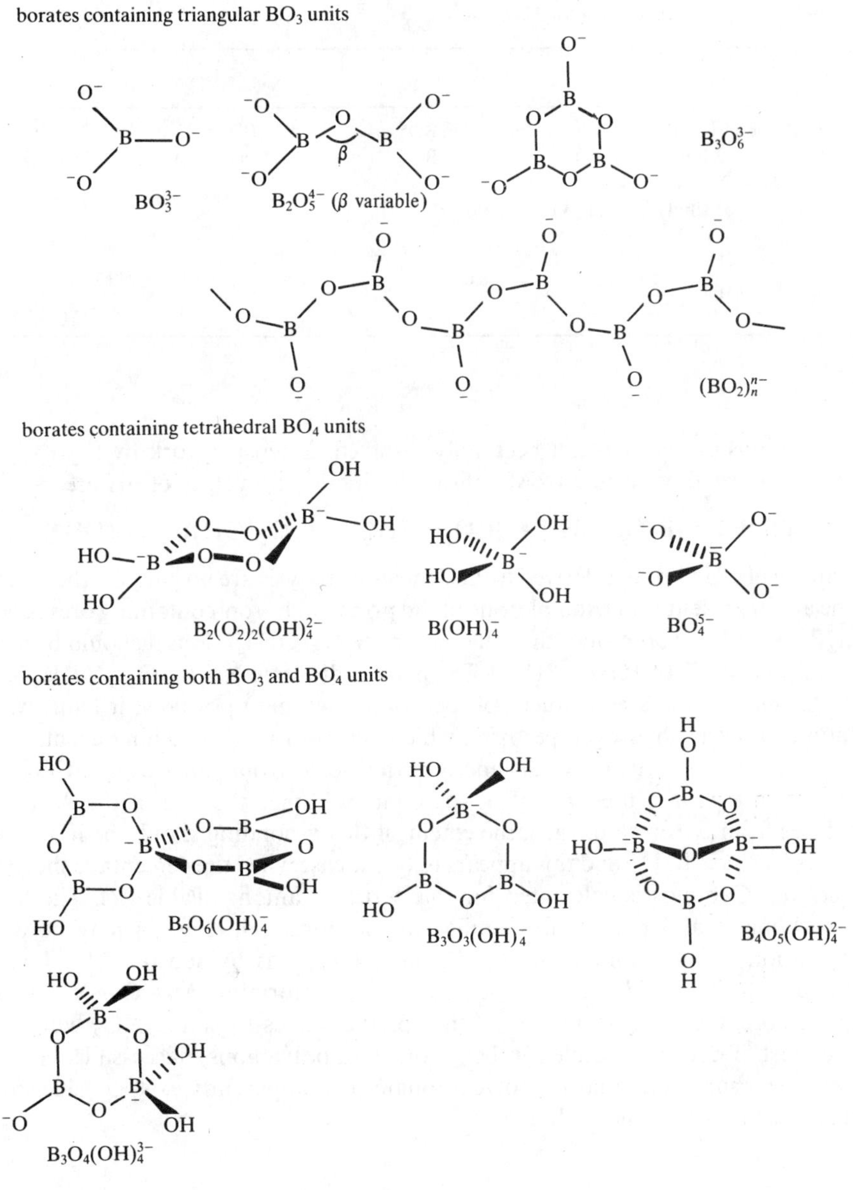

BO_3^{3-} $B_2O_5^{4-}$ (β variable) $B_3O_6^{3-}$

$(BO_2)_n^{n-}$

borates containing tetrahedral BO_4 units

$B_2(O_2)_2(OH)_4^{2-}$ $B(OH)_4^-$ BO_4^{5-}

borates containing both BO_3 and BO_4 units

$B_5O_6(OH)_4^-$ $B_3O_3(OH)_4^-$ $B_4O_5(OH)_4^{2-}$

$B_3O_4(OH)_4^{3-}$

Fig. 54—Some common polyborate anions.

pyroborates, $B_2O_5^{4-}$, variations can occur in the B—O—B angle, the relative orientation of the two BO_3 planes and, of course, the accompanying cations. The overall charge on a given borate ion is determined by the total number of terminal oxygens and tetrahedral boron atoms, *each of which* contributes one unit of negative charge.

The composition of aqueous borate solutions is complex and depends on pH, temperature and concentration; even after half a century of investigation the identity

Table 27—Bond length data for MO_3^{n-} ions

	B	C	N
M—O bond length (Å)	1.37 (BO_3^{3-})	1.29 (CO_3^{2-})	1.22 (NO_3^-)
Single M—O bond length (Å)	1.48 (BO_4^{4-})	1.43 (average)	1.39 (NO_4^-)
$\Delta = [(\text{M—O in } MO_3^{n-}) - (\text{M—O for single bond})]$ (Å)	0.11	0.14	0.17
$\dfrac{\Delta}{\text{M—O in } MO_3^{n-}} \times 100\%$	8.0	10.9	13.9

of all the ions present has not been fully resolved. Extensive work by Ingri suggests that the species present in a 0.4M solution at increasing values of pH are:

$$B(OH)_3 \rightarrow B_5O_6(OH)_4^- \rightarrow B_3O_3(OH)_5^{2-} + B_4O_5(OH)_4^{2-} \rightarrow B(OH)_4^-$$

Formulae of solid borates derived from elemental analysis are no guide to the structure of the anion present. For example, one of the principal boron-containing ores is *borax*, '$Na_2B_4O_7.10H_2O$', but since the salt contains $B_4O_5(OH)_4^{2-}$ ions it should be written correctly as $Na_2B_4O_5(OH)_4.8H_2O$. The peroxo-diborate $Na_2B_2(O_2)_2(OH)_4$ is used in large amounts as a constituent of 'perborate' washing powders; it hydrolyses in solution to release hydrogen peroxide which acts as a mild bleaching agent.

The relatively short M—O distances in the isoelectronic ions BO_3^{3-}, CO_3^{2-} and NO_3^- are due to the presence of $p_\pi-p_\pi$ bonding. Since the size of the 2p orbitals decreases from boron to oxygen, the extent of this π bonding should be in the order $NO_3^- > CO_3^{2-} > BO_3^{3-}$ and this appears to be the case when the percentage shortening of the M—O bonds is calculated for this series of anions (Table 27). The lack of polycarbonates and polynitrates similar to the borates in Fig. 54 may be due to stabilization of the 'monomeric' CO_3^{2-} and NO_3^- ions by substantial π bonding, some of which would have to be sacrificed on forming M—O—M linkages. Furthermore, the ready formation of the volatile oxides CO_2 and NO_2 will tend to make most, if not all, possible synthetic routes to polyanions inaccessible. However, it does seem surprising that no polycarbonates can apparently exist considering the rather wide range of known borates.

Sulphides

All the elements react on heating with an excess of sulphur to give the trisulphides M_2S_3. One of the lower sulphides of gallium, GaS, has a layer structure containing $[Ga—Ga]^{4+}$ units and is therefore not a derivative of gallium(II); the sulphide TlS is a completely different type of compound in that half the thallium is present as thallium(I) and half as thallium(III): $Tl^I[Tl^{III}S_2]$. Gallium exists as gallium(I) in the sulphide Ga_2S.

Nitrides

The boron–nitrogen bond is isoelectronic with a carbon–carbon bond and hence it might be anticipated that boron nitride, $(BN)_x$, could exist in two forms: one having a layer lattice like graphite and consisting of sheets of hexagonal $(3B, 3N)$ rings (Fig. 55), the other having a giant, three-dimensional lattice like diamond (see Fig. 63). Both types of boron nitride have been realized. The graphitic form is the ultimate product formed by heating a wide variety of boron–nitrogen compounds. On the application of high temperatures and high pressures, the layer form of boron nitride is converted to the diamond-like variety called borazon. Borazon is exceedingly hard and is the only known compound capable of cutting diamond.

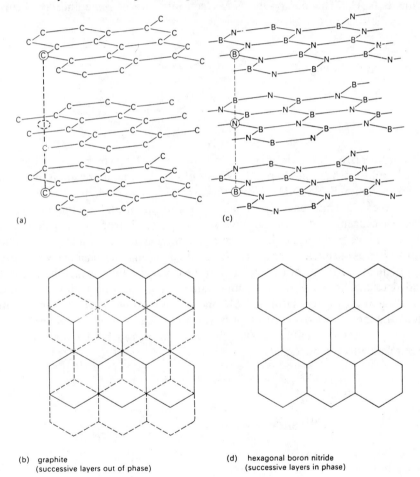

(a)

(c)

(b) graphite
(successive layers out of phase)

(d) hexagonal boron nitride
(successive layers in phase)

Fig. 55—(a), (b) The structure of graphite (successive layers out of phase). (c), (d) The structure of hexagonal boron nitride (successive layers in phase). Both structures consist of layers of hexagonal (either C_6 or B_3N_3) rings. In graphite the rings are in phase only at every other layer and hence each carbon atom has only carbon atoms from the same plane as its neighbours. All the layers of boron nitride have their rings in phase, each boron atom having two nitrogen atoms (from neighbouring layers) above and below it. Because of this weak boron–nitrogen interaction, the layers in boron nitride do not slide over each other as readily as those in graphite.

The nitrides of aluminium, gallium and indium are known; aluminium nitride, AlN, produces ammonia to treatment with water or acid and for a time was considered as a means of 'fixing' atmospheric nitrogen as ammonia for use in agriculture:

$$\text{liquid air} \xrightarrow[\text{distillation}]{\text{fractional}} N_2 \xrightarrow[\text{heat}]{\text{Al}} \text{AlN} \xrightarrow{H_2O} NH_3$$

Borazine

Although the final product of heating diborane with ammonia is hexagonal boron nitride, an intermediate in the decomposition is a colourless, volatile liquid called borazine, $B_3N_3H_6$. This compound is isoelectronic with benzene and, like benzene, it has a cyclic structure:

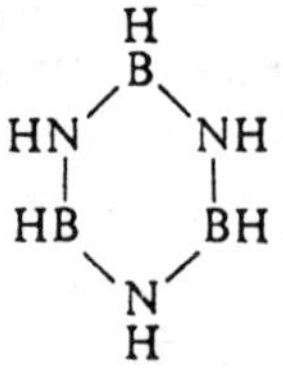

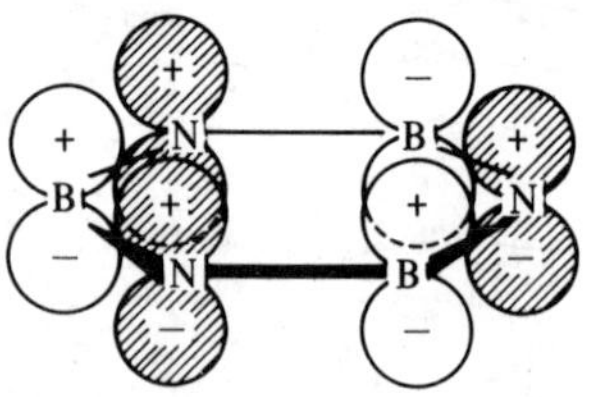

All the boron–nitrogen bonds are equal in length at 1.44 Å, which is to be compared with the boron–nitrogen single-bond distance of 1.60 Å found in the ammonia–boron trifluoride complex $F_3B \leftarrow NH_3$. From this evidence it is concluded that π bonding occurs between filled nitrogen $2p_z$ orbitals and empty $2p_z$ orbitals on the boron atoms.

Although benzene and borazine have similar physical properties, their chemical properties are very different. For example, borazine readily undergoes addition reactions whereas such reactions are relatively uncommon in the chemistry of benzene. Typically, three molecules of hydrogen chloride (or methyl iodide, methanol, water) are added to the borazine ring at room temperature, the ring losing its aromatic character and reverting to a saturated, cyclohexane-like cyclic system. In such addition reactions the most electronegative half of the attacking molecule becomes attached to the boron atoms, probably because the boron atoms in borazine are positively charged relative to the nitrogens:

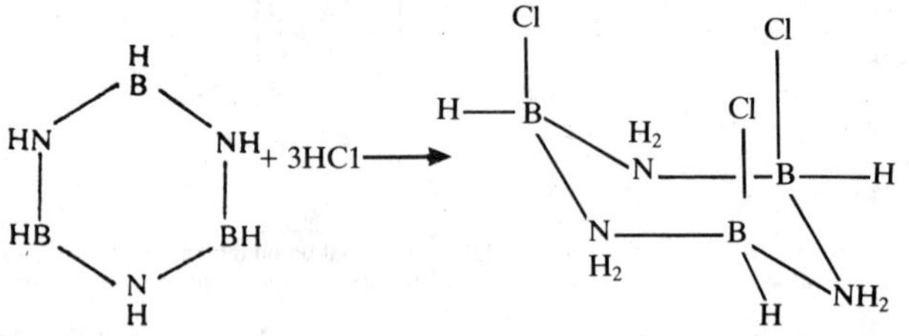

(planar) (probably 'chair' conformation more stable than 'boat')

It is of interest to note that similar additions involving benzene are *thermodynamically* favoured but, because high activation energies are involved, there is a *kinetic* block to their occurrence. As would be expected using the isoelectronic principle, multi-ring boron–nitrogen systems and even a borazine–chromium 'sandwich' compound can be prepared:

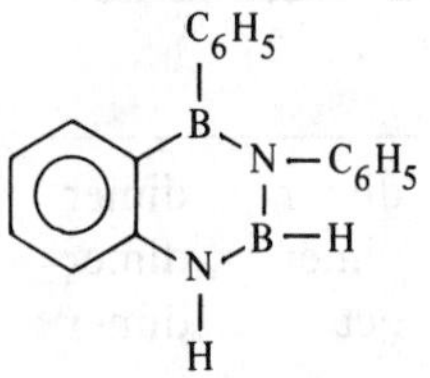

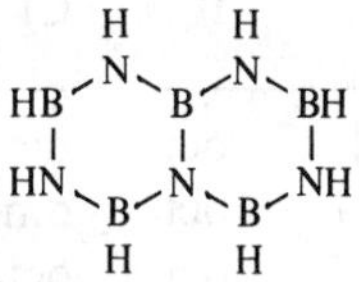

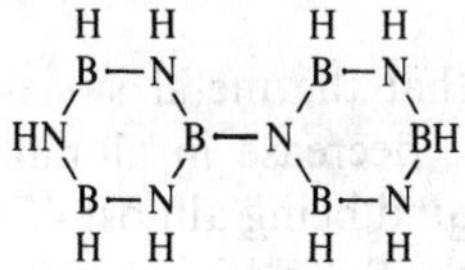

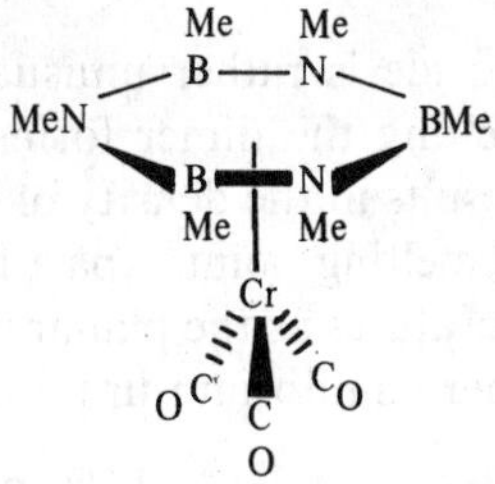

Salts of oxo-acids

The metals aluminium, gallium and indium dissolve in the stronger oxo-acids to give, for example, the M^{III} nitrates, sulphates, selenates, halates and perhalates, which may be crystallized from solution as hydrates. Thallium dissolves to give thallium(I) salts.

Aluminium sulphate forms a series of essentially isomorphous double sulphates (called 'alums') with a wide variety of sulphates containing monovalent cations. The alums have the general formula $M^I Al(SO_4)_2.12H_2O$, where M^I is Na^+, K^+, Rb^+, Cs^+, NH_4^+ or Tl^+; gallium and indium can be substituted for aluminium, but the Tl^{3+} ion appears to be too large for inclusion in the alum lattice. The water of crystallization is divided equally between the two cations, the coordination sphere of the univalent ion being completed by oxygens from the sulphate groups. Presumably Li^+ cannot form alums because it is too small to accommodate these extra sulphate oxygens.

Trihalides

All the trihalides of this group, with the exception of TlI_3, may be prepared by treating the respective element with the free halogen (or HF may be used in place of fluorine). The product formed on treating thallium with an excess of iodine has the formula TlI_3 but, like the isomorphous RbI_3, it contains an almost linear $[I{-}I{-}I]^-$ ion and hence is thallium(I) tri-iodide. It is interesting to note that thalium can be oxidized to thallium(III) by iodine in the presence of iodide ions, when the tetrahedral complex ion $[Tl^{III}I_4]^-$ is formed.

The trihalides of Al, Ga, In and Tl, unlike the monomeric boron halides, have crystal structures which contain either discrete dimer molecules or octahedrally coordinated metal atoms:

	F	Cl	Br	I
Al	oct	oct	dimer	dimer
Ga	oct	dimer	dimer	dimer
In	oct	oct	oct	dimer
Tl	oct	oct	—	—

Aluminium trichloride is rather unusual in that the metal is six-coordinate in the crystal but on melting the dimer forms. This decrease in aluminium coordination from six to four results in the density of the liquid being almost 45% lower than that of the solid at the melting point. Apart from the fluorides, which enter the gas phase only at high temperatures as the planar monomers, most of the other halides undergo a dimer⇌monomer equilibrium in the vapour on moderate heating: e.g.

$$Ga_2Br_6(g) \rightleftharpoons 2GaBr_3(g),$$

14% monomer at the melting point (122°C)
30% monomer at the boiling point (279°C)

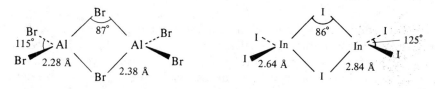

Two descriptions have been given for the bonding in the bridge region of these dimers. Three-centre, four-electron bonding similar to that in diborane (but with a pair of electrons occupying the non-bonding molecular orbital) would naturally account, at least in part, for the relatively long M—X (bridge) bonds. More usually it is assumed that a lone pair of electrons from the halogen on one metal is donated to an empty orbital of the other:

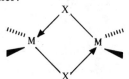

Steric congestion in the four-membered ring and d_π–p_π bonding between M and the terminal halogens may also contribute to the difference in M—X bond lengths.

Lower halides

All the Group III elements form diatomic halides in the gas phase at elevated temperatures which, with the exception of Tl^+F^-, are covalently bonded:

$$2M + MX_3 \xrightarrow{\text{heat}} 3MX$$

Unlike the other members of this group, boron has several volatile lower halides which contain catenated molecules:

diboron tetrahalides (decompose slowly at room temperature)

triboron pentahalides (only the fluoride has been isolated and this decomposes at $-30°C$)

The most extensively studied boron subhalide is B_2Cl_4, which can be made either by passing boron trichloride through an electrical discharge between mercury electrodes or, on a larger scale, by condensing copper atoms with BCl_3 vapour:

$$2BCl_3 + 2Hg \xrightarrow[\text{pressure}]{\text{low}} Hg_2Cl_2 + B_2Cl_4$$

$$BCl_3 + Cu(\text{atoms}) \xrightarrow[\text{warm to room temperature}]{\text{co-condense at } -196°C,} CuCl + B_2Cl_4$$

Rotation about the boron–boron bond occurs easily: in the gaseous and liquid states at room temperature the B_2Cl_4 molecules adopt the 'staggered' structure, whereas crystal forces in the solid make the molecules assume a planar shape for ease of packing:

Boron also forms three crystalline monochlorides, B_4Cl_4, B_8Cl_8 and B_9Cl_9, all of which possess closed cages of boron atoms (Fig. 56). The bonding in these molecules is complex and probably involves multicentred σ bonds extending over all the boron atoms in the cage. Low yields of B_4Cl_4 are obtained by passing B_2Cl_4 through an electric discharge between mercury electrodes, whereas B_8Cl_8 and B_9Cl_9 result when B_2Cl_4 slowly decomposes at room temperature:

$$B_4Cl_4 \xleftarrow[\text{discharge}]{\text{mercury}} B_2Cl_4 \xrightarrow[\text{decomposition}]{\text{thermal}} B_8Cl_8 + B_9Cl_9$$

$$+ [B_{10}Cl_{10} + B_{11}Cl_{11} + B_{12}Cl_{12}]$$

The structures of the red compounds $B_{10}Cl_{10}$, $B_{11}Cl_{11}$ and $B_{12}Cl_{12}$ are not yet known.

The halides of empirical formula MX_2 ($X \neq F$), which exist for gallium, indium and thallium, are not volatile dimers like those of boron but are ionic solids having the composition $M^I[M^{III}X_4]$. The two main methods used for their preparation are either the reduction of the trihalide with the free metal:

$$MCl_3 + M \rightarrow M^+[M^{III}Cl_4]^-$$

or stoichiometric halogenation of the element using mercury(II) halides:

$$2M + 2HgX_2 \rightarrow M^+[M^{III}X_4]^- + 2Hg$$

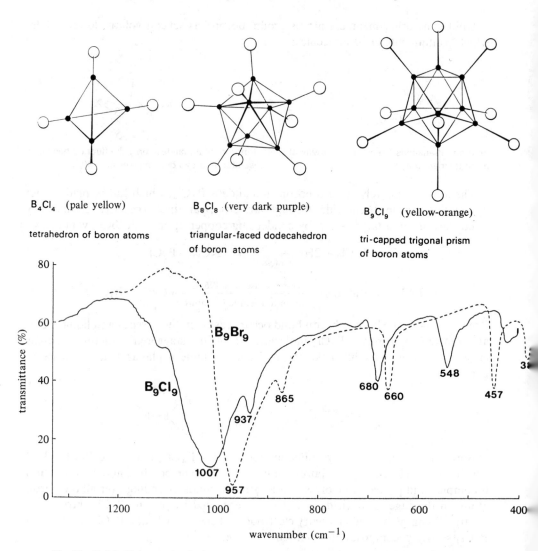

B_4Cl_4 (pale yellow)

tetrahedron of boron atoms

B_8Cl_8 (very dark purple)

triangular-faced dodecahedron
of boron atoms

B_9Cl_9 (yellow-orange)

tri-capped trigonal prism
of boron atoms

Fig. 56—Polyhedral cages in the boron monochlorides. Infrared spectroscopy is widely used to confirm the molecular structures of specimens by comparing their spectra with those of authentic compounds. It is sometimes possible to deduce the structure of a *new* compound by comparison of its infrared spectrum with that of a closely related molecule. Here the close similarity of the spectra of B_9Cl_9 (whose structure is known from single-crystal X-ray studies) and B_9Br_9 strongly suggests that the bromide has an identical cage structure. In fact, the X-ray powder patterns of the two halides show them to be isostructural.

The latter reaction can also be carried out in the presence of aluminium (for M = Ga or In), when Ga^+ or In^+ derivatives of the tetrahaloaluminate ion are produced: e.g.

$$Ga + Al + 2HgCl_2 \rightarrow Ga^+AlCl_4^- + 2Hg$$

Treatment of these 'dihalides' with a variety of ligands results in the formation of complexes containing M—M bonds (see page 179), e.g.

$$\text{'InBr}_2\text{'} + \text{PEt}_3 \longrightarrow$$

which are analogous to the derivatives obtained by addition of Lewis bases to diboron halides:

$$\text{B}_2\text{Cl}_4 + 2\text{NMe}_3 \longrightarrow \text{Cl}_2\text{B}-\text{BCl}_2$$

$$\begin{array}{cc} \uparrow & \uparrow \\ \text{N} & \text{N} \\ \text{Me}_3 & \text{Me}_3 \end{array}$$

Reduction of the trihalides of gallium, indium and thallium with an excess of the free element results in the formation of the solid monochlorides MX. (See page 201 for a brief summary of thallium(I) chemistry, including that of the halides.) The monohalides of gallium and indium are rather intractable and insoluble in many solvents. However, it has recently been shown that solutions of InCl, InBr and InI can be stabilized at low temperatures in toluene containing tetramethylethylene diamine; disproportionation occurs above 0 °C.

ORGANOMETALLIC COMPOUNDS

Grignard and lithium reagents readily react with Group III trihalides to give the alkyls or aryls: e.g.

$$3\text{CH}_3\text{MgI} + \text{BF}_3 \xrightarrow{\text{ether}} (\text{CH}_3)_3\text{B}$$

$$3\text{C}_6\text{H}_5\text{Li} + \text{GaCl}_3 \rightarrow (\text{C}_6\text{H}_5)_3\text{Ga}$$

$$4\text{CH}_3\text{Li} + \text{AlCl}_3 \rightarrow \text{LiAl}(\text{CH}_3)_4$$

$$3\text{C}_6\text{H}_5\text{Li} + \text{BF}_3 \rightarrow (\text{C}_6\text{H}_5)_3\text{B} \xrightarrow{\text{LiC}_6\text{H}_5} \text{LiB}(\text{C}_6\text{H}_5)_4$$

Aluminium trialkyls, which are valuable intermediates in the polymerization of olefins, are prepared industrially by a 'direct synthesis' from aluminium, hydrogen and olefin; preformed AlR_3 is added to start the reaction since AlH_3 is not formed directly: e.g.

$$\text{Al} + (\text{AlEt}_3)_2 + \tfrac{3}{2}\text{H}_2 \rightarrow (\text{AlEt}_2\text{H})_3 \xrightarrow{3\text{C}_2\text{H}_4} \tfrac{3}{2}(\text{AlEt}_3)_2$$

Aluminium appears to occupy a unique place within the group in that its triphenyl and lower trialkyls are dimeric in the solid, liquid and gaseous states (dissociation of the monomers occurs on heating the dimers in the vapour). The bonding

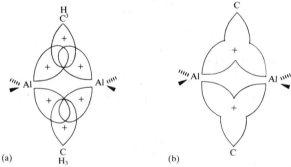

in the Al—C—Al bridges is thought to involve three-centre molecular orbitals formed from essentially sp^3 orbitals on the carbon and aluminium atoms (Fig. 57). This type of three-centre interaction in the bridge necessarily results in an acute Al—C—Al angle (75°); the fact that the boron alkyls and aryls are monomeric in all phases is presumably a direct result of the small size of the boron atom, which does not allow the formation of a stable four-membered ring in the bridge system. (For example, the two carbon atoms would have to approach to within about 2.5 Å of each other, which is much less than twice the van der Waals radius for carbon, ~ 3.8 Å.) The organo-derivatives of gallium, indium and thallium, although apparently weakly associated in the solid state, do not form dimers like the aluminium compounds; this may be due to adverse metal–metal steric interaction which would occur in the M—C—M bridge system.

Fig. 57—The trimethylaluminium dimer $Al_2(CH_3)_6$: (a) the sp^3 atomic orbitals used in bonding; (b) the approximate shape of the resulting two bonding molecular orbitals, each of which holds two electrons.

As discussed on page 87, boron alkyls can be made via the addition of olefins to diborane (hydroboration). The mechanism of this reaction probably proceeds via the initial interaction of the C=C double bond with either BH_3 or the mono-bridged species H_2B—H—BH_3; there are several equivalent ways of describing this three-centre bonding:

Formation of the boron–carbon bond can then proceed via the involvement of one B—H bond in a similar three-centred intermediate:

$$
\begin{array}{ccc}
\overset{\displaystyle C-C}{\underset{\displaystyle B-H}{\diagdown\diagup}} & \longrightarrow & \overset{\displaystyle C-C}{\underset{\displaystyle B\quad H}{|}} & \longrightarrow & \overset{\displaystyle C-C}{\underset{\displaystyle B}{|}}\diagdown H
\end{array}
$$

Probably using a similar mechanism, aluminium alkyls add ethene across the Al—C bonds:

$$
(AlEt_3)_2 \; + \; C_2H_4 \quad \xrightarrow[\text{100 atm}]{100\text{–}160\ ^\circ C} \quad
\underset{Et(C_2H_4)_y \quad (C_2H_4)_z Et}{\overset{(C_2H_4)_x Et}{\overset{|}{Al}}}\diagup\diagdown
$$

The values of x, y and z, which can be controlled by varying the temperature, pressure and reaction time, are usually about 6 but can be 100 or more under extreme conditions. The $\sim C_{14}$ groups are removed from the aluminium by controlled oxidation (to give alkoxide) followed by hydrolysis; the resulting primary alcohols are converted into sulphonic acids, $ROSO_3H$, for use as detergents:

$$
\diagdown\!\!\diagup Al(C_2H_4)_x Et \xrightarrow{O_2} \diagdown\!\!\diagup Al-O(C_2H_4)_x Et \xrightarrow{H_2O} HO(C_2H_4)_x Et \; + \; \diagdown\!\!\diagup AlOH
$$

(direct hydrolysis of the aluminium alkyls without prior oxidation simply gives the corresponding alkanes).

When the polymerization of propene is attempted, the addition of one mole occurs smoothly but the resulting aluminium alkyl is unstable:

$$
\diagdown\!\!\diagup Al-Pr + CH_3CH=CH_2 \longrightarrow \diagdown\!\!\diagup Al-CH_2-\underset{Me}{\overset{Pr}{CH}} \longrightarrow \diagdown\!\!\diagup AlH + CH_2=\underset{Me}{\overset{Pr}{C}}
$$

$+CH_3CH=CH_2$ (regeneration of catalyst)

Thermal degradation of the product yields *isoprene*, required for the production of synthetic rubber:

$$
CH_2=\underset{CH_3}{\overset{CH_2CH_2CH_3}{C}} \xrightarrow[-CH_4]{\Delta} CH_2=\underset{CH_3}{\overset{H}{C}}-C=CH_2 \xrightarrow{\text{catalyst}} \diagdown\underset{H}{\overset{CH_2CH_2}{C}}=\underset{CH_3}{\overset{}{C}}\diagup \underset{H}{\overset{CH_2CH_2}{C}}=\underset{CH_3}{\overset{}{C}}\diagdown
$$

(isoprene)   (rubber)

THE CHEMISTRY OF THALLIUM(I)

As discussed at the beginning of the chapter, thallium is more stable in the univalent rather than the trivalent state, a fact reflected in the standard oxidation potentials:

$$Tl^{3+} + 3e^- \rightleftharpoons Tl(s), \quad E^{\ominus} = +1.26 \text{ V}$$

$$Tl^+ + e^- \rightleftharpoons Tl(s), \quad E^{\ominus} = -0.34 \text{ V}$$

The Tl^+ ion is only very weakly hydrated in solution, self-diffusion measurements suggesting that no water molecules actually move through the solution with it.

As expected from the negative value of the Tl^+/Tl potential, thallium metal dissolves in acids to give hydrogen and Tl^I derivatives, although the reaction can be sluggish if the resultant salt is insoluble and covers the metal surface. Dissolution of the oxide, hydroxide or carbonate in acid is also a useful, and quite general, method of salt synthesis:

$$Tl_2CO_3 + 2HF(40\% \text{ aqueous solution}) \xrightarrow[\text{crucible}]{Pt} 2TlF + CO_2 + H_2O$$

$$TlOH + HNO_3 \rightarrow TlNO_3 + H_2O$$

Other salts can be made by double decomposition reactions if they are relatively insoluble:

$$TlNO_3 + NaCl \rightarrow TlCl\downarrow + NaNO_3$$

$$2TlNO_3 + Na_2S_2O_3 \xrightarrow[\text{solution}]{\text{conc}} Tl_2S_2O_3\downarrow + 2NaNO_3$$

(thiosulphate)

The thallium(I) ion (radius 1.64 Å) is unique in having an $(n-1)d^{10}ns^2$ outer electronic structure and hence there are no other singly charged ions to which its chemistry might be compared directly. The nearest approaches to such an electronic structure in ions of comparable size are to be found in Ag^+ ($4d^{10}$ core, ionic radius 1.14 Å) and Rb^+ ($5s^5 5p^6$ core, ionic radius 1.66 Å). In much of its chemistry the colourless thallium(I) ion resembles the heavier alkali metals and, for example, thallium(I) chloride and thallium(I) bromide are ionic halides which normally have the caesium chloride structure. On the other hand, thallium(I) halides show a remarkable resemblance in colour and solubility to the silver halides: thallium(I) fluoride (like silver fluoride) is very soluble in water, whereas thallium(I) chloride, bromide and iodide are insoluble and darken on exposure to light. Thallium(I) hydroxide is a strong base, is soluble in water and absorbs carbon dioxide from the air to give the carbonate Tl_2CO_3 (compare potassium hydroxide and rubidium hydroxide). The closeness of the ionic radii of Tl^+ and Rb^+ makes many of their salts (e.g. sulphate, chromate, nitrate, chloride, bromide and iodide) isomorphous. Also Tl^+ is able to replace Rb^+ in many double salts such as the alums.

SIMILARITIES BETWEEN GALLIUM AND ZINC

The presence of the $3d^{10}$ electron core at zinc results in an unexpectedly high effective nuclear charge on the Zn^{2+} ion and makes the cation much more strongly polarizing than the corresponding ion of Group IIA, Ca^{2+}. Thus zinc tends to have many of the properties associated with tripositive Group IIIB ions such as Al^{3+} and Ga^{3+}.

Property or reaction	Ga	Zn
Metal in air	Stable at room temperature but burns to Ga_2O_3 on heating	Stable at room temperature but burns to ZnO on heating
Metal plus acid	Dissolves with evolution of H_2 to give Ga^{3+} salts	Dissolves with evolution of H_2 to give Zn^{2+} salts
Metal plus alkali	Dissolves with evolution of H_2 to give gallates containing the $Ga(OH)_4^-$ ion	Dissolves with evolution of H_2 to give zincates containing the $Zn(OH)_4^{2-}$ ion
Other preparations of common salts	Dissolution of oxide or hydroxide in acids	Dissolution of oxide, hydroxide or carbonate in acids
Common soluble salts	Halides (except F^-), sulphate, nitrate, perchlorate	Halides (including the slightly soluble F^-), sulphate, nitrate, perchlorate
Common insoluble compounds	Oxide, hydroxide, fluoride and phosphate	Oxide, hydroxide, carbonate, phosphate, cyanide and oxalate
Heat hydroxide or nitrate	Oxide, Ga_2O_3, formed	Oxide, ZnO, formed; yellow when hot, white when cold

Reactions of M^{n+} ions in aqueous solution

1 Addition of aqueous NaOH	White precipitate of $GaO.OH$, soluble in excess to give gallate ions, $Ga(OH)_4^-$	White precipitate of $Zn(OH)_2$, soluble in excess to give zincate ions, $Zn(OH)_4^{2-}$
2 Addition of aqueous ammonia	White precipitate of $GaO.OH$, soluble in excess of reagent owing to formation of $Ga(OH)_4^-$	White precipitate of hydroxide, soluble in excess of reagent owing to formation of ammines such as $Zn(NH_3)_4^{2+}$
3 Addition of aqueous Na_2HPO_4	White precipitate of mixed phosphate, $Na_2HGa(PO_4)_2$	White precipitate of zinc phosphate

Property or reaction	Ga	Zn
4 Addition of aqueous alkali metal carbonate	Carbonate unknown	White precipitate of basic carbonate, $2ZnCO_3.3Zn(OH)_2$
5 Addition of aqueous sodium oxalate	No precipitate; oxalate soluble	White precipitate of zinc oxalate, ZnC_2O_4, in neutral solution
6 Addition of aqueous KCN	Cyanide unknown	White precipitate of $Zn(CN)_2$ which dissolves in excess reagent to give $Zn(CN)_4^{2-}$

COMPARISON OF ALUMINIUM AND SCANDIUM

The Group IIIA elements (Sc, Y and La) possess a d^1s^2 outer electronic configuration and, like the Group IIIB metals, are normally trivalent. Therefore some of the chemical properties of scandium and aluminium can be expected to be similar.

Property	Aluminium	Scandium
Metal in air	Protected by a strongly adherent oxide film	Rapidly attacked by damp air
Metal in water	Unaffected by water	Reduces water with the evolution of hydrogen
Metal in alkali	Dissolves with evolution of H_2	Dissolves with evolution of H_2
Metal in dilute acid	Dissolves with evolution of H_2 unless highly pure	Dissolves with evolution of H_2
Common soluble salts	Halides (except fluoride), nitrate, sulphate	Halides (except fluoride), nitrate, sulphate

Reactions in aqueous solution

1 $M(H_2O)_6^{3+}$ ions	Partial hydrolysis to give species such as $AlOH^{2+}$, $Al_2(OH)_2^{4+}$ and higher polymers	Partial hydrolysis to give species such as $ScOH^{2+}$, $Sc_2(OH)_2^{4+}$ and $Sc_3(OH)_5^{4+}$

Property	Aluminium	Scandium
2 Aqueous NaOH	White precipitate of $Al(OH)_3$, soluble in excess reagent to give $Al(OH)_4^-$; precipitate also soluble in acids to give salts	White precipitate of hydrous oxide, $Sc(OH)O$, soluble in excess of concentrated reagent to give $Sc(OH)_6^{3-}$; precipitate also soluble in acids to give salts
3 Aqueous NaF	White precipitate of AlF_3, soluble in excess reagent owing to formation of fluoro-complexes, ultimately AlF_6^{3-}	White precipitate of ScF_3, soluble in excess reagent to give ScF_6^{3-}
4 Aqueous sodium oxalate	Precipitate of aluminium oxalate, soluble in excess owing to formation of complex anions, e.g. $Al(C_2O_4)_2^-$ and $Al(C_2O_4)_3^{3-}$	Precipitate of scandium oxalate, soluble in excess owing to formation of complex anions such as $Sc(C_2O_4)_2^-$
5 Aqueous sodium hydrogen phosphate	White precipitate of $AlPO_4$	White precipitate of $ScPO_4$

Chapter 7

Group IV. Carbon, Silicon, Germanium, Tin and Lead

INTRODUCTION

The ionization energies in Table 28 follow the same trends as were noted for the Group III elements; the irregular changes down the group are due to inner d orbitals (germanium, tin, lead), inner f orbitals (lead) and relativistic effects (lead). The poor shielding characteristics of the 3d electrons of germanium are also responsible for the relatively small increase in covalent radius between silicon and germanium (0.05 Å) because the other electrons in germanium are more strongly attracted to the nucleus than might otherwise have been expected; the ns and np electrons of germanium and tin are *both* outside filled $(n-1)$d shells and hence the change in covalent radius between the two elements (0.19 Å) is much more normal. Likewise, the difference in radius between tin and lead of only 0.13 Å is due to the presence of the filled 4f shell and relativistic effects at lead.

The exceedingly high energies required to form M^{4+} ions, coupled to the fact that such ions would be small and highly polarizing, means that the existence of M^{4+} cations is highly unlikely, but they may just conceivably be present in SnO_2 or PbO_2. In favourable cases the high ionization energy can be offset by complexing ligands round the M^{4+} ion; as well as recouping energy by M–ligand bond formation, this increases the effective ionic size and so reduces the polarizing power of the highly charged cation, thus making its salts more stable:

$$CX_4 + 2 \quad \overset{As(CH_3)_2}{\underset{As(CH_3)_2}{\bigcirc}} \quad \longrightarrow \quad \left[\left(\begin{array}{c} As \\ \diagdown \; C^{4+} \text{-}As \\ \diagup \quad \diagdown As \\ As \end{array} \right) \right] \left[X^- \right]_4$$

(X = Br, I)

(this compound could also be regarded as a complex arsonium salt)

Table 28—The Group IV elements

Element	Configuration	1st IE (kJ mol^{-1})	2nd IE (kJ mol^{-1})	(3rd IE (kJ mol^{-1})	4th IE (kJ mol^{-1})	Covalent radius (Å)
6 Carbon, C	$1s^2 2s^2 2p^2$	1086	2352	4619	6222	0.77
15 Silicon, Si	$[Ne]3s^2 3p^2$	786.1	1576	3227	4355	1.17
32 Germanium, Ge	$[Ar]3d^{10}4s^2 4p^2$	761.5	1537	3302	4393	1.22
50 Tin, Sn	$[Kr]4d^{10}5s^2 5p^2$	708.5	1412	2943	3929	1.41
82 Lead, Pb	$[Xe]4f^{14}5d^{10}6s^2 6p^2$	715.5	1450	3081	4083	1.54

$$SiI_4 + 6DMF \rightarrow [Si(DMF)_6][I^-]_4$$

(the dimethyl formamide is bound to the Si by its oxygen atom)

$$SiI_4 + 6Opy \rightarrow [Si(Opy)_6][I^-]_4$$

(Opy = pyridine-N-oxide)

TETRAVALENCY IN GROUP IV

From their outer electron configuration of ns^2np^2, one might expect the Group IV elements to have covalent compounds in which they bond only to two other atoms using their half-filled p orbitals; this would lead to the formation of MX_2 molecules in which the angle X—M—X would be aproximately $90°$. However, this is contrary to observation because, in the vast majority of their compounds, these elements (and more especially carbon) are tetrahedrally surrounded by four other groups. This occurs because it is relatively easy for an ns electron to be 'promoted' to the unfilled np orbital; for carbon the $ns \rightarrow np$ promotion energy is 405.8 kJ mol^{-1}, that of the other elements probably being somewhat smaller.

However, although the $ns \rightarrow np$ promotion certainly leaves the element with four unpaired electrons, the electrons have *all their spins parallel*. To obtain the element in its *valence* (i.e. reacting) *state*, work must be done to 'randomize' the electron spins—and for carbon this absorbs about a further 250 kJ mol^{-1}. (The hypothetical process of hybridizing one s and three p orbitals to give four sp^3 tetrahedral orbitals is simply a mathematical step and requires no energy absorption on the part of the reacting element.) Therefore, to form CX_4 from X_2 molecules and graphite, the standard state of carbon, the following energy steps must be considered:

$$C(s) \xrightarrow[715 \text{ kJ mol}^{-1}]{\Delta H_{sub}} C(g) \xrightarrow{P} C(g) \ (s^{\uparrow}p_x^{\uparrow}p_y^{\uparrow}p_z^{\uparrow}) \xrightarrow{R} C(g) \text{ (valence state)}$$

$$2X_2 \xrightarrow{\Delta H_{diss}} 4X \searrow CX_4$$

where P is the promotion energy and R is the energy required to randomize the electron spins. The total energy input is thus $\Delta H_{sub} + P + R + 2\Delta H_{diss}$, which for the production of methane is about 2260 kJ mol^{-1}; this energy has to be regained by the formation of four strong C—X bonds. The production of CX_2 requires fewer energy-consuming steps:

$$C(s) \xrightarrow{\Delta H_{sub}} (Cg) \xrightarrow{R'} C(g) \text{ (valence state)}$$

$$X_2 \xrightarrow{\Delta H_{diss}} 2X \searrow CX_2$$

but only two C—X bonds are formed to compensate for this energy.

The question arises as to which of these two processes leads to the more thermodynamically stable molecule. At room temperature the former process resulting in the formation of MX_4 molecules is almost always the more satisfactory both for carbon and the other members of the group; for example, the enthalpies of formation at 25°C for CH_2 and CH_4 are about $+343$ and -74.9 kJ mol^{-1} respectively. Thus although the utilization of the ns^2 electrons requires a higher initial input of energy, this is more than offset by the formation of two extra M—X bonds. This does not mean that MX_2 molecules cannot be made at all. Those of C and Si often result as highly reactive intermediates from either the reduction of MX_4 molecules by the free element at high temperature:

$$SiF_4 + Si \xrightarrow{1150°C} [SiF_2]$$

or by the photolysis of unstable M^{IV} compounds:

$$CH_2N_2 \xrightarrow{hv} [CH_2] + N_2$$

(diazomethane)

In marked contrast many divalent species of the heavier elements are stable at room temperature:

$$GeBr_4 + Ge \text{ (or Zn)} \xrightarrow{heat} GeBr_2$$

$$Sn + I_2 \xrightarrow[\text{2M HCl}]{\text{heat in}} SnI_2$$

$$Sn + HX(g) \xrightarrow{heat} SnX_2 \quad (X = F \text{ or } Cl)$$

The M—X covalent bond strength decreases as the atomic number of M increases (Table 29), with the result that at lead not all Pb—X bonds are capable of supplying the energy required to stabilize the lead(IV) state with respect to lead(II). For example, lead tetrafluoride, tetrachloride and tetrabromide decompose readily on heating:

$$PbCl_4 \xrightarrow{50°C} PbCl_2 + Cl_2$$

whilst lead tetraiodide is too unstable to exist at room temperature:

$$\therefore \quad Pb + I_2 \xrightarrow{heat} PbI_2$$

Table 29—Bond energies (kJ mol^{-1}) of the Group IV elements

	Carbon	Silicon	Germanium	Tin
M—M	347	205	188	151
M=M	611	—	—	—
M≡M	841	—	—	—
M—H	411	323	285	251
M—F	490	598	473	—
M—Cl	326	402	339	314
M—Br	272	331	280	268
M—I	—	—	213	—

Gaseous tin and lead dihalides are non-linear molecules in which the X—M—X bond angles are close to the value of 90° expected if p orbitals on M are used in bonding. However, CH_2 ('carbene') is found to be linear in the ground state and contains two unpaired electrons; the carbon atom is sp-hybridized and the two

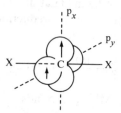

unpaired electrons occupy the remaining p orbitals. This triplet ground state of carbene has only slightly less energy than a singlet state in which the CH_2 molecule is bent, the H—C—H angle being 102.4°, and has all its electrons spin-paired. Substituents other than hydrogen on the carbon can invert the stability sequence of the triplet and singlet states, an example being CF_2 which is a bent, diamagnetic molecule having an F—C—F angle of 105°; SiF_2 is similar and has an F—Si—F angle of 102°.

THE DIVALENT IONIC STATE

An alternative to the covalent M^{II} state is of course the formation of M^{2+} by ionization of the np electrons. However, there are no solid compounds known of carbon, silicon, germanium or tin in which M^{2+} ions have been reported with certainty, and of the lead(II) derivatives perhaps only lead(II) fluoride, which has the rutile structure, can be said to be a typical ionic compound.

The *hydrated* Sn^{2+} and Pb^{2+} ions which occur in aqueous solutions of tin and lead perchlorates suffer extensive and complex hydrolysis; some or all of the following species are probably present, with the last ion in each case being thought to predominate:

$$Sn_{aq}^{2+} \rightarrow [SnOH]_{aq}^{+}, [Sn_2(OH)_2]_{aq}^{2+}, [Sn_3(OH)_4]_{aq}^{2+}$$

$$Pb_{aq}^{2+} \rightarrow [PbOH]_{aq}^{+}, [Pb_3(OH)_4]_{aq}^{2+}, [Pb_3(OH)_5]_{aq}^{+}, [Pb_6(OH)_8]_{aq}^{4+}, [Pb_4(OH)_4]_{aq}^{4+}$$

As an example of the behaviour of dihalides in aqueous solution, $SnCl_2$ dissolves in small quantities of water as *undissociated* $[SnCl_2]_{aq}$, but on dilution the basic chloride SnClOH precipitates as a white solid; addition of hydrochloric acid results in the formation of the pyramidal trichlorostannate(II) ion:

$$SnCl_2 + Cl^- \rightarrow SnCl_3^-$$

CATENATION

Carbon is unique among the Group IV elements in its ability to catenate (i.e. to make chains with itself as in the paraffin hydrocarbons). Although the C—C bond is one of the strongest homonuclear bonds apart from that in H_2, the strength of the other M—M bonds (Table 29) decreases rapidly down the group. Thus whereas many hundreds of carbon atoms may be linked together in 'polythene', thermal stability usually limits the number of M—M bonds occurring in catenated hydrides and halides of the other Group IV elements to eleven or twelve for silicon, nine or ten for germanium and only two for tin.

Even disilane is thermodynamically unstable towards decomposition into SiH_4 and silicon via the equation

$$Si_2H_6(g) \rightarrow SiH_4(g) + H_2 + Si(s)$$

for which $\Delta G^\ominus$ is -70 kJ mol^{-1}. Thus it, and the higher silanes, must fortuitously be *kinetically* stable towards the above type of decomposition. However, as attempts are made to purify the higher hydrides by either distillation or gas chromatography, thermal decomposition does increasingly become a problem as the chains grow longer.

MULTIPLE p_π–p_π BONDING

Although there are a large number of compounds known in which carbon forms π bonds with itself or with nitrogen and oxygen, it is only very recently that similar species have been prepared for silicon and the other members of Group IV. The unsaturated derivatives of these heavier elements are usually highly unstable, unless bulky groups are present to block possible decomposition routes to more thermodynamically favoured products containing single bonds. This instability is partially due to the inherent weakness of multiple bonds involving overlap of np orbitals when $n > 2$ (page 62), but there is probably a kinetic factor involved also because ethene, a molecular normally considered 'stable', is actually *thermodynamically unstable* relative to the singly bonded system of polythene.

Some species are so fragile that they can only be studied transiently in the gas phase using spectroscopy, e.g.

triazido(phenyl)silane   phenyl sila*iso*cyanide

or are investigated using the method of *matrix isolation*. In this technique the compound is often made by pyrolysis at high temperature and very rapidly cooled to the temperature of boiling helium (4K) in the presence of a large excess of inert gas such as argon. Under these conditions the highly reactive molecules are effectively separated from each other by the solid matrix of argon and hence cannot undergo polymerization (of course, thermal decomposition is impossible at such low temperatures). The identity of the specimen can normally be established beyond reasonable doubt by investigating the matrix mixture using a variety of spectroscopic methods:

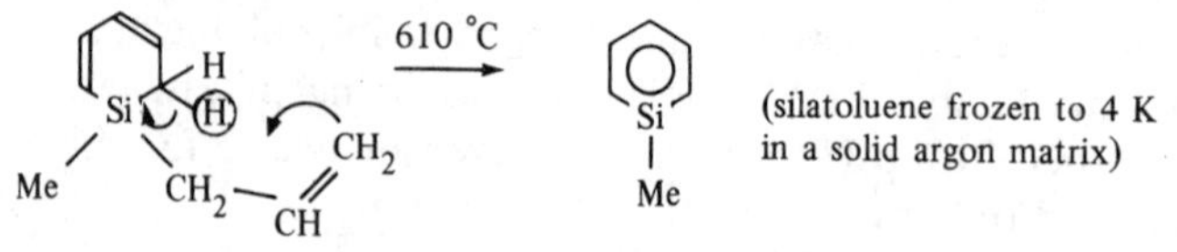

(silatoluene frozen to 4 K
in a solid argon matrix)

Slow and controlled warming of the matrix whilst it is still under spectroscopic observation sometimes allows the original compound to decay into a known product which can provide further confirmatory structural evidence:

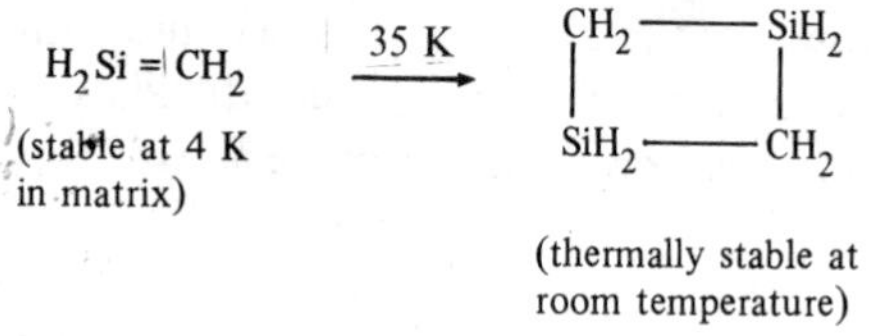

$H_2Si = CH_2$

(stable at 4 K
in matrix)

$\xrightarrow{35\ K}$

(thermally stable at
room temperature)

Many M_2R_4 species of the heavier Group IV elements are now known which are thermally stable at room temperature, and some of the unsymmetrical silenes have even been isolated in both the *trans* and *cis* forms. However, the M—M bonds behave less and less like double bonds as the atomic weight of M increases (page 63). Two typical syntheses of silenes are shown below:

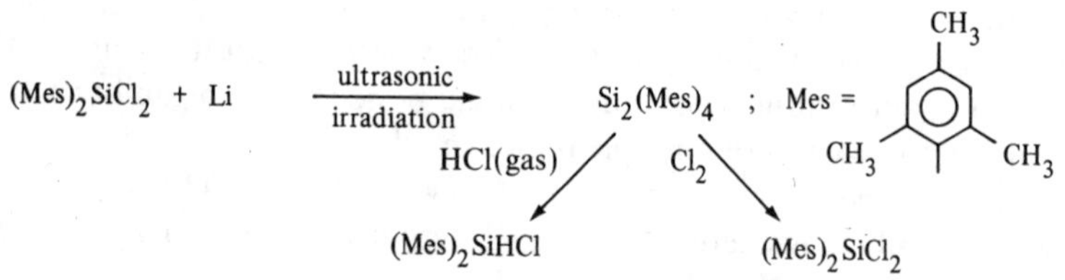

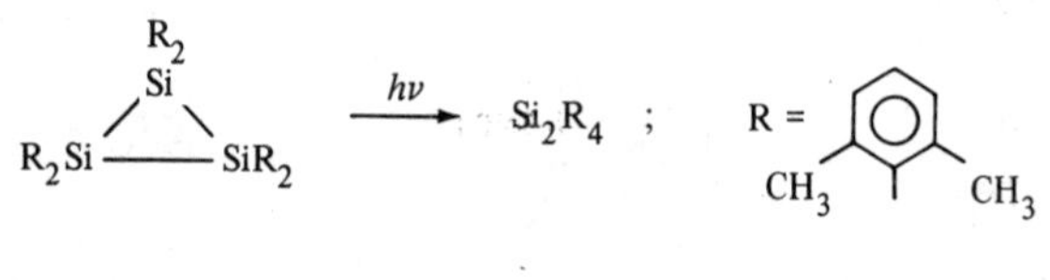

$(Mes)_2SiCl_2$ + Li $\xrightarrow[\text{irradiation}]{\text{ultrasonic}}$ $Si_2(Mes)_4$; Mes =

HCl(gas)

$(Mes)_2SiHCl$

Cl_2

$(Mes)_2SiCl_2$

$\xrightarrow{h\nu}$ Si_2R_4 ; R =

M–HALOGEN π BONDING IN THE TETRAHALIDES

The filled halogen p orbitals in the Group IV tetrahalides have the same symmetry as empty d orbitals on the central Si, Ge, Sn and Pb atoms. There now seems to be ample evidence that considerable π bonding occurs by interaction between these two types of orbital; in particular, electron spectroscopy shows that the halogen np lone pairs in all the tetrahalides studied, *except those of carbon*, are stabilized by this interaction, the effect being greatest for SiF_4 (Fig. 58).

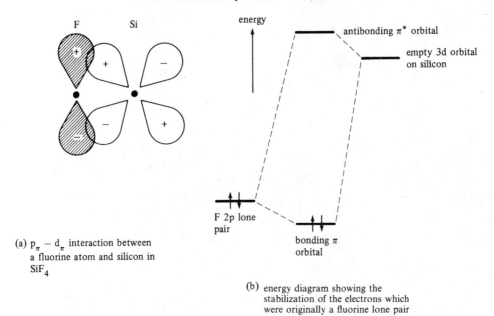

(a) $p_\pi - d_\pi$ interaction between a fluorine atom and silicon in SiF$_4$

(b) energy diagram showing the stabilization of the electrons which were originally a fluorine lone pair

Fig. 58—Fluorine–silicon π bonding in SiF$_4$.

Theoretical calculations have demonstrated that the normally large valence shell nd orbitals on a non-transition element shrink considerably when very electronegative groups such as halogens are attached to that element. Such a d orbital contraction will result in better p–d overlap and enhance the above type of π bonding.

No secondary π bonding can occur in the carbon tetrahalides because carbon has no d orbitals in its valence shell. (Compare the C—X and Si—X bond energies given for the halides in Table 29).

STEREOCHEMISTRY OF M^{II} DERIVATIVES

The divalency of Group IV elements is normally rationalized in terms of the presence of an inert (or 'reluctant') ns^2 electron pair. However, it is clear from a vast amount of structural data that this electron pair is very often *sterically active* and reduces the stereochemical symmetry of many M^{II} derivatives. Not all compounds are affected in this way and some retain perfectly regular octahedral arrangements; unfortunately there is no simple theory which permits us to predict those systems which will be distorted. In naive terms *all* M^{II} compounds should have regular structures if the lone pair remained in a spherically symmetrical s orbital, but it appears that in many systems s–p mixing occurs to give the lone pair orbital considerable p (i.e. directional) character.

The simplest manifestation of the lone pair's stereochemical activity is to be found in the non-linearity of the gaseous, monomeric dihalides, the structures of which can

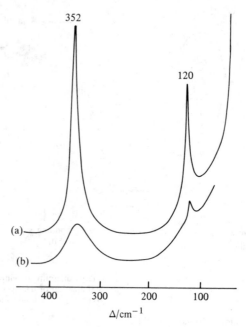

Fig. 59 — The Raman spectrum of gaseous $SnCl_2$ monomer. For a non-linear, triatomic molecule two polarized bands are expected, one arising from the symmetric stretch and the other from a bending deformation:

The polarized nature of the bands can be seen by comparing the intensities when a polaroid screen is placed first parallel and then perpendicular to the plane of polarization of the incident exciting radiation; for a polarized band the intensity decreases when this is done as in spectrum (b). Only one band, which occurs at 361 cm^{-1}, is expected for the symmetric stretch in a linear $ZnCl_2$ molecule.

be deduced from Raman spectra (Fig. 59) or electron diffraction:

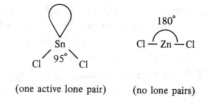

(one active lone pair)   (no lone pairs)

In crystals, steric activity of a lone pair is usually deduced from the presence of irregular coordination about the central atom, the electrons being presumed to occupy apparently empty spaces within the coordination polyhedra defined by the ligands. For example, in tin(II) compounds the normally observed geometries are trigonal pyramidal and distorted trigonal bipyramidal with the lone pair in the hemisphere opposite the ligand plane (there are several much more distant neighbours on the lone pair side of the Ge, Sn and Pb atoms which, if they are assumed to be weakly bonded, result in total coordination numbers of 6–9):

(CsSnCl$_3$)   (solid SnCl$_2$)   (polymeric SnCl$_2$.dioxane)

The presence of the lone pair in SnCl$_3^-$ can also be detected chemically because the ion acts as an effective Lewis base towards both transition metals and main group elements:

$$BX_3 + SnCl_3^- \rightarrow [X_3BSnCl_3]^- \quad (X = halogen)$$

$$PtCl_4^{2-} + 2SnCl_3^- \rightarrow cis\text{-}[Cl_2Pt(SnCl_3)_2]^{2-} + 2Cl^-$$

C—O VERSUS Si—O BONDING: THE SILICONES

Carbon is the only Group IV element which is able to form strong π bonds with oxygen. The effect of this can be seen in the physical properties of many carbon compounds which differ in a dramatic way from their silicon, germanium, tin and lead counterparts, although both may have the same stoichiometry; two typical examples are carbon dioxide, which has a sublimation point of $-78\,°C$, and silicon dioxide, which has a melting point of $1710\,°C$. In the carbon dioxide molecule the carbon atom is bound to the two oxygens by both σ and $p_\pi-p_\pi$ bonds (see Fig. 70, page 237) and hence the solid contains molecules of CO_2 held in the lattice only by weak van der Waals forces. Silicon, however, gains more energy by forming four Si—O single bonds than by being multiply bonded to two oxygen atoms only, and this results in a polymeric structure for solid SiO_2 based on O—Si—O bridges; on melting, many strong Si—O bonds must be broken and this requires a very high temperature.

The formation of Si—O—Si single bonds in preference to Si=O bonds is the fundamental basis of the 'silicone' industry. Silicone polymers, which can be either cyclic or long chain molecules, have the R_2SiO group as the unit building block:

(R is normally methyl)

They are formed by hydrolysis of dialkysilicon dichlorides, the intermediate dihydroxyl derivative being highly unstable:

$$n R_2SiCl_2 + 2nH_2O \longrightarrow n \left[\begin{array}{c} R \\ \diagdown \\ Si \\ \diagup \\ R \end{array} \begin{array}{c} \text{\tiny{wwww}} OH \\ \\ OH \end{array} \right] \xrightarrow{-nH_2O} \left\{ \begin{array}{c} R \\ \diagdown \\ Si \\ \diagup \\ O \end{array} \begin{array}{c} \text{\tiny{wwww}} O_{\text{\tiny{wwww}}} \\ \\ R \end{array} \right\}_n$$

(The analogous carbon diol $R_2C(OH)_2$ is also unstable and loses a molecule of water, but in this case the product is a ketone $R_2C{=}O$ in which the carbon and oxygen atoms are linked by a $\sigma + \pi$ double bond.)

To limit the length of the polymer to some desired value of n, the above hydrolysis is carried out in the presence of R_3SiCl, which stops polymer growth by 'sealing off' the end of the chain with an $-SiR_3$ group:

$$R_2SiCl_2 + R_3SiCl \xrightarrow{H_2O} R_3SiO\underbrace{H + HO}Si-O\left[\begin{array}{c}R \\ | \\ Si-O \\ | \\ R\end{array}\right]_n \longrightarrow R_3Si-O\left[\begin{array}{c}R \\ | \\ Si-O \\ | \\ R\end{array}\right]_{n+1}$$

The simple chain polymer produced in this type of reaction is, depending on the degree of polymerization, either a liquid or a waxy solid. If it is desired to have a rigid, solid polymer this can be achieved by 'cross-linking' the chains of silicon and oxygen atoms by adding a controlled quantity of $RSiCl_3$ to the hydrolysis reaction mixture:

$$RSi\begin{array}{c}OH \\ -OH \\ OH\end{array} + \begin{array}{c}HO\left[\begin{array}{c}R \\ | \\ Si-O \\ | \\ R\end{array}\right]_n \\ HO\left[\begin{array}{c}R \\ | \\ Si-O \\ | \\ R\end{array}\right]_m \\ HO\left[\begin{array}{c}R \\ | \\ Si-O \\ | \\ R\end{array}\right]_o\end{array} \longrightarrow RSi\begin{array}{c}O\left[\begin{array}{c}R \\ | \\ Si-O \\ | \\ R\end{array}\right]_n \\ -O\left[\begin{array}{c}R \\ | \\ Si-O \\ | \\ R\end{array}\right]_m \\ O\left[\begin{array}{c}R \\ | \\ Si-O \\ | \\ R\end{array}\right]_o\end{array}$$

The starting materials for these polymers are normally produced by the 'direct synthesis' technique, in which silicon is heated with an alkyl chloride, e.g.

$$Si + CH_3Cl \xrightarrow{Cu} CH_3SiCl_3 + (CH_3)_2SiCl_2 + (CH_3)_3SiCl$$

and the mixture of alkylsilicon chlorides separated by fractional distillation.

FIVE- AND SIX-COORDINATE COMPOUNDS

The covalency maximum for carbon is four, but the other members of the group are able to increase their coordination numbers to five and six. This is especially true in the case of the tetrahalides, which react with a wide variety of ligands to form trigonal bipyramidal $MX_4.L$ and octahedral $MX_4.2L$ complexes:

$$
\begin{array}{c}
L \\
| \\
X - M\cdots X \\
| \quad\searrow X \\
X
\end{array}
\qquad\qquad
\left[
\begin{array}{c}
X \\
| \\
X - M\cdots X \\
| \quad\searrow X \\
X
\end{array}
\right]^{-}
$$

e.g. $GeF_4.N(CH_3)_3$ e.g. $SiF_5^-,\ GeF_5^-,\ SnCl_5^-$

$$
\begin{array}{c}
L \\
X\cdots\!\!\underset{|}{M}\!\!\cdots X \\
X\swarrow | \searrow X \\
L
\end{array}
\qquad
\begin{array}{c}
L \\
X\cdots\!\!\underset{|}{M}\!\!\cdots L \\
X\swarrow | \searrow X \\
X
\end{array}
\qquad
\left[
\begin{array}{c}
X \\
X\cdots\!\!\underset{|}{M}\!\!\cdots X \\
X\swarrow | \searrow X \\
X
\end{array}
\right]^{2-}
$$

trans   *cis*

e.g. $SiCl_4.2P(CH_3)_3$ e.g. $SiF_4.2bipy$ e.g. SiF_6^{2-}
$SnCl_4.2DMF$ $SnCl_4.2DMF$ $PbCl_6^{2-}$
$SnBr_4.2DMF$ $SnBr_4.2DMF$ $SnCl_nBr_{6-n}^{2-}\ (n=0\text{--}6)$
$SiF_4.2pyridine$ $SnCl_4.(MeCN)_2$ $GeX_6^{2-}\ (X=F\text{ or }Cl)$
$GeCl_4.2pyridine$ SnI_6^{2-}
$(DMF = $ dimethyl
formamide$)$;
$bipy = 2,2'$-bipyridine

These higher coordination numbers are achieved without a major bonding role being assumed by d orbitals on the central atom. All six ligands in octahedral complexes, and the axial pair in trigonal bipyramidal derivatives, are held by three-centre, four-electron bonds of the type described on page 69; the three equatorial groups of the trigonal bipyramidal complexes are attached by typical covalent bonds involving sp^2 hybrid orbitals on the central Group IV atom. Carbon is probably too small to accommodate even the one extra ligand required for five coordination (page 68). The fact that silicon, germanium, tin and lead are able to increase their coordination number in this way probably accounts for the ready hydrolysis of their tetrahalides on treatment with water:

$$
MX_4 + OH^- \longrightarrow
\left[
\begin{array}{c}
X \\
| \\
HO\overset{\delta-}{----}M\overset{\delta-}{----}X \\
\diagup\!\!\Vert \\
X \quad X
\end{array}
\right]^{-}
\longrightarrow X_3MOH \longrightarrow \longrightarrow
$$

This associative mechanism helps to explain the high kinetic stability (as opposed to thermodynamic stability) of CCl_4, SiF_6^{2-} and SF_6 towards hydrolysis. Such compounds are coordinatively saturated and a high steric activation energy barrier would thus be presented to an attack by OH^- ions. Acyl chlorides and SF_4 on the other hand undergo ready hydrolysis because the carbon and sulphur atoms can now satisfactorily accommodate the incoming OH^- ion thereby causing a substantial lowering of the hydrolysis activation energy.

POLYATOMIC ANIONS

Following the trend for polyatomic ions noted at mercury, the heavier Group IV elements form a number of highly coloured, polyatomic anions when alkali metal alloys with Ge, Sn and Pb are dissolved in ammonia or 1,2-diaminoethane ($H_2NCH_2CH_2NH_2$, en). Normally the species revert back to the starting alloys on removal of solvent, but it has proved possible to crystallize out a few solvates such as $Na_4Sn_9.7en$. More recently it has been found that addition of 2,2,2-crypt ligand (page 117) to the amine solutions allows complexation of the alkali cations as $M(crypt)^+$ and so prevents electron transfer from the anions back to M^+ on solvent removal. In this way alkali crypt derivatives of M_4^{2-}, M_5^{2-}, M_9^{2-} and M_9^{4-} anions have been isolated and studied by X-ray crystallography; the idealized structures of these anions are shown in Fig. 60.

Mixed metal anions have also been made by dissolving, for example, Na–Sn–Pb alloys in 1,2-diaminoethane; ^{119}Sn NMR spectra show that the solutions contain $Pb_2Sn_7^{4-}$, $Pb_3Sn_6^{4-}$, $Pb_4Sn_5^{4-}$, $Pb_5Sn_4^{4-}$ and $Pb_6Sn_3^{4-}$. Other ions made in a similar fashion include $Sn_{9-x}Ge_x^{4-}$, $TlSn_8^{5-}$ and $SnTe_4^{4-}$.

(a) Tetrahedron: Ge_4^{2-}; Sn_4^{2-}

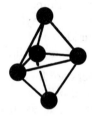

(b) Trigonal bipyramid
Sn_5^{2-}; Pb_5^{2-}

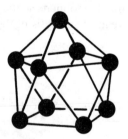

(c) Mono-capped square anti-prism:
Ge_9^{4-}; Sn_9^{4-}

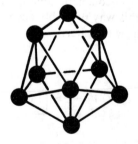

(d) Tri-capped trigonal prism:
Ge_9^{2-}

Fig. 60—Idealized structures of polyatomic anions containing Group IV elements.

^{13}C, ^{29}Si, ^{119}Sn AND ^{207}Pb NMR SPECTROSCOPY

The nuclei of carbon-13 (1.1%), silicon-29 (4.7%), tin-119 (8.7%) and lead-207 (22.6%) all possess a spin of $\frac{1}{2}$ and may be used in nuclear magnetic resonance studies; two other tin isotopes, ^{115}Sn (0.35% abundance) and ^{117}Sn (7.7%), also have a nuclear spin but the former isotope is the one normally preferred. Structure determinations using ^{13}C NMR are now routine in both organic chemistry and organometallic chemistry and will not be discussed further.

Owing to the low abundance of ^{29}Si and ^{119}Sn it is sometimes advantageous to improve the signal-to-noise ratio by using samples enriched in the chosen isotope. Thus by using ^{29}Si-enriched silicates it can be shown that aqueous solutions with 1:1 ratios of K:Si contain at least 22 different soluble silicate anions in an extremely complex mixture. Chloroform solutions of a perchloropentasilane, Si_5Cl_{12}, made by photolysis of $HSiCl_3$ show a ^{29}Si NMR spectrum which consists of just two lines with an intensity ratio of 4:1. The spectrum is therefore consistent with the *neo*-structure for Si_5Cl_{12} since the *n*- and *iso*-isomers would show more than two peaks; the *extreme sharpness* of the observed lines virtually rules out the possibility of accidental overlap of resonances which would, of course, have given rise to a deceptively simple spectrum:

$$Cl_3Si \overset{SiCl_3}{\underset{SiCl_3}{\underset{\diagdown}{\overset{\diagup}{Si}}}} \quad \text{Cl}_3Si$$

neo-dodecachlorosilane, Si_5Cl_{12}

In a recently published paper, ^{119}Sn NMR spectroscopy was used to study a mixture of fluorochlorostannates(IV), $SnF_xCl_{6-x}^{2-}$ ($x = 0-6$), resonance peaks being observed for all the species present in the CH_2Cl_2 solution of the mixture, including *cis–trans* and *mer–fac* isomers. The presence of ^{19}F—^{119}Sn coupling provided a useful aid in component assignments because the multiplicity of lines making up the various peaks was equal to $(2nI + 1)$, where n is the number of fluorine atoms (which have a nuclear spin I of $\frac{1}{2}$) present in the ion: e.g.

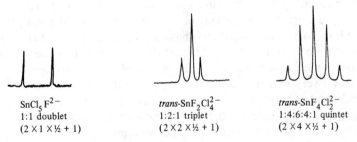

$SnCl_5F^{2-}$
1:1 doublet
$(2 \times 1 \times \frac{1}{2} + 1)$

trans-$SnF_2Cl_4^{2-}$
1:2:1 triplet
$(2 \times 2 \times \frac{1}{2} + 1)$

trans-$SnF_4Cl_2^{2-}$
1:4:6:4:1 quintet
$(2 \times 4 \times \frac{1}{2} + 1)$

The observed chemical shifts of ^{207}Pb cover an immense range of almost 4000 ppm, the standard zero being the shift of lead-207 in $Pb(CH_3)_4$; compare these shifts to the 10 ppm range which covers most of the proton resonances in organic compounds. Very small structural changes in the environment of the ^{207}Pb induces relatively large changes in the chemical shift, which is useful, for example, in studying the Pb^{2+} ion in aqueous solution. When $Pb(ClO_4)_2$ dissolves in water, the lead-207 chemical shift

(-2880 ppm) changes little with concentration because no complexation takes place between Pb_{aq}^{2+} and the perchlorate anion; the number of water molecules surrounding the Pb^{2+} ion was thought to be about eight. In contrast, the shift of lead in 0.01M solutions of lead nitrate was almost identical (-2854 ppm) to that in the perchlorate system, demonstrating the presence of mainly free Pb_{aq}^{2+} ions, but as the concentration was steadily increased to 1.0M, the chemical shift began to change until it finally reached a value of -2961 ppm owing to the build-up of the nitrato-complex $[Pb(NO_3)]_{aq}^+$ in solution. Lead ethanoate was found to undergo only partial dissociation in water:

$$[Pb(OOCCH_3)_2]_{aq} \rightleftharpoons [Pb(OOCCH_3)]_{aq}^+ + CH_3COO^-, \quad K_{diss} = 1.4 \times 10^{-2} \ (30°)$$
$$(-1175 \text{ ppm}) \qquad\qquad (-2410 \text{ ppm})$$

MÖSSBAUER, OR NUCLEAR GAMMA RESONANCE, SPECTROSCOPY

This technique, which differs from other types of spectroscopy in that the transitions from an excited state to the ground state occur within the nucleus, can be used successfully for only a relatively few nuclei, including ^{119}Sn, ^{57}Fe, ^{127}I and ^{129}Xe. It involves the recoilless emission and resonant absorption of γ radiation, achieved by using solid samples in which the recoil energy of the emitter (considerable for loss of high-energy γ rays) is absorbed by the crystal lattice; for tin compounds the recoil effect is minimized further by holding the samples at the boiling point of liquid nitrogen, $-196°C$.

The energy difference between the excited and ground states of the nucleus, and hence the frequency of the γ radiation, is slightly sensitive to the environment of the nucleus, thus providing information about the chemical state of the atom. In ^{119}Sn Mössbauer spectrometry it has been possible in many cases to use 'chemical shifts' in the γ ray frequency to determine the oxidation state of tin in new compounds.

A tin-119 Mössbauer spectrum consists of a sharp, single peak when the metal has regular tetrahedral or octahedral coordination. Hence *tetravalent* SnX_4 and SnX_6^{2-} species will show only a single line, in contrast to most *divalent* species which usually have a grossly distorted arrangement of ligands about the tin owing to the steric effect of the lone pair; such compounds have a spectrum showing two lines of equal intensity which are separated by an amount Δ called the quadrupole splitting. The $Sn^{II}Br_6^{4-}$ ion in the caesium salt Cs_4SnBr_6 is one of the few divalent tin complexes which exhibits a very narrow, single line, suggesting that the tin atom in the ion has regular octahedral symmetry and hence a sterically inactive lone pair. The perfectly octahedral shape of the anion predicted from the Mössbauer spectrum has since been confirmed by an X-ray crystal structure determination.

X-RAY PHOTOELECTRON SPECTROSCOPY (XPS)

It is now possible by using XPS to measure the inner-core electron energies of the various atoms making up a molecule. Contrary to the predictions of simple valence

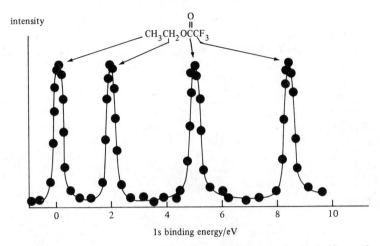

Fig. 61—The 1s electron binding energies of the various carbon atoms in ethyl trifluoroethanoate determined by X-ray photoelectron spectroscopy. The energy 'zero' corresponds to the 1s electron binding energy of the carbon atom in the CH_3 group, 291.2 ev (28 096 kJ); the energy of the 1s electrons in the other carbon atoms is obtained by adding the 'chemical shifts', in eV, to 291.2. Peak intensities, which correspond to the number of photoejected electrons having the stated energy, are equal within experimental error because there is one carbon in each of the four different chemical environments.

theories, the 1s electrons of carbon, for example, *are quite markedly* affected by bonding, and to differing degrees depending on the appended atoms (Fig. 61). These 'chemical shifts' arise owing to penetration of the 2s orbital into the space occupied by the 1s electrons, which results in electrostatic repulsion and thus has a moderating effect on the energy of the core electrons. The higher the electronegativity of an atom attached to carbon, the more the 2s density will be drawn away from the core region so causing an *increase* in the binding energy of the 1s electrons.

OCCURRENCE OF THE GROUP IV ELEMENTS

Carbon

About 0.8% of the Earth's crust; crude oil and coal provide the highest natural concentrations of carbon.

Silicon

27.6%; the second most common element after oxygen in the Earth's crust; occurs widely in silicates and as silica, SiO_2, in quartz and sandstone.

Germanium

About 7 ppm; some zinc and silver ores contain considerable amounts of germanium, but there appears to be no widely occurring germanium ore.

Tin

0.004%; small amounts of tin have been reported to occur native; the oxide SnO_2 is mined as 'tinstone'.

Lead

About 16 ppm of the Earth's crust; chief ore is galena, PbS.

EXTRACTION

Carbon

Artificial graphite is made by heating either powdered anthracite or coke with sand in an electric furnace for about 24 h. This is the thermodynamically stable form of carbon at room temperature, but the more dense diamond can be made directly from graphite by the application of 125 000 atm pressure at about 3000 K; in the presence of a transition metal as catalyst, the transformation can be accomplished at 70 000 atm and 2000 K (see Fig. 62).

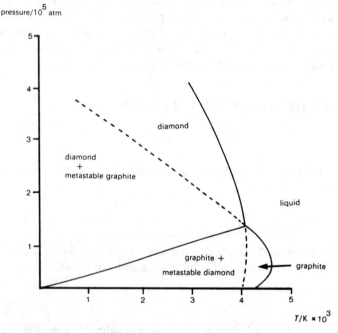

Fig. 62—Sketch of the phase diagram for carbon [see *Journal of Chemical Physics* vol. 38 (1963) page 631 for full details].

Silicon, germanium and tin

Reduction of the dioxides with carbon at high temperatures gives the free elements; silicon and germanium may be made ultra-pure for use as semiconductors by zone refining (see Fig. 68).

Lead

Galena, PbS, is roasted in air to give the oxide, which is then reduced by carbon monoxide in a blast furnace.

THE ELEMENTS

There is an increase in metallic character of the elements in going down the group: carbon is a typical non-metal; silicon and germanium are 'metalloidal', having properties between the true metals and non-metals; tin and lead are essentially metallic. The behaviour is reflected in the structures adopted by the elements. Carbon, silicon and germanium adopt diamond-like structures, the crystals of which are actually giant molecules with each atom tetrahedrally surrounded by four neighbours (Fig. 63). Tin exists in three forms, the low-temperature α form having the non-metallic diamond structure, whereas the two high-temperature modifications each have a typically metallic, close-packed structure. The transition temperatures for the tin allotropes are:

$$\alpha\text{-Sn} \overset{13.2\,^{\circ}\text{C}}{\rightleftharpoons} \beta\text{-Sn} \overset{161\,^{\circ}\text{C}}{\rightleftharpoons} \gamma\text{-Sn}$$

(grey   (white   (brittle
tin)   tin)   tin)

Grey tin, having the more open diamond structure, has a lower density than either of the two metallic allotropes.

As already stated, only carbon of the Group IV elements is able to form strong p_π–p_π bonds with itself and hence the graphite structure is unique to carbon. In this

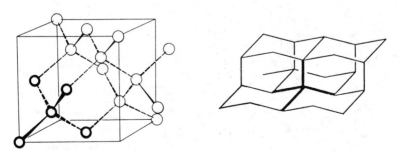

Fig. 63—The structure of diamond. Each atom is tetrahedrally surrounded by four neighbours. The structure is adopted by carbon, silicon and germanium.

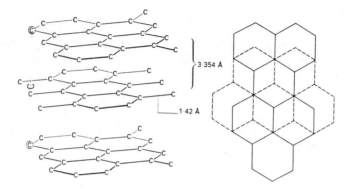

Fig. 64 The structure of crystalline graphite. It is only alternate layers in graphite which are 'in phase'. Such an ababab ... arrangement of the layers reduces the steric interaction between the carbon atoms in adjacent layers to a minimum.

allotropic form the carbon atoms are linked in a planar hexagonal network and the separate layers stacked on each other at about the van der Waals distance (Fig. 64). Hence there are no strong forces holding the layers in place, which gives rise to the lubricating properties of graphite; recent work indicates that adsorbed gases play some part in the lubrication mechanism as shown by the fact that graphite does not act as a lubricant under high-vacuum conditions. Within the layers each sp^2-hybridized carbon atom forms σ bonds with three neighbouring atoms in the same plane (see Fig. 65). Every atom has a half-filled 2p orbital above and below this plane of atoms, which will interact to give a large number of molecular orbitals extending over the whole layer. The bonding π molecular orbitals are full and the antibonding molecular orbitals are empty, but because their energy separation is slight, excitation of electrons from the bonding to the antibonding orbitals occurs relatively easily (e.g. by absorption of light quanta or on the application of an electric field).

The high melting points of carbon (3930°C), silicon (1420°C) and germanium (959°C) are due to the strong interatomic (covalent) forces which have to be broken in order to break up the lattice; the lowering of the melting points from carbon to germanium is due mainly to the decreasing strength of the M—M bonds down the group. The low melting points of tin (232°C) and lead (327.5°C) indicate that these

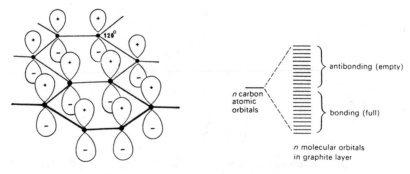

Fig. 65—π Bonding within a single layer of graphite.

metals do not use all their outer electrons in binding the metallic lattice together. (Compare the above figures with the higher melting points of the alkaline earth metals, which use only two electrons per atom in binding their metallic lattices.)

CHEMISTRY OF GRAPHITE

Compared with diamond, the graphitic form of carbon is quite reactive. The weak van der Waals forces holding the graphite layers together allow the penetration of a wide variety of reactants into the lattice giving, broadly, two types of product: (a) those in which the carbon layers are buckled owing to the destruction of the delocalized π molecular orbital system; (b) those in which the graphite layers retain their π electron system but become separated from each other by as much as 10 Å (in pure graphite in the interlayer distance is only 3.35 Å). In compounds of type (a) the graphite loses its electrical conductivity and often much of its colour, whereas those of type (b) retain the electrical conductivity.

Graphite compounds of type (a)

At 400°C, pure graphite undergoes a quiet reaction with fluorine to give a product containing approximately equal amounts of carbon and fluorine $(CF_x)_n$; when x approaches 1.0 the colour is white. An X-ray examination of the product shows that the layers of carbon atoms have swelled apart to a distance of 6.6 Å. Fig. 66 illustrates one of the possible structures of $(CF)_n$, the fluorines being covalently bound to the carbon atoms. The latter are now sp^3-hybridized and adopt the boat or chair conformation; because the delocalized π molecular orbitals have been destroyed, the colour and electrical conductivity of the original graphite largely disappear.

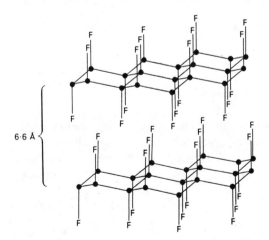

Fig. 66—One of the suggested structures of 'graphite fluoride', showing the cyclohexane chair type of configuration adopted by the carbon atoms; it is possible that the rings may, instead, take up the boat conformation. The evidence at present available is contradictory regarding the ring shape.

Very strong oxidizing agents, such as mixtures of nitric acid and potassium chlorate, cause graphite to swell enormously, during which the colour changes to brown. Analysis of the dried product indicates a carbon:oxygen ratio of about 2:1, but a small amount of combined hydrogen is always retained; the chemical properties of 'graphite oxide' suggest that at least three variations of carbon–oxygen linkage is present:

(1) Ether groups, which are thought to bridge mainly the 1,3 positions:

(2) Tertiary alcohols, which can be methylated and esterified:

(3) Carbonyl groups, for which it has been suggested that keto–enol tautomerism is possible:

enol form   keto form

Graphite compounds of type (b)

Graphite readily absorbs the vapours of the alkali metals to give highly coloured solids in which the original graphite structure is largely undisturbed except that the layers are farther apart to allow access of the metal atoms; on strong heating the alkali metal volatilizes unchanged out of the graphite lattice: e.g.

$$C + K \rightarrow C_{60}K \xrightarrow{K} C_{48}K \xrightarrow{K} C_{36}K \xrightarrow{K} C_{24}K \xrightarrow{K} C_{8}K$$

deep   steel   bronze
blue   blue

In the C_8M compounds there are alternate layers of carbon and metal atoms, the latter being arranged so that they lie above and below centre of two C_6 rings in the neighbouring planes:

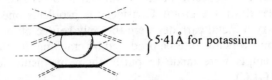

$5\cdot41\text{Å}$ for potassium

Each metal atom is equidistance from twelve carbon atoms; note that unlike pure graphite the carbon layers are superimposed.

The stoichiometry of the several compounds arises from the number of carbon planes between each metal layer and the arrangement of the metal atoms within their layers as shown in Fig. 67. These metal–graphite intercalation compounds conduct electricity more readily than graphite due to the transference of electrons from the metal atoms into the π-orbitals on the carbon layers, where they are relatively 'free' to move on the application of an electric field.

It is also possible for a wide variety of metal halides, usually transition metal chlorides in their higher oxidation states, to be incorporated into the graphite structure. Unlike the alkali metal systems the graphite layers remain in the unaltered ABABAB sequence, and the structures of the metal halides are largely preserved. One of the most studied systems is that of graphite with iron(III) chloride, for which a range of compositions has been given by various workers; some reports suggested that partial reduction to Fe(II) takes place on intercalation, but the *single* [57]Fe Mössbauer resonance given by the compound is characteristic of an iron(III) atom in the tetrahedral environment of an Fe_2Cl_6 dimer molecule.

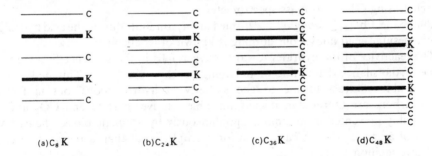

Fig. 67—Type (b) lamellar compounds of graphite; a further compound, $C_{60}K$, has five carbon layers between the potassium atoms. The metal atoms in each layer do not necessarily lie directly over each other. For example, in the C_8K system the graphite layers (G) and the metal layers $(\alpha, \beta, \dots)$ might stack $G\alpha G\beta G\gamma G\delta$ or $G\alpha G\gamma G\delta G\beta$ or $G\alpha G\delta G\beta G\gamma$.

HYDRIDES OF THE GROUP IV ELEMENTS

The paraffin hydrocarbons belong to a study of organic chemistry and will not be discussed here, although it should be remembered that they can be in the form of either straight or branched chains.

Relatively few hydrides are known for the other elements, partly because suitable syntheses are difficult to conceive or carry out but also because of the decreasing stability of hydrides down the group. Consider the problems one would encounter in organic chemistry if metal carbides were the only source of methane and all the higher hydrocarbons, which had to be produced by cracking this methane, tended to decompose as attempts were made to purify them by distillation or gas–liquid chromatography (GLC).

Classical methods used for preparing silanes and germanes used the acid hydrolysis of magnesium silicide and magnesium–germanium alloys, but now a considerable range of syntheses is available, including:

$$SiCl_4 + LiAlH_4 \xrightarrow{\text{ether}} SiH_4 \quad \text{(a general method for hydrides)}$$

$$MO_2 + LiAlH_4 \xrightarrow{\text{heat}} MH_4 \quad (M = Si \text{ or } Ge)$$

$$Si_3Cl_8 + LiAlH_4 \rightarrow Si_3H_8$$

$$SnCl_4 + LiAlH_4 \rightarrow SnH_4$$

$$+SiF_2+_n + HF_{aq} \rightarrow Si_nH_{2n+2} \quad (n = 1-6)$$

$$Mg/Si + Mg/Ge + H_{aq}^+ \rightarrow \text{mixed hydrides,} \quad \text{e.g. } H_3Si-GeH_3$$

$$GeH_4 \text{ (pass through a silent electrical discharge)} \rightarrow Ge_nH_{2n+2} \quad (n = 2-9)$$

In this last synthesis, Ge_2H_6, Ge_3H_8, n- and iso-Ge_4H_{10} and n-, iso- and neo-Ge_5H_{12} were isolated and characterized by ^1H NMR; the germanes from $n = 6$ to $n = 9$ were detected using GLC and mass spectrometry.

Plumbane, PbH_4, is thought to be formed in tiny amounts during the acid hydrolysis of magnesium–lead alloys, but nothing is known of its chemistry.

The reactivity of the hydrides towards oxygen falls in the sequence $Si > Ge > Sn$, some of the silanes detonating or inflaming on exposure to air. This may seem a considerable contrast to the oxygen stability of hydrocarbons, but in fact all hydrocarbons are thermodynamically unstable relative to H_2O and CO_2 and are presumably prevented from igniting spontaneously by a kinetic block: the catalytic action of a spark, as in a car engine for example, is all that is required to cause explosive ignition.

TETRAHALIDES OF THE GROUP IV ELEMENTS

All the tetrahalides except lead tetraiodide are known; the strength of the M—X bond decreases with increasing atomic number of M (see Table 29), so that at lead the lead–iodine bond is too weak to support the promotion of lead to lead(IV); even lead tetrachloride and tetrabromide are very unstable thermally. Although carbon–halogen bond strengths are high, the steric crowding round the small carbon atom in carbon tetraiodide makes the molecule decompose rather readily to tetraiodoethylene either on heating or in the presence of ultraviolet light; in this way the I—C—I bond angle is increased, so reducing the iodine–iodine steric interaction:

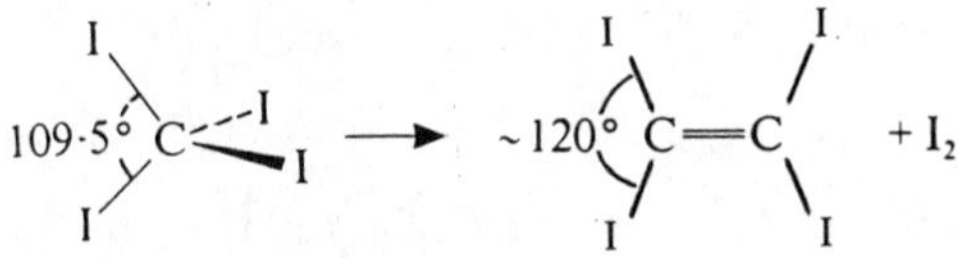

The tetrahalides are, except for the tetrafluorides of tin and lead, very volatile compounds, suggesting that the molecules are covalently bonded and have only weak van der Waals forces operating *between* the molecules in the solid and liquid phases. Those of the larger silicon, germanium, tin and lead are able to increase their coordination numbers to five or six by complex formation: e.g.

$$GeF_4 + N(CH_3)_3 \longrightarrow$$

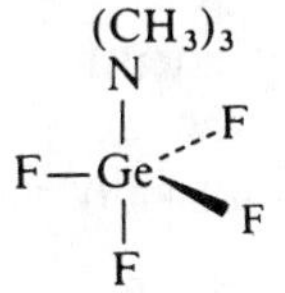

trigonal bipyramidal coordination round Ge atom

$$SiF_4 + 2L \longrightarrow \qquad or$$

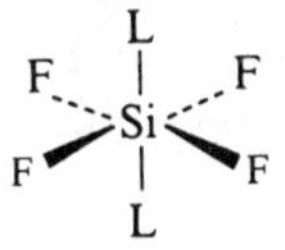

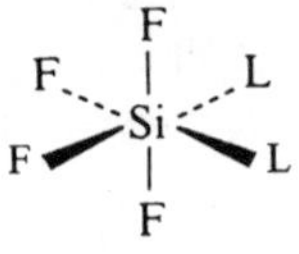

trans
SiF_4(pyridine)

cis
SiF_4(bipy)

$$MX_4 + 2X^- \rightarrow MX_6^{2-} \text{ (octahedral)} \quad SiF_6^{2-}, GeF_6^{2-}, GeCl_6^{2-}, SnI_6^{2-}, PbCl_6^{2-}$$

$$SnCl_4 + 2Me_2SO \xrightarrow{20\,°C} cis\text{-}(Me_2SO)_2SnCl_4 \xrightarrow[20\ h]{135\,°C} trans\text{-}(Me_2SO)_2SnCl_4$$

$$(Me_2SO = \text{dimethyl sulphoxide})$$

Unlike the tetrahalides of carbon, those of the other elements hydrolyse on exposure to water, although it is possible to isolate a few hydrates of the tin(IV) halides. The pentahydrate has only two water molecules directly coordinated to the metal and should be written $cis\text{-}(H_2O)_2SnCl_4.3H_2O$; it readily undergoes partial hydrolysis to give a hydroxo-bridged dimer:

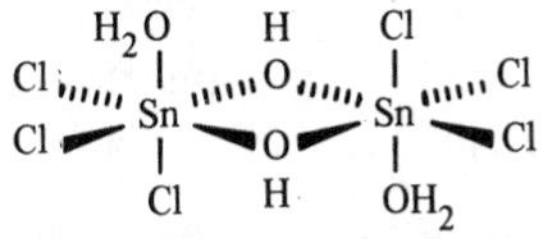

Other hydrates, such as [$cis\text{-}(H_2O)_2SnBr_4$.dioxane], [$cis\text{-}(H_2O)_2SnCl_4$.18-crown-6] and [$trans\text{-}(H_2O)_2SnCl_4$.15-crown-5], are also known in which the organic ether molecules are simply hydrogen-bonded to the coordinated waters and do not interact with the tin.

A wide variety of tetravalent halides exists in which some of the halogen atoms have been substituted either by hydrogen or, more usually, by organic groupings, some examples being H_3SiCl, H_3GeCl, $HGeCl_3$, $(CH_3)_2PbCl_2$, $(CH_3)_2SnCl_2$ and $C_6H_5GeCl_3$:

$$GeCl_2 + HCl(g) \xrightarrow{40\,°C} HGeCl_3 \quad \text{(the Ge analogue of chloroform)}$$

$$Sn(CH_3)_4 + SnCl_4 \xrightarrow[1:1]{heat} 2(CH_3)_2SnCl_2$$

CATENATED HALIDES

Carbon, silicon and germanium are capable of extensive catenation in their halides. This is especially true of carbon, which forms the industrially important polytetrafluoroethylene ('Fluon', 'Teflon') when tetrafluoroethylene is subjected to pressure:

$$C_2F_4 \rightarrow -CF_2CF_2CF_2CF_2- \quad \text{(chain length measured in hundreds of carbon atoms)}$$

Only carbon of the Group IV elements is capable of forming multiple bonds with itself, so that $X_2M{=}MX_2$ and $XM{\equiv}MX$ *halogen* derivatives are unique to carbon; so, also, are halogen derivatives of aromatic compounds containing delocalized π bonds. A new branch of organic chemistry has grown up in recent years based on highly fluorinated aromatic molecules such as hexafluorobenzene:

hexachlorobenzene   hexafluorobenzene

The van der Waals radius of a fluorine atom bound to carbon is only slightly larger than that of hydrogen and hence virtually all the known aromatic derivatives are capable of having fluorinated analogues; their chemistry is, however, considerably modified owing to the presence of the highly electronegative fluorine atoms.

Polymeric $-SiF_2-$ and $-SiCl_2-$ chains have also been prepared by passing either silicon tetrafluoride or tetrachloride over heated silicon:

$$Si + SiF_4 \rightarrow [SiF_2] \rightarrow -SiF_2SiF_2SiF_2-$$

When the polymeric fluoride is destructively distilled in a vacuum, many polyfluorosilanes are produced, of which SiF_4, Si_2F_6, Si_3F_8 and Si_4F_{10} have been isolated and species up to $Si_{16}F_{34}$ could be detected using mass spectrometry. Methods used to synthesize catenated chlorides and bromides include the following:

$$Cl_2 + \text{calcium–silicon mixture} \rightarrow SiCl_4(65\%) + Si_2Cl_6(30\%) + Si_3Cl_8(4\%)$$
$$+ Si_4\text{–}Si_6 \text{ species } (1\%)$$

$$HSiCl_3 \xrightarrow{photolysis} SiCl_4 + Si_2Cl_6 + Si_3Cl_8 + Si_4Cl_{10} + neo\text{-}Si_5Cl_{12}$$

$$GeBr_2 + GeBr_4 \xrightarrow[\text{toluene}]{\text{reflux in}} Ge_2Br_6$$

$$cyclo\text{-}Si_5(C_6H_5)_{10} + HBr(\text{anhydrous}) \rightarrow cyclo\text{-}Si_5Br_{10}$$

$$(SiF_2) + Cl_2 \rightarrow Cl_2FSi\text{—}SiCl_3 + F_3Si\text{—}SiCl_3 + ClF_2Si\text{—}SiF_2\text{—}SiCl_3$$

$$(SiF_2) + Br_2 \rightarrow Si_2Br_6 + Si_2F_5Br + F_3Si\text{—}SiBr_2\text{—}SiF_3 + F_3Si\text{—}SiF_2\text{—}SiBr_3$$

The paucity of known higher halides for germanium, tin and lead may be due, at least in part, to the stability of solid MX_2 species for these elements, which makes available a facile decomposition path not possible for carbon and silicon: e.g.

$$Ge_2Br_6 \xrightarrow{80\,^\circ C} GeBr_4(l) + GeBr_2(s)$$

DIVALENT HALIDES OF Ge, Sn AND Pb

Germanium dihalides may be obtained by reduction of the tetrahalides with elemental germanium on heating:

$$GeX_4 + Ge \rightarrow 2GeX_2$$

The most stable and best characterized is GeF_2 which has a fluorine-bridged chain structure with further weak bridging *between* the chains giving rise to cross-linking. The stereochemistry about the germanium atom is approximately that of a trigonal bipyramid with an equatorial position probably occupied by a lone pair of electrons:

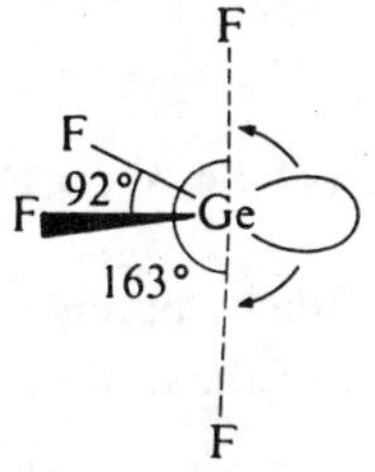

Gaseous hydrogen halides attack tin to give the readily hydrolysable divalent halides. The orthorhombic form of tin difluoride has three near neighbours with F—Sn—F angles close to 90°, and three other more distance fluorines complete a distorted octahedral coordination about the metal; this basic trigonal pyramidal structure is fairly common among Sn(II) compounds, being found for example in $SnCl_2$, $SnCl_2.2H_2O$, $SnCl_3^-$ and SnS:

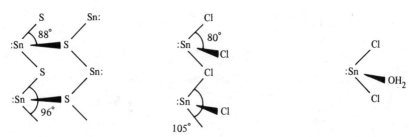

SnS; zigzag chains
cross-linked by Sn—S
interactions; the selenide is
isostructural

SnCl$_2$; chain polymer

SnCl$_2$.2H$_2$O, SnCl$_2$.H$_2$O;
second H$_2$O in the dihydrate is
hydrogen-bonded to the
coordinated water molecule

Lead dihalides are easily made by metathesis using lead nitrate as the source of soluble lead(II). Many complexes are formed by the Group IV dihalides, including the above SnCl$_2$.2H$_2$O, PbBr$_2$(dimethyl sulphoxide) and complex anions such as SnBr$_3^-$, SnBr$_6^{4-}$ and PbCl$_6^{4-}$.

TETRAVALENT ETHANOATES, NITRATES AND SULPHATES OF Sn AND Pb

These water-sensitive derivatives have not been studied very extensively except for lead(IV) tetraethanoate which is widely used as an oxidizing agent in organic chemistry; a few complexes are known, including salts such as Cs$_2$Sn(NO$_3$)$_6$ and K$_2$Pb(SO$_4$)$_3$ as well as adducts like the octahedral Sn(NO$_3$)$_4$(pyridine)$_2$:

$$Pb_3O_4 + CH_3COOH(anhydrous) \rightarrow Pb(OOCCH_3)_4 \quad \text{(readily hydrolyses to PbO$_2$)}$$

$$Pb(OOCCH_3)_4 + \text{concentrated } H_2SO_4 \rightarrow Pb(SO_4)_2$$

$$SnO_2 + \text{hot dilute } H_2SO_4 \rightarrow Sn(SO_4)_2 \quad \text{(hydrolyses to SnO$_2$)}$$

$$SnCl_4 + N_2O_5 \rightarrow Sn(NO_3)_4 \quad \text{(volatile in a vacuum)}$$

$$Cs_2SnCl_6 + N_2O_4 \rightarrow Cs_2Sn(NO_3)_6$$

OXO-ACID SALTS OF DIVALENT Sn AND Pb

Salts of divalent tin and lead are made by purely conventional syntheses, the method chosen depending on the solubility of the desired product:

$$Pb(OOCH_3)_2 \text{ (or } Pb(NO_3)_2) + Na_2CO_3 \xrightarrow[\text{of CO}_2]{\text{in presence}} PbCO_3 \downarrow$$

$$PbCO_3 \text{ (or PbO)} + \text{acid} \rightarrow \text{soluble Pb(II) salt} \quad \text{(ideal for HNO$_3$, CH$_3$COOH,}$$
$$\text{HClO$_4$, HClO$_3$, HBrO$_3$, HBF$_4$)}$$

$$Pb(NO_3)_2 + K_2SO_4 \rightarrow PbSO_4 \downarrow$$

$$Pb(NO_3)_2 + K_2SeO_4 \rightarrow PbSeO_4 \downarrow$$

$$Pb(NO_3)_2 + K_2CrO_4 \rightarrow PbCrO_4 \downarrow$$

$$Pb(NO_3)_2 + 2KIO_3 \rightarrow Pb(IO_3)_2 \downarrow$$

These preparations of lead(II) salts are highly typical of the two metals, but it should be remembered that tin(II) compounds in aqueous solution are oxygen-sensitive and, unless strict precautions are taken to exclude air, some contamination by Sn(IV) will always occur.

Two unusual methods for generating tin(II) solutions take advantage of the relative positions of tin and copper in the electrochemical series:

$$Cu(ClO_4)_2 + Sn/Hg \xrightarrow[\text{atmosphere}]{\text{inert}} Sn(ClO_4)_2 + Cu \downarrow$$

$$CuSO_4 + Sn \xrightarrow[\text{atmosphere}]{\text{inert}} SnSO_4 + Cu \downarrow$$

This synthetic method is obviously capable of extension to other parts of the periodic table.

CARBIDES

Many elements form carbides M_xC_y when the free element is heated with either carbon or a hydrocarbon; in some cases the element may be made *in situ* by the carbon reduction of an oxide or some other suitable compound. The carbides of non-metals such as boron (B_4C) or silicon (SiC) contain covalently bonded lattices and are both hard and chemically inert. Silicon carbide, more commonly known as the abrasive carborundum, has a diamond-like structure in which alternate carbon atoms have been replaced by silicon. Many of the transition metals form 'interstitial' carbides in which carbon atoms occupy holes within a (usually) close-packed metallic lattice; they are high-melting, extremely hard and chemically inert.

The carbides of the metals in Groups I–III are more salt-like and contain essentially C^{4-} and C_2^{2-} anions; carbides containing the former ions (Na_4C, Be_2C, Al_4C_3) liberate methane on hydrolysis, whereas those containing C_2^{2-} ions (MgC_2, CaC_2) form mainly ethyne when treated with water or dilute acids. The crystal structure of calcium carbide is very similar to that of sodium chloride, except that all the $[C \equiv C]^{2-}$ ions are lined up in parallel to each other, resulting in one crystal axis being slightly longer than the other two (see Fig. 42); the carbon–carbon distance of 1.20 Å is that expected for two triply bonded carbon atoms.

Magnesium carbide, MgC_2, loses some carbon on heating to give Mg_2C_3; the fact that this new carbide liberates methylacetylene, $CH_3C \equiv CH$, on hydrolysis suggests the possible presence of C_3^{4-} ions in the crystal.

ORGANO-DERIVATIVES OF SILICON, GERMANIUM, TIN AND LEAD

The tetra-organo-derivatives are prepared in the usual way by treating halides with an excess of Grignard or lithium reagents:

$$SiCl_4 + 4LiR \rightarrow SiR_4$$

$$2PbCl_2 + 4RMgX \rightarrow [2PbR_2] \rightarrow Pb + PbR_4$$

When the tetrahalide is in excess the products are the alkyl or aryl halides, e.g.

$$GeCl_4 + nLiR \rightarrow R_nGeCl_{4-n}$$

although these halides are more easily prepared by heating stoichiometric amounts of MR_4 with the corresponding tetrahalide:

$$GeBr_4 + GeR_4 \rightarrow 2R_2GeBr_2$$

$$GeBr_4 + 3GeR_4 \rightarrow 4R_3GeBr$$

The alkylsilicon chlorides required by the silicone industry are made on a large scale by 'direct synthesis' using silicon and an alkyl chloride:

$$Si + CH_3Cl \xrightarrow{Cu} CH_3SiCl_3 + (CH_3)_2SiCl_2 + (CH_3)_3SiCl$$

The most commercially important of the Group IV organo-derivatives is tetraethyllead, $(C_2H_5)_4Pb$, which is used widely as an 'anti-knock' in petrol. It is made on a large scale by treating ethyl chloride with a sodium–lead alloy:

$$Na/Pb + C_2H_5Cl \rightarrow NaCl + (C_2H_5)_4Pb$$

The products are treated with water to remove excess sodium and the tetraethyllead is purified by steam distillation.

Catenated organo-derivatives can be made via Würtz-type coupling reactions or by the action of alkylating or arylating agents on divalent halides. The products are very much more stable than the corresponding hydrides; for example, $Si_{12}(CH_3)_{26}$ can be distilled in air at temperatures approaching $300\,°C$:

$$Me_2SiCl_2 + Na/K \rightarrow (Me_2Si)_n \quad (n = 5, 6, 7; \text{ non-planar rings})$$

$$Me_3SiCl + Me_2SiCl_2 \xrightarrow{Na/K} Me_3Si(SiMe_2)_nSiMe_3 \quad (n = 1-16)$$

$$GeI_2 + (Me_3Al)_2 \rightarrow Me_3Ge(GeMe_2)_nGeMe_3 \quad (n = 1-22)$$

$$SnCl_2 + C_6H_5MgBr \rightarrow [Sn(C_6H_5)_2]_6 \quad \text{(chair form of ring)}$$

For examples of 'unsaturated' $(R_2M)_2$ species see page 62.

SEMICONDUCTOR PROPERTIES OF THE GROUP IV ELEMENTS

Silicon and germanium are being increasingly used in semiconductor devices, and for this application they have to be obtained in a state of high purity. This is relatively easily accomplished using the technique of zone refining (Fig. 68).

Very pure silicon and germanium are not good conductors of electricity (Fig. 69); for example, the room temperature resistivity of germanium is $4700\ \Omega\ m^{-1}$. With

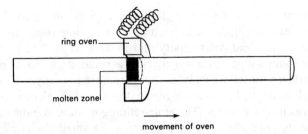

ring oven

molten zone

movement of oven

Fig. 68—Zone refining; a section through the apparatus.

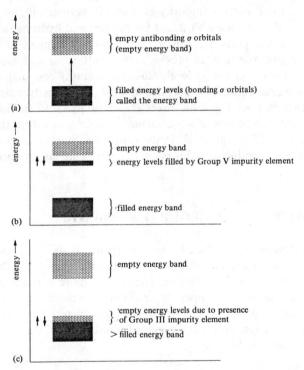

(a)
} empty antibonding σ orbitals
} (empty energy band)

} filled energy levels (bonding σ orbitals)
} called the energy band

(b)
} empty energy band
} energy levels filled by Group V impurity element

} filled energy band

(c)
} empty energy band

} empty energy levels due to presence
} of Group III impurity element
> filled energy band

Fig. 69—(a) A schematic representation of the energy levels in pure germanium. There is a relatively large gap between the completely filled energy levels and the vacant energy levels in the crystal (~ 0.72 eV) and few electrons are excited to the empty levels at room temperature. The lack of these mobile electrons results in a high resistance. (b) Energy levels in germanium containing a Group V element as an impurity. The addition of the Group V element results in more electrons than are necessary to fill the bonding energy levels in the crystal. These extra electrons are accommodated in the normally empty, high-energy levels, resulting in a mobile electron system able to carry the electric current. (c) Energy levels in germanium containing a Group III element as an impurity. A Group III element as an impurity in the germanium crystal results in too few electrons being available to fill all the normal energy levels. This gives mobility to the electrons when under the influence of an electric field. Basically, it appears that these empty levels ('holes') are causing electrical conduction, but in actual fact the holes only confer mobility upon the electrons, which, of course, carry the electric current.

increasing temperature their resistivity decreases rapidly owing to electrons being thermally excited into an empty conduction band; trace impurities also cause a very marked increase in electrical conductivity and as little as $10^{-6}\%$ of arsenic atoms will decrease the room temperature resistivity of germanium to 400 Ω m^{-1}. Thus, to control the electrical properties of silicon and germanium to any useful degree, their initial purity must be controlled to a staggering degree and the element handled under 'sterile' conditions in glove boxes. The zone-refining method is indicated in Fig. 68; a rod of reasonably pure silicon or germanium has a small electrically-heated ring oven placed round it at one end and a narrow strip of the rod is melted; the ring oven then slowly traverses along the rod, moving the molten zone along with it. The impurities (usually) concentrate in the molten phase and are gradually moved towards the far end of the rod when the oven completes its journey. The oven is placed at the beginning of the rod again and the process repeated several times. By using an inert atmosphere the concentration of impurity elements in germanium, for example, can be made as low as 10^{-7} ppm each (except for oxygen and hydrogen).

The addition of a few parts per million of a Group V element to the silicon or germanium lattice gives rise to a 'surplus' of electrons (one for each Group V atom), which are accommodated in energy levels close to an empty conduction band of the host material, giving rise to an n-type semiconductor (n for negative electrons, which are the current carriers); see Fig. 69. Conversely, a few parts per million of a Group III element (e.g. indium) results in a 'deficiency' of electrons within the host lattice and not all the available energy levels in the (normally filled) energy band are full; this gives a p-type semiconductor (p for positive holes acting as the current carriers); see Fig. 69.

OXYGEN COMPOUNDS OF GROUP IV ELEMENTS

The chemistry of compounds formed between oxygen and the Group IV elements is both extensive and of a rather different nature to that involving the derivatives discussed in the first part of the chapter. Hence it is more convenient to describe briefly this aspect of the group's properties separately. Furthermore, strong π bonding between carbon and oxygen so completely divides the chemistry of carbon from that of the other members of Group IV that there is virtually no point at which any direct or useful comparisons can be made.

COMMENTS ON RELEVANT ISOELECTRONIC SYSTEMS

When the relative positions of carbon, nitrogen and oxygen in the periodic table are compared it is immediately apparent that CO and N_2 are isoelectronic. Therefore the bonding in carbon monoxide can be described in terms of a molecular orbital diagram very similar to that shown on page 269 for N_2; thus the fact that there is a triple bond between carbon and oxygen, $C\equiv O$, explains the very short C---O distance (1.128 Å) found in this oxide. Carbon monoxide forms may complexes with the transition metals and it might be anticipated that N_2, as its isoelectronic partner, could behave in a similar manner. Although chemists had made this prediction many years ago, it was only after a prodigious amount of research that metal–N_2 complexes were finally prepared—the first one being made accidentally:

$$[(NH_3)_5Ru^{II}H_2O]^{2+} + N_2 \xrightarrow[\text{aqueous solution}]{\text{bubble through}} [(NH_3)_5Ru^{II}-N\equiv N]^{2+} + H_2O$$

The cyanide ion CN^- is also isoelectronic with carbon monoxide, and its complexes with the transition metals are, of course, very well known.

In the linear carbon dioxide molecule there are two delocalized π bonds at right angles to each other (Fig. 70); the *three* atomic p orbitals which interact in each plane form *three* molecular orbitals, one of which is bonding, one non-bonding and one antibonding. There are sufficient electrons available in carbon dioxide for the degenerate pairs of both bonding *and* non-bonding orbitals to be filled as shown in the energy level diagram of Fig. 70.

The active ion NO_2^+ present in HNO_3–H_2SO_4 nitrating mixtures used in organic chemistry is isoelectronic with carbon dioxide and can be predicted, correctly, to be a linear ion and to have an electronic structure similar to that shown for CO_2 in Fig. 70.

The isoelectronic principle links together some surprisingly diverse compounds and often allows chemists to make quite astonishing correlations about structures. A particularly good example is aluminium orthophosphate, $AlPO_4$, which, being isoelectronic with silica ('Si_2O_4' $\equiv AlPO_4$), is predictably able to adopt *all six* crystal forms of silica shown on page 246. The pseudo-isoelectronic compounds BPO_4, $BAsO_4$ and $AlAsO_4$ also exist in at least one form of the silica structures.

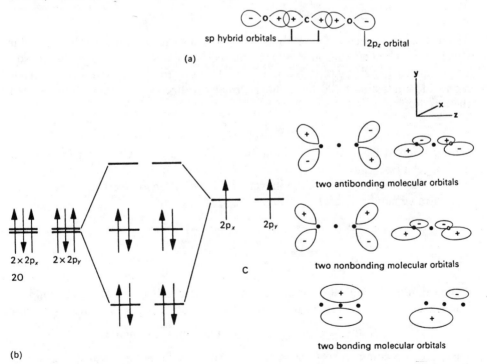

Fig. 70—Bonding in the carbon dioxide molecule: (a) σ bonding interaction; (b) π bonding interaction and π energy level diagram.

OXIDES OF CARBON

Carbon has three main oxides, CO, CO_2 and C_3O_2, all of which contain strong carbon–oxygen π bonds and therefore have no stable counterparts among the other elements in the group. Since these oxides differ so markedly from those of silicon, germanium, tin and lead, they will be discussed separately.

CARBON MONOXIDE

General Chemistry

Carbon monoxide is isoelectronic with N_2 and, not surprisingly, the molecules have several physical properties in common: they are colourless, diamagnetic gases with very low melting points (N_2, $-210\,°C$; CO, $-205\,°C$) and boiling points (N_2, $-196\,°C$; CO, $-190\,°C$). Carbon monoxide, however, is much the more reactive of the two and combines directly with the halogens (except iodine), cyanogen, oxygen, the alkali metals and sulphur: e.g.

$$CO + X_2 \rightarrow O{=}CX_2 \quad (X = F, Cl \text{ or } Br)$$

$$CO + K \text{ (in liquid ammonia solution)} \rightarrow K_2[OC{\equiv}CO] \text{ (potassium ethynediolate)}$$

$$CO + S \rightarrow OCS$$

The iron atom of haemoglobin binds strongly to CO to form carboxyhaemoglobin which can no longer transport oxygen; thus carbon monoxide is a powerful poison, being particularly dangerous because it is odourless. Fortunately, although carbon monoxide binds about 100 times more strongly to haemoglobin than does oxygen, the reaction

$$HbO_2 + CO \rightleftharpoons HbCO + O_2$$

can be reversed by removing the victim from the source of CO and administering pure oxygen (see page 472).

From its laboratory preparation via the dehydration of methanoic acid using concentrated sulphuric acid,

$$HC\overset{\displaystyle O}{\underset{\displaystyle OH}{\diagup\diagdown}} \xrightarrow{-H_2O} CO$$

carbon monoxide could be considered as methanoic anhydride. However, unlike a true acid anhydride it is very insoluble in water and only reacts with alkalis on heating:

$$NaOH + CO \rightarrow NaOOCH \quad \text{(sodium methanoate)}$$

There are two commercial sources of carbon monoxide: *producer gas* (c. 25% CO, 70% N_2, 4% CO_2; made by blowing air through incandescent coke) and *water gas* (c. 40% CO, 50% H_2, 5% CO_2, 5% N_2; made by blowing superheated steam through white hot coke). On a much smaller scale, very pure carbon monoxide can be obtained by decomposing nickel tetracarbonyl:

$$Ni(CO)_4 \rightarrow Ni + 4CO$$

Carbon monoxide as a ligand

The carbon and oxygen atoms in CO are linked by a triple bond: a σ bond plus two π bonds at right angles to each other formed by sideways overlap of p orbitals. The lone pair of electrons remaining on the carbon are in an orbital with considerable p character (in the simplest terms, an sp hybrid orbital), thus making carbon monoxide a potential Lewis base:

$$C \equiv O$$

carbon–oxygen distance 1.12 Å

Surprisingly, however, very few CO complexes of the main group elements are known; the first example to be found was the highly unstable borane-carbonyl produced when diborane reacts with carbon monoxide under pressure:

$$B_2H_6 + 2CO \rightarrow 2[H_3B \leftarrow :CO]$$

Very much more stable complexes are formed between carbon monoxide and the transition metals, a typical example being the colourless liquid nickel tetracarbonyl prepared by direct synthesis from the metal:

$$Ni + 4CO \xrightarrow{50\,°C}$$

(nickel tetracarbonyl)

tetrahedral

More generally, the finely divided transition metal is generated *in situ* under a high pressure of carbon monoxide in a heated autoclave: e.g.

$$CrCl_3 + Na \xrightarrow[\text{200 atm CO}]{\text{tetrahydrofuran}} Cr(CO)_6 \quad (MoCl_5 \rightarrow Mo(CO)_6; WCl_6 \rightarrow W(CO)_6)$$

$$OsO_4 \xrightarrow[\text{50 atm CO}]{\text{100 °C}} Os(CO)_5 \quad \text{(CO acts as reducing agent)}$$

$$Re_2O_7 \xrightarrow[\text{300 atm CO}]{\text{300 °C}} Re_2(CO)_{10} \quad \text{(CO acts as reducing agent)}$$

$$CoCO_3 + H_2 + CO \xrightarrow[\text{250 atm}]{\text{150–200 °C}} Co_2(CO)_8$$

The oxidation state of the transition metal in these simple metal carbonyls is zero, and in all but a very few cases the metal surrounds itself by sufficient carbon monoxide groups (each donating two electrons) to attain the same number of electrons as the next inert gas and hence fill up all the available low-energy orbitals; this is called the effective atomic number (EAN) rule: e.g.

$$Ni^0 \quad 1s^2 2s^2 2p^6 3s^2 3p^6 3d^8 4s^2$$

Nickel(0) requires eight electrons to gain the same number of electrons as the next inert gas, krypton; these can be supplied by four carbon monoxide ligands as in $Ni(CO)_4$. A rhenium(0) atom (seven outer electrons) requires eleven further electrons to obey the EAN rule. This can be achieved if two 17-electron $Re(CO)_5$ fragments form a dimeric molecule containing a metal-to-metal bond:

$$2[\cdot Re(CO)_5] \quad \rightarrow \quad (OC)_5 Re\!-\!Re(CO)_5$$

(one unpaired electron   (electrons shared to form a covalent
on each Re atom)   Re—Re bond)

Similar metal bonds occur in $Mn_2(CO)_{10}$, $Tc_2(CO)_{10}$ and $Co_2(CO)_8$ (when the latter is in solution; the solid has a different structure). In other metal carbonyls the carbon monoxide is also able to form 'bridges' between metal atoms; in such cases the carbon atom is better regarded as being sp^2-hybridized, two of the sp^2 orbitals being used in the formation of M—C σ bonds. *Each* metal atom thus receives *one* electron from the bridging carbonyl group:

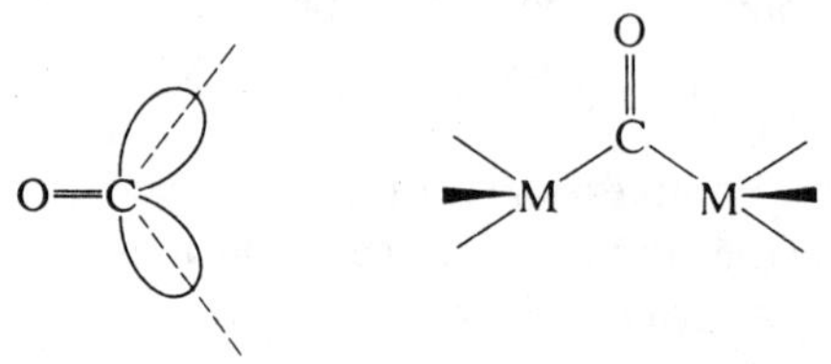

Di-iron enneacarbonyl, $Fe_2(CO)_9$, is the simplest metal carbonyl which shows all three types of interaction described above: it has three terminal C≡O groups, three bridging $>$C=O groups and an iron–iron bond. The presence of this latter bond is assumed from the diamagnetism of the complex; without it each iron atom would possess one impaired electron:

$$Fe + CO \xrightarrow[\text{100 atm}]{\text{200 °C}} Fe(CO)_5 \quad \text{(volatile yellow-brown liquid)}$$

$$Fe(CO)_5 \xrightarrow[\text{25°C}]{\text{sunlight}} Fe_2(CO)_9$$

di-iron
enneacarbonyl

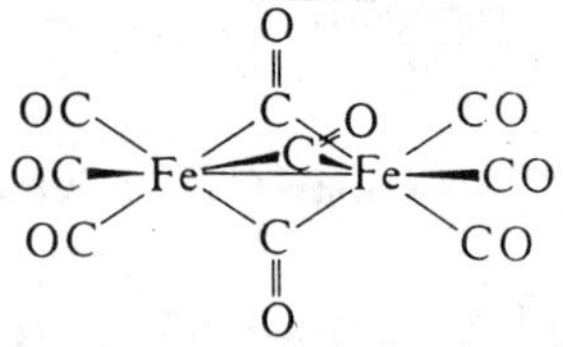

Each iron atom in $Fe_2(CO)_9$ requires ten electrons to gain the effective atomic number of krypton and this is achieved as follows: six electrons from the three terminal CO molecules, one electron from each of the three carbonyl bridges and a final electron from the other iron via the Fe—Fe bond.

Many carbonyls are now known which contain complex clusters of metal atoms held together either by metal–metal bonds or a mixture of metal bonds and doubly (sometimes, even triply) bridging CO groups. The more common cluster derivatives

Table 30—A survey of the more common transition metal carbonyls

Electrons required by EAN rule:	12	11	10	9	8
3d metals	$Cr(CO)_6$	$Mn_2(CO)_{10}$	$Fe(CO)_5$ $Fe_2(CO)_9$ $Fe_3(CO)_{12}$	$Co_2(CO)_8$ $Co_4(CO)_{12}$ $Co_6(CO)_{16}$	$Ni(CO)_4$
4d metals	$Mo(CO)_6$	$Tc_2(CO)_{10}$	$Ru(CO)_5$ $Ru_3(CO)_{12}$	$Rh_4(CO)_{12}$ $Rh_6(CO)_{16}$	—
5d metals	$W(CO)_6$	$Re_2(CO)_{10}$	$Os(CO)_5$ $Os_3(CO)_{12}$	$Ir_4(CO)_{12}$ $Ir_6(CO)_{16}$	—

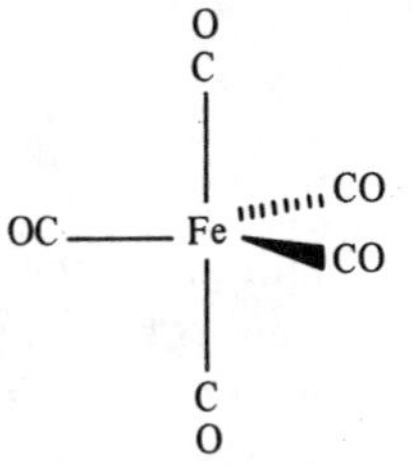

trigonal bipyramidal

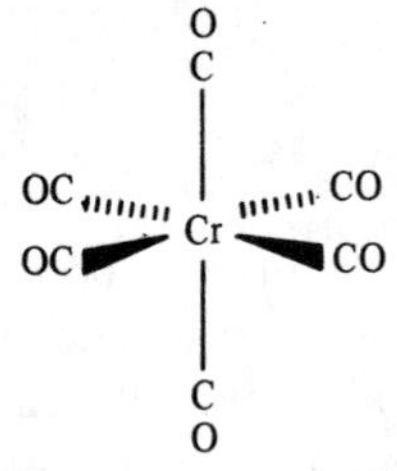

octahedral

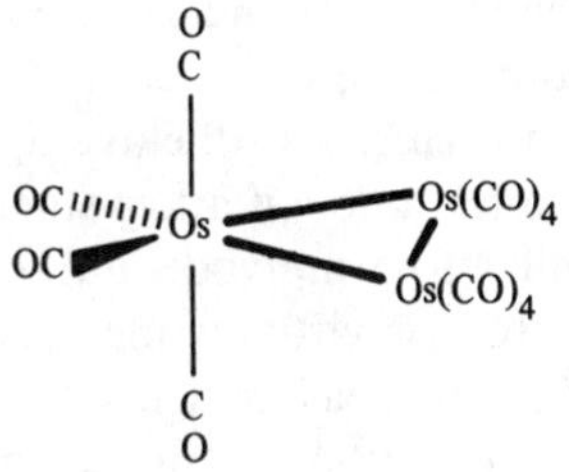

$Os_3(CO)_{12}$
$[Ru_3(CO)_{12}]$

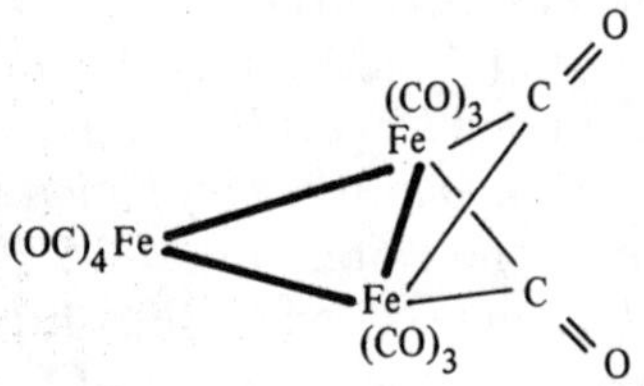

$Fe_3(CO)_{12}$

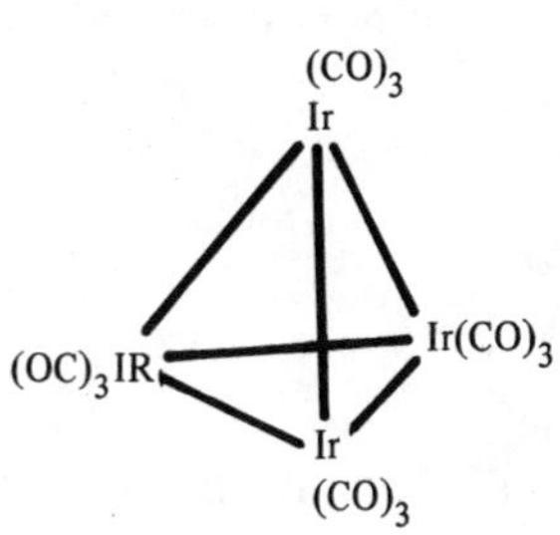

$Ir_4(CO)_{12}$
$[Rh_4(CO)_{12}]$

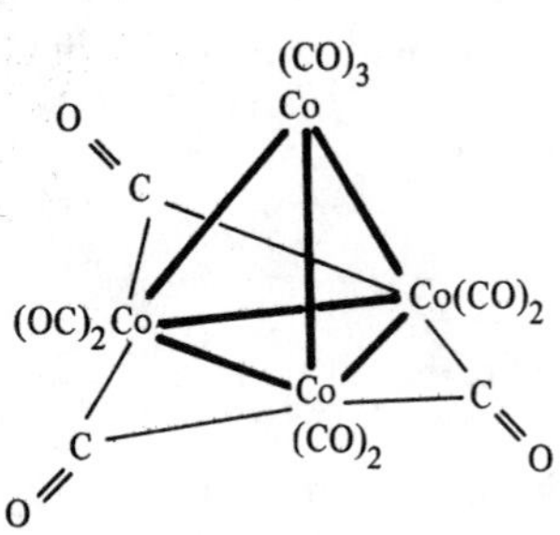

$Co_4(CO)_{12}$

contain metal atom triangles $[Fe_3(CO)_{12};\ Ru_3(CO)_{12};\ Os_3(CO)_{12}]$, tetrahedra $[Co_4(CO)_{12};\ Rh_4(CO)_{12}]$ and octahedra $[Co_6(CO)_{16};\ Rh_6(CO)_{16}]$ (see Table 30).

Bonding in transition metal carbonyls

The high thermal stabilities of the transition metal carbonyl compounds compared with those formed by elements like boron are probably due to a secondary bonding effect in which filled metal d orbitals interact with the empty antibonding π orbitals on the carbon monoxide ligand:

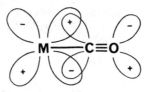

The σ electron donation from the carbon monoxide results in a build-up of negative charge on the metal which has been shown by theoretical calculation to be a highly undesirable condition for a metallic element; the 'back-bonding' from metal to carbon monoxide tends to relieve this charge build-up.

This bonding scheme implies some multiple character in the metal–carbon interaction and should therefore lead to a shortening of the M—CO bond. By choosing a molecule such as $C_5H_5Mo(C_2H_5)(CO)_3$, a direct comparison of an Mo—C_2H_5 single-bond length (2.38 Å) can be made, in the same molecule, with the average of three Mo—CO bond lengths (1.97 Å). (Although we do not know if a hypothetical Mo—CO bond *without a π contribution* would be similar to a molybdenum–ethyl bond, it has been shown that the B—C distance in H_3BCO is virtually identical to that in $B(CH_3)_3$ and its derivatives.) From these and other measurements using a variety of metals it is obvious that M—CO bonds are particularly 'short', in agreement with the suggested scheme of back-bonding; the instability of main group carbonlys suggests that carbon monoxide is a rather poor σ donor and that the existence of the vast range of transition metal derivatives may be due entirely to the secondary π bonding.

CARBON DIOXIDE

This linear molecule is prepared on a huge scale industrially, mainly via the thermal decomposition of limestone heated by producer gas; the carbon dioxide, amounting to about 30% of the effluent gases, is removed by absorption with aqueous potassium carbonate, from which it is liberated by warming. Exhaled human breath contains about 4% of CO_2.

Solid carbon dioxide does not melt under atmospheric pressure but sublimes directly to the vapour state at $-78\,°C$, and because of this it makes a very useful commercial refrigerant: 'dry ice'. It is the most thermodynamically stable oxide of carbon at $25\,°C$, but in the presence of carbon an equilibrium is set up:

$$CO_2 + C \rightleftharpoons 2CO$$

The generation of an extra mole of gas on the right-hand side of this equation gives a large, positive entropy change to the reaction accompanying the formation of carbon monoxide (ΔS at 25°C is $+42.9$ entropy units), thus ensuring that the formation of carbon monoxide is progressively favoured by increasing temperature. The reaction can be a nuisance in certain cases: the use of carbon dioxide gas as a heat exchanger in nuclear reactors constructed of graphite is hampered by the take-up of carbon in hot zones and the deposition of carbon in the colder pipelines, when the carbon monoxide disproportionates back to carbon and carbon dioxide.

Although carbon dioxide is the anhydride of carbonic acid, its aqueous solution contains few bicarbonate or carbonate ions, the bulk being present simply as solvated molecules (the ratio $CO_2(aq):H_2CO_3$ is about 600 at 25°C):

$$CO_2(g) \rightleftharpoons CO_2(aq) \rightleftharpoons H_2CO_3 \rightleftharpoons H^+(aq) + HCO_3^- \rightleftharpoons H^+(aq) + CO_3^{2-}$$

Hydration of CO_2 to carbonic acid is a surprisingly slow process, and nature has had to develop the catalyst carbonic anhydrase to speed up the reaction during animal respiration (see page 462).

The *true* first ionization constant for carbonic acid, $K_a = [H^+][HCO_3^-]/[H_2CO_3]$, is 2.5×10^{-4} at 25°C. However, the *apparent* constant $K = [H^+][HCO_3^-]/[CO_2 + H_2CO_3]$ is only 4.5×10^{-7} at 25°C owing to presence of large amounts of $CO_2(aq)$ in the solution; the pH of a saturated solution of carbon dioxide at 1 atm is 3.7.

CARBONATES

The carbonate ion CO_3^{2-} is planar, the carbon orbitals involved in the σ bonding being sp^2-hybridized. The remaining p orbital, perpendicular to the trigonal plane, interacts with similar oxygen orbitals to give a delocalized π system extending over the whole ion, resulting in C—O bond lengths (1.29 Å) which are considerably shorter than that of a single bond (1.43 Å):

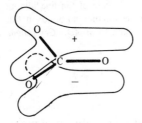

Two very general methods are available for the preparation of metal carbonates: (1) the reaction of carbon dioxide with a soluble base (e.g. Na, Ca) or (2) precipitation via addition of a soluble carbonate or bicarbonate to an aqueous solution of a metal salt. Since all carbonates, except those of the ammonium ion and the alkali metals, are insoluble in water, the second method has the wider applicability. However, its

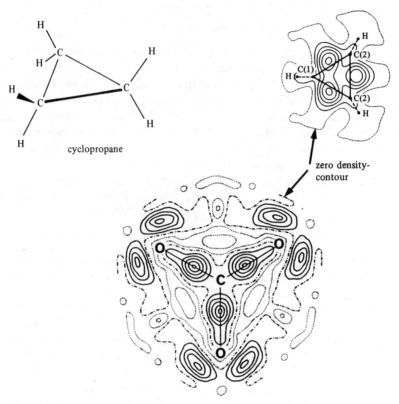

cyclopropane

zero density-contour

Fig. 71—Electron density distributions in *cyclo*propane and the carbonate ion. By using accurate X-ray crystal diffraction data it is possible to calculate the electron density in a molecule. When the calculated electron density of the free *atoms* is subtracted from the experimental density for the *molecule*, a 'difference map' is obtained which emphasizes the build-up of electron density, as a contour diagram, in the bonding and lone pair regions (shown by solid lines). The map of the carbonate ion shows that as expected from valence theory, there is a concentration of electron density *between* the bonded atoms; lone pairs on the oxygen atoms are also visible. (The dotted lines indicate negative contours of reduced electron density.) Cyclopropane has CCC angles of only 60° and as a consequence the carbon orbitals cannot overlap along the C—C directions. The electron density in the carbon–carbon bonds of cyclopropane is seen to occur to one side of the line joining each pair of carbon atoms, in agreement with the theoretical suggestion of 'bent bonds' in this molecule. [The electron density plots in these two species were taken from *Acta Crystallographica* vol. B44 (1988) pp. 289 and 362.]

true scope is somewhat limited by instability of many carbonates, and basic salts or even hydroxides sometimes result; this is particularly true for metals with tripositive cations such as Al.

The high symmetry of the carbonate ion is not always maintained in crystals owing to distortions arising from lattice forces, including hydrogen-bonding effects. Transition metal complexes can be made which contain either monodentate or bidentate carbonate groups:

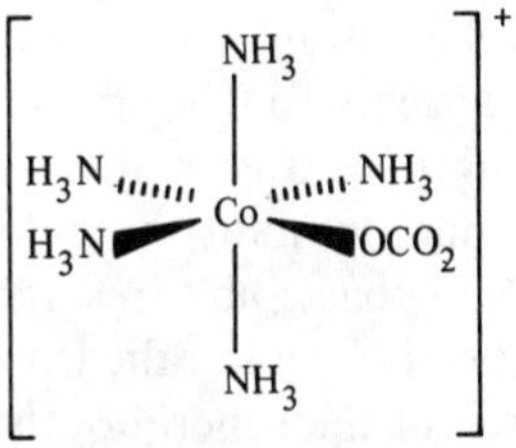

$$[Co^{III}(NH_3)_5CO_3]^+ \qquad\qquad [Co^{III}(NH_3)_4CO_3]^+$$

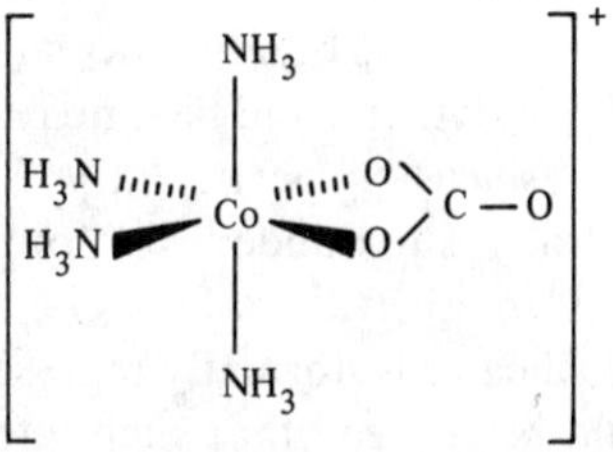

Hydrogen carbonates (bicarbonates) are formed when CO_2 is added to alkali or alkaline earth carbonates:

$$Na_2CO_3 + CO_2 + H_2O \xrightarrow[\text{solution}]{\text{aqueous}} 2NaHCO_3 \quad \text{(soluble)}$$

$$CaCO_3(s) + CO_2 + H_2O \xrightarrow[\text{suspension}]{\text{aqueous}} Ca(HCO_3)_2 \quad \text{(soluble; stable only in solution)}$$

In solid alkali metal hydrogen carbonates the planar HCO_3^- ions are linked together into polymeric chains by hydrogen bonding involving the 'acid' hydrogen atoms:

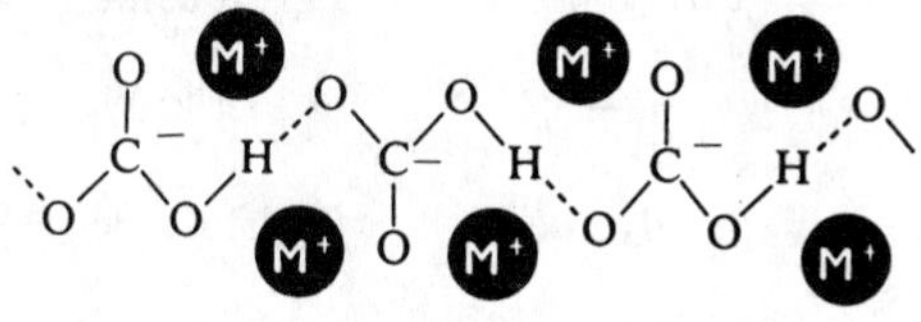

The weakness of carbonic acid results in the occurrence of appreciable hydrolysis in solutions of alkali carbonates, the hydroxyl ion concentration being about 0.01M in 0.5M sodium carbonate.

CARBON SUBOXIDE, C_3O_2

This linear oxide, boiling point $6\,^\circ C$, is prepared by dehydrating malonic acid with phosphorus pentoxide at $300\,^\circ C$. In its reactions with water, ammonia and hydrogen chloride it behaves as the true anhydrous of malonic acid by giving the acid, the amide and the chloride respectively:

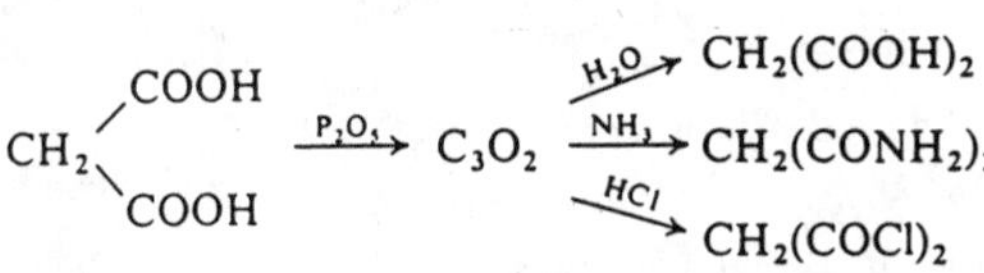

OXIDES OF SILICON, GERMANIUM, TIN AND LEAD

Since strong p_π–p_π bonding does not occur between these elements and oxygen, their oxides are polymeric and bear no resemblance to those of carbon. The crystal chemistry of SiO_2 (silica) is particularly complex, but in all the various forms silicon is tetrahedrally surrounded by four oxygen atoms, and infinite, three-dimensional networks are formed via Si—O—Si bridging. The SiO_4 basic unit in silicon dioxide and the silicates is normally represented diagrammatically by a tetrahedron (Fig. 72). It should be realized that such tetrahedra only represent the general outline defined by the van der Waals spheres of the four oxygen atoms, and where two tetrahedra join at an apex there is almost always a *bent* Si $\overset{O}{\diagup\diagdown}$ Si bridge.

To fully appreciate the various polymer structures the reader is urged to cut out small equilateral triangles (to simulate the projected outline of a tetrahedron in plan) and assemble them loosely on a flat surface or overhead projector; two or three cardboard tetrahedra also help one to understand the steric implications of the two-dimensional models and diagrams.

The stable form of SiO_2 under normal conditions is α-quartz, the other crystal modifications having the following transition temperatures:

$$\alpha\text{-quartz} \underset{}{\overset{870\,°C}{\rightleftharpoons}} \alpha\text{-tridymite} \underset{}{\overset{1470\,°C}{\rightleftharpoons}} \alpha\text{-cristobalite} \underset{}{\overset{1710\,°C}{\rightleftharpoons}} \text{liquid}$$

$$\Big\updownarrow 573\,°C \qquad\qquad \Big\updownarrow 120\text{–}160\,°C \qquad\qquad \Big\updownarrow 200\text{–}275\,°C$$

$$\beta\text{-quartz} \qquad\qquad \beta\text{-tridymite} \qquad\qquad \beta\text{-cristobalite}$$

The structures of quartz, tridymite and cristobalite differ in the way in which the SiO_4 tetrahedra are linked together; in all three forms each oxygen atom is shared between two tetrahedra (i.e. the crystals have 4:2 coordination) so that the overall composition is SiO_2. The interconversion of these three crystalline phases requires the rupture of many strong silicon–oxygen bonds and the reformation of others, so that the transitions occur only very slowly. Hence α-tridymite and α-cristobalite are

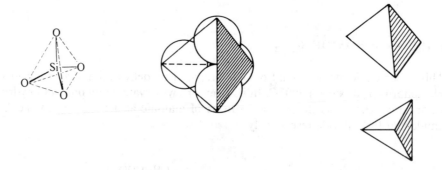

Fig. 72—The tetrahedral SiO_4 group found in silicates and silica.

found in nature as metastable crystals. The difference between the α and β forms of quartz, tridymite and cristobalite is slight and only involves the rotation of the SiO_4 tetrahedra relative to each other; this is reflected in the relative rapidity of the α–β transitions.

Quartz and cristobalite are true polymorphs of silica, but the formation of tridymite needs nucleation catalysts and hence this form of silica is always 'impure' owing to the presence of foreign ions. (For a discussion of the problem see *American Mineralogy* vol. 52 (1967) p. 1233.)

The crystallization of molten silica is very difficult to achieve, and even on slow cooling a vitreous, non-crystalline solid is normally formed. This amorphous silica softens at about 1500 °C, when it becomes plastic and can be worked and blown in either an oxy-hydrogen or oxy-acetylene flame to give transparent apparatus which is highly insensitive to thermal shock owing to the low coefficient of thermal expansion of silica. Quite complex pieces of blown-ware may be heated to redness and plunged into cold water without harm. Silica is also transparent in the ultraviolet region of the spectrum, which makes it useful in the construction of sample cells, prisms and lenses for use with ultraviolet light.

The dioxides of germanium, tin and lead normally adopt the rutile structure (6:3 coordination), although germanium dioxide transforms to the α-quartz structure above 1033 °C and also readily forms an amorphous glass like silica.

Divalent oxides of tin and lead, MO, are well known and adopt an unusual layer structure in which each metal is bonded to four oxygens arranged in a square to one side of it; the lone pair of electrons present in the M^{II} state is stereochemically responsible for this metal coordination:

The oxide Pb_3O_4 contains lead in two oxidation states, $Pb_2^{II}Pb^{IV}O_4$. The lead(IV) atoms are at the centres of PbO_6 octahedra which are fused into chains by sharing two opposite edges; these chains are cross-linked by the lead(II) atoms which are pyramidally coordinated to three oxygens. Again the lone pair of electrons on the lead(II) atom appear to be influencing the stereochemistry of the metal.

SILICATES

The dioxides of silicon, germanium, tin and lead have acidic properties and give salts with a variety of metallic ions. The stannates and plumbates, which contain $M(OH)_6^{2-}$ octahedral anions, have been comparatively little studied, whereas the silicates occur widely, mainly owing to both the preponderance of silicon and oxygen in nature (together they account for more than 74% of the Earth's crust) and the strength of the silicon–oxygen bond. No silicon analogue of the carbonate ion is known because silicon and oxygen are unable to form strong p_π–p_π bonds with each other; all the silicon–oxygen bonds are single and hence the simplest possible silicate ion is the tetrahedral orthosilicate SiO_4^{4-}, derived from the hypothetical orthosilicic acid $Si(OH)_4$:

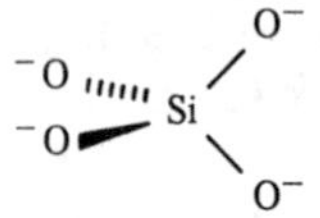

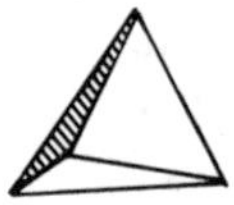

$$SiO_4^{4-}$$

In all the known silicates each silicon is tetrahedrally attached to four oxygen atoms. The wide variety of structures adopted by the silicates arises solely from the different ways in which the SiO_4 tetrahedra are linked together by sharing one, two or three oxygen atoms. Electrical neutrality in the silicates is maintained by the presence of a suitable number of cations, which may be monovalent, divalent or trivalent or mixtures of differently charged ions.

Silicates with one oxygen atom shared

When just two SiO_4 tetrahedra share a single oxygen atom they create a pyrosilicate anion, $Si_2O_7^{6-}$. In this and other polysilicate ions each terminal oxygen atom contributes *one unit* to the overall negative charge whilst the bridge oxygens bestow none:

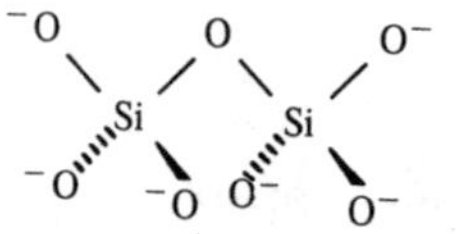

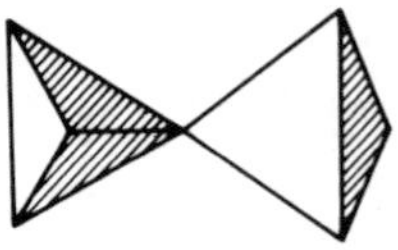

As with the orthosilicates, a diversity of salts is possible since many different cations may, in theory, be used to neutralize the anionic charge; a further variation noted in pyrosilicates is that the Si—O—Si angle can range from about $130°$ to $180°$.

Silicates with two oxygen atoms shared

Cyclic or infinite linear polymers result when each SiO_4 unit shares two oxygens with its immediate neighbours; these anions, which are still sometimes known by their old name 'metasilicates', obviously must have the general formula $(SiO_3^{2-})_n$. Some of the possible structures are shown below in two, or three, different representations; the linear polymers can have their tetrahedra arranged in a surprising variety of relative orientations, the *pyroxene* chain (b) being by far the most common.

Cyclic silicates, $(SiO_3^{2-})_n$

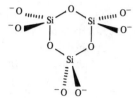

$Si_3O_9^{6-}$

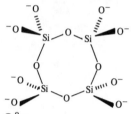

$Si_4O_{12}^{8-}$

$Si_6O_{18}^{12-}$

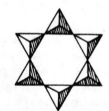

$Si_8O_{24}^{16-}$

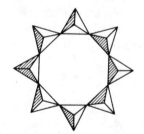

A few linear silicates, $(SiO_3^{2-})_\alpha$

(a)

(known only in a metagermanate)

(b)

very common pyroxene structure

(c)

(d)

Silicates with three oxygen atoms shared

In the simplest type of these silicates two parallel pyroxene chains are linked together to form the double-stranded *amphibole* system. Each alternate SiO_4 tetrahedron shares three oxygens with its neighbours whereas the other half of the tetrahedra share only two. Thus the repeating unit $[(SiO_{2.5}^-)(SiO_3^{2-})(SiO_{2.5}^-)(SiO_3^{2-})]$ gives amphibole silicates the general formula $Si_4O_{11}^{6-}$:

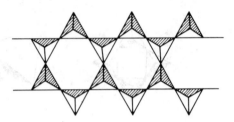

The pyroxene and amphibole chains are held together by the electrostatic attraction of the counter-cations and hence crystal cleavage can be expected to occur along the chain direction since it will be easier to break the non-directional ionic links rather than strong Si—O covalent bonds. Not surprisingly, amphiboles, for example, constitute the *fibrous* asbestos minerals.

When all the tetrahedra in parallel pyroxene chains share three oxygen atoms, two-dimensional, sheet-like polymers arise having the general formula $(SiO_{2.5}^-)_{2n}$ or, in whole numbers, $(Si_2O_5^{2-})_n$. If cations lie between each polymer layer, ready cleavage into the thin sheets characteristic of micas will occur; cations between alternate layers leave only van der Waals forces between the corresponding layers, resulting in soft minerals such as talc:

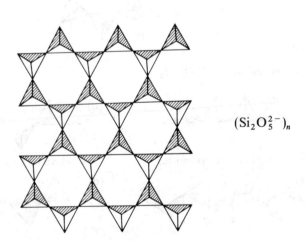

$$(Si_2O_5^{2-})_n$$

Silicates with four oxygen atoms shared

Three-dimensional structures can be built up if half the tetrahedra in a layer structure point up and the other half point down and all then share four oxygen atoms; however, such an arrangement has no negatively charged, terminal oxygens and hence the whole network is neutral. Such structures constitute the various forms of silica since the general formula is now $(SiO_{4/2})_n$ or $(SiO_2)_n$.

Aluminium is able to replace certain silicon atoms to give AlO_4^-·tetrahedra which generate an overall negative charge on the three-dimensional polymers; when a quarter of the silicon atoms are replaced the polymer formula is $(AlSi_3O_{16/2}^-)_n$ or $(AlSi_3O_8^-)_n$, and if half are replaced the formula is $(Al_2Si_2O_8^{2-})_n$. Alumino-silicates of this type make up the important class of minerals called feldspars which alone account for about 60% of the Earth's crust.

Zeolites are commercially important alumino-silicates which possess an open, three-dimensional structure and are able to absorb a number of small gaseous molecules, such as water, ammonia, alcohol and even mercury, into vacant cavities within the network. The cations are held more strongly in the zeolite lattice the higher their charge; that is, $M^{3+} > M^{2+} > M^+$. Hence on passing a solution of calcium ions (e.g. hard water) over a sodium zeolite, an exchange of sodium ions for calcium ions occurs within the zeolite structure (i.e. giving soft water). The calcium zeolite may be changed back to the sodium form by treatment with a concentrated solution of sodium chloride, when the law of mass action for the reaction

$$CaZ + 2Na^+ \rightleftharpoons Na_2Z + Ca^{2+}$$

ensures the exchange of calcium ions for sodium ions. This is the basis of the Permutit water softeners.

Silicates in solution

The structures discussed above are only stable in the solid state; dissolution of the silicates in strong base breaks down the polymers into smaller, more soluble species. The constitution of such solutions is very complex, as highlighted by a recent ^{29}Si NMR study of potassium silicate which showed the presence of at least 22 different anions. (See *Journal of the Chemical Society, Dalton Transactions* (1988) p. 1457.)

REDUCTION OF METAL OXIDES BY CARBON

It has been stated that the whole of the modern metallurgical industry depends entirely upon the existence and thermodynamic stability of carbon monoxide. Most metals are extracted from their oxides, which either occur naturally or can be obtained by roasting an ore containing, say, the sulphide or carbonate. Hence for a reduction process based on carbon (and essentially all elements can be isolated by such an oxide reduction provided that the temperature is sufficiently high) a knowledge is

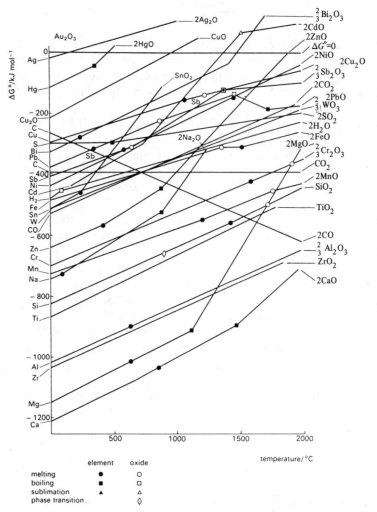

Fig. 73—Plot of $\Gamma G^{\ominus}$ against temperature for a variety of oxides. [Reproduced with permission from *Principles of the Extraction of metals* by D. J. G. Ives (Royal Institute of Chemistry, 1960) p. 21.] The reaction $C + O_2 \rightarrow CO$ involves no change in the number of molecules of gaseous products or reactants, so there is only a small entropy change ($\Delta S^{\ominus} = 2.93$ J K^{-1} mol^{-1}) and hence $\Delta G^{\ominus}$ is essentially independent of temperature as shown by the practically horizontal $\Delta G^{\ominus}/T$ graph. On the other hand, the reaction $2C + O_2 \rightarrow 2CO$ has $\Delta S^{\ominus} = 179.5$ J K^{-1} mol^{-1}, because one mole of gaseous reactant produces two moles of gaseous product; hence the stability of carbon monoxide increases with temperature (downward slope of $\Delta G^{\ominus}/T$ graph).

required of the thermodynamic functions relating to processes such as

$$2M + O_2 \rightarrow 2MO$$

$$2C + O_2 \rightarrow 2CO$$

$$C + O_2 \rightarrow CO_2$$

The approximately linear plots of $\Delta G^{\ominus}$ (the Gibbs free energy change) against temperature for these reactions are perhaps the most useful thermodynamic quantities in predicting possible reduction processes. The slope of such a plot would be equal to $-\Delta S$, the change in entropy for the reaction. This, of course, follows from the relation

$$\Delta G^{\ominus} = \Delta H^{\ominus} - T\Delta S^{\ominus}$$

the slope of a graph of $\Delta G^{\ominus}$ against T being $\partial(\Delta G^{\ominus})/\partial T$, i.e. $-\Delta S$.

All such $\Delta G^{\ominus}/T$ plots for the formation of element oxides (Fig. 73) rise with increasing temperature except that representing the formation of carbon monoxide from carbon, which has a negative slope; this *positive* entropy change is due to the net generation of one mole of gas during the oxidation of one mole of carbon to carbon monoxide. At the temperature where the carbon monoxide plot intersects another oxide plot, the equilibrium constant for the reaction

$$MO + C \rightleftharpoons CO + M$$

is unity and above that temperature the reduction of MO by carbon will be increasingly in thermodynamic and also, owing to the high temperatures involved, kinetic favour of metal formation. A point to note is that the $\Delta G^{\ominus}/T$ plot for carbon dioxide follows the same pattern as the oxides of the other elements, emphasizing that the versatility of carbon as reducing agent does indeed rest upon the stability of carbon monoxide and hence, ultimately, the unique strength of $2p_{\pi}-2p_{\pi}$ bonds.

Chapter 8

Group V: The Pnictides. Nitrogen, Phosphorus, Arsenic, Antimony and Bismuth

INTRODUCTION

The shielding effect of d^{10} and f^{14} inner shells on the ionization energies in this group is not as marked as in Group IIB, the point in the periodic table where the phenomenon first becomes important. However, the trend towards unexpectedly high ionization energies and relatively small covalent radii for the heavier elements is still apparent from Fig. 74; relativistic effects make the fourth and fifth ionization energies of bismuth even higher than those of antimony (Table 31).

No M^{5+} cations can be expected because of the high energies involved, and even the presence of free Sb^{3+} and Bi^{3+} ions in crystalline derivatives is questionable: in the more well-known salts $Sb_2(SO_4)_3$ and $Bi(NO_3)_3 . 5H_2O$ the metals are strongly coordinated to the anions. Highly acidic solutions of bismuth perchlorate possibly contain hydrated Bi^{3+} ions, but with only slight increases in pH the cation $[Bi_6(OH)_{12}]^{6+}$ rapidly becomes the predominant soluble species. This ion consists of a central bismuth octahedron with hydroxo bridges between each pair of metal atoms:

● Bi

• OH

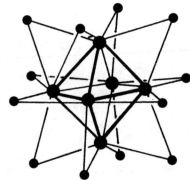

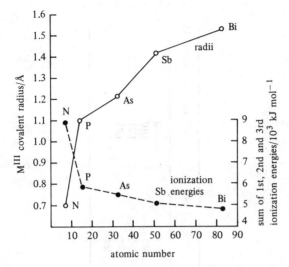

Fig. 74—Variation of M^{III} covalent radii, and sum of the first, second and third ionization energies, with atomic number.

Polycations containing similar clusters of bismuth atoms occur in a variety of crystalline basic salts. For example, the first solid hydrolysis product appearing from bismuth nitrate solutions at a pH of about 1.2 has the formula $[Bi_6O_4(OH)_4]$ $(NO_3)_6.4H_2O$.

Attempts have been made to generate chelated cations of antimony and bismuth by reacting their trichlorides with crown ethers. However, the isolated complexes contain only discrete, pyramidal MCl_3 molecules whose lone pair electrons appear to be accommodated within the central cavity of the coordination ether thus allowing the oxygens to interact with the antimony atom:

It is possible for the Group V elements to achieve a rare gas electron configuration by accepting three electrons to form M^{3-} anions. Again this process is not very energetically favourable and, owing to strong inter-electron repulsions, the formation of N^{3-} requires a huge 2130 kJ mol^{-1}. As a consequence, ionic nitrides are only possible when small cations such as Li^+ and Mg^{2+} are present to give crystals stabilized by high lattice energies. Electrons have more space on phosphorus, which lowers their mutual repulsion and results in the formation of P^{3-} requiring only about 1450 kJ.

Table 31—The Group V elements

Element		1st IE (kJ mol^{-1})	2nd IE (kJ mol^{-1})	3rd IE (kJ mol^{-1})	M$^{(III)}$ covalent radius (Å)
7 Nitrogen, N	$1s^2 2s^2 2p^3$	1402	2857	4578	0.70
15 Phosphorus, P	[Ne]$3s^2 3p^3$	1012	1902	2910	1.10
33 Arsenic, As	[Ar]$3d^{10} 4s^2 4p^3$	947	1798	2735	1.21
51 Antimony, Sb	[Kr]$4d^{10} 5s^2 5p^3$	834	1794	2443	1.41
83 Bismuth, Bi	[Xe]$4f^{14} 5d^{10} 6s^2 6p^3$	703	1610	2467	1.52

In this group most of the compounds are covalent with the elements having an oxidation state of either $+III$ or $+V$. The MR_3 derivatives are pyramidal and have a sterically active lone pair which can be used to form complexes with both main group Lewis acids and transition metals: e.g.

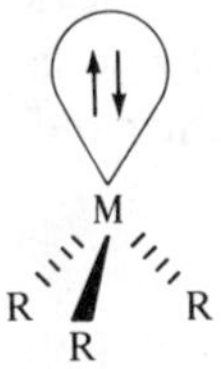

$$F_3B - PMe_3$$

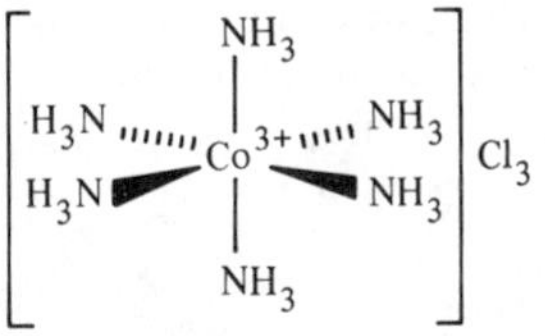

It is possible to resolve a pyramidal $:MRR'R''$ species into its optical isomers if inversion, which leads to racemization, takes place slowly enough:

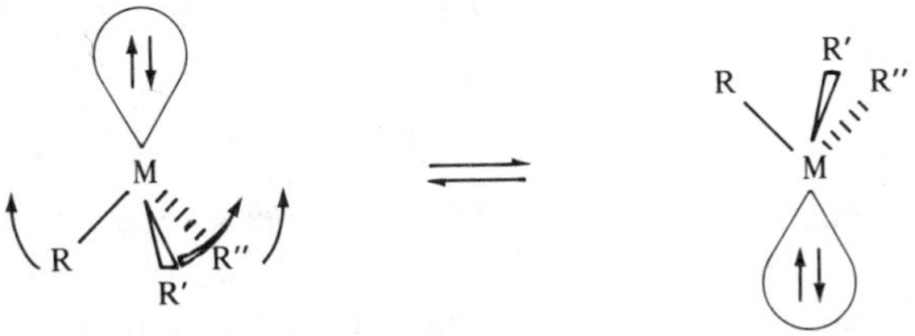

The energy barriers to inversion for $M(CH_3)_3$ derivatives are, in kJ mol^{-1}, 34 (N), 133 (P), 122 (As) and 112 (Sb), hence the process will be particularly rapid at 20°C for ammonia (10^{11} times per second) and the amines. Calculations have suggested that inversion barriers need to be above 100 kJ mol^{-1} if pure optical isomers are to be stable under normal conditions, and hence asymmetric tertiary phosphines and arsines are often readily resolvable. As an example, methylethylphenylarsine, AsMeEtPh, has an inversion energy of 177 kJ mol^{-1} and racemizes slowly at 218°C with a half-life of 740 h; the optical isomers have an $[\alpha]_D^{20}$ of 3.1°.

STEREOCHEMISTRY AND FLUXIONAL BEHAVIOUR OF MR₅ COMPOUNDS

Many of the pentavalent MR_5 molecules of Group V have a trigonal bipyramidal structure in which the axial bonds are slightly longer than the equatorial bonds, typical examples being PF_5 and gaseous PCl_5:

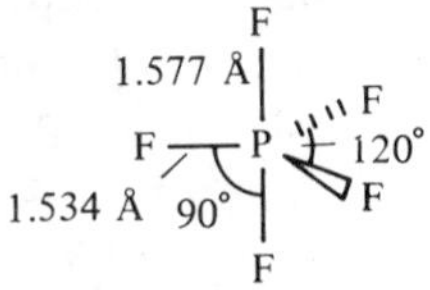

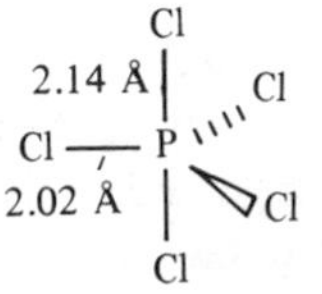

(the three atoms lying in the trigonal plane are known as the equatorial ligands since they lie on the 'equator' of the molecule; the two axial ligands are at right angles to this plane)

Common exceptions are SbF_5 and BiF_5, which have polymeric F—M—F bridged structures, and $Sb(C_6H_5)_5$, which is square pyramidal in the solid; the solid halides PCl_5 and PBr_5 are not molecular but have ionic crystals: $PCl_4^+ PCl_6^-$ and $PBr_4^+ Br^-$.

Although phosphorus pentafluoride has a trigonal bipyramidal structure it shows only one ^{19}F NMR resonance peak, when two might have been expected to arise from the differing environments of the axial and equatorial fluorine atoms. This simplified spectrum does not arise from accidental overlap of resonances but is due to the fluorine atoms undergoing rapid interchange of their positions via the 'pseudo-rotation', or Berry twist, mechanism:

During this process, which occurs about 100 000 times per second at room temperature, two of the F—P—F angles *increase* from 90° to 120° while two others *decrease* from 120° to 90°. The resulting 'new' molecule has an orientation at 90° to that which had been adopted by the original molecule, hence the origin of the name pseudo-rotation ('false rotation') used to describe this mechanism. It will be noted that the *pivot* atom F^1 remains unchanged in position throughout the rearrangement. In PF_5 the fluorines change places too rapidly for the NMR spectrometer to recognize the two different fluorine positions, and only a resonance describing their averaged position is observed.

The pseudo-rotation of PF_5 is essentially an intramolecular 'reaction' and, like all reactions, it has an activation energy barrier (calculated to be 16 kJ mol^{-1}) which determines the rate. Such reaction rates in pseudo-rotations depend markedly on the ligand R, and, for example, the chlorine atoms in a PCl_5 molecule change places only about once every three or four hours.

THE RELATIVE INSTABILITY OF AsCl₅

The non-metals in the row immediately following the first series of transition metals are relatively difficult to oxidize to their highest oxidation state owing to the shielding properties of the underlying $3d^{10}$ shell. (The effect of the $4d^{10}$ shell on the next row of elements is partially offset by the tendency of heavier atoms, in the absence of shielding effects, to be oxidized more easily, as was seen in Groups I and IIA; thus the changing resistance to oxidation is found to be more abrupt when the first d shell is involved.) In Group V this results in arsenates being considerably more oxidizing than phosphates and also leads to difficulties in synthesizing pentahalides. Although

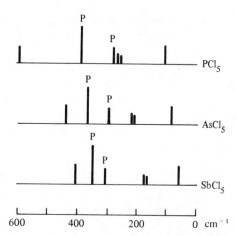

Fig. 75—Comparison of the Raman spectra of Group V pentachlorides (P shows those peaks which are polarized). [Data taken from *Angewandte Chemie* (International Edition in English) (1976) p. 377.]

AsF_5 has been known for a long time, $AsCl_5$ resisted all early attempts to make it; $AsBr_5$ and AsI_5 remain unknown.

Arsenic pentachloride was finally obtained by irradiating an $AsCl_3$–Cl_2 mixture at $-105\,^\circ C$, but it was found to rapidly lose chlorine above $-50\,^\circ C$ and revert back to the trichloride. The compound was characterized by elemental analysis and by comparing its Raman spectrum with those of liquid PCl_5 and $SbCl_5$. As shown in Fig. 75 there is an excellent correlation of peak intensities, peak positions and peak polarization characteristics which leaves no doubts concerning either the identity of $AsCl_5$ or the fact that it has a trigonal bipyramidal structure like the other two pentachlorides: a remarkable achievement for so thermally fragile a molecule.

BiF_5 is the only pentahalide formed by bismuth; in this case relativistic effects and the presence of both $5d^{10}$ and $4f^{14}$ inner shells contribute to the instability of bismuth(V) halides and the highly oxidizing nature of other Bi^V compounds. The existence of AsF_5 and BiF_5 reflects the well-known fact that fluorine is able to oxidize most elements to their highest valence state. This is due to a combination of factors, including the small size of the fluorine atom (which allows high coordination numbers to be achieved), the abnormally low dissociation energy of F_2 and the very high dissociation energies of M—F bonds. The last two factors contribute markedly to the thermodynamic stability of fluorides relative to other halides.

p_π–p_π MULTIPLE BONDS

These elements follow a similar trend to that noted in Group IV: nitrogen at the head of the group forms many multiply bonded compounds which have no stable analogues among the other members. However, there is a rapidly growing number of phosphorus and arsenic derivatives which contain double bonds. Such compounds are thermodynamically unstable with respect to singly bonded decomposition products, but it is often possible to kinetically block pathways which lead to decomposition by using bulky substituents:

More transient species containing triple bonds between carbon and phosphorus have also been prepared recently and are of particular interest in being the analogues of cyanides:

$$PH_3 \text{ (pass through carbon arc)} \rightarrow HC\equiv P$$

$$CF_3PH_2 \text{ (pass vapour over solid KOH at } 20°C) \xrightarrow{-HF} F_2C=PH \xrightarrow{-HF} FC\equiv P$$

The dipole moment of HCP (0.39 D) is much less than that of hydrogen cyanide (3.00 D), mainly owing to the lower electronegativity of phosphorus relative to nitrogen.

Surprisingly, the N—N single bond energy (160 kJ mol^{-1}) is *lower* than that for a P—P bond (200 kJ), which presumably arises from the same factors discussed on page 58 for the labile bond in F$_2$. However, the double- and triple-bond energies for these two elements are 419 (N=N), 945 (N≡N), 310 (P=P) and 490 (P≡P) kJ mol^{-1}. These figures correlate well with the existence of elemental nitrogen as a diatomic molecule, N≡N, while phosphorus forms a series of solid allotropes containing single bonds. Heating white phosphorus, P$_4$, to about 1700°C under vacuum results in partial dissociation into P$_2$ molecules which have a bond length of only 1.89 Å, that of a P—P single bond being 2.20 Å. This suggests the presence of a triple bond in P$_2$ corresponding to that in N$_2$. The dissociation process P$_4 \rightarrow 2P_2$ is presumably entropy-driven by the $T\Delta S$ term at the high temperatures involved, because the P$_2$ molecule is thermodynamically unstable under ambient conditions.

d$_\pi$–p$_\pi$ BONDING

In phosphine oxides, R$_3$PO, the filled oxygen 2p orbitals have the correct symmetry to interact with empty 3d orbitals on phosphorus, which would give the P—O bond some multiple character of the partial triple-bond type. In such a situation the energy of the oxygen 'lone pairs' will appear to become more stable:

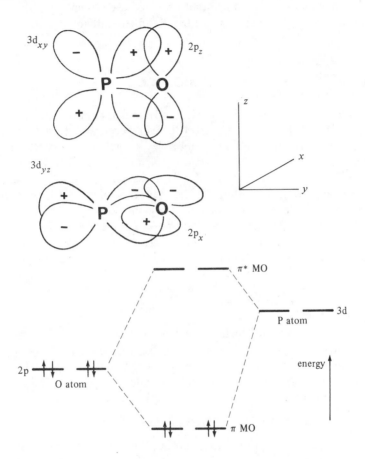

This energy-lowering effect on the lone pairs has been shown by X-ray photoelectron spectroscopy to be in the order OPF$_3$ > OPCl$_3$ > OPMe$_3$; Jolly presumes this order to be due to the increasing size of the phosphorus d orbitals as the electronegativity of the ligands drops. (Large, diffuse 3d orbitals will overlap less effectively with small oxygen 2p orbitals.)

Doubts about the relative importance of such d$_\pi$-p$_\pi$ bonding arose when a very short N—O bond of 1.16 Å (0.26 Å less than a single bond) was found in trifluoramine oxide, ONF$_3$. Nitrogen 3d orbitals cannot reasonably be expected to contribute to the bonding in this molecule. However, theoretical calculations show that the oxygen 2p orbitals holding its lone pairs are able to π-bond with the filled N—F σ orbitals, giving substantial p$_\pi$-p$_\pi$ character to the N—O bond. Such an interaction depends on the ligands being highly electronegative, and accordingly the N—O bond length (1.39 Å) in trimethylamine oxide, ONMe$_3$, is virtually equal to that of a single bond.

From this description of the bonding in ONF$_3$ it would appear unnecessary to always imply the involvement of d orbitals when discussing the reason for abnormally short bonds, and significant p$_\pi$-p$_\pi$ bonding is now considered to be present in OPF$_3$. However, theoretical studies have shown that d orbitals *also* make a substantial contribution to the bonding. For example, in OPF_3 the P—O bond shortens by 0.13 Å (to a value almost identical to that observed experimentally) when d orbitals are included in calculations, and part of this bond strengthening is thought to be of the $d_\pi-p_\pi$ type.

BONDING IN PF_5 AND RELATED MOLECULES

The assumption of sp^3d hybrids in PF_5 and similar pentavalent compounds of Group V is not even approximately correct, although accurate calculations do require the inclusion of d orbital functions in the relevant equations: in PF_5 the total d orbital population is only about 0.6 electrons and this arises mainly via $d_\pi-p_\pi$ back-bonding from filled fluorine 2p orbitals rather than from σ bonding.

The two axial fluorines in PF_5 are held to the phosphorus by three-centre, four-electron bonds as described on page 69, which probably accounts for the axial P—F bonds being slightly longer than the equatorial ones; in the PF_6^- anion, which is isoelectronic with SF_6, all the fluorine atoms are involved in 3c–4e bonds, resulting in a plausible explanation of the ion's octahedral shape (see page 69).

FORMATION OF 'ONIUM' SALTS

Since N^+ is isoelectronic with carbon, the ammonium ion NH_4^+, like methane, has a tetrahedral structure. As described on page 122, virtually all ammonium salts are soluble in water and their chemistry is very similar to that of the heavier alkali metals; of the other Group V elements only phosphorus and arsenic form similar MH_4^+ ions but they do so much less readily than nitrogen. For a reaction such as

$$MH_3 + HX(g) \rightarrow MH_4^+X^- \quad (X = halogen)$$

there are several energy terms which are important when assessing the possible stability of the product salt:

(a) The strength of the H—X bond.
(b) The energy required to promote M to the valence M^+ state.
(c) The strength of the M—H bonds.
(d) The lattice energy of $MH_4^+X^-$.

The balance of these appears to occur at M = As and X = Br or I because it is just possible to prepare $AsH_4^+Br^-$ and $AsH_4^+I^-$ but the strength of the H—Cl and H—F bonds is too high to allow the formation of arsonium chloride or fluoride. For antimony and bismuth the M—H bonds are so weak that no SbH_4^+ or BiH_4^+ salts can be synthesized.

Tetra-substituted onium salts are well known for all the group and, although the substituents are usually alkyl or aryl groups, halogen atoms can also be used:

$$AsF_3 + AsCl_3 + Cl_2 \rightarrow AsCl_4^+ AsF_6^-$$

$$MR_3 + RI \rightarrow MR_4^+ I^- \quad (M = N, P, As \text{ or } Sb)$$

$$PCl_3 + Cl_2 \rightarrow PCl_4^+ PCl_6^-$$

$$NF_3 + F_2 + SbF_3 \xrightarrow[\text{100 atm}]{\text{200 °C}} NF_4^+ SbF_6^-$$

Primary, secondary and tertiary amines are quite basic and will react directly with aqueous acids to give salts containing partially substituted NH_4^+ cations:

$$H_2NR + HCl \rightarrow NH_3R^+ Cl^-$$

$$HNR_2 + HCl \rightarrow NH_2R_2^+ Cl^-$$

$$NR_3 + HCl \rightarrow NHR_3^+ Cl^-$$

No quaternary ammonium salts containing silyl groups result when silyl chloride or iodide react with ammonia; instead, a smooth reaction occurs to give the *planar* molecule trisilylamine:

$$3H_3SiCl + NH_3 \rightarrow N(SiH_3)_3 + [3HCl]$$

COMPLEXES OF GROUP V PENTAHALIDES

The pentahalides behave as strong Lewis acids and many *octahedral* derivatives are known, including some of the very unstable arsenic pentachloride. In the general preparative methods shown below note how the $AsCl_5$ and BiF_5 are prepared *in situ*:

$$AsCl_3 + Cl_2 + NEt_4Cl \rightarrow NEt_4^+ AsCl_6^-$$

$$AsCl_3 + Cl_2 + R_3PO \rightarrow AsCl_5(OPR_3)$$

$$PF_5 + NH_3 \rightarrow H_3NPF_5$$

$$CsF + SbF_5 \rightarrow Cs^+ SbF_6^-$$

$$KCl + SbCl_5 \rightarrow K^+ SbCl_6^-$$

$$BiF_3 + F_2 \xrightarrow[\text{solvent}]{BrF_3} BrF_2^+ BiF_6^- \xrightarrow[BrF_3]{Ag} Ag^+ BiF_6^-$$

For the pentafluorides the Lewis acidity is in the order $SbF_5 > BiF_5 > AsF_5 > PF_5$, which is partially reflected in their structures: PF_5 and AsF_5 are monomeric whereas SbF_5 and BiF_5 are associated via M—F—M bridging. Not surprisingly, liquid antimony pentafluoride has a high viscosity (40 cP at 20 °C) and is rather similar to glycerol in consistency.

COMPLEXES OF MR₃ MOLECULES

Trivalent derivatives of the Group V elements are capable of behaving as either Lewis acids or Lewis bases; as the latter they have acted as workhorses for several generations of chemists and still remain indispensable. Transition metal complexes of ammonia are well known as *ammines*, of which $[Cu(NH_3)_4]SO_4$ and $[Co(NH_3)_6]Cl_3$ are typical examples; ammonia differs markedly from triorgano-phosphines and -arsines in that nitrogen, unlike P and As, has no outer d orbitals which can participate in supplementary back-bonding (see page 62).

One most unusual ligand which forms a wide range of zero-valent transition metal complexes is phosphorus trifluoride. It might be anticipated that the σ donor properties of PF_3 would be rather poor owing to the presence of three highly electronegative fluorines; thus the occurrence of back-bonding must be a very important factor in stabilizing its derivatives. Typical syntheses include the following:

$$OsCl_3 + PF_3 \xrightarrow[\text{autoclave}]{\text{Cu}} Os(PF_3)_5$$

$$ReCl_5 + PF_3 \xrightarrow[\text{autoclave}]{\text{Cu}} (PF_3)_5 Re\text{---}Re(PF_3)_5$$

$$Ni(\text{vapour}) + PF_3 \xrightarrow[-196\,^\circ C]{\text{cool to}} Ni(PF_3)_4$$

$$Fe(CO)_5 + PF_3 \rightarrow Fe(PF_3)_5$$

$$Ni(PCl_3)_4 + SbF_3 \rightarrow Ni(PF_3)_4 + SbCl_3$$

When the Group V trihalides act as Lewis acids towards halide ions, interest usually focuses on the stereochemical influence of the central element's lone pair of electrons. However, care has to be exercised because the stoichiometry of a product does not always reflect the shape of the anion owing to the frequent occurrence of halogen bridging:

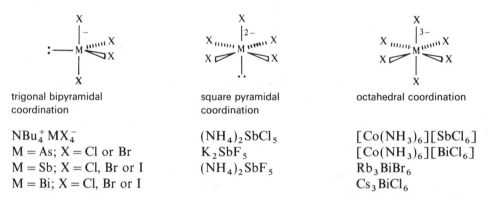

trigonal bipyramidal coordination	square pyramidal coordination	octahedral coordination
$NBu_4^+ MX_4^-$	$(NH_4)_2SbCl_5$	$[Co(NH_3)_6][SbCl_6]$
M = As; X = Cl or Br	K_2SbF_5	$[Co(NH_3)_6][BiCl_6]$
M = Sb; X = Cl, Br or I	$(NH_4)_2SbF_5$	Rb_3BiBr_6
M = Bi; X = Cl, Br or I		Cs_3BiCl_6

When pyridine is added to $SbCl_3$ dissolved in concentrated hydrochloric acid, the pyridinium salt $pyH^+SbCl_4^-$ is obtained. In this not untypical complex there are three different Sb—Cl distances in the distorted octahedral coordination around antimony, *but* all the Cl—Sb—Cl angles are close to 90°. Thus, as the authors commented, there appears to be no reasonable space to assign to the lone pair, which must then be assumed to be in an essentially spherical orbital like those in the $SbCl_6^{3-}$ and BiX_6^{3-} ions.

CATENATION WITHIN GROUP V

Catenation in compounds of the Group V elements is much less extensive than was noted for carbon, silicon and germanium, but layer structures derived from M_6 rings are present in some allotropes of elemental P, As and Sb; page 274. The limit for nitrogen appears to be an eight-atom chain, but often such compounds are *extremely* explosive. Organic substituents tend to increase the stability of polyaza derivatives to the extent that their chemistry may be studied in relative safety:

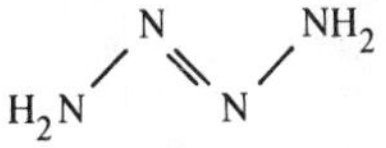

(decomposes below 0°C)

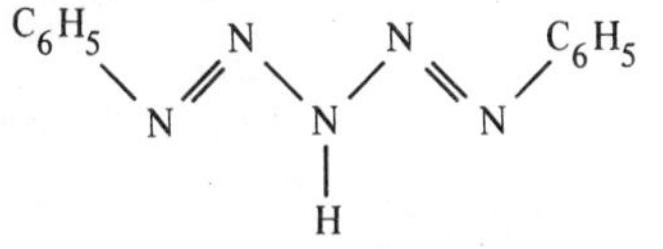

(explodes just above room temperature)

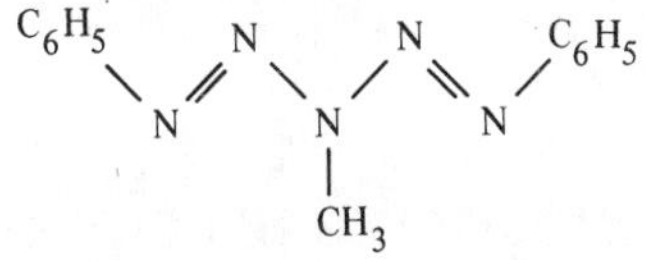

(decomposes at 122°C)

The heavier elements form a few catenated hydrides such as M_2H_4 and M_3H_5 (M = P or As), with larger chains having been detected for phosphorus. When organic groups are present, two-atom chains or non-planar cyclic species are possible:

$$Et_2PCl + Na \rightarrow Et_2P—PEt_2$$

$$RPCl_2 + Mg \rightarrow (PR)_n \quad (n = 4, 5, 6; R = Me \text{ or } C_6H_5)$$

$$CF_3AsCl_2 + Hg \rightarrow (CF_3As)_n \quad (n = 4, 5)$$

$$(CF_3)_2SbI + Hg \rightarrow (CF_3)_2Sb—Sb(CF_3)_2$$

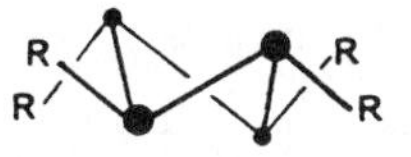

(R = CF$_3$)

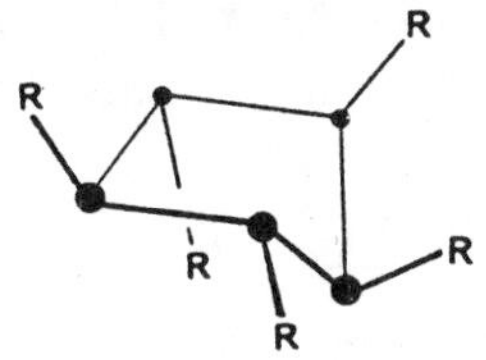

(R = CF$_3$, CH$_3$)

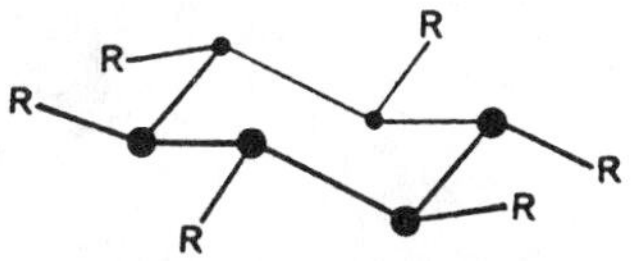

(R = C$_6$H$_5$; equatorial)

Many more-complex ring systems are known, especially for phosphorus. These include the polyatomic ions discussed on page 267, an adamantane-like structure derived wholly from phosphorus atoms and, more recently, transition metal 'sandwich' compounds containing planar P_5 and P_6 rings:

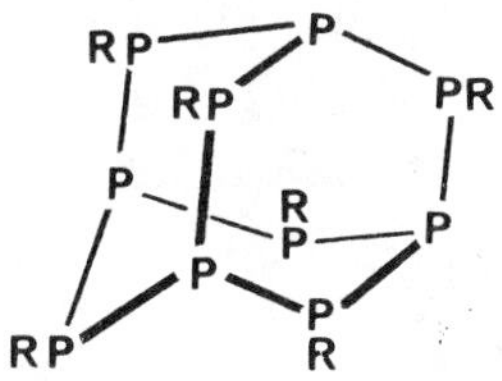

$P_4(P\phi)_6$; $\phi = C_6H_5$
(adamantane structure)

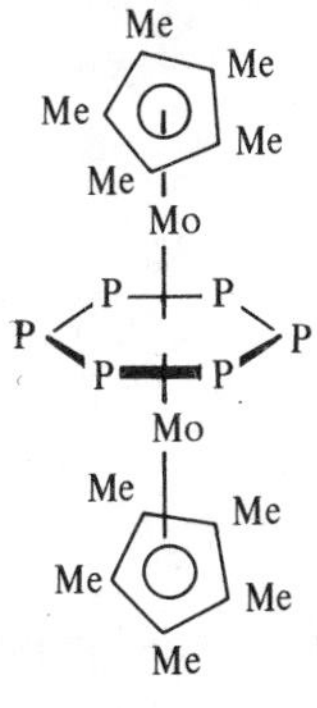

$$P_4 + [C_5Me_5Fe(CO)_2]_2 \xrightarrow{\text{heat}}$$

PHOSPHAZENE POLYMERS

Although nitrogen and phosphorus themselves do not catenate appreciably, together they form an interesting series of polymers called the phosphazenes. The chlorophosphazenes are synthesized by heating phosphorus pentachloride with ammonium chloride and can take the form of both rings, $(Cl_2PN)_n$, and open chains, $Cl_4P(NPCl_2)_nNPCl_3$. The rings usually contain either six or eight atoms, the six-membered rings being planar and the eight-membered ones assuming both 'boat' and 'chair' conformations. In both types of ring all the phosphorus–nitrogen bond lengths are equal (1.56–1.59 Å) and are considerably shorter than the accepted bond length of 1.77 Å for a phosphorus–nitrogen single bond. This suggests that delocalized (i.e. 'aromatic') π bonding occurs between the filled nitrogen 2p orbitals and empty 3d orbitals on the phosphorus atoms:

'boat'   'chair'

linear polymer

The phosphorus–chlorine bonds on the rings are reactive and can be substituted by treatment with suitable reagents. For example, secondary alkylamines give dialkyl-amino-derivatives:

$$\text{Cl}_2\text{P}\diagdown \text{Cl} + (\text{CH}_3)_2\text{NH} \longrightarrow \text{Cl}_2\text{P}\diagdown \text{N(CH}_3)_2 + \text{HCl}$$

and substitution of chlorine by aryl groups takes place on treatment with LiAr or ArMgX. These substitution reactions can, of course, give rise to a variety of isomers dependent both on the position of the substituted phosphorus atoms relative to each other in the ring and on the spatial arrangement of the appended groups. For a six-membered cyclic chlorophosphazene in which two of the chlorine atoms have been substituted by a group R, the three possible isomers are

POLYATOMIC IONS

Although simple cations of Group V are unknown, several polyhedral cations containing clusters of bismuth atoms have been isolated. They are formed either by reduction of $BiCl_3$ with bismuth or by oxidation of elemental bismuth with AsF_5 in liquid sulphur dioxide:

$$4Bi + BiCl_3 + 3AlCl_4 \xrightarrow{\text{fuse}} Bi_5(AlCl_4)_3$$

$$Bi + AsF_5 \xrightarrow{\text{SO}_2(\text{l})} Bi_8^{2+} \longrightarrow Bi_5^{3+} \quad \text{(isolated as hexafluoroarsenates)}$$

$$Bi + BiCl_3 \xrightarrow{\text{heat}} {}^{\text{'}}Bi_{24}Cl_{28}{}^{\text{'}}$$

The compound $Bi_{24}Cl_{28}$ consists of a mixture of Bi_9^{5+}, Bi^+, $BiCl_5^-$ and $Bi_2Cl_8^{2-}$ ions and is of considerable interest in that it contains the first authenticated Bi^+ cation as well as the Bi_9^{5+} cluster:

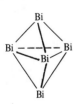

trigonal bipyramid
Bi_5^{3-}

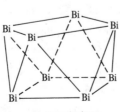

square antiprism
Bi_8^{2+}

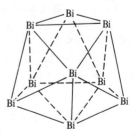

tri-capped trigonal
prism Bi_9^{5+}

Alkali metal 'alloys' of P, As, Sb and Bi react in 1,2-diaminoethane (en) to give a variety of unusual anions which can be isolated in crystalline solids on addition of 2,2,2-crypt ligand (page 116) to the mixture. The crypt forms complexes with the alkali metal cations and so prevents decomposition via transfer of electrons from the anions to M^+ on removal of the solvent. Mixed tetrahedral clusters such as $Pb_2Sb_2^{2-}$ and $Sn_2Bi_2^{2-}$ result when ternary alloys are used: e.g.

$$K_5Bi_4 + crypt \xrightarrow{en} [K(crypt)]_2Bi_4$$

$$KAs_2 + crypt \xrightarrow{en} [K(crypt)]_3As_{11}$$

$$NaSb_3 + crypt \xrightarrow{en} [Na(crypt)]_3Sb_7$$

$$KSnBi + crypt \xrightarrow{en} [K(crypt)]_2[Sn_2Bi_2]$$

square planar

$As_4^{2-}, Sb_4^{2-}, Bi_4^{2-}$ As_7^{3-}, Sb_7^{3-} $P_{11}^{3-}, As_{11}^{3-}$

NITROGEN FIXATION

Reduction of N_2 to ammonia under ambient or near-ambient conditions has been a serious goal of chemists for more than three decades. The realization that nitrogenase, the nitrogen-fixing agent in biological systems, contained molybdenum (or vanadium) and iron suggested that the first step in the reaction might be a metal–N_2 complex and much subsequent research appears to support this idea. Initial coordination of the nitrogen molecule to molybdenum is probably followed by its reduction to an intermediate hydrazido(2-) species $Mo{=}N{-}NH_2$ and finally to NH_3. The strong $\underset{\alpha}{}\ \underset{\beta}{}$ triple bond in N_2 is weakened by progressive build-up of multiple-bond character between Mo and the α-nitrogen atom with concomitant protonation of the β-nitrogen:

$$\sim\!\sim\!\sim Mo{-}N_2 \rightarrow MoNNH \rightarrow Mo{=}NNH_2 \xrightarrow{\hspace{1cm}} \overset{NH_3}{\nearrow} Mo{\equiv}N \rightarrow \rightarrow NH_3$$

It originally took a great deal of effort to find efficient ways of making similar nitrogen complexes in the laboratory, but now there is a surprising array of ligand possibilities realized by N_2, the two most common being demonstrated in the same reaction medium:

$$(Ru(NH_3)_5H_2O)^{2+} \xrightarrow[\text{aqueous solution}]{N_2 \text{ bubbled through}} (Ru(NH_3)_5N_2)^{2+}$$
$$\sim 50\% \text{ yield}$$

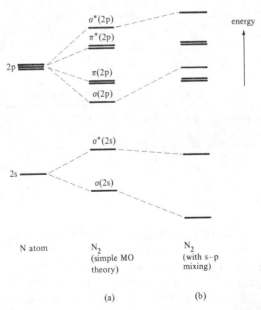

(back-bonding interaction involving a π^* antibonding orbital on N_2)

$$Ru(NH_3)_5H_2O^{2+} + (H_3N)_5RuN_2^+ \rightleftharpoons [(H_3N)_5Ru\text{—}NN\text{—}Ru(NH_3)_5]^{4+}$$

(linear bridge)

THE PHOTOELECTRON SPECTRUM OF N₂

In the photoelectron technique, monochromatic electromagnetic radiation interacts with an atom or molecule to eject an electron whose kinetic energy is measured. On subtracting this kinetic energy from the energy of the incident radiation, the energy of the orbital from which the electron came can be estimated. By scanning a range of kinetic energies, the energy of all the orbitals within the species under study can be obtained. Several separate experiments using a variety of 'light' sources are required in order to cover the wide energy range of orbitals involved, from the strongly bound

Fig. 76—Orbital energy diagram for the N_2 molecule. The electron configuration is $\sigma(2s)^2\sigma^*(2s)^2\pi(2p)^4\sigma(2p)^2$.

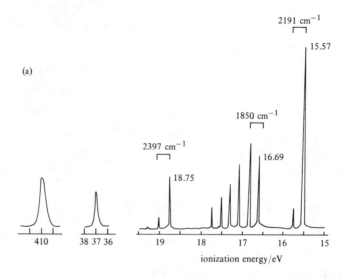

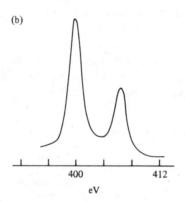

Fig. 77—Photoelectron spectra of N_2 and N_3^-. (a) The photoelectron spectrum of the N_2 molecule. (b) The X-ray photoelectron spectrum of sodium azide. The spectrum shows the energy of the nitrogen 1s orbitals in the linear azide ion $[N—N—N]^-$. Two peaks, with an intensity ratio of 2:1, are observed which is in keeping with the presence of two different nitrogen environments within the ion (400 eV ≡ 38 600 kJ).

inner core to the outer valence shell. Fig. 77 shows the photoelectron spectrum of the N_2 molecule; the ionization energies of the various orbitals are shown on the x axis while the ordinate gives the relative numbers of electrons ejected.

The product which arises on extraction of an electron from N_2 is the molecular ion N_2^+, and some of the absorbed photon's energy may be used to leave the ion in a vibrationally excited state. When this happens a series of peaks, separated by the vibration frequency of N_2^+, will be observed:

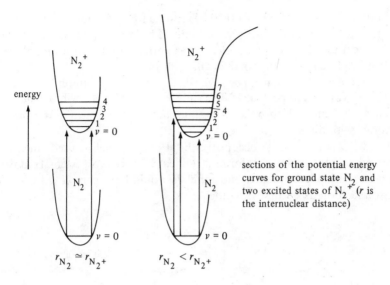

sections of the potential energy curves for ground state N_2 and two excited states of N_2^+ (r is the internuclear distance)

When the bond lengths in N_2 and N_2^+ are almost identical, there will be very little fine structure since few vibrationally excited ions will be produced (owing to the Franck-Condon principle; see above diagram). However, if the bond length in N_2^+ is either shorter or longer than that in N_2, considerable fine structure will be observed. In those N_2^+ ions arising from the ejection of essentially non-bonding electrons there is little fine structure in the spectrum which usually consists of one strong peak and one or two much less intense ones. When a bonding or antibonding electron is removed, the bond length in N_2^+ will differ from that of N_2 and the resulting spectrum will have much vibrational fine structure. Loss of an antibonding electron will strengthen the N—N bond and give a higher vibration frequency than occurs in N_2 (as measured by the peak separation in the fine structure); the opposite occurs when a bonding electron is lost because the bond in N_2^+ is now weaker and vibrates with a lower frequency than N_2.

Several interesting features emerge from the photoelectron spectrum of N_2 shown in Fig. 77. The peak at 410 eV (39 500 kJ) corresponds to the loss of a tightly bound 1s electron; its singlet nature shows that the 1s orbitals are too small to overlap significantly and hence do *not* give two well-separated molecular orbitals, one bonding and one antibonding.

The almost complete lack of fine structure on the lowest-energy peak at 15.57 eV (1502 kJ) shows that the highest occupied orbital in N_2 is essentially non-bonding. This is confirmed by the relatively small change in the vibration frequency of 2191 cm^{-1} compared to N_2 (2345 cm^{-1}). Calculations show that this peak corresponds to ejection of an electron from the $\sigma(2p)$ orbital; *simple* molecular orbital theory puts $\sigma(2p)$ below the degenerate pair of $\pi(2p)$ orbitals, but this neglects the fact that $\sigma(2p)$ and $\sigma(2s)$ have the same symmetry and will mix. Such s–p mixing ('hybridization') lowers the energy of $\sigma(2s)$ to about 37 eV and increases the energy of $\sigma(2p)$ so much that it rises above the $\pi(2p)$ orbitals and becomes almost non-bonding. (This is a general phenomenon: two orbitals of *similar energies* and identical symmetry will always mix such that the lower-energy orbital becomes more stable and the higher orbital is destabilized.)

The group of peaks at 16.69 eV (1610 kJ) results from ejection of an electron from one of the degenerate bonding $\pi(2p)$ orbitals. The product ion will thus be more weakly bound than the nitrogen molecule, as verified by the vibrational frequency change from 2345 cm^{-1} in N_2 to 1850 cm^{-1} in N_2^+.

Finally, loss of an electron from $\sigma*(2s)$, which will be less antibonding after mixing with $\sigma*(2p)$, gives a main peak at 18.75 eV (1809 kJ) with slight vibrational structure of frequency 2397 cm^{-1}. Hence this orbital has actually become almost non-bonding as a result of s–p mixing.

Presumably it is the highest occupied orbital $\sigma(2p)$ which is used when nitrogen acts as ligand to transition metals forming a linear M—N≡N system. Had the $\pi(2p)$ orbitals been the highest occupied most metals would have complexed with N_2 in a sideways-on fashion as occurs with ethene.

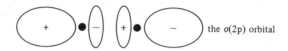

the $\sigma(2p)$ orbital

BONDING IN THE P_4 MOLECULE

Although the ring strain in tetrahedrane, C_4H_4, is thought to be of the order of 500 kJ mol^{-1}, calculations suggest that such strain is almost absent in the isostructural P_4 molecule. To a first approximation the photoelectron spectrum of P_4 shows that, because of the 60° P—P—P angle in the tetrahedron, phosphorus 3p orbitals are virtually non-bonding and the P_4 molecule relies on the omnidirectional character of the 3s orbitals for its existence: their interaction is strongly bonding, giving rise to a high electron density build-up within each triangular face. If s bonding had not been so successful in holding the P_4 tetrahedron together one might have anticipated the existence of a P_8 cube-shaped molecule since p overlap would then be maximized.

It is interesting to note the effects of shielding and penetration on the inner-core orbitals of phosphorus as measured in the photoelectron spectrum of P_4. The 1s orbital, closest to the nucleus, has an energy of 208 000 kJ, with the 2s (18 700 kJ) and 2p (13 500 kJ) separated by about 5000 kJ owing to the penetration properties of the 2s orbital. Shielding by the 1s electrons, the increase in principal quantum number (which places the electrons further from the nucleus) and mutual inter-electron repulsions reduce the energy of the $n = 2$ shell by nearly 200 000 kJ relative to the 1s electrons. In contrast, the valence orbitals have a binding energy of barely 1000 kJ.

OCCURRENCE OF THE GROUP V ELEMENTS

Nitrogen

The gas is the principal constituent of air, constituting about 78.1% by volume. Combined nitrogen occurs as ammonia, nitrites and nitrates (e.g. 'Chile saltpetre', $NaNO_3$); as an essential element to life it occurs in organic matter (e.g. in proteins, which average about 16% nitrogen).

Phosphorus

About 0.1% in the lithosphere, almost exclusively as phosphates; principal ores are phosphorite, $Ca_3(PO_4)_2$, and apatite, $3Ca_3(PO_4)_2.Ca(F,Cl)_2$.

Arsenic

About 5 ppm of the Earth's crust; the minerals realgar, As_2S_2, and orpiment, As_2S_3, were known to the ancients. Principal present source is the flue gases obtained in the extraction of nickel, iron and cobalt.

Antimony

0.5 ppm of the Earth's crust; most important ore is stibnite, Sb_2S_3.

Bismuth

0.1 ppm abundance; normally extracted from the flue gases obtained during the roasting of lead sulphide ores.

EXTRACTION

Nitrogen

Pure liquid nitrogen, boiling point $-196°C$ (1 atm), is obtained by fractional distillation of liquid air; liquid nitrogen is a useful refrigerant. Nitrogen required for ammonia and cyanamide production is made in huge quantities using the producer gas reaction in which air is blown over red-hot coke:

$$C + O_2 + N_2 \rightarrow 2CO + N_2$$

Phosphorus

Calcium phosphate is heated to $1300°C$ with silica and carbon in an electric furnace; the calcium silicate, $CaSiO_3$, formed under these conditions remains behind in the furnace while the phosphorus distils away and may be condensed in suitable receivers as white phosphorus:

$$2Ca_3(PO_4)_2 + 6SiO_2 \rightarrow 6CaSiO_3 + P_4O_{10}$$
$$P_4O_{10} + 10C \rightarrow P_4 + 10CO$$

Arsenic and bismuth

The oxides obtained from flue gas deposits ('flue dusts') are reduced by heating with carbon.

Antimony

The sulphide ore is roasted in air to give the oxide which is reduced by carbon monoxide in a blast furnace.

THE ELEMENTS

Nitrogen is the only member of the group which solidifies (melting point $-210\,°C$) to a molecular lattice composed of diatomic molecules; under normal conditions of temperature and pressure it exists as a rather inert diatomic gas constituting slightly more than 78% by volume of the Earth's atmosphere. The inert nature of nitrogen gas is undoubtedly due to the strong bonding in the N_2 molecule which requires the input of some 941.4 kJ before it can be broken up into atoms and undergo reaction; as a result of this only lithium of all the elements combines directly with nitrogen at room temperature, and then only slowly. By passing nitrogen through an electrical discharge at low pressure it can be dissociated into atoms which are much more reactive and combine directly at room temperature with such elements as mercury, sulphur, iodine, phosphorus and arsenic. The high dissociation energy of the N_2 molecule makes many compounds of nitrogen endothermic and hence potentially unstable; for example, all the nitrogen oxides (except dinitrogen pentoxide, which is just exothermic), cyanogen and nitrogen trichloride are endothermic. This often means that a nitrogen derivative is thermally more unstable than its phosphorus analogue (e.g. NCl_3, $\Delta H_f = +230$ kJ mol^{-1}; PCl_3, $\Delta H_f = -259$ kJ mol^{-1}).

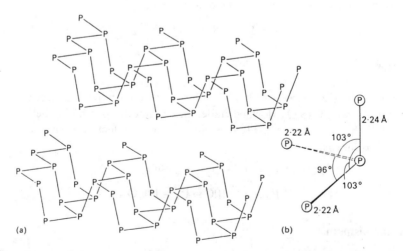

Fig. 78—The structure of black phosphorus. (a) The structure is composed of layers of puckered six-membered rings (only two partial layers are shown). (b) Each phosphorus atom is pyramidally coordinated to its three nearest neighbours.

Table 32—Inter-atomic distances in Group V elements

	Three nearest neighbours at	Three next-nearest neighbours at
Arsenic	2.51 Å	3.15 Å
Antimony	2.87 Å	3.37 Å
Bismuth	3.10 Å	3.47 Å

Phosphorus, arsenic and antimony have several allotropes, the most stable form of all three (and for bismuth the only form) being that of a layer structure made up of puckered six-membered rings in the 'chair' conformation (see Fig. 78 for the structure of black phosphorus which is of this type). The layer forms of arsenic, antimony and bismuth differ slightly from black phosphorus in that a degree of metallic bonding helps to hold the layers together (instead of purely van der Waals forces as in black phosphorus), allowing the solids to conduct both heat and electricity. This metallic interaction increases from arsenic to bismuth as shown by the fact that the three next-nearest neighbours from other layers become structurally significant and relatively more close (Table 32).

Two other important forms of phosphorus are the white and the red modifications. Pyrophoric white phosphorus results whenever phosphorus vapour below about 1000°C is condensed; it consists of a molecular crystal produced by the packing of P_4 tetrahedra. The action of light or heat converts the white form into the red. The latter is considerably less volatile and less soluble in organic solvents than white phosphorus and undoubtedly contains larger units than P_4, but its structure has not yet been determined. The highly unstable yellow forms of arsenic and antimony, prepared in a similar fashion to white phosphorus, appear to have a structure containing M_4 molecules like that of white phosphorus.

The vapours of phosphorus, arsenic and antimony contain M_4 tetrahedra until quite high temperatures are reached; it is only at about 1000°C that the P_4 molecules begin to dissociate slightly into P_2, and even at 1700°C about 50% of the tetrahedra remain intact. At very high temperatures the elements are present, at least partly, as atoms. Bismuth vapour contains only an equilibrium mixture of Bi_2 molecules and bismuth atoms, although Bi_2 molecules persist at high temperatures and can still be detected at 2000°C.

HYDRIDES OF THE GROUP V ELEMENTS

Nitrogen forms four hydrides—hydrazoic acid, N_3H; di-imide, N_2H_2; hydrazine, N_2H_4; ammonia, NH_3—of which the first two contain p_π–p_π multiple bonds and hence have no analogues among the other elements of the group.

MH$_3$ derivatives

The small 1s orbital of hydrogen becomes less compatible with the orbitals of the Group V elements as their atomic number increases. Hence the strength of the M—H bond decreases down the group, making the hydrides more thermally unstable: ammonia only begins to break down appreciably into its elements at temperatures around 2000 °C, whereas bismuthine, BiH$_3$, has a half-life of only a few minutes at room temperature.

Among the more spectacular differences between the trihydrides of this group are the bond angles shown in Table 33, and many attempts have been made to explain the trend. The distances between adjacent hydrogens in NH$_3$ and PH$_3$ molecules when the bond angles are held at 90° are about 1.4 and 2.5 Å respectively; hence a classical explanation for the observed H—N—H angle would be that electrostatic repulsion between similarly charged hydrogen atoms is greater in ammonia than in phosphine. However, this electrostatic force is apparently negligible compared with *Pauli repulsion* between the bonding electron pairs. This repulsion arises from Pauli's principle that electrons of like spin occupy different regions of space; bonding pairs of electrons are constrained to be close to their two neighbouring nuclei and hence cannot avoid each other as readily as lone pairs.

The *dominating influence* on molecular geometry thus arises from Pauli repulsion between bond pairs, which varies exponentially with distance. Relatively long bonds with considerable space between them, as in PH$_3$, will repel each other less than the bonds in ammonia. Hence if it is assumed that the central atom in the hydrides uses pure p orbitals for bonding, the differences in H—M—H bond angles for the group may be rationalized.

Although this is obviously a very simplified description of the bonding, it is nevertheless advantageous to keep as much electron density in the low-lying s orbitals as possible.

Ammonia is produced industrially in huge quantities, the capacity of some of the larger plants being about 1500 tons per day. The process involves the reversible reaction between hydrogen and nitrogen,

$$N_2 + 3H_2 \underset{200-300 \text{ atm}}{\overset{500\,°C}{\rightleftharpoons}} 2NH_3, \quad \Delta H(25\,°C) = -92.4 \text{ kJ mol}^{-1}$$

which is favoured by high pressure (owing to the volume decrease) and low temperature (since ΔS is negative); nitrogen is obtained from fractional distillation of liquid air and hydrogen from either water-gas or the petroleum industry. To increase the reaction rate at the operating temperatures, a catalyst, usually finely divided iron, must be used.

Table 33—Bond angles (°) in MH$_3$ molecules

NH$_3$	PH$_3$	AsH$_3$	SbH$_3$	BiH$_3$
106.6	93.5	91.8	91.3	—

The general method of preparing a covalent hydride, reduction of the corresponding halide with $LiAlH_4$, is applicable for the other trihydrides, although the yields are low for the unstable bismuthine. Acid hydrolysis of alloys containing Group V elements and Na, Mg or Ca is also a useful source of the hydrides, and in the case of phosphine the reaction may be scaled up to the kilogram level. Traces of other hydrides, $M_2H_4(P, As, Sb)$ and $M_3H_5(P, As)$, are obtained as by-products; mixtures of these lower hydrides are formed when PH_3 or AsH_3 is passed through an ozonizer (a silent electric discharge apparatus used for making ozone from O_2).

Hydrazine, N_2H_4

In hydrazine each nitrogen carries a sterically important lone pair of electrons which dictate, in part, the *gauche* conformation which the molecule adopts in the gaseous phase:

angle of twist α from
the *cis* configuration is about 90°

Commercially, hydrazine is manufactured by the Raschig process, in which ammonia is oxidized by sodium hypochlorite in aqueous solution:

$$NH_3(\text{large excess}) + NaOCl \rightarrow NH_2Cl + NaOH$$
$$(\text{chloramine})$$

$$NH_2Cl + 2NH_3 \rightarrow N_2H_4 + NH_4Cl$$

Glue or gelatine is added to the mixture to inhibit the reaction

$$N_2H_4 + 2NH_2Cl \rightarrow N_2 + 2NH_4Cl$$

It is thought that the glue complexes with traces of heavy metal ions present in the solution which catalyse this loss of hydrazine. A dilute ($\sim 2\%$) solution of hydrazine is thus obtained which can be concentrated by distillation to hydrazine hydrate, $N_2H_4.H_2O$; azeotropic distillation with aniline gives anhydrous hydrazine. It is a weaker base than ammonia but does form two series of salts, $N_2H_5^+X^-$ and $N_2H_6^{2+}(X^-)_2$; for example, $X = Cl$ or Br.

M_2H_2 derivatives

A hydride of this formulation requires a double bond between the two M atoms, which is only possible when M is nitrogen. Di-imide, N_2H_2, is the hydrogen analogue of the organic azo-derivatives $RN{=}NR$; it is highly unstable and cannot be isolated at room temperature but is considered to be an intermediate in a few reactions. The infrared spectrum of di-imide, prepared by passing an electrical discharge through hydrazine and freezing out the di-imide at liquid nitrogen temperatures to arrest the decomposition, suggests that the molecule probably has the *cis* configuration

$$
\begin{array}{c}
H \qquad\qquad H \\
\diagdown \qquad\quad \diagup \\
N=N \\
\text{(lone pairs)}
\end{array}
$$

Hydrazoic acid, HN$_3$

Hydrazoic acid is interesting because it is one of the very few molecules in which more than two nitrogen atoms are linked directly together, the three nitrogens being collinear:

$$
\underset{\substack{\diagup \\ H}}{N} \overset{1.34\text{Å} \; 1.34\text{Å}}{-\!N\!-\!-\!N}
$$

$$111°$$

Hydrazoic acid is only slightly dissociated in aqueous solution ($pK_a \sim 5$); electropositive metals dissolve to give azides, but hydrogen is not released because it reduces part of the remaining hydrazoic acid to ammonia and nitrogen:

$$4Li + 6HN_3 \rightarrow 4LiN_3 + 2NH_3 + 2N_2$$

$$Zn + 3HN_3 \rightarrow Zn(N_3)_2 + NH_3 + N_2$$

The azides of Group I and Group II elements are salts and contain the symmetrically linear azide ion N_3^- (see Fig. 77):

$$N \overset{1.15\text{ Å}}{-\!-\!-\!-\!-} N \overset{1.15\text{ Å}}{-\!-\!-\!-\!-} N$$

MH$_5$ derivatives

No binary hydrides have yet been isolated in which the central Group V element is pentacovalent probably due to two major factors: (a) the high dissociation energy of H_2 will markedly destabilize a pentahydride relative to PF_5 and (b) the electronegativity of hydrogen is possibly too low to allow satisfactory 3-centre, 4-electron axial bonding (page 69). It is presumably significant that the known hydrides PHF_4, PH_2F_3 and PH_3F_2 have fluorine atoms in the axial positions:

$$
\begin{array}{ccc}
\begin{array}{c}
F \\
| \\
H\!-\!P\!\cdots\!F \\
| \quad\diagdown F \\
F
\end{array}
&
\begin{array}{c}
F \\
| \\
H\!-\!P\!\cdots\!F \\
| \quad\diagdown H \\
F
\end{array}
&
\begin{array}{c}
F \\
| \\
H\!-\!P\!\cdots\!H \\
| \quad\diagdown H \\
F
\end{array}
\end{array}
$$

Hydroxylamine, HONH$_2$

This molecule is thought to have a structure in which the lone pair electrons on nitrogen and oxygen are *trans* to each other:

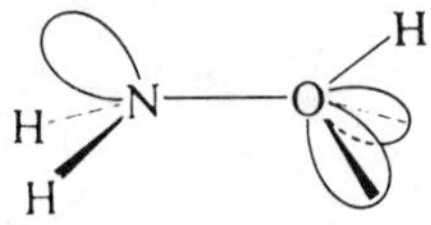

Commercially, hydroxylamine is isolated from its bisulphate made by adding concentrated sulphuric acid to boiling nitromethane:

$$CH_3NO_2 + H_2SO_4 \rightarrow NH_3OH^+HSO_4^- + CO\uparrow$$

The anhydrous compound is unstable and decomposes without melting into water, ammonia, dinitrogen monoxide and nitrogen monoxide; like hydrazine, it is a very weak base.

OXIDES AND OXO-ACIDS OF THE GROUP V ELEMENTS

Oxides of nitrogen

As previously mentioned, in all the oxides of nitrogen there is $p_\pi-p_\pi$ multiple bonding between the nitrogen and oxygen atoms, and consequently the oxides have no phosphorus, arsenic, antimony or bismuth analogues; therefore it is convenient to discuss the oxides of nitrogen separately.

Dinitrogen monoxide, N$_2$O

This colourless and rather inert gas may be prepared by heating ammonium nitrate; it has an asymmetric, linear structure:

$$N \underset{\text{1.13 Å}}{\rule{2cm}{0.4pt}} N \underset{\text{1.19 Å}}{\rule{2cm}{0.4pt}} O$$

Nitrogen monoxide, NO

Nitrogen monoxide is a rather unusual molecule in that it contains one unpaired electron and yet shows no signs of dimerization until it is cooled to very low temperatures. On dimerization, nitrogen monoxide becomes diamagnetic, showing that all the electrons become paired, but the association forces are much weaker than normal single bonds, as shown by the low heat of dimerization (15.5 kJ mol^{-1}). From the energy level diagram shown in Fig. 79 it can be seen that the unpaired electron is in an antibonding orbital (π^*2p). This can be demonstrated rather dramatically by comparing the nitrogen–oxygen bond length in nitrogen monoxide (1.15 Å) with that in the nitrosonium ion NO$^+$ (1.06 Å). To form the nitrosonium ion, an electron

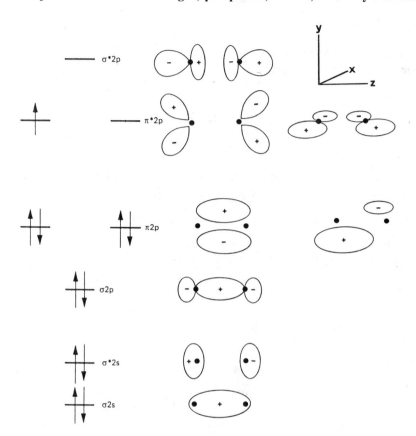

Fig. 79—Energy level diagram for the nitrogen monoxide molecule.

must be removed from the highest-energy orbital in nitrogen monoxide which is the antibonding π^*2p; this will have the effect of binding the nitrogen and oxygen atoms more strongly together, so causing a shortening of the nitrogen–oxygen bond in the nitrosonium ion relative to that in nitrogen monoxide. The infrared stretching frequency (which is some measure of a bond's strength) of the nitrogen–oxygen bond will therefore increase from 1877 cm^{-1} for nitrogen monoxide to about 2320 cm^{-1} for the nitrosonium ion.

Nitrogen monoxide forms a rather large number of compounds ranging from the covalent nitrosyl derivatives through transition metal complexes to nitrosonium salts. Nitrosyl halides (except the iodide, which is unknown) are prepared simply by allowing nitrogen monoxide to react with the respective halogen: e.g.

$$2NO + Cl_2 \xrightarrow{-80°C} 2Cl—NO$$

Their stability decreases with increasing atomic weight of the halogen, so that nitrosyl fluoride is stable at 20 °C whereas nitrosyl chloride, ClNO (0.5%), and nitrosyl bromide, BrNO (7%), are partially dissociated into nitrogen monoxide and halogen under the same conditions. The molecules are bent:

$$X \diagdown \overset{N-O}{\underset{\alpha}{\diagup}}$$

$$X = F: \alpha = 110°; \; X = Cl: \alpha = 116°; \; X = Br: \alpha = 117°$$

Many of the transition metal complexes of nitrogen monoxide can be envisaged as arising from the donation of *three* electrons from nitrogen monoxide to the metal. Since back-donation from metal d orbitals to antibonding orbitals on the nitrogen monoxide molecule apparently contributes towards the stability of these metal nitrosyls, the transition metals are often in low oxidation states, so that the maximum electron density is present in the d orbitals to facilitate such back-bonding. The effective atom number rule (see under metal carbonyls, page 239) is normally found to apply to these complexes, e.g. for iron:

$$Fe_3(CO)_{12} + NO \longrightarrow Fe(NO)_2(CO)_2 \quad \text{(iso-electronic with } Fe(CO)_5)$$
$$\downarrow {\scriptstyle I_2}$$
$$FeI_2 + NO \xrightarrow{100\,°C} Fe(NO)_2I$$

$$Fe(CO)_5 + NO \xrightarrow[\text{pressure}]{50\,°C} Fe(NO)_4$$

tetrahedral   iron–iron bond   ionic

Although the π^*2p orbital in nitrogen monoxide is antibonding relative to the free nitrogen and oxygen atoms, it still requires about 912 kJ mol^{-1} to remove an electron from that orbital. The simplest method of synthesizing compounds containing the nitrosonium ion is to treat a nitrosyl halide with a halide of an element which is capable of forming a complex halo-anion e.g.

$$NOCl + AlCl_3 \rightarrow NO^+AlCl_4^-$$
$$NOF + PF_5 \rightarrow NO^+PF_6^-$$

Many of these nitrosonium salts are isomorphous with the corresponding ammonium salts because the cations are of similar size. The nitrosonium ion is isoelectronic with N_2.

Dinitrogen trioxide, N_2O_3

An equimolar mixture of nitrogen monoxide, NO, and nitrogen dioxide, NO_2, when condensed at low temperatures, gives dinitrogen trioxide, N_2O_3, as a blue solid. Dinitrogen trioxide melts to a blue liquid, but in the gas phase it completely dissociates back into nitrogen monoxide and nitrogen dioxide. Recent spectroscopic data suggest that the molecule has a planar structure with non-equivalent nitrogen atoms and a very long N—N bond:

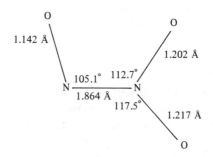

Dinitrogen tetroxide, N_2O_4

On the laboratory scale, the thermal decomposition of lead nitrate is a satisfactory source of dinitrogen tetroxide, N_2O_4; the gas is also formed when certain metals (e.g. copper) dissolve in concentrated nitric acid.

Solid dinitrogen tetroxide is both colourless and diamagnetic; it melts at $-10\,^\circ C$ to a pale yellow liquid and boils at $21\,^\circ C$, the vapour at the boiling point containing about 16% of the dark brown nitrogen dioxide, NO_2. At $150\,^\circ C$ the vapour contains only paramagnetic nitrogen dioxide molecules and is almost black. Dinitrogen tetroxide is a planar molecule and has a rather unexpectedly long nitrogen–nitrogen bond of 1.64 Å (cf. the nitrogen–nitrogen bond length of 1.47 Å in hydrazine).

Nitrogen dioxide, NO_2 is an angular molecule. A satisfactory approximation to the bonding may be obtained by assuming that the nitrogen atom is originally sp^2-hybridized, one sp^2 hybrid containing the unpaired electron and the other two making σ bonds with p orbitals on the oxygen atoms; the remaining nitrogen p orbital combines with similar oxygen orbitals to make a delocalized π-type molecular orbital system extending over the whole molecule:

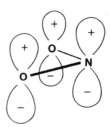

2p atomic orbitals available for π bonding in the nitrogen dioxide molecule

On this idealized model the O—N—O bond angle would, of course, be $120\,^\circ$; because only one electron is in an sp^2 orbital, the mutual repulsion between electrons in the two nitrogen–oxygen bonds will force the O—N—O angle to open up to more than $120\,^\circ$. Treatment of nitrogen dioxide with fluorine or chlorine results in the formation of nitryl halides, XNO_2, the previously unpaired nitrogen dioxide electron now being used in bonding to the halogen atom (nitryl bromide and iodide appear to be unknown):

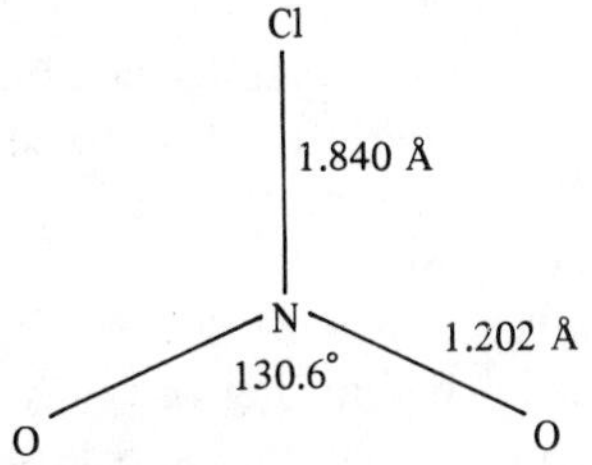

Dinitrogen pentoxide, N_2O_5

In the crystalline state this oxide is a salt, nitronium nitrate ($NO_2^+ NO_3^-$), but the vapour contains *covalent* N_2O_5 molecules; it is the anhydride of nitric acid:

$$N_2O_5 + H_2O \rightarrow 2HNO_3$$

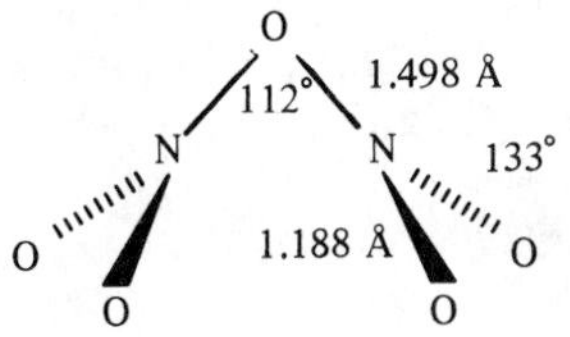

Oxides of phosphorus, arsenic, antimony and bismuth

Trioxides

All four elements react directly with oxygen to form the trioxides, but a limited supply of oxygen must be used with phosphorus to stop the formation of phosphorus pentoxide. In the solid state, phosphorus trioxide and the high-temperature modifications of arsenic trioxide and antimony trioxide contain M_4O_6 molecules, the structures of which are not entirely unrelated to the M_4 tetrahedral molecules found in some allotropes of these elements, the tetrahedra being somewhat expanded and having M—O—M bridges linking the Group V atoms. The trioxides sublime directly as M_4O_6 molecules, although at high temperatures they begin to dissociate into pyramidal M_2O_3 molecules:

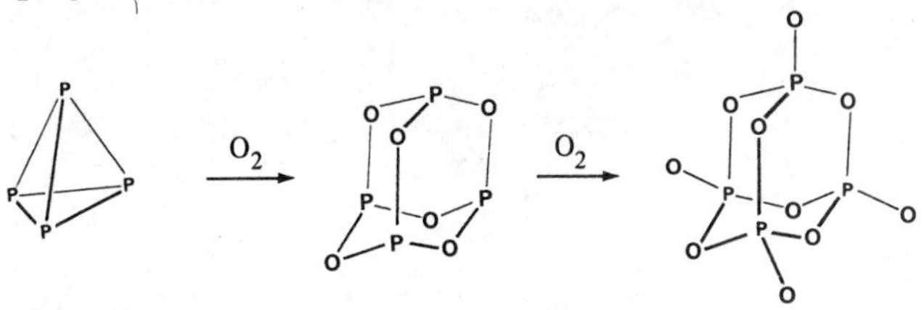

Arsenic and antimony trioxides also exist in low-temperature, polymeric forms in which pyramidal MO_3 units are linked together via oxygen bridges. Unlike the other oxides, Bi_2O_3 is entirely basic and dissolves in acids to give bismuth(III) salts.

Pentoxides

Phosphorus trioxide takes up further oxygen on heating to give the pentoxide, which is present in the vapour state as dimeric P_4O_{10} molecules, the structure of which is shown above. Condensation of the vapour produces phosphorus pentoxide as a white solid containing discrete P_4O_{10} molecules. This particular form of phosphorus pentoxide is actually metastable, and two other modifications have been described in which PO_4 tetrahedra link up by sharing oxygens to give polymeric structures; these polymeric forms are much less volatile and less reactive than the metastable P_4O_{10}. Other, mixed valence oxides are known in which one, two or three P^{III} atoms in the P_4O_6 structure have been oxidized to P^V giving the stoichiometries P_4O_7, P_4O_8 and P_4O_9.

The pentoxides of arsenic, antimony and bismuth are less well defined and have not been studied structurally. They cannot be made by heating the trioxides in oxygen; careful dehydration of arsenic acid, H_3AsO_4, and the so-called 'antimonic acid' is said to give arsenic pentoxide and antimony pentoxide respectively. Bismuth pentoxide is obtained as an ill-defined solid by oxidizing bismuth trioxide with a strong oxidizing agent such as chlorine, potassium chlorate or a persulphate.

The oxo-acids of nitrogen

The bonding in the oxo-acids of nitrogen involves $p_\pi-p_\pi$ nitrogen–oxygen bonding and, since phosphorus, arsenic, antimony and bismuth are unable to form strong $p_\pi-p_\pi$ bonds with oxygen, there can be no direct analogues for these elements. The relationship between the nitrogen oxides and acids is shown below:

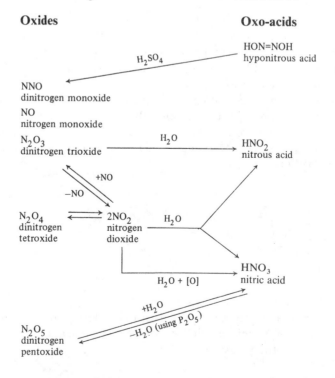

Hyponitrous acid, $H_2N_2O_2$

Although hyponitrous acid is dehydrated by sulphuric acid to give dinitrogen monoxide, this oxide is not the acid anhydride because it dissolves quite readily in water but gives a neutral solution. It is probable that hyponitrous acid has the *trans* configuration rather than *cis* because the hyponitrite ion $N_2O_2^{2-}$ has Raman and infrared spectra entirely consistent with the *trans* structure. Sodium hyponitrite is made by reducing $NaNO_2$ with sodium amalgam in aqueous solution; silver hyponitrite and other insoluble derivatives can be precipitated from solutions of their nitrates:

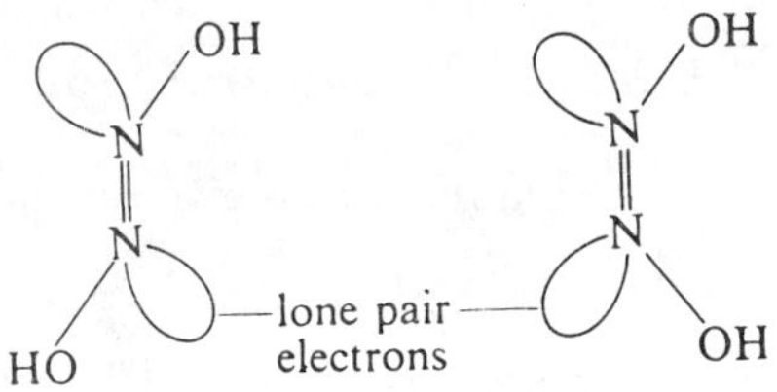

trans   *cis*

Nitrous acid, HNO_2

The free acid is very unstable; dilute aqueous solutions may be prepared by adding a mineral acid to an alkali metal nitrite. Dinitrogen tetroxide is a 'mixed' acid anhydride, producing nitrous and nitric acids when dissolved in water:

$$N_2O_4 + H_2O \rightarrow HNO_2 + HNO_3$$

Organic and metallic derivatives of nitrous acid are known in which the NO_2 residue is bonded either through nitrogen, as $-NO_2$, or through oxygen, $-O-N-O$:

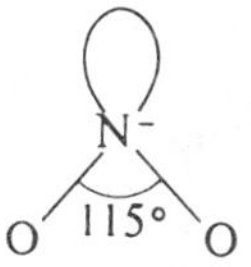

nitrite ion

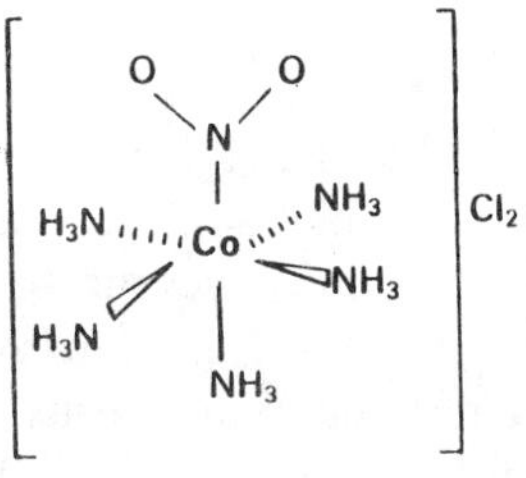

nitropentamminecobalt(III) chloride

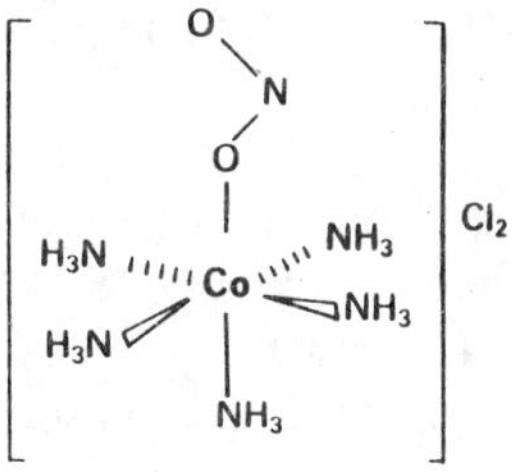

nitritopentamminecobalt(III) chloride

Nitric acid, HNO_3

The anhydride of nitric acid is dinitrogen pentoxide:

$$H_2O + N_2O_5 \rightarrow 2HNO_3$$

The reverse reaction, that of dehydration of nitric acid to give the oxide, may be accomplished using phosphorus pentoxide. On the industrial scale nitric acid is manufactured by the direct oxidation of ammonia using platinum as a catalyst.

The nitrate ion is planar with all the $O-N-O$ bond angles equal to $120°$, which suggests that the nitrogen atom uses sp^2 hybrid orbitals to form σ bonds to each of the three oxygen atoms; this σ interaction is augmented by π bonding involving the remaining nitrogen p orbital and similar orbitals on oxygen:

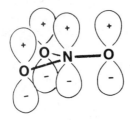   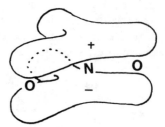

interacting atomic orbitals   *bonding molecular orbital formed*

Nitric acid dissolves in pure sulphuric acid to give the nitronium ion,

$$HNO_3 + 2H_2SO_4 \rightarrow NO_2^+ + H_3O^+ + 2HSO_4^-$$

which is the active agent when this acid mixture is used for nitrating aromatic systems in organic chemistry:

The sulphuric acid is therefore a necessary constituent of the 'nitrating mixture' because it ensures that all the nitric acid is converted into the active species of the slow step in the nitration mechanism (NO_2^+), thus maximizing the rate of nitration.

The oxo-acids of phosphorus

The oxo-acids of nitrogen have no phosphorus analogues, and although metaphosphoric acid, HPO_3, has the same empirical formula as nitric acid, it is a highly polymeric material quite unlike HNO_3. The difference lies in the fact that strong $p_\pi-p_\pi$ bonding does not occur between phosphorus and oxygen. However, in all the phosphorus acids and their salts the terminal phosphorus–oxygen bonds are shorter than the accepted value for a phosphorus–oxygen single bond; from this it is often assumed that the σ bonding between phosphorus and oxygen is augmented by $d_\pi-p_\pi$ bonding between empty 3d orbitals on phosphorus and filled 2p orbitals on oxygen, so giving the phosphorus–oxygen linkages multiple-bond character.

The phosphorus oxo-acids can be divided into two classes: the *phosphorous* acids, in which the formal oxidation state of phosphorus is +1 or +3, and the *phosphoric* acids, which have phosphorus in formal oxidation states of +4 or +5.

Phosphorous acids

In the phosphorous acids the tetrahedrally coordinated phosphorus atoms have either one or two non-acidic hydrogen atoms attached directly to them.

Table 34(a) — Reactions of nitrites and nitrates

Test	Nitrite, NO_2^-	Nitrate, NO_3^-
1 Solubility in water	All nitrites soluble; least soluble is $AgNO_2$, 3.3 gm dm^{-3}	All normal nitrates soluble; basic nitrates of **Bi** and **Hg** insoluble but dissolve in dilute HNO_3
2 Dilute HCl	Release of unstable $3HNO_2 \rightarrow H_2O + HNO_3 + 2NO$ (brown fumes from $NO + O_2 \rightarrow NO_2$)	No reaction
3 Fe^{2+} + dilute acid	Brown coloration due to presence of $FeNO(H_2O)_5^{2+}$	No reaction in presence of dilute acid
4 Fe^{2+} + concentrated H_2SO_4	Brown $FeNO(H_2O)_5^{2+}$ formed	Brown $FeNO(H_2O)_5^{2+}$ formed
5 Acidic $KMnO_4$	Decolorized without gas evolution; nitrate formed by oxidation	No reaction
6 Boil with NH_4^+	Nitrogen released: $NH_4^+ + NO_2^- \rightarrow N_2 + 2H_2O$	No reaction
7 Iodide + dilute acid	I_2 formed together with NO_2 (from $NO + O_2$)	No reaction
8 $AgNO_3$ solution	White precipitate of $AgNO_2$ from concentrated solutions	No reaction
9 $BaCl_2$ solution	No reaction	No reaction
10 Zn + ethanoic acid	No reaction	Nitrite formed
11 Zn or Al plus hot NaOH	Nitrite reduced to ammonia	Nitrate reduced to ammonia
12 Action of heat on anhydrous compounds	All decompose; alkali nitrites most stable (see below:)	All decompose (see below)

Nitrite:

$$3HNO_2 \xrightarrow[\text{soln}]{\text{in}} HNO_3 + 2NO + H_2O$$
$$NH_4NO_2 \rightarrow N_2 + 2H_2O$$
$$2LiNO_2 \rightarrow Li_2O + NO + NO_2$$
$$2NaNO_2 \xrightarrow{>800\,°C} Na_2O + NO + NO_2$$
$$Pb(NO_2)_2 \rightarrow PbO + NO + NO_2$$
$$AgNO_2 \rightarrow Ag + NO_2$$

Nitrate:

$$4HNO_3 \rightarrow 2H_2O + 4NO_2 + O_2$$
$$NH_4NO_3 \rightarrow N_2O + 2H_2O$$
$$4LiNO_3 \rightarrow 2Li_2O + 4NO_2 + O_2$$
$$2NaNO_3 \xrightarrow{500\,°C} 2NaNO_2 + O_2$$
$$2Pb(NO_3)_2 \rightarrow 2PbO + 4NO_2 + O_2$$
$$2AgNO_3 \rightarrow 2Ag + 2NO_2 + O_2$$

Hypophosphorous acid, $PH_2(O)(OH)$

This monobasic acid may be prepared by boiling white phosphorus with barium hydroxide and then adding H_2SO_4 to precipitate barium sulphate:

$$Ba(OH)_2 + P(white) \rightarrow Ba(H_2PO_2)_2 \xrightarrow{H_2SO_4} PH_2(O)(OH) + BaSO_4\downarrow$$

Two of the hydrogen atoms in hypophosphorus acid are directly linked to the central phosphorus, making it and its salts active reducing agents:

Orthophosphorous acid, $PH(O)(OH)_2$

This dibasic acid results from the reaction between water and phosphorus trichloride at $0°C$:

$$PCl_3 + H_2O \rightarrow PH(O)(OH)_2$$

Pyrophosphorous acid, $P_2H_2(O_3)(OH)_2$

This is unknown but its salts can be made by heating monohydrogen orthophosphites, when a condensation reaction occurs with extrusion of a molecule of water:

Phosphoric acids

In the phosphoric acids the phosphorus atoms are tetrahedrally surrounded by four oxygen atoms, and the variations which occur within this class arise via the sharing of no corner, one corner or two corners of the PO_4 tetrahedra. If three corners are shared then the entity which is formed is electrically neutral and has the composition P_2O_5. Thus it is not possible to obtain polymeric phosphates with layer structures similar to those found in some of the polysilicates.

Orthophosphoric acid, P(O)(OH)₃

This is a tribasic acid prepared industrially by treating phosphate rock with sulphuric acid:

$$Ca_3(PO_4)_2 + H_2SO_4 \rightarrow CaSO_4\downarrow + H_3PO_4$$

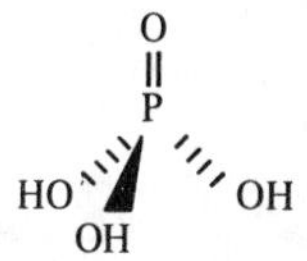

Pyrophosphoric acid, H₄P₂O₇

This is a tetrabasic acid obtained by dehydrating orthophosphoric acid between 213 and 316 °C; pyrophosphates result when monohydrogen othophosphates are heated:

$$H_3PO_4 \xrightarrow{\;-H_2O\;} H_4P_2O_7$$

$$2HPO_4^{2-} \xrightarrow{\;-H_2O\;} P_2O_7^{4-}$$

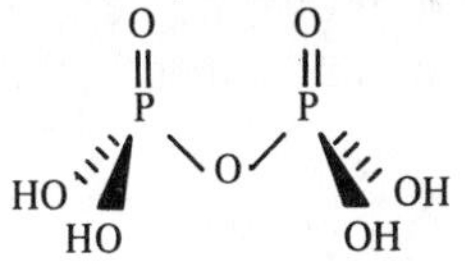

Condensed phosphates

In addition to the pyrophosphates there is a wide variety of polyphosphates in which some or all of the PO_4 tetrahedral units share two corner oxygens to give short chains $(P_nO_{3n+1})^{(n+2)-}$ ($n = 1-16$), infinite chains $(PO_3^-)_\infty$ or cyclic species $(PO_3)_n^{n-}$ ($n = 3-10$, but commonly 3 or 4). The thermal condensations used in the preparation of polyphosphates often give rise to very complex mixtures which have to be carefully separated by chromatography. The pyrolysis of monosodium orthophosphate serves as an example:

$$NaH_2PO_4 \xrightarrow{>160\,°C} \underset{\substack{\text{dihydrogen}\\\text{pyrophosphate}}}{Na_2[H_2P_2O_7]} \xrightarrow{>240\,°C} \text{cyclo-}Na_3P_3O_9 \xrightleftharpoons[300-500\,°C\,(\text{chain})]{625\,°C} (NaPO_3)_n$$

$P_3O_9^{3-}$

$P_4O_{12}^{4-}$

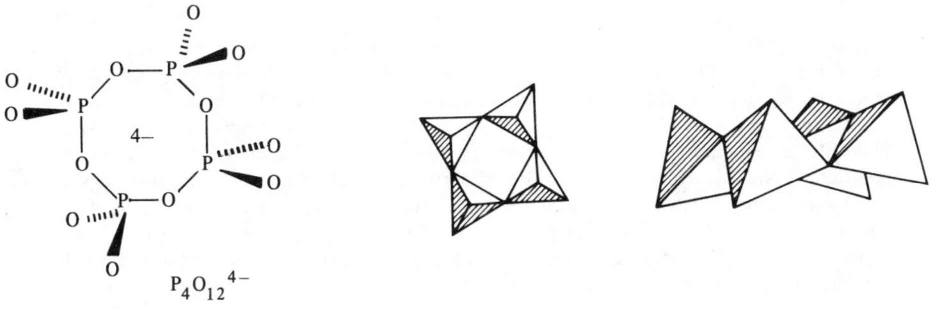

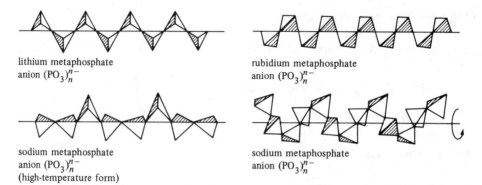

lithium metaphosphate
anion $(PO_3)_n^{n-}$

rubidium metaphosphate
anion $(PO_3)_n^{n-}$

sodium metaphosphate
anion $(PO_3)_n^{n-}$
(high-temperature form)

sodium metaphosphate
anion $(PO_3)_n^{n-}$

Note that each phosphorus atom with n terminal oxygen atoms contributes only $(n-1)$ units of negative charge to a phosphate anion because one oxygen is formally double-bonded and so carries no charge.

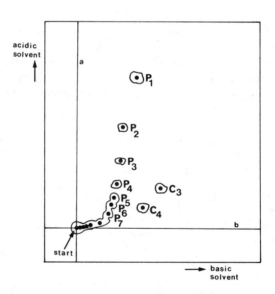

Fig. 80—Separation of linear (P) and cyclic (C) polyphosphates using two-dimensional paper chromatography. The sample mixture is carefully placed as a single drop on a square of chromatography paper at the position arrowed. The chromatogram is then developed along line b by placing the left-hand edge in an aqueous–alcoholic mixture containing ammonia. After about 10 h the paper is removed from the tank and the bottom edge placed in an aqueous–alcoholic mixture made acid with trichloroacetic acid; this causes the phosphates to move towards the top edge. Using two-dimensional chromatography allows the linear and cyclic species to be resolved. P_1 is orthophosphate, P_2 is diphosphate (i.e. pyrophosphate), P_3 is linear triphosphate, etc.; C_3 and C_4 are cyclic tri- and tetraphosphates respectively. [For full details see *Analytical Chemistry* vol. 28 (1956) p. 1093.]

Table 34(b) — Reactions of phosphites and orthophosphates in aqueous solution

The presence of a P—H bond in the phosphite ion gives the latter reducing properties not shared by orthophosphates. Most metal derivatives of the phosphorus oxo-acids, except those of Na, K, Rb and Cs, are insoluble in water.

Reaction	Phosphite ion (using $NaHPO_3$)	Orthophosphate ion (using Na_2HPO_4)
1 Addition of $BaCl_2$	White precipitate, $BaHPO_3$	White precipitate, $BaHPO_4$, in neutral solution; in presence of dilute NH_3, $Ba_3(PO_4)_2$ is precipitated
2 Addition of $AgNO_3$	Initial white precipitate of unstable Ag_2HPO_3; turns black due to metallic Ag	Yellow precipitate, Ag_3PO_4
3 $KMnO_4$ acidified with ethanoic acid	Decolorized on warming as PO_4^{3-} formed	No reaction
4 Addition of $HgCl_2$ solution	White precipitate, Hg_2Cl_2; on warming with excess of HPO_3^{2-}, metallic Hg formed	Orange-yellow precipitate, $HgNaPO_4$
5 Concentrated H_2SO_4	On warming, SO_2 evolved as PO_4^{3-} is formed by oxidation	No reaction
6 Zn + dilute H_2SO_4	PH_3 formed by reduction	No reaction
7 Addition of $CuSO_4$	Light blue precipitate, $CuHPO_3$	Pale blue, $CuNaPO_4$
8 Concentrated NaOH	PO_4^{3-} formed with evolution of H_2	No reaction

Hypophosphites, containing the $H_2PO_2^-$ ion with two P—H bonds, have properties very similar to phosphites.

Oxo-acids of arsenic and antimony

The arsenates are similar to their phosphorus analogues except that the pyro- and meta-arsenic acids are not known as clearly defined compounds; their salts, however, can be made in a similar manner to pyrophosphates and metaphosphates by heating the monohydrogen or dihydrogen ortho-arsenates respectively.

Various hydrated forms of antimony trioxide and pentoxide have been described, but their structures are unknown. In no case is there any structural similarity between the oxo-acid salts of phosphorus and antimony; all the known antimonates contain the octahedral $Sb(OH)_6^-$ ion. Other antimony–oxygen derivatives are known which contain SbO_6 octahedra but no hydrogen: e.g.

$$MSb(OH)_6 \xrightarrow[-H_2O]{heat} MSbO_3 \quad (M = alkali\ metal)$$

However, these are best considered purely as mixed oxides and not antimonates.

FURTHER COMPOUNDS OF THE GROUP V ELEMENTS

Sulphides

Of the wide variety of sulphur–nitrogen compounds which are now known, perhaps the most interesting is tetrasulphur tetranitride, S_4N_4, made by treating a sulphur chloride with ammonia; it has a cradle-like structure in which all four nitrogen atoms lie in the same plane. The nitrogen–sulphur bond lengths are all equal at 1.62 Å (the theoretical length for a single bond is 1.76 Å):

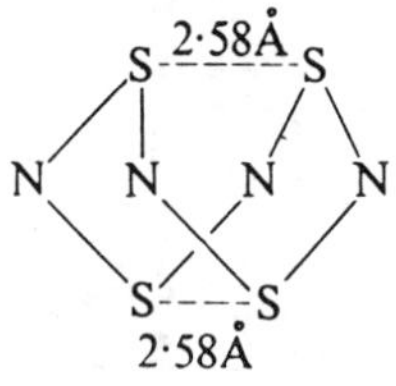

The pairs of sulphur atoms above and below the plane of four nitrogen atoms are within 2.58 Å of each other (the sum of the van der Waals radii for two sulphur atoms is 3.7 Å and the normal sulphur–sulphur single-bond length is 2.04 Å), which suggests that there is a weak sulphur–sulphur interaction. Delocalized π bonding involving sulphur 3d and nitrogen 2p orbitals is thought to occur over the whole molecule, which would account for the rather short sulphur–nitrogen bonds.

Depending on the reagents employed, groups can be placed on either the nitrogen or sulphur atoms of tetrasulphur tetranitride, with resultant loss of delocalized π bonding and a change in structure to a puckered eight-membered sulphur–nitrogen ring:

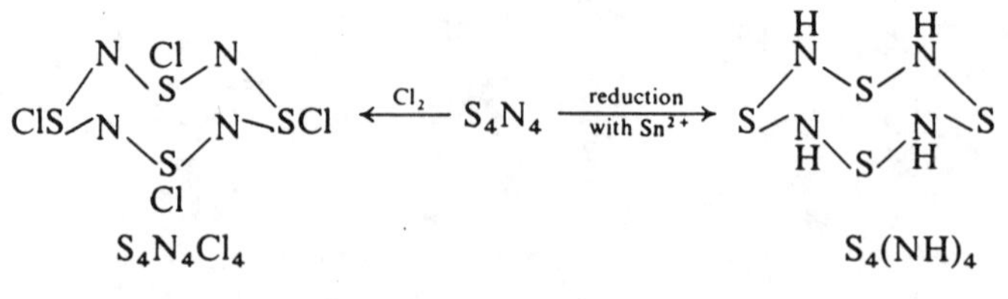

Tetrasulphur tetraimide, $S_4(NH)_4$, is another of the products arising from the reaction of ammonia with sulphur chlorides. It can be considered as arising from a puckered S_8 ring by substitution of four sulphur atoms by nitrogen. All the sulphur imides in the series $S_{8-n}(NH)_n$, where $n = 1, 2, 3$ or 4, have been isolated and separated into their respective isomers (no isomers exist with two nitrogen atoms linked directly to each other in the ring). For example, $S_6(NH)_2$ has three isomers:

| 1,3-isomer | 1,5-isomer | 1,4-isomer |

By treating the sulphur imides with S_2Cl_2, the rings can be linked together with —S—S— bridges and as many as five rings have been joined in this way using the isomers of $S_6(NH)_2$ and S_2Cl_2: e.g. for heptasulphur imide

$$2S_7NH + S_2Cl_2 \longrightarrow$$ $$+ 2HCl$$

Polythiazyl $(SN)_2$

When tetrasulphur tetranitride vapour is passed over silver gauze inside an evacuated tube, the four-membered ring compound S_2N_2 is first formed, but, on standing, this transforms into a bronze-coloured solid $(SN)_x$. This polymer has received considerable attention in recent years because, as well as having a metallic-like electrical conductivity under normal conditions, it becomes superconducting at 0.26 K:

$$\underset{\text{(orange)}}{S_4N_4} \xrightarrow[\text{gauze}]{\text{Ag}} \underset{\text{(white)}}{S_2N_2} \longrightarrow \underset{\text{(bronze)}}{(SN)_x}$$

| (square planar) | (almost planar chain) |

Other Group V sulphides

Phosphorus forms several P_4S_x sulphides of which P_4P_3, P_4S_7, P_4S_9 and P_4S_{10} can be made by direct combination from stoichiometric amounts of red phosphorus and sulphur. Their structures may be considered as being derived from a P_4 molecule by either the insertion of sulphur into the tetrahedron's edges to give P—S—P bridges or the formation of terminal P=S bonds. Somewhat surprisingly P_4S_6 has yet to be obtained in the pure state but is thought to adopt the P_4O_6 structure as does As_4S_6.

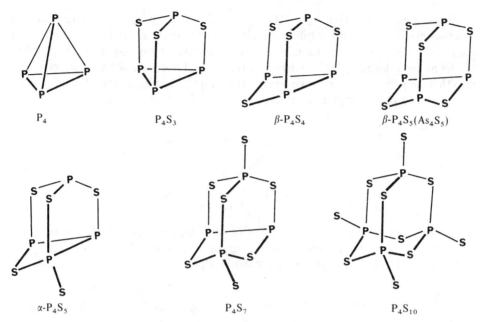

P_4 P_4S_3 β-P_4S_4 β-$P_4S_5(As_4S_5)$

α-P_4S_5 P_4S_7 P_4S_{10}

Arsenic forms the isostructural As_4S_3, As_4S_5 and As_4S_{10}, together with one form of As_4S_4 which has a cradle-like shape very similar to that of tetrasulphur tetranitride. Hydrogen sulphide passed into aqueous, acidified arsenites or acidified salts of antimony and bismuth precipitates the corresponding trisulphide M_2S_3. On heating, arsenic trisulphide sublimes as the dimer As_4S_6 which has the same structure as P_4O_6:

2.49 Å

2.23 Å

As_4S_4 As_4S_6

Halides

All the Group V elements form the normal *trihalides*, although nitrogen tribromide and tri-iodide are very unstable and are not known free of ammonia. Nitrogen is unable to achieve a coordination number of five and hence has no MX_5 derivatives; the M—X bond strength decreases with increasing atomic weight of M (and X) and becomes less able to stabilize the M^V state, with the result that the *pentahalides* become fewer in number from phosphorus to bismuth (Table 35).

All the molecular chloro-fluorides MCl_xF_{5-x} are known for phosphorus and arsenic, but those of arsenic are extremely unstable. When several arrangements are possible, the isomer having the more electronegative fluorine in axial positions is

Table 35—Halides of the Group V elements

	F	Cl	Br	I
N	NF_3	NCl_3	(NBr_3)	(NI_3)
	N_2F_4	—	—	—
	N_2F_2 (cis and trans)	—	—	—
P	PF_3	PCl_3	PBr_3	PI_3
	P_2F_4	P_2Cl_4	P_2Br_4	P_2I_4
	PF_5	PCl_5	PBr_5	PI_5
As	AsF_3	$AsCl_3$	$AsBr_3$	AsI_3
	AsF_5	$AsCl_5$	—	—
Sb	SbF_3	$SbCl_3$	$SbBr_3$	SbI_3
	SbF_5	$SbCl_5$	—	—
Bi	BiF_3	$BiCl_3$	$BiBr_3$	BiI_3
	BiF_5	—	—	—

always the one isolated in the gas phase, although *ionic* species can exist for some of the compounds in the solid state:

$$PCl_3F_2$$

$$AsCl_2F_3 \text{ (g)}$$
$$[AsCl_4^+ \ AsF_6^- \text{ (s)}]$$

$$PCl_4F \text{ (g)}$$
$$[PCl_4^+ \ PCl_4F_2^- \text{ (s)}]$$

Crystalline phosphorus pentachloride is also ionic, $PCl_4^+ PCl_6^-$; this can be visualized as a salt in which phosphorus pentachloride has ionized to the tetrahedral PCl_4^+ cation and a chloride ion, the latter being 'solvated' by another phosphorus pentachloride molecule to give the octahedral PCl_6^- anion. Phosphorus pentabromide on the other hand forms an ionic salt $PBr_4^+ Br^-$; apparently the energy released by forming the solvated ion PBr_6^- is not enough to compensate for the loss of lattice energy accompanying the increase in anionic radius as the bromide ion is changed into PBr_6^-. In some solvents phosphorus pentachloride ionizes to give PCl_4^+ and chloride ion, the latter now being solvated by the solvent and not with a further molecule of phosphorus pentachloride.

The trihalides of phosphorus, arsenic, antimony and bismuth can be made by direct reaction, whereas those of nitrogen are produced by treating ammonia or ammonium salts with the respective halogen. In the case of nitrogen trifluoride, the fluorine is produced *in situ* by electrolysing fused ammonium bifluoride:

$$NH_4F.HF \xrightarrow{\text{electrolysis}} (F_2) \xrightarrow{NH_4^+} NF_3$$

Cis- and *trans-*difluorodiazine are minor products in this reaction. The instability of nitrogen trichloride, tribromide and tri-iodide is largely due to the high bond energy of the nitrogen N_2 molecule which makes these halides very endothermic compounds (e.g. ΔH_f for NCl_3 is $+230$ kJ mol^{-1}).

Only nitrogen is able to form strong $p_\pi-p_\pi$ bonds, and hence *cis*- and *trans*-difluorodiazine, which have a double bond between the nitrogen atoms, have no analogues among other elements in the group:

difluorodiazine

cis   *trans*

It is worth noting the different products formed on hydrolysis of nitrogen and phosphorus trichlorides:

$$NCl_3 + 3OH^- \rightarrow 2NH_3 + 3OCl^-$$

$$PCl_3 + 3OH^- \rightarrow H_2P(H)O_3 + 3Cl^-$$

This has been attributed to the electronegativity sequence $N > Cl > P$, which results in differing polarities in the M—Cl bonds: $\overset{\delta-}{N}—\overset{\delta+}{Cl}$ and $\overset{\delta+}{P}—\overset{\delta-}{Cl}$. The negatively charged hydroxyl ion will attack the chlorine atoms in nitrogen trichloride and the phosphorus atom in phosphorus trichloride.

Although nitrogen has a high electronegativity, that of fluorine is even higher, which results in $\overset{\delta+}{N}—\overset{\delta-}{F}$ bond polarities in the nitrogen fluorides. The very small electric dipole moment of 0.2 Debye possessed by nitrogen trifluoride (compare this to the dipole moment of ammonia, which is 1.44 Debye) is thought to be due to the partial cancelling of the dipole moment due to the lone pair by the nitrogen–fluorine bond polarities, which act in the opposite direction:

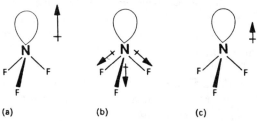

(a)   (b)   (c)

The dipole moment of nitrogen trifluoride: (a) lone pair moment; (b) three nitrogen–fluorine bond moments; (c) small resultant moment.

By passing nitrogen trifluoride or phosphorus trichloride through an electric discharge between mercury electrodes, one halogen atom is removed from each molecule by excited mercury atoms. The MX_2 fragments so formed couple together to give tetrafluorohydrazine, $F_2N—NF_2$, or diphosphorus tetrachloride, $Cl_2P—PCl_2$, respectively. In gaseous tetrafluorohydrazine at room temperature there are about equal amounts of the *trans* and *gauche* forms:

cis   trans   gauche

In the solid state, P_2I_4, made by heating stoichiometric amounts of phosphorus and iodine, has the *trans* conformation.

Phosphorus trihalides react with oxygen to produce the phosphoryl halides,

$$2PX_3 + O_2 \rightarrow 2OPX_3 \quad (\text{cf. } PX_3 + S \rightarrow SPX_3)$$

which also result from the partial hydrolysis of the pentahalides:

$$PX_5 + H_2O \xrightarrow{-2HX} OPX_3 \xrightarrow{3H_2O} H_3PO_4 + 3HX$$

The phosphoryl halides possess a tetrahedral structure, the phosphorus–oxygen bond being considerably shortened possibly by d_π–p_π bonding (see page 62). They are powerful donor molecules and form complexes with compounds of both transition and non-transition elements, the *oxygen* atom being the donor site:

$$Cl_3PO \rightarrow BCl_3 \qquad Cl_3PO \rightarrow SbCl_5 \qquad Cl_3PO \rightarrow FeCl_3 \qquad \begin{array}{c} Cl_3PO \searrow \\ \\ Cl_3PO \nearrow \end{array} TiCl_4$$

Although $OAsCl_3$ cannot be made either by careful hydrolysis of the highly unstable $AsCl_5$ or by the action of oxygen on arsenic trichloride, it is readily precipitated from a variety solvents by the low-temperature reaction between ozone and $AsCl_3$. Slow decomposition begins to occur at $-25\,°C$:

$$OAsCl_3 \rightarrow AsCl_3 + Cl_2 + \text{polymeric } As_2O_3Cl_4$$

Salts of oxo-acids

The relative stability of such salts is well illustrated by the reaction of N_2O_4 with Group V elements in dimethyl sulphoxide, Me_2SO. Arsenic does not react, antimony gives a basic nitrate as a solvate, $SbO(NO_3).Me_2SO$, and bismuth forms the normal nitrate $Bi(NO_3)_3.3Me_2SO$.

Dissolution of bismuth or Bi_2O_3 in concentrated nitric acid gives bismuth nitrate, which may be crystallized out as the pentahydrate. The bismuth atom is coordinated to four water molecules and three bidentate nitrate groups to give very distorted ten-coordination; Raman studies suggest that bidentate nitrate groups still persist when the salt is dissolved in aqueous solution. Other salts, including $Bi(ClO_4)_3 \cdot 5H_2O$, $Bi_2(SO_4)_3 \cdot nH_2O$ and $Sb_2(SO_4)_3$ as well as complexes such as $KBi(SO_4)_2$, are also known; all these species are hydrolysed very readily to basic salts.

ORGANO-DERIVATIVES OF THE GROUP V ELEMENTS

A discussion of the amines, R_nNH_{3-n}, belongs to organic chemistry and will not be given here.

The halides of phosphorus, arsenic, antimony and bismuth react readily with an excess of either Grignard reagents or lithium reagents to form organo-derivatives:

$$MCl_3 + 3RMgX \rightarrow MR_3$$

$$OPCl_3 + 3LiR \rightarrow OPR_3$$

$$MX_5 + 5LiR \rightarrow MR_5$$

A deficiency of the Grignard or lithium reagent normally leads to the formation of intermediate organo-halides, which are useful for synthetic work:

$$PCl_3 + LiR \rightarrow RPCl_2$$

$$PCl_3 + 2LiR \rightarrow R_2PCl \xrightarrow{Na} R_2P{-}PR_2$$

Direct synthesis of the organo-derivatives by simply heating a Group V element with an alkyl or aryl halide is sometimes possible:

$$CF_3I + P \xrightarrow{heat} P(CF_3)_3 + IP(CF_3)_2 + I_2PCF_3$$

$$\Big\downarrow \begin{smallmatrix} heat \\ Hg \end{smallmatrix} \qquad \Big\downarrow \begin{smallmatrix} heat \\ Hg \end{smallmatrix}$$

$$(CF_3)_2P{-}P(CF_3)_2 \qquad (PCF_3)_n$$

$$(n = 4, 5 \text{ or } 6; \text{ cyclic derivatives})$$

(1,6-distibatriptycene)

The Group V triorgano-derivatives are excellent donor molecules and have been employed as ligands to transition metals in particular for many years. In some cases skilful design of complex polydentate species has allowed chemists to force specific stereochemistries onto the acceptor metal atom. Two simple organo-arsenicals illustrate the typical synthetic pathways which have been adopted:

$$Me_2AsI + Na \xrightarrow{\text{THF}} NaAsMe_2$$

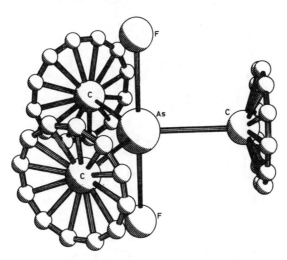

Some of the pentavalent organo-halides of phosphorus can exist as several different species. Thus tetraphenylphosphorus fluoride has been prepared as the trigonal bipyramidal molecule $(C_6H_5)_4PF$ and also in two ionic forms: $P(C_6H_5)_4^+ F^-$ and the 'dimeric' $P(C_6H_5)_4^+ P(C_6H_5)_4F_2^-$.

We are normally accustomed to thinking of molecules as having static groups present in them so that we are able to draw their structures or make ball-and-stick models. Obviously, though, this is not the actual case since bonds will be vibrating or bending and whole groups may be spinning around a bond. This latter situation is well illustrated in Fig. 81, which shows the model used in an electron diffraction study of gaseous Me_3AsF_2 to take into account the fact that the three methyl groups were undergoing virtually independent free rotation.

Fig. 81—Perspective view of the molecule $(CH_3)_3AsF_2$ illustrating the skeletal geometry and the model used to define the free rotation of the methyl groups. [Reproduced with permission from *Journal of the Chemical Society, Dalton Transactions* (1988) p. 451.]

Pyridine, C_5H_5N, is a well-known aromatic heterocycle, being isoelectronic with benzene. Although the other Group V elements do not form particularly strong $p_\pi-p_\pi$ bonds to carbon, it has proved possible to make the higher analogues of pyridine and also to form transition metal 'sandwich' complexes similar to those of benzene:

$$C_5H_5M + Cr(CO)_6 \xrightarrow{\text{THF}} \qquad (M = N, P, As, Sb \text{ or } Bi)$$

$$C_5H_5As + Cr(\text{atoms}) \xrightarrow[\text{at } -196\ ^\circ C]{\text{co-condense}}$$

Chapter 9

Group VI: The Chalcogens. Oxygen, Sulphur, Selenium, Tellurium and Polonium

INTRODUCTION

This is the last main group containing an element which can be truly classed as metallic, polonium. Since tin in Group IV, the penultimate elements have shown less and less metallic character until at the halogens, iodine and astatine behave entirely as non-metals. Physically, the increasing metallic character down Group VI is reflected in the electrical properties of the elements: oxygen and sulphur are insulators, selenium and tellurium behave as semiconductors, whereas polonium shows typical metallic conduction.

Bismuth is the last element in the periodic table to possess any stable isotopes, so that a study of polonium is hampered by the radioactivity of its most accessible isotope ^{210}Po, made in milligram-to-gram amounts by an (n,γ) reaction using a bismuth target:

$$^{209}_{83}Bi \xrightarrow{\;(n,\gamma)\;} {}^{210}_{83}Bi \xrightarrow[t_{1/2}\,=\,5\ \text{days}]{\;\beta^-\;} {}^{210}_{84}Po \quad (\alpha\text{-active};\ t_{1/2} = 138.4\ \text{days})$$

The intense α radiation not only presents obvious problems concerning the adequate protection of personnel, but can also seriously affect the chemistry in rather subtle ways: the nuclear disintegration generates considerable sample heating (since it involves the production of about 140 W per gram), causes very extensive radiation damage in solids and, in solution, produces reactive radicals via α-induced solvent decomposition. All these effects can accelerate the decomposition of polonium compounds as described later on page 314.

The ionization energies fall steadily down the group owing to the decreased shielding effects along the Se, Te and Po periods of the d^{10} and f^{14} inner shells; however, the covalent radii of selenium and tellurium are still considerably less than expected by simple extrapolation from O and S, which have no d^{10} cores (Table 36). The high

Table 36—The Group VI elements

Element		1st IE (kJ mol^{-1})	2nd IE (kJ mol^{-1})	3rd IE (kJ mol^{-1})	4th IE (kJ mol^{-1})	M^{II} (covalent radius (Å))
8 Oxygen, O	$1s^2 2s^2 2p^4$	1314	3388	5296	7469	0.66
16 Sulphur, S	$[Ne]3s^2 3p^4$	1000	2253	3358	4564	1.04
34 Selenium, Se	$[Ar]3d^{10}4s^2 4p^4$	941	2076	3086	—	1.17
52 Tellurium, Te	$[Kr]4d^{10}5s^2 5p^4$	870	1472	2954	3649	1.37
84 Polonium, Po	$[Xe]4f^{14}5d^{10}6s^2 6p^4$	813	—	—	—	—

ionization energies obviously rule out the possibility of a rare gas configuration produced by loss of all six outer electrons to give M^{6+} ions; furthermore, even M^{4+} ions (having an ns^2 structure) probably do not exist in any compounds with the possible exception of one form of PoO_2 which has the fluorite structure normally associated with ionic species.

The more usual ways these elements attain a rare gas electronic structure include the formation of (1) M^{2-} ions as in the alkali metal oxides, sulphides, selenides and tellurides, (2) two covalent bonds, e.g. H_2O and $(CH_3)_2S$, (3) MR^- ions such as OH^- and HS^-, and (4) 'onium' salts typified by H_3O^+ and R_3S^+. In addition, many hypervalent MR_4 and MR_6 derivatives are known for all the elements except oxygen.

The electronegativity of oxygen is very high, being second only to that of fluorine. This leads to a large degree of ionic character in M—O bonds, to the limit where many oxides actually have ionic structures. For the same reason the phenomenon of hydrogen bonding (page 98) within this group is almost completely restricted to hydroxy-compounds.

Oxygen also differs markedly from the rest of its group in forming strong $p_\pi-p_\pi$ bonds with itself, carbon and nitrogen. Thus the two π-bonded allotropes of oxygen, O_2 and O_3, are unique in type since the other group members either form rings and chains containing single bonds (S, Se, Te) or metallic lattices (Po). Multiple bonds between hypervalent sulphur and carbon apparently occur in a number of species as judged from the particularly short C—S distances; it seems likely that these are of the $d_\pi-p_\pi$ type (see page 62):

$$F_5SCH_2Br + C_4H_9Li \xrightarrow{-70\,°C} F_5SCH_2Li \xrightarrow{-LiF} \underset{1.554\ \text{Å}}{F_4S{=}CH_2}$$

$$CF_3CH{=}SF_4 \xrightarrow[60\,°C;\ -HF]{\text{solid KOH}} \underset{1.395\ \text{Å}}{CF_3C{\equiv}SF_3} \quad (\text{dimerizes above } -30\,°C)$$

CATENATION WITHIN GROUP VI

One of the most interesting properties of these elements is their exceptional ability to link together and form rings or chains. This capacity is usually limited to two for oxygen, as in the peroxides and superoxides, although O_3 and the exceedingly unstable O_4F_2 are also known. On the other hand, elemental sulphur is unique within the whole periodic table for the immense variety of its allotropes and the ease with which S—S bonds are broken and remade. If the appropriate separation and identification techniques were available it would probably be found that S_n rings and, in some cases, chains can exist for every value of n from 2 onwards; already rings up to $n = 23$ have been isolated from rapidly chilled molten sulphur using HPLC, and higher homologues are known to be present in the mixtures (Figs. 82 and 83). Some modifications of solid selenium and tellurium contain infinitely long chains and, on melting, these break down into smaller rings or chains, each one of which, if it could be obtained in pure form, represents a new allotrope. However, the catenation

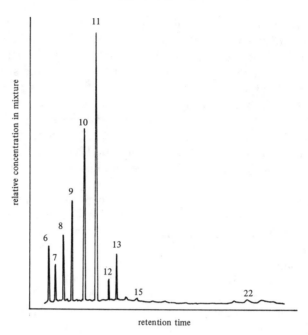

Fig. 82—HPLC chromatogram of cyclic sulphur species, S_n. This mixture was obtained by treating $(C_5H_5)_2TiS_5$ with S_xCl_2 where $x \simeq 6$. The number of sulphur atoms in each eluted ring is shown above the corresponding peak; as expected, S_{11} is the main species formed in the reaction. (Idealized chromatogram; for typical experimental curves see *Zeitschrift für Naturforschung* vol. 41b (1986) p. 958.)

characteristics of Se and Te are slightly more limited than those of sulphur since their chains and rings do not increase in size during heating of the melts; see page 317 for a discussion on the behaviour of molten sulphur.

Successful attempts have been made in recent years to form 'mixed' rings containing both sulphur and selenium. NMR studies, using ^{77}Se, on molten mixtures of the two elements allowed the detection of 29 $S_{8-n}Se_n$ cyclic species, including positional isomers. The most abundant isomer in any one group was that in which the selenium atoms remained linked to each other in the mixed rings; thus NMR peaks due to $1,2,3\text{-}Se_3S_5$ were the most intense of triseleniumpentasulphur rings and consisted of a doublet (intensity 2) and a triplet (intensity 1):

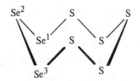

^{77}Se, nuclear spin $\frac{1}{2}$; therefore number of lines observed due to spin–spin coupling is $(2n \times \frac{1}{2} + 1)$, i.e. $(n+1)$ where n is the number of neighbouring selenium atoms.

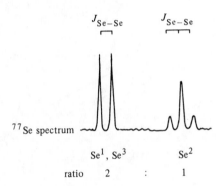

J_{Se-Se} J_{Se-Se}

^{77}Se spectrum

Se^1, Se^3 Se^2

ratio   2   :   1

Some individual members of the mixed series containing six- and seven-membered rings have been isolated by using a preformed M$_5$ fragment stabilized as a titanium

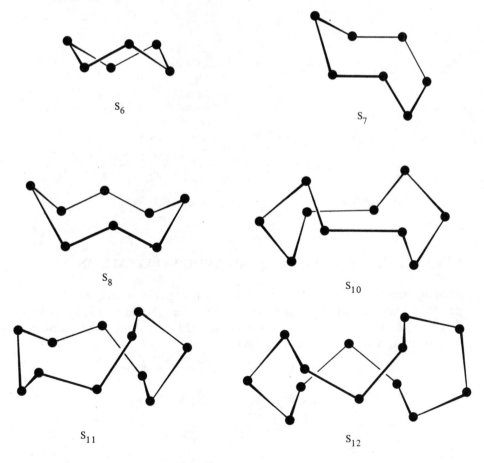

S_6

S_7

S_8

S_{10}

S_{11}

S_{12}

Fig. 83—Rings present in some sulphur allotropes.

heterocycle:

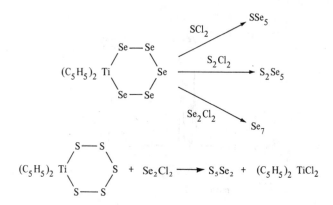

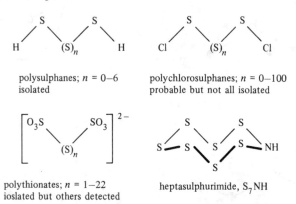

Catenation is not restricted to the elemental forms of these elements. Many compounds, particularly of sulphur, contain rings and chains of which the latter might be regarded as oligomers (i.e. small polymers) stabilized by end-stopping groups such as —H, —Cl or —SO_3^-:

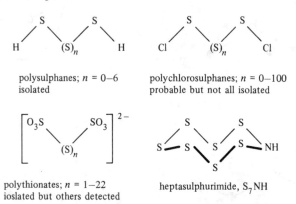

polysulphanes; $n = 0$–6
isolated

polychlorosulphanes; $n = 0$–100
probable but not all isolated

polythionates; $n = 1$–22
ioslated but others detected

heptasulphurimide, S_7NH

CATIONIC CLUSTERS CONTAINING GROUP VI ELEMENTS

Although simple cations are unknown within this group, several highly coloured, polyatomic cations have been isolated in recent years (Table 37 and Fig. 84). Since they are all unstable in water they have to be prepared either directly or in non-aqueous solvents using a variety of oxidizing agents:

$$O_2 + PtF_6 \rightarrow O_2^+ PtF_6^- \quad (O_2^+ \text{ is paramagnetic})$$

$$S + SbF_5(\text{excess}) \xrightarrow{1.SO_2} S_4^{2+}(SbF_6^-)_2$$

$$S + AsF_5 \xrightarrow{1.SO_2} S_8^{2+}(AsF_6^-)_2$$

$$Se + AsF_5 \xrightarrow{1.SO_2} Se_{10}^{2+}(AsF_6^-)_2$$

$$\text{Te} + \text{AsF}_5 \xrightarrow{\text{1.SO}_2} \text{Te}_4^{2+}(\text{AsF}_6^-)_2$$

$$\text{Te} + \text{TeCl}_4 \xrightarrow{\text{AlCl}_3} \text{Te}_4^{2+}(\text{AlCl}_4^-)_2$$

$$\text{Te} + \text{oleum} \rightarrow \text{Te}_6^{4+}$$

Table 37—Polyatomic cations formed by the chalcogens

O_2^+	—	—	—	—
—	S_4^{2+}	S_8^{2+}	—	S_{19}^{2+}
—	Se_4^{2+}	Se_8^{2+}	Se_{10}^{2+}	—
—	Te_4^{2+}	—	—	—

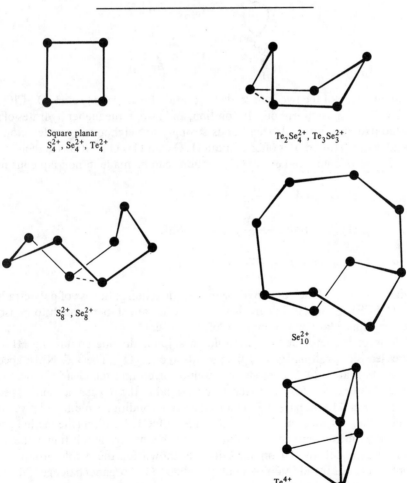

Square planar
S_4^{2+}, Se_4^{2+}, Te_4^{2+}

$Te_2Se_4^{2+}$, $Te_3Se_3^{2+}$

S_8^{2+}, Se_8^{2+}

Se_{10}^{2+}

Te_6^{4+}

Fig. 84—Some polychalcogen cations. Those atoms linked by dashed lines are very close and may be bonded together.

When mixtures of chalcogens are oxidized with AsF_5 or SbF_5 in liquid sulphur dioxide, species such as $Te_2Se_6^{2+}$, $Te_3S_3^{2+}$ and $Te_2Se_8^{2+}$ are formed. It will be noticed from the list of cations that the most highly oxidized cluster is formed by tellurium (Te_6^{4+}) and that cations with progressively less charge per atom occur through selenium (Se_{10}^{2+}) to sulphur (S_{19}^{2+}).

ONIUM CATIONS

The pyramidal oxonium ion H_3O^+, present in many crystalline monohydrates of strong acids, is isoelectronic with NH_4^+:

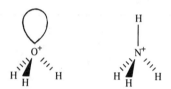

The similar shape of the two ions leads to several of their salts (e.g. $H_3O^+ClO_4^-$ and $NH_4^+ClO_4^-$) being isomorphous. In solution, and also in the higher hydrates of many acids, the oxonium ion hydrogen-bonds strongly to neighbouring water molecules; thus acid di- and tri-hydrates often contain $H_5O_2^+$ and $H_7O_3^+$ cations in their crystals. Oxonium, peroxonium and even H_3S^+ cations can be made in non-aqueous media:

$$H_2O + AsF_5 \xrightarrow{\text{l.HF}} H_3O^+AsF_6^-$$

$$H_2O_2 + SbF_5 \xrightarrow{\text{l.HF}} H_3O_2^+SbF_6^-$$

$$H_2S + SbF_5 \xrightarrow{\text{l.HF}} H_3S^+SbF_6^-$$

The high reactivity of the liquid hydrogen fluoride requires the use of polytetrafluoro-ethene (PTFE) apparatus, but the resulting species are stable enough to be isolated in the crystalline state on evaporation of the solvent.

Addition of ^{17}O-enriched water to liquid sulphur dioxide containing $HF–SbF_5$ mixtures leads to protonation and the formation of H_3O^+. The ^{17}O NMR spectrum of the solution at $-20°C$ consists of a well-resolved quartet, clearly indicating the presence of three *equivalent* hydrogen atoms bound to the oxygen; a similar spectrum is shown in Fig. 110 on page 399. Under the same conditions, methanol gives rise to a triplet in the ^{17}O spectrum resulting from the $MeOH_2^+$ cation (the methyl protons are evidently too far from the oxygen to cause further hyperfine splitting of the triplet).

Many organo-substituted onium salts are known for the whole group and are generally prepared by addition of an alkyl iodide to a diorganochalcogen; the iodide may then be exchanged for other groups using typical reactions:

$$Me_2Se + MeI \rightarrow Me_3Se^+I^-$$

$$Me_3Se^+I^- + Ag_2O(\text{moist}) \rightarrow AgI + Me_3Se^+OH^-(\text{basic})$$

$$Me_3Te^+OH^- + HF \rightarrow Me_3Te^+F^- + H_2O$$
$$\quad\;\text{base} \qquad\quad \text{acid} \quad\; \text{salt} \qquad\;\; \text{water}$$

$$Me_3Se^+Cl^- + AgF \xrightarrow[\text{solution}]{\text{aqueous}} AgCl\downarrow + Me_3Se^+F^-$$

The onium ions are optically active when substituted by three different organic groups and several have been separated into their respective isomers; however, the OR_3^+ species undergo very rapid inversion and cannot be resolved.

The triphenyloxonium cation $(C_6H_5)_3O^+$, like its isoelectronic partner triphenylamine, is virtually planar with the phenyl rings twisted at such an angle that there can be only very minimal π bonding between the central atom's 2p orbital and the empty π^* orbitals of the aromatic rings. Presumably this planarity arises solely because of steric crowding on the small oxygen and nitrogen atoms (compare trisilylamine, page 67).

BONDING IN THE OXYGEN MOLECULE

The energy separation between the 2s and 2p orbitals in the oxygen atom is much larger than that in nitrogen; for the reasons, see page 52. This results in comparatively little mixing of the $\sigma(2s)$ and $\sigma(2p)$ orbitals in the O_2 molecule so that the energy of $\sigma(2p)$ remains below that of the degenerate pair of $\pi(2p)$ orbitals (see Fig. 85). Two

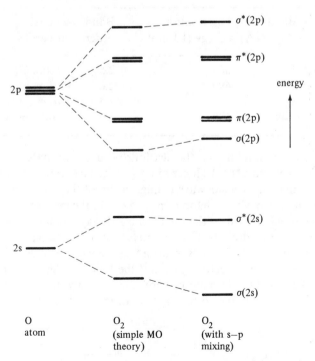

Fig. 85—Orbital energy diagram for the O_2 molecule. The electron configuration is $\sigma(2s)^2\sigma^*(2s)^2\pi(2p)^4\pi^*(2p_x)^1\pi^*(2p_y)^1$

unpaired electrons (Hund's rule) singly occupy the degenerate $\pi^*(2p)$ orbitals and account for the paramagnetism of O_2. (Paramagnetic compounds are attracted by a magnetic field, and hence a tube of liquid oxygen, suspended by a cotton thread, moves *towards* a magnet placed close by and sticks to the pole faces.)

From the electron configuration of O_2 the bond order (defined as *half* the difference in the number of electrons occupying bonding orbitals compared to those in antibonding orbitals) is two, normally shown simply as $O=O$. Three other species which contain two oxygen atoms linked together are the dioxygenyl cation (O_2^+), the superoxide anion (O_2^-) and the peroxide anion (O_2^{2-}). These arise via the addition of electrons to, or the subtraction of an electron from, the two $\pi^*(2p)$ orbitals:

$$O_2^+ \quad \sigma(2s)^2\sigma^*(2s)^2\sigma(2p)^2\pi(2p)^4\pi^*(2p_x)^1\pi^*(2p_y)^0 \quad \text{bond order 2.5}$$

$$O_2 \quad \sigma(2s)^2\sigma^*(2s)^2\sigma(2p)^2\pi(2p)^4\pi^*(2p_x)^1\pi^*(2p_y)^1 \quad \text{bond order 2.0}$$

$$O_2^- \quad \sigma(2s)^2\sigma^*(2s)^2\sigma(2p)^2\pi(2p)^4\pi^*(2p_x)^2\pi^*(2p_y)^1 \quad \text{bond order 1.5}$$

$$O_2^{2-} \quad \sigma(2s)^2\sigma^*(2s)^2\sigma(2p)^2\pi(2p)^4\pi^*(2p_x)^2\pi^*(2p_y)^2 \quad \text{bond order 1.0}$$

Because of the antibonding nature of the π^* orbitals we have the rather unusual situation that *removal* of an electron from O_2 to give O_2^+ results in stronger bonding between the oxygen atoms as seen from the dissociation energies and bond lengths; on the other hand, *addition* of electrons to form O_2^- and O_2^{2-} weakens the O---O bonding and lowers the bond orders relative to O_2 (see Fig. 86):

	Bond length (Å)	Dissociation energy (kJ mol^{-1})	Bond order	Magnetic properties
O_2^+	1.12	623	2.5	Paramagnetic
O_2	1.12	490	2.0	Paramagnetic
O_2^-	1.32	—	1.5	Paramagnetic
O_2^{2-}	1.49	398	1.0	Diamagnetic

Unlike the situation found in N_2, the last four occupied orbitals in O_2 all have π symmetry, being $\pi(2p)$ and $\pi^*(2p)$. Hence dioxygen–metal complexes can be expected to have either an angular or a sideways configuration of the O_2 ligand; an example of the former is found in oxyhaemoglobin (page 471). In the majority of its complexes the O_2 molecule attaches itself to the metal in a sideways manner using one of the filled $\pi(2p)$ orbitals to donate electrons to the metal, the interaction being strengthened by back-bonding from metal d orbitals to the $\pi^*(2p)$ oxygen orbitals. Obviously both the partial donation of the π electron pair and the filling of the antibonding $\pi^*(2p)$ orbitals will weaken the O---O bonding; not unexpectedly, the O—O distances in these metal complexes are found to be larger than that in free O_2:

$$Pt[P(C_6H_5)_3]_4 + O_2 \longrightarrow$$

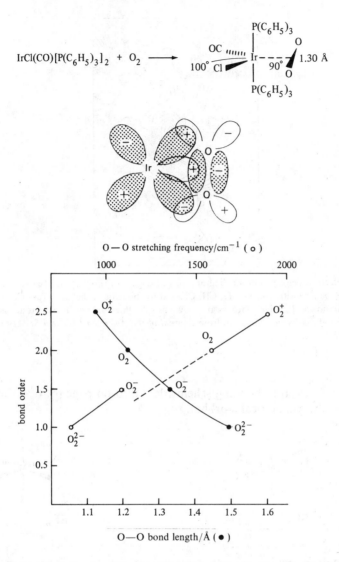

Fig. 86—Bond order versus bond length and bond stretching frequency for O—O species. As expected, the bond length decreases with bond order. The O—O stretching frequency is also very sensitive to bond order, the graph actually consisting of two part-curves, the discontinuity being due to the loss of some binding energy as the electrons are forced to pair up in the π^* orbitals.

HYPERVALENT SPECIES: MX$_4$, MX$_6$ AND THEIR COMPLEXES

The unused pair of electrons present in the valence shell of a tetravalent chalcogen is very sterically active and imposes a trigonal bipyramidal geometry on all MX$_4$ species; in the solid state additional groups from neighbouring molecules may weakly interact with M giving higher coordination numbers. The presence of the equatorial lone pair in Me$_2$TeCl$_2$ is clearly visible in a difference electron density map (Fig. 87). MX$_6$ derivatives have an undistorted octahedral structure which implies, using the

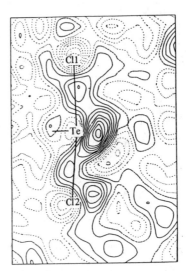

Fig. 87—Electron density contour diagram for $(CH_3)_2TeCl_2$ showing the Te—Cl bond and Te(IV) lone pair densities. Atoms Te, Cl1, Cl2 and the bisector of the C1—Te—C2 angle are in the plane shown. (Positive contours are shown as solid lines and negative contours as dotted lines). [Reproduced with permission from *Journal of the American Chemical Society* vol. 105 (1983) p. 229.]

three-centre, four-electron bonding scheme described on page 69, that the lone pair is in a spherically symmetrical s orbital:

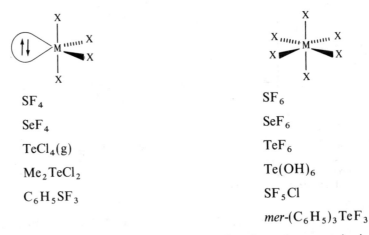

SF_4	SF_6
SeF_4	SeF_6
$TeCl_4(g)$	TeF_6
Me_2TeCl_2	$Te(OH)_6$
$C_6H_5SF_3$	SF_5Cl
	mer-$(C_6H_5)_3TeF_3$

Oxygen is too small an atom to accommodate more than four electron pairs in its valence shell and hence differs from the rest of the group by not forming OX_4 or OF_6 derivatives.

Interest in complexes of tetravalent chalcogens has centred on the steric properties of the lone pair electrons but, although many complexes have been studied spectroscopically, relatively few structures have been substantiated by X-ray crystallography. In octahedral hexahalo-anions, *trans*-$SeCl_4(pyridine)_2$ and *trans*-

$TeCl_4$(tetramethyl thiourea)$_2$ the structures are undistorted, suggesting that the lone pair occupies an s orbital; on the other hand, R_3M^+ cations are pyramidal with a lone pair taking up the fourth position of a tetrahedral environment around M:

SC(NMe$_2$)$_2$ = tetra
methyl thiourea

$$2KCl + TeCl_4 \rightarrow K_2TeCl_6$$

$$MO_2 + HX(conc) \rightarrow MX_6^{2-} \rightarrow \text{precipitate with large cation}$$

$$TeO_2 + HCl(conc) + SC(NMe_2)_2 \rightarrow Cl_4Te[SC(NMe_2)_2]_2$$

Tellurium hexafluoride forms $MTeF_7$ and M_2TeF_8 species with the heavier alkali metal fluorides which probably contain TeF_7^- or TeF_8^{2-} anions, but nothing is known about their structures or those of the corresponding amine complexes $TeF_6.2R_3N$.

It is significant that the only complexes formed by a Group VI hexafluoride are those containing the largest atom, tellurium. Sulphur hexafluoride is a highly inert compound and, for example, although being thermodynamically unstable towards hydrolysis it is unaffected even by superheated steam at 500 °C. This is obviously not due to an intrinsic stability of S—F bonds because SF_4 is extremely sensitive to even traces of water. Presumably the inertness of SF_6 arises from a kinetic origin: sulphur is too small to bind a seventh ligand required for the formation of an activated intermediate in a reaction sequence. A similar situation was noted when the hydrolytic stability of carbon tetrahalides was discussed on page 68. Tellurium hexafluoride on the other hand slowly hydrolyses at room temperature.

d ORBITALS AND S—O BONDING

One of the most controversial issues of theoretical sulphur chemistry over the past few years has been the importance or otherwise of 3d orbitals in bonding. Many recent *ab initio* studies have shown that although d orbitals are relatively large and diffuse, it is essential to include them to some extent in calculations on second-row elements. (The calculated position of the amplitude maximum in 3d orbitals on the S^+ ion is about 1.4 Å from the nucleus compared to values somewhat less than 1 Å for the 3s and 3p orbitals.)

An important contributing factor to the shortness of sulphur–oxygen bonds is their polar character, $S^{\delta+}$—$O^{\delta-}$, and this is augmented by some d_π–p_π bonding. However, the latter is by no means as substantial as the p_π–p_π interactions which occur in, say, carbonyl (C=O) groups; thus although the d orbitals do have a considerable stabilizing effect, there is apparently little justification for depicting sulphate or sulphite groups with S—O double bonds.

Electron density maps calculated from X-ray diffraction work on crystalline sulphates, thiosulphates, sulphites and phosphoric acid have been interpreted as showing direct evidence for some 3d–2p interaction between the central element and the oxygen ligands. Such interactions will be enhanced by the positive character of the sulphur atom in a polar bond because this will reduce the size of the d orbitals and result in improved overlap. The increasing effective nuclear charge which occurs along the second row from Si to Cl will also contract the 3d orbitals to give increasingly strong back-bonding in the MO_4^{n-} anions.

THE BISULPHITE ION IN SOLUTION

The behaviour of the bisulphite anion HSO_3^- is quite complex because it can exist as three separate species depending on the conditions. In solid rubidium and caesium bisulphites the hydrogen atom is attached to sulphur but in aqueous solution an equilibrium is set up between the S—H and O—H isomers:

$$H—SO_3^- \rightleftharpoons H—OSO_2^-$$

As the concentration increases, a second equilibrium involving the pyrosulphite ion begins to establish itself:

$$2HSO_3^- \rightleftharpoons [O_2S—SO_3]^{2-} + H_2O$$

(The name pyrosulphite for the $S_2O_5^{2-}$ ion is somewhat unfortunate because it suggests the presence of an S—O—S bridge rather than a direct S—S bond.) In contrast, it is thought that the biselenite ion exists only as the $H—OSeO_2^-$ isomer.

PROBLEMS ASSOCIATED WITH THE RADIOACTIVITY OF POLONIUM

There are no stable isotopes of polonium and hence research into its chemistry is greatly hampered by the associated radioactivity. Like all elements, polonium has several isotopes, two of which, $^{208}Po(\alpha; t_{1/2}, 2.9$ years) and $^{209}Po(\alpha; t_{1/2}, 42$ years), have quite long half-lives; unfortunately they are less easy to make and hence much more expensive than $^{210}Po(\alpha; t_{1/2}, 138.4$ days) which can now be isolated in gram quantities by an (n,γ) reaction on ^{209}Bi:

$$^{209}Bi \xrightarrow{(n,\gamma)} {}^{210}_{83}Bi \xrightarrow[t_{1/2}=5 \text{ days}]{-\beta^-} {}^{210}_{84}Po$$

To put the half-life of ^{210}Po into perspective, it represents 10^{13} disintegrations per minute* for each milligram of metal, or 10 000 C (3.7×10^{14} Bq) per litre for a 0.01 M solution. The resulting decomposition of the water due to the α activity present in such a 0.01M solution gives a visible and continuous evolution of gaseous products. In air, the α radiation induces the direct combination of N_2 and O_2 to produce nitrogen oxides which, in a closed system, results in the metal or its compounds being

---

* There are approximately 10^{13} seconds in 30 000 years.

coated by a layer of basic nitrate. Glass is so weakened by irradiation that containers can disintegrate altogether when left in contact with polonium compounds.

Intense α activity causes charring of organic material. This can lead to problems when large amounts of polonium are being separated on ion exchange columns, and organo-polonium derivatives also char so rapidly that it is difficult to obtain accurate elemental analyses for new products. Personnel are obviously at risk from similar radiation damage and hence all work has to be carried out in very efficient glove boxes (Fig. 109, page 396) provided with *stout* gloves; the maximum safe level for ingestion is considered to be about 7×10^{-12} g, which is barely 2×10^{10} atoms.

There are two ways to study the chemistry of polonium: visible quantities may be used in conventional reactions, or tiny invisible amounts are mixed with a carrier and used in 'tracer' experiments. In the latter, ^{210}Po might be mixed with macro-amounts of tellurium and checks made to see if the α radiation follows the tellurium through a synthesis; if it does then one may assume that polonium and tellurium have formed an identical compound.

It is more satisfactory to use larger, visible specimens of polonium, but the radiation may sometimes adversely affect the chemistry. For example, polonium *metal* results from addition of aqueous ammonia to its compounds because of the reducing action of atomic hydrogen formed by disruption of NH_3 on α bombardment. Normally all isotopes of an element show the same chemical properties, but this is not always the case for radioactive species because the intense radiation from a short-lived isotope may prevent the formation of specific compounds. Thus PoF_6 is formed when ^{208}Po is heated in fluorine, but the hexafluoride cannot be made using ^{210}Po; even with the longer-lived isotope, rapid radiation-induced reduction to the tetrafluoride occurs.

OCCURRENCE OF THE GROUP VI ELEMENTS

Oxygen

Most abundant of all the elements, making up about 49.4% of the Earth's crust and 23% (by weight) of the atmosphere.

Sulphur

About 0.05% abundance; occurs in the free state, as sulphides and as sulphates.

Selenium

0.09 ppm of the Earth's crust; occurs as selenides of lead, copper, silver, mercury and nickel, and the main source is as a by-product from the treatment of the sulphide ores of these elements, especially those of copper.

Tellurium

At 0.002 ppm abundance it is about as rare as gold; found as metallic tellurides, often of silver and gold; main source is anode slime from electrolytic purification of copper.

Polonium

One of the rarest naturally occurring elements (about 10^{-10} ppm of the Earth's crust); main natural source is uranium ores, which contain about 10^{-4} g of polonium per tonne. All isotopes are radioactive.

EXTRACTION

Oxygen

The main industrial source of oxygen is from the fractional distillation of liquid air. The material may be transported either as the highly compressed gas or as the liquid in vacuum-jacketed tanks.

Sulphur

Native sulphur of high purity (about 99.5%) is mined using the Frasch process in which a boring is made down to the deposit and lined with three concentric pipes (Fig. 88). Superheated water, when pumped down the outer pipe, melts the sulphur which is then pumped to the surface by applying compressed air to the central tube. Large quantities of sulphur are also recovered from both the refining of crude oil and the roasting of metal sulphides.

Selenium and tellurium

The 'anode mud' resulting from the electrolytic purification of copper contains 3%–28% selenium and about 8% tellurium. The dried mud is fused with soda ash and silica and blown with air to oxidize the selenium and tellurium to their dioxides. Addition of sodium hydroxide to the melt converts the dioxides to sodium selenite and sodium tellurite. After cooling and leaching with water, sulphuric acid is added to neutralize the solution, when tellurium dioxide is precipitated leaving selenous acid in solution. Tellurium dioxide is reduced to tellurium with carbon at red heat and selenium precipitated from the mother liquor by adding sulphur dioxide:

$$H_2SeO_3 + 2SO_2 + H_2O \rightarrow Se\downarrow + 2H_2SO_4$$

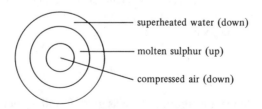

Fig. 88—Cross-section through the concentric pipes of the Frasch process.

Polonium

All 27 isotopes of polonium are radioactive and many have very short half-lives. Natural sources of the metal are not used for extraction purposes because of the minute amounts present; the most accessible isotope, polonium-210, is made artificially by the irradiation of natural bismuth (100% ^{209}Bi):

$$^{209}_{83}Bi \xrightarrow{\text{n,}\gamma\text{ reaction}} {}^{210}_{83}Bi \xrightarrow[t_{1/2}=5\text{ days}]{\beta^- \text{ decay}} {}^{210}_{84}Po$$

THE ELEMENTS

The increasing metallic character down Group VI finally manifests itself with the typically metallic lattices adopted by the two allotropic forms of elemental polonium. Both forms have a polonium coordination of six, the low-temperature α-polonium being cubic and the high-temperature β-polonium rhombohedral:

$$\alpha\text{-Po} \overset{36°C}{\rightleftharpoons} \beta\text{-Po}$$

The vapour at the boiling point (962°C) contains Po_2 molecules and presumably represents the maximum chain length attainable by polonium.

Oxygen normally exists as the diatomic molecule O_2, the electronic structure of which was discussed on page 309. The action of a silent electric discharge converts the paramagnetic O_2 into diamagnetic ozone, O_3, which is a non-linear molecule isoelectronic with the nitrite ion:

$$\underset{O}{\overset{O}{\diagdown}}\underset{117°}{\overset{}{\diagup}}\underset{O}{\overset{\text{1.28 Å}}{}}$$

Rhombic sulphur (S_α), the most common allotropic form of sulphur, contains S_8 puckered rings; in the monoclinic form (S_β) identical S_8 rings pack into the crystals in a slightly different manner to those in S_α. When S_8 molecules are dissolved in polar solvents such as methanol or acetonitrile, a slowly established equilibrium is set up,

$$\underset{(98.9\%)}{S_8} \rightleftharpoons \underset{(0.3\%)}{S_7} + \underset{(0.8\%)}{S_6}$$

the components of which can be separated by high-pressure liquid chromatography. In hexane at room temperature S_8 rings are stable for weeks, but above 100°C the smaller rings begin to appear within a few hours.

Liquid sulphur is a much more complex system. On melting at about 115°C, a yelow, mobile liquid is formed which consists almost wholly (93.6%) of S_8 molecules, although tiny amounts of other cyclic species from S_6 to at least S_{30} can be detected in the quenched melt using reverse-phase HPLC. The viscosity of molten sulphur falls on further heating, which is typical of any normal liquid. However, at 159°C the viscosity begins to rise very sharply, reaching a maximum at 170°C (Fig. 89); this viscosity increase is accompanied by a darkening of the colour to deep red.

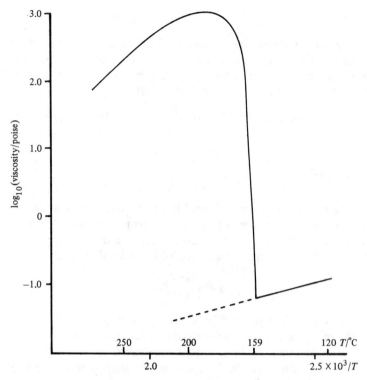

Fig. 89—A plot of log (viscosity) against $1/T$ for liquid sulphur. The dotted part of the curve is the extrapolated viscosity for S_8.

HPLC analysis of the quenched melts shows that the concentration of rings other than S_8 passes through a maximum at $159\,^{\circ}C$ and insoluble polymeric material (S_μ) is present to the extent of 3%; the maximum concentration of S_μ (40%) occurs at about $250\,^{\circ}C$, as shown in Fig. 90.

The high viscosity of liquid sulphur arises from the presence of polymeric material which mainly consists of large rings, but some chains, which necessarily possess an unpaired electron on both terminal sulphur atoms, must also be present because the melt is slightly paramagnetic. The number of these free radicals increases nearly two-hundred-fold between 200 and $375\,^{\circ}C$ as some of the large cyclic molecules split into small chains. After S_7 and S_8, the most frequently occurring cyclic molecules are S_6, S_9 and S_{12}, although their concentrations always remain below 1%.

The vapour at the boiling point $(444\,^{\circ}C)$ contains mainly S_7 (40%), S_6 (30%) and S_8(20%) together with minor amounts of S_2, S_4 and S_5.

Mixtures containing relatively high proportions of the smaller cyclic allotropes can be made by treating $(C_5H_5)_2TiS_5$ with chlorosulphanes. Unfortunately the latter are normally obtained as mixtures, Cl_2S_x, so that the resulting sulphur molecules have to be isolated using preparative-scale HPLC; a typical chromatogram is shown in

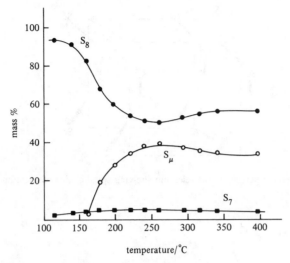

Fig. 90—Variation in amounts of S_8, S_7 and S_μ present in sulphur rapidly chilled from several different temperatures. The cyclic species soluble in CS_2 were separated using HPLC and varied from S_6 to about S_{22}; the insoluble S_μ was a complex mixture of components having high molecular weights. [Data taken from *Angewandte Chemie* (International Edition in English) vol. 24 (1985) p. 59.]

Fig. 82 for mixtures obtained from the reaction sequence

$$S_6 + Cl_2 \longrightarrow \text{'}S_6Cl_2\text{'}$$

$$(C_5H_5)_2Ti \begin{array}{c} S-S \\ \diagup \quad \diagdown \\ S \\ \diagdown \quad \diagup \\ S-S \end{array} + \text{'}S_6Cl_2\text{'} \longrightarrow S_x \ (x \text{ mainly 10 and 11})$$

Crystalline grey Se and Te contain long spiral molecules as shown in Fig. 91. Typically, when selenium melts at 220 °C the viscosity of the liquid decreases rapidly with temperature, indicating that the long chains originally present in the solid are breaking up into smaller units; at the boiling point (685 °C) the vapour consists of Se_2 and Se_6 molecules. Crystallization of selenium dissolved in carbon disulphide yields two modifications of red selenium, Se_α and Se_β, both of which contains Se_8 rings; highly polymeric selenium is made by pouring molten selenium into cold water.

SEMICONDUCTOR PROPERTIES OF SELENIUM

Grey selenium and tellurium have electrical conductivities intermediate between those of insulators and those of metals and which increase with rise in temperature. These semiconductor properties are explained by assuming that at low temperatures all the

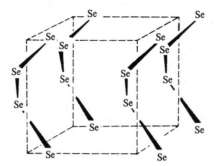

Fig. 91—The crystal structure of grey selenium showing spiral chains of selenium atoms.

available low-energy orbitals are completely filled so that no electrons are mobile (or 'free') to carry an electric current; under these conditions selenium and tellurium behave as insulators (Fig. 92(a)). As the temperature rises, thermal energy promotes a small number of electrons to low-lying empty orbitals which allows some movement of electrons in the now incompletely filled orbitals when an electric field is applied (Fig. 92(b)).

Illumination of selenium, especially with light having a wavelength of about 7000 Å, also induces promotion of electrons into the empty orbitals and allows the passage of an electric current. In this way the conductivity of selenium can be increased by as much as a thousandfold and the phenomenon forms the basis of selenium photocells used in the measurement of light intensity.

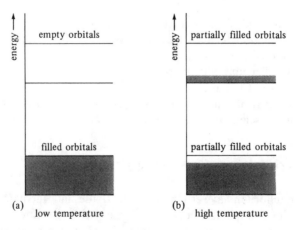

Fig. 92—Energy level diagrams of grey selenium.

HYDRIDES OF THE GROUP VI ELEMENTS

All the MH_2 hydrides can be made directly from the elements, except for polonium hydride. A better synthesis for hydrogen sulphide, selenide or telluride is to treat a metal chalcogenide with aqueous acid. The small hydrogen 1s orbital becomes less and less compatible with the bonding orbitals of the heavier Group VI elements, making the M—H bonds decrease in strength down the group, which is reflected in the heats of formation of the hydrides listed in Table 38. Consequently hydrogen selenide and hydrogen telluride are relatively unstable and strong reducing agents.

A series of higher hydrides having the general formula H_2M_x (oxygen, $x = 2$; sulphur, $x = 2-8$; selenium, $x = 2, 3$; tellurium, $x = 2$) are known. Except for hydrogen peroxide (see page 324), these higher hydrides are very unstable and readily deposit the free element:

$$H_2M_x \rightarrow (x-1)M + H_2M \qquad \begin{cases} \Delta H_f^\ominus, \ H_2S_2 = -8.4 \text{ kJ mol}^{-1} \\ \Delta H_f^\ominus, \ H_2Se_2 = +105 \text{ kJ mol}^{-1} \\ \Delta H_f^\ominus, \ H_2Te_2 = +166 \text{ kJ mol}^{-1} \end{cases}$$

A mixture of polysulphanes may be prepared by treating sodium polysulphide with hydrochloric acid, and it is separated into the various components ($x = 2-8$) by simultaneous distillation and thermal cracking; the purified sulphanes must be stored at low temperatures to prevent decomposition. Although the rapid decomposition of the sulphanes to give free sulphur might suggest otherwise, it has been established that the polysulphanes do not contain branched sulphur chains.

The molecular structure of disulphane is similar to that of hydrogen peroxide, showing that the lone pair electrons in the sulphur 3p orbitals exert considerable steric influence and cause the molecule to adopt a configuration in which the lone pairs are farthest apart and hence interact least with each other:

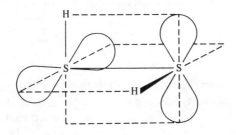

Table 38—Heats of formation and bond angles of the Group VI hydrides

	Heat of formation (kJ mol^{-1})	Bond angle (deg)
Water, H_2O	−241.8	104.5
Hydrogen sulphide, H_2S	−20.1	92.1
Hydrogen selenide, H_2Se	+85.8	91
Hydrogen telluride, H_2Te	+154.4	90

Water

The graphical plot of boiling point against molecular weight given on page 100 for the Group VI hydrides shows that water is anomalous. Direct extrapolation from hydrogen sulphide gives the 'expected' boiling point for water as about $-80\,°C$, a value almost $200\,°C$ too low. This behaviour is due to hydrogen bonding (page 98) which augments the normal van der Waals-type forces within the liquid; the solubility of many alcohols, sugars, carboxylic acids and phenols is due to their ability to participate in this hydrogen bonding.

As described in Chapter 12, ions are strongly hydrated in aqueous solution and very often several water molecules remain closely associated with a cation when salts crystallize out on evaporation. The extent of hydration in crystals depends on ionic size and ionic charge. Since the water molecules are held mainly by ion–dipole forces,

$$H^{\delta+}$$
$$\diagdown$$
$$O^{\delta-}\!-\!-\!-\!-\,M^{n+}$$
$$\diagup$$
$$H_{\delta+}$$

the interaction will be greatest for small ions of high charge and, for cations of similar size, is in the order $M^{3+} > M^{2+} > M^{+}$. Size has a secondary effect in that it restricts the number of water molecules which are able to pack round the ion in a crystalline hydrate:

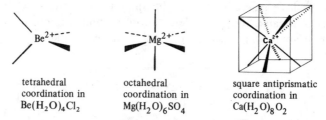

tetrahedral coordination in $Be(H_2O)_4Cl_2$	octahedral coordination in $Mg(H_2O)_6SO_4$	square antiprismatic coordination in $Ca(H_2O)_8O_2$

The water molecules in beryllium sulphate tetrahydrate, which are among the most strongly coordinated in any crystal, are also hydrogen-bonded to sulphate oxygen atoms. In such a situation it might reasonably be assumed that, apart from atomic vibrations, the system would be 'static'. However, proton NMR studies have shown that each of the four water molecules is undergoing a rapid $180°$ flip motion between its two sulphate partners:

$$Be^{2+}\!\!-\!\!O\begin{array}{c} {}^{2}H\!\cdots OSO_3^{2-} \\ \\ {}^{1}H\!\cdots OSO_3^{2-} \end{array} \quad\rightleftharpoons\quad Be^{2+}\!\!-\!\!O\begin{array}{c} {}^{1}H\!\cdots OSO_3^{2-} \\ \\ {}^{2}H\!\cdots OSO_3^{2-} \end{array}$$

In solution, beryllium and magnesium bind strongly to four and six water molecules over a wide concentration range, but the *larger* calcium ion hydrates much more weakly and the number of coordinated solvent molecules falls from about ten at 1M to barely six at 4.5M. Large univalent cations like rubidium have such a poorly defined hydration sphere that accurate values of the coordination number are virtually impossible to measure. As shown in Fig. 93, rapid exchange occurs between coordinated water and bulk solvent even for a small, doubly charged ion like Be^{2+}.

The water molecules in the immediate vicinity of an ion are strongly polarized and their protons become positively charged relative to those in the free solvent; hence it can be expected that more water molecules will be quite strongly hydrogen-bonded to these positive protons forming a second (and then a third, etc.) hydration layer around the cation. However, the influence of the cationic charge will decrease rapidly with each successive layer of solvent. Thus in solution the total number of water molecules associated with a particular ion is difficult to define since it is actually a matter of degree, and the value obtained depends markedly on the particular method used to determine it. When sugar is added to a salt solution prior to electrolysis it is possible to measure the amount of water associated with the electrically diffusing cations (assuming that the uncharged sugar distributes evenly between bulk water molecules and those in the various hydration spheres of the cation). Such experiments show that a total of about 15–20 solvent molecules are carried about in solution by beryllium and magnesium cations.

The same polarization effect which gives rise to the tiny 'iceberg' round the cation also induces the ionization of protons from the coordinated water:

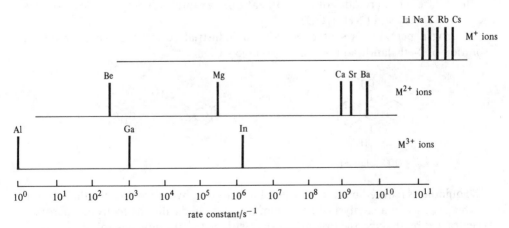

Fig. 93—The rate of water substitution in hydrated M^+, M^{2+} and M^{3+} ions. The rate is proportional to the strength of the water–cation interaction and is clearly in the order $M^{3+} > M^{2+} > M^+$ for ions of the same quantum shell.

This effect will be most marked for small, highly charged ions (e.g. Be^{2+}, Al^{3+}) since they will polarize the protons to the greatest extent. Such ions are very readily hydrolysed, giving rise to either basic salts or a precipitate of the metal hydroxide; hydrolysis is catalysed by base because the added hydroxyl ions disturb the above equilibria by removing protons from solution. Conversely, singly charged cations do not polarize the protons very strongly and aqueous solutions of their salts are stable to hydrolysis (e.g. the alkali metals).

Hydrogen peroxide, H_2O_2

The gauche structure adopted by hydrogen peroxide shows small angular variations between the isolated molecules found in the gaseous phase and those present in the hydrogen-bonded solid:

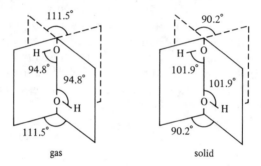

| gas | solid |

Like water, hydrogen peroxide has a high dipole moment (2.01 Debye), a high dielectric constant (89 at $0°C$) and also undergoes a slight self-ionization,

$$2H_2O_2 \rightleftharpoons H_3O_2^+ + HO_2^-$$

for which $[H_3O_2^+][HO_2^-] = 1.5 \times 10^{-12}$. Several salts will crystallize from solution with 'hydrogen peroxide of crystallization', examples being the carbonates $K_2CO_3.3H_2O_2$ and $(NH_4)_2CO_3.H_2O_2$.

Hydrogen peroxide is synthesized on an industrial scale by passing air into a solution of 2-ethylanthraquinol in an organic solvent:

(10% solution)

The quinone is reduced back to the quinol by bubbling hydrogen through the solution in the presence of a catalyst of palladium metal. Periodically the hydrogen peroxide is extracted by shaking the organic solvent with water; the aqueous solution may be concentrated up to about 90% by fractional distillation at low pressure.

On the laboratory scale hydrogen peroxide can be prepared by heating barium oxide in air or oxygen and treating the barium peroxide which results with sulphuric acid

$$\text{BaO} \xrightarrow{\;O_2\;} \text{BaO}_2 \xrightarrow{\;H_2SO_4\;} \text{BaSO}_4\downarrow + H_2O_2$$

Although the heat of formation of liquid hydrogen peroxide is -187.8 kJ mol^{-1}, indicating high stability with respect to decomposition into hydrogen and oxygen, it is thermodynamically unstable with respect to water and oxygen:

$$H_2O_2(l) \rightarrow H_2O(l) + \tfrac{1}{2}O_2 \quad \Delta H = -98.3 \text{ kJ mol}^{-1}, \quad \Delta G = -121.4 \text{ kJ mol}^{-1}$$

At room temperature, oxygen evolution is very slow in the absence of catalysts (silver, platinum or several transition metal oxides, such as manganese dioxide) because of the reaction's high activation energy, but the heated liquid and vapour will explode well below the extrapolated boiling point (150 °C).

Peroxides and peroxo-salts

The alkali and alkaline earth metals form solid ionic peroxides containing the $[O—O]^{2-}$ ion in which the oxygen–oxygen distance, 1.49 Å, is typical of a single bond; the alkali metals also form hydroperoxides M^+OOH^- akin to the hydroxides of the water system. Addition of cold aqueous acid to both peroxides and hydroperoxides liberates hydrogen peroxide. To a limited extent it is possible to substitute —OOH for —OH groups in ortho oxo-acids and —O—O— for —O— bridges of pyro oxo-acids:

sulphuric acid
H_2SO_4

peroxomonosulphuric acid
H_2SO_5 (Caro's acid)

pyrosulphate ion
$S_2O_7^{2-}$

peroxodisulphate ion
$S_2O_8^{2-}$

Peroxophosphoric acid (H_3PO_5), peroxodicarbonates ($O_2CO—OCO_2^{2-}$) and peroxodiphosphates ($O_3PO—OPO_3^{4-}$) can be envisaged as arising via similar substitutions. The preparation of peroxomonosulphuric acid is quite logical and involves the addition of hydrogen peroxide to chlorosulphonic acid, a reaction very similar to the ready hydrolysis of chlorosulphonic acid to sulphuric acid:

$$\underset{O}{\overset{O}{\underset{\diagup}{\diagup}}}S\underset{OH}{\overset{Cl}{<}} + H_2O \longrightarrow HCl + \underset{O}{\overset{O}{\underset{\diagup}{\diagup}}}S\underset{OH}{\overset{OH}{<}}$$

However, salts of the peroxo-diacids have to be synthesized by means of electrolytic oxidation of the corresponding simple salts:

$$CO_3^{2-} \xrightarrow{-e} \left[\underset{O}{\overset{O}{\underset{\diagup}{\diagup}}}CO\cdot \right]^{-} \xrightarrow{\text{dimerization}} C_2O_6^{2-} \text{ (also for } PO_4^{3-}, SO_4^{2-})$$

possible intermediate
radical ion

METALLIC OXIDES, SULPHIDES, SELENIDES AND TELLURIDES

All the Group VI elements form alkali metal derivatives which contain X^{2-} ions, but, owing to the increasing metallic character down the group, the stability of the X^{2-} ions decreases from oxygen to tellurium. For example, treatment of these compounds with water produces XH^- with oxygen and sulphur, but the hydrides XH_2 with selenium or tellurium. On warming, aqueous solutions of hydrosulphides evolve hydrogen sulphide, demonstrating that the hydrosulphide ion HS^- is much less stable than hydroxide.

The most common metallic oxide types are:

(a) M_2^IO. Alkali metal oxides (also sulphides, selenides and tellurides) crystallize with the *anti*-fluorite structure (8:4 coordination), the O^{2-} ions taking the place of Ca^{2+} ions and M^+ replacing F^- in the normal fluorite lattice; see Fig. 41.

(b) $M^{II}O$. The Group II elements (with the exception of beryllium), cadmium, vanadium, managanese, cobalt and nickel have oxides (sulphides and selenides) which have this stoichiometry; they adopt the 6:6 coordination of the sodium chloride lattice; see Fig. 149.

(c) $M_2^{III}O_3$. Oxides of this stoichiometry often have the corundum (Al_2O_3) structure, in which the metal is octahedrally coordinated by O^{2-} ions and each O^{2-} ion is tetrahedrally surrounded by M^{3+} ions.

(d) $M^{IV}O_2$. Typical structures adopted by these oxides are those of rutile (TiO_2, Fig. 152) and fluorite (CaF_2. Fig. 41).

(e) M_3O_4. These oxides are unusual in that the metal is present in two oxidation states: either $M^{II}M_2^{III}O_4$, e.g. $Fe^{II}Fe_2^{III}O_4$; or $M^{IV}M_2^{II}O_4$, e.g. $Pb^{IV}Pb_2^{II}O_4$.

HIGHER CHALCOGENIDES

Ionic species containing unbranched chains with 1–6 chalcogen atoms are known for the alkali and alkaline earth metals; they are prepared either from H_2X or by direct union:

$$Ba(OH)_2 + Cs_2SO_4 + H_2S \xrightarrow[\text{solution}]{\text{aq}} Cs_2S$$

$$Cs_2S + 4S \xrightarrow[\text{solution}]{\text{aq}} Cs_2S_5$$

$$Na + Te \xrightarrow[\text{NH}_3]{\text{liq}} Na_2Te_2 + Na_2Te_3$$

$$K + (2,2,2\text{-crypt}) + Te \xrightarrow[\text{ethane}]{\text{1,2-diamino-}} [K(crypt)]_2Te_3$$

$$Na_2S_5 + NMe_4Cl \xrightarrow[\text{solution}]{\text{aq}} (NMe_4)_2S_6$$

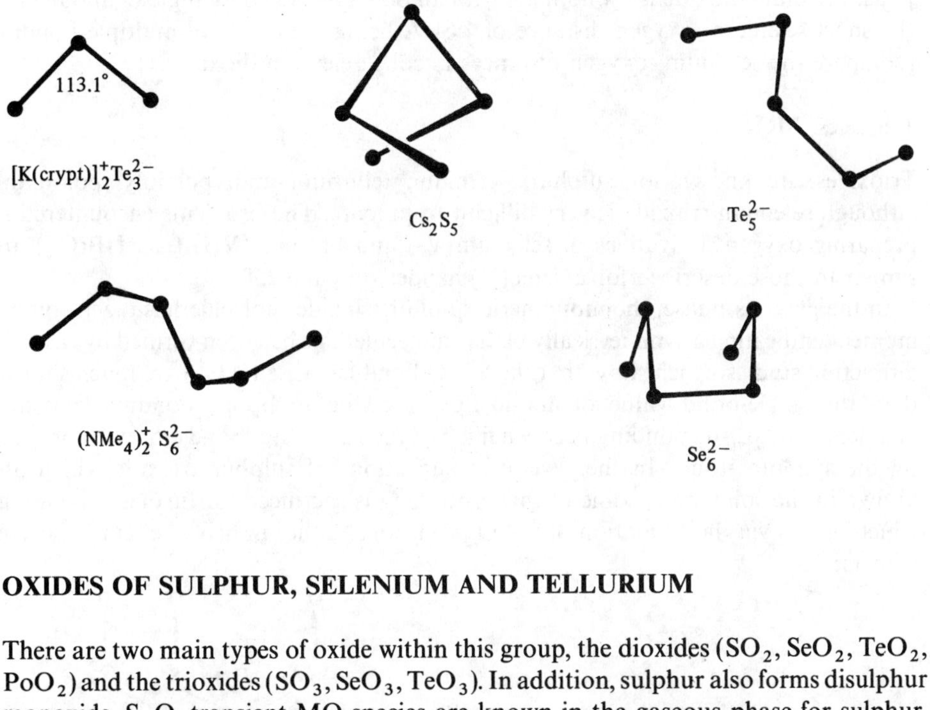

113.1°

$[K(crypt)]_2^+ Te_3^{2-}$

Cs_2S_5

Te_5^{2-}

$(NMe_4)_2^+ S_6^{2-}$

Se_6^{2-}

OXIDES OF SULPHUR, SELENIUM AND TELLURIUM

There are two main types of oxide within this group, the dioxides (SO_2, SeO_2, TeO_2, PoO_2) and the trioxides (SO_3, SeO_3, TeO_3). In addition, sulphur also forms disulphur monoxide, S_2O, transient MO species are known in the gaseous phase for sulphur, selenium and tellurium, and polonium forms a black monoxide, PoO.

Disulphur monoxide, S_2O

This oxide is prepared by passing an electrical discharge through a mixture of sulphur vapour and sulphur dioxide. It is a bent molecule, being closely related structurally and electronically to ozone and sulphur dioxide:

S

1.88 Å   1.46 Å

S   118°   O

O

O   117°   O

S

1.43 Å

O   119.5°   O

Dioxides, MO_2

The dioxides may be prepared directly by heating the elements in air or oxygen. The decreasing M—O π bond strength and the increasing metallic character from sulphur to polonium are reflected in the structures of the solid dioxides: solid sulphur dioxide contains discrete SO_2 molecules, selenium dioxide consists of infinite chains,

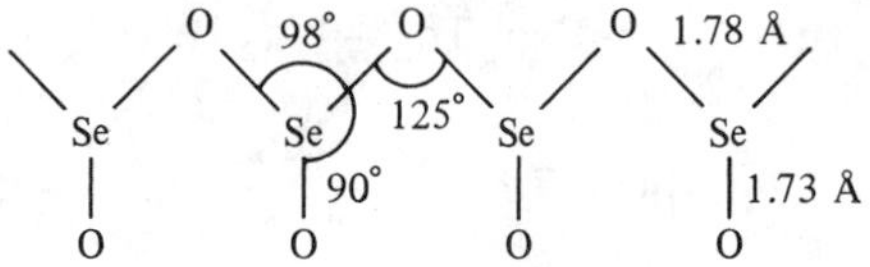

whereas tellurium dioxide and polonium dioxide are essentially ionic solids having rutile and face-centred cubic structures (Figures 49 and 26) respectively. In the gaseous phase selenium dioxide is monomeric with an Se—O—Se bond angle of about 125°, the short selenium–oxygen distance of 1.61 Å being indicative of multiple bonding (compare the selenium–oxygen distances in solid selenium dioxide).

Trioxides, MO_3

Trioxides are known for sulphur, selenium, tellurium and, possibly, polonium, although selenium trioxide is very difficult to prepare. The problems encountered in preparing oxygen derivatives of selenium(VI) and bromine(VII) (e.g. $HBrO_4$) are similar to those described for arsenic(V) halides on page 258.

In the gaseous phase, the monomeric sulphur trioxide molecule has a zero dipole moment, indicating a symmetrically planar molecule; this has been verified by electron diffraction studies which show that the S—O bond lengths (1.43 Å) are much shorter than the single-bond value of about 1.60 Å owing to p_π-p_π bonding (possibly augmented by p_π-d_π bonding between filled oxygen 2p orbitals and empty 3d orbitals on the sulphur atom). In the several modifications of sulphur trioxide which are known in the solid state, some of this π bonding is sacrificed for further σ bonding, which occurs via the formation of S—O—S bridges either in helical chains or cyclic trimers:

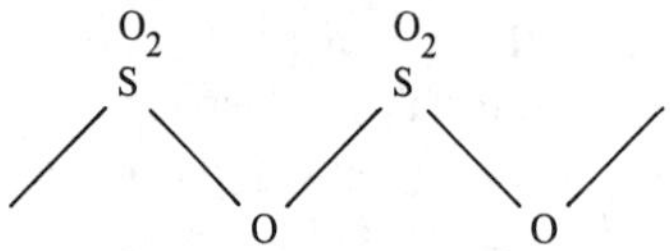

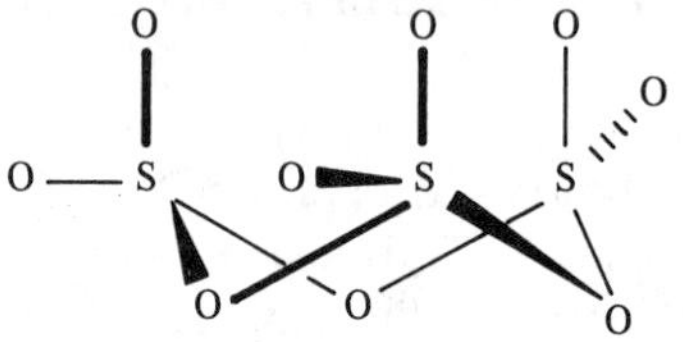

The strongly oxidizing selenium trioxide can be made by either dehydrating selenic acid (H_2SeO_4) with P_2O_5 at 150°C or refluxing potassium selenite (K_2SeO_3) with SO_3; it exists as an Se—O—Se bridged cyclic tetramer in the solid state. TeO_3, formed by dehydrating telluric acid ($Te(OH)_6$) at about 350°C, is only *slowly* rehydrated by water, in contrast to SO_3 and SeO_3.

OXO-ACIDS OF SULPHUR, SELENIUM AND TELLURIUM

Space does not permit a complete discussion on all of the many oxo-acids particularly of sulphur, which exist for these elements.

Sulphurous acid, H_2SO_3

Sulphur dioxide, although formally the anhydride of H_2SO_3, dissolves in water as $(SO_2)_{aq}$ with little or no formation of the free acid. However, many salts containing sulphite and bisulphite ions are known, for example Na_2SO_3 and $NaHSO_3$.

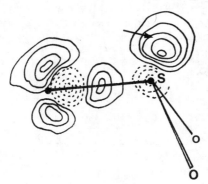

The lone pair of electrons on a sulphite ion, SO_3^{2-}, shows up clearly as a peak (arrowed) on an electron density map plotted in the plane of an S—O bond. The build-up of electronic charge (solid contours) at the mid-point of the S—O bond and the oxygen lone pairs are also prominent features. This map shows the *difference* in experimentally determined electron density in the ion and that calculated for the un-bonded atoms placed in the same positions.

Alkali metal sulphites react slowly with sulphur on boiling in aqueous solution to give thiosulphates (see Fig. 94); iodine oxidizes thiosulphates to tetrathionates as in iodine–thiosulphate titrations:

Sulphuric acid, H_2SO_4

This is prepared on an industrial scale via the catalytic oxidation of sulphur dioxide to sulphur trioxide:

$$SO_2 + O_2 \xrightarrow[\text{catalyst}]{V_2O_5} SO_3 \xrightarrow[\text{in } H_2SO_4]{\text{dissolved}} H_2S_2O_7 \xrightarrow{H_2O} 2H_2SO_4$$

The anhydrous acid, which is frequently used as a non-aqueous solvent, has a high electrical conductivity due to extensive self-ionization:

$$\left. \begin{array}{l} 2H_2SO_4 \rightleftharpoons H_3SO_4^+ + HSO_4^- \\ 2H_2SO_4 \rightleftharpoons H_3O^+ + HS_2O_7^- \end{array} \right\} \text{ both reactions occur simultaneously}$$

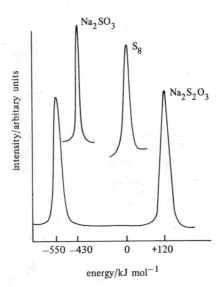

energy/kJ mol^{-1}

Fig. 94—Photoelectron spectrum of the thiosulphate ion. Monochromatic X-radiation of known energy hv is used to eject a 1s electron from the sulphur atoms in $S_2O_3^{2-}$ and the kinetic energy (KE) of these electrons is measured; the 1s electron binding energy for sulphur can then be calculated using the equation

$$hv = \text{binding energy of electron} + \text{KE}(+ \text{apparatus constant})$$

Because the wave functions of the outer electrons penetrate to the interior of the atom, the 1s binding energy is found to be sensitive to the oxidation state of the sulphur. This photoelectron spectrum shows the presence of two sulphur atoms in different oxidation states (formally $+$VI and $-$II); since the peaks are of equal intensities, the two types of sulphur must be present in the ratio 1:1 required by the formula O_3SS^{2-} of the thiosulphate ion. The photoelectron spectra of elemental sulphur (formal oxidation state 0) and sulphite ($+$IV) are also shown on the diagram for comparison purposes; the very large energy differences for the 1s electrons of sulphur in various oxidation states are notable, particularly when simple valence theories assume that the inner electrons are unaffected by bonding.

It is widely used in organic chemistry both as a sulphonating agent and as a solvent for nitric acid in nitration reactions (when NO_2^+ is the electrophile):

$$HNO_3 + 2H_2SO_4 \rightarrow NO_2^+ + H_3O^+ + 2HSO_4^-$$

Oxo-acids of selenium

In general these are similar to the sulphur acids, although slight differences do occur. For example, selenium dioxide is (like sulphur dioxide) very soluble in water, but the selenous acid H_2SeO_3 which is formed is only negligibly dissociated; however, the $HSeO_3^-$ and SeO_3^{2-} ions are readily formed when alkali is added to aqueous selenous acid.

Selenite and selenate are isostructural with the pyramidal sulphite and tetrahedral sulphate ions respectively: as a result, many metal sulphates and selenates are found to be isomorphous; for example, the alums and seleno-alums $M_2SeO_4 \cdot Al_2(SeO_4)_3 \cdot 24H_2O$, where M is an alkali metal.

Oxo-acids of tellurium

Tellurous acid, H_2TeO_3, is similar to but weaker than selenous acid; both normal and acid salts are known, e.g. Cs_2TeO_3 which contains pyramidal TeO_3^{2-} ions.

Unlike sulphur and selenium, tellurium(VI) forms a hexahydroxy acid $Te(OH)_6$, which is prepared by the oxidation of tellurium dioxide in aqueous nitric acid using potassium permanganate. On titration with standard alkali using phenolphthalein as indicator, telluric acid behaves only as a monobasic acid giving salts $M^ITeO(OH)_5$, but other salts can be prepared in which two or six protons have been replaced; for example, $Li_2TeO_2(OH)_4$ and Ag_6TeO_6.

Other oxo-acids of sulphur(VI)

Many acids and salts exist which can be considered to arise from H_2SO_4 and sulphates via *isoelectronic* substitutions. A similar situation was found in Groups IV and V where the wide variety of silicates and phosphates arose from the replacement of —OH groups by isoelectronic —O— bridges.

To be quite correct, sulphuric acid should be called ortho-sulphuric acid since it contains the maximum known number of hydroxyl groups bound to one sulphur(VI) atom. The pyro-acid may be thought of as derived from sulphuric acid by loss of water between two molecules of the ortho-acid:

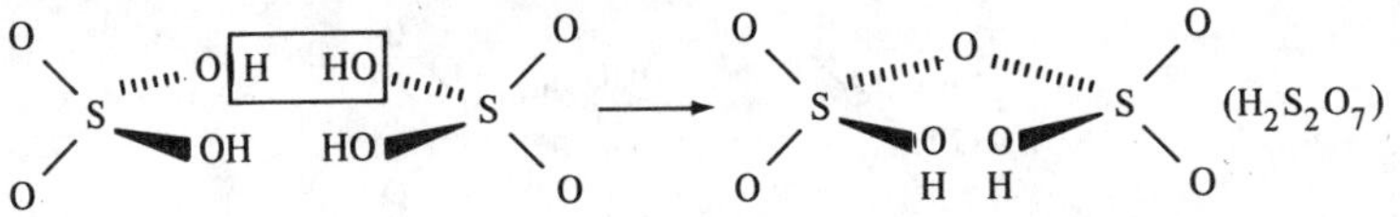

pyrosulphuric acid
(disulphuric acid)

In the pyro-acid the sulphur atoms share one oxygen atom; attempts to make 'meta-sulphuric acid', in which each sulphur shares two oxygen atoms, result only in the formation of sulphur trioxide:

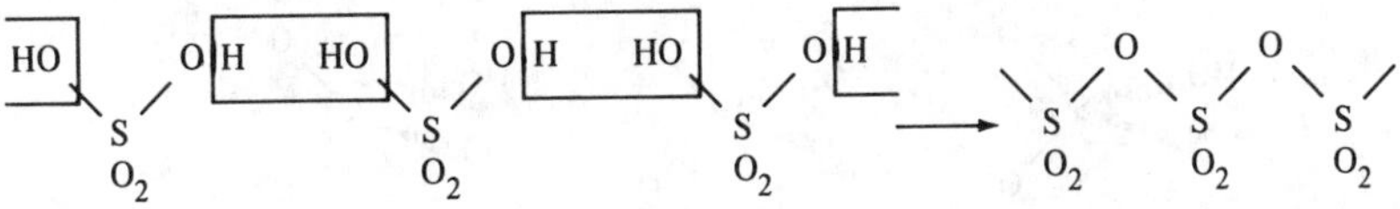

Although polysulphuric acid and polysulphates containing infinite linear chains cannot exist because neutral species result, the oligomeric ions $S_3O_{10}^{2-}$, $S_4O_{13}^{2-}$ and $S_5O_{16}^{2-}$ are known:

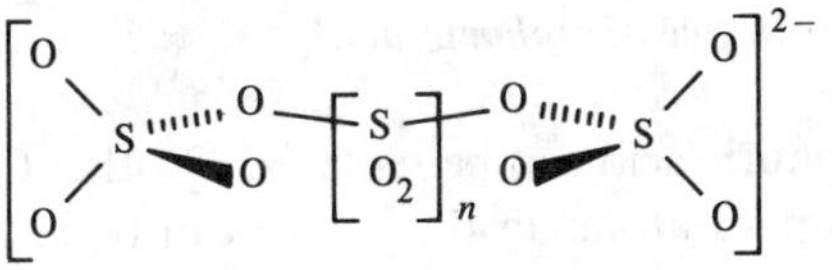

$n = 1, 2, 3$ (but other values are obviously possible)

Replacement of a free oxygen on H_2SO_4

Other elements of Group VI should be capable of substitution for the two free oxygen atoms in the sulphuric acid molecule. Examples of type A are thiosulphuric acid and selenosulphate salts derived from selenosulphuric acid. No derivatives are yet known of type B:

type A:

; examples

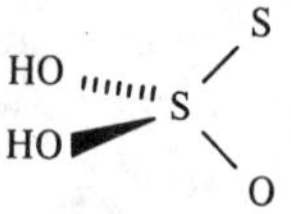

thiosulphuric
acid

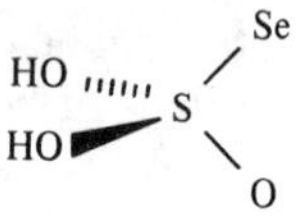

selenosulphuric
acid

type B:

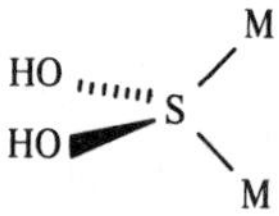

; no example yet known

Replacement of a hydroxyl group on H_2SO_4

The hydroperoxide group —OOH is very similar to the hydroxyl group —OH and should be capable of at least partial replacement. Peroxomonosulphuric acid, also known sometimes as Caro's acid, does indeed exist, but derivatives of the doubly substituted acid have yet to be isolated:

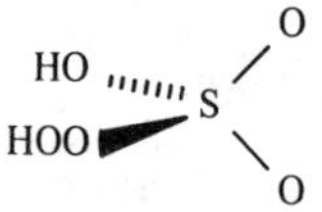

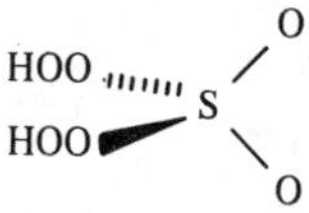

peroxomonosulphuric   (not known)
acid

Groups isoelectronic to hydroxyl may also be substituted into sulphuric acid. For example, an amino group may be substituted for hydroxyl to give sulphamic acid, and substitution of fluorine and chlorine yields fluorosulphonic acid and chlorosulphonic acid respectively:

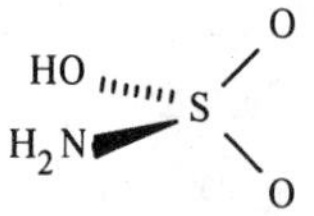

sulphamic
acid

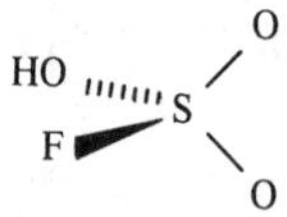

fluorosulphonic
acid

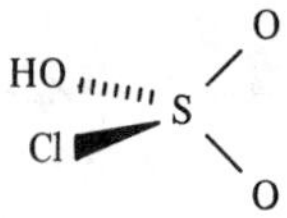

chlorosulphonic
acid

Replacement of the bridge oxygen atom in pyro-sulphuric acid

The bridging oxygen atom in pyro-sulphuric acid can be replaced by other Group VI elements and also the number of Group VI atoms in the bridge can be increased:

trithionic acid

peroxodisulphuric acid   tetrathionic acid

pentathionic acid $S(Se)_n S$ $(n = 1\text{--}6)$

Very recently, polythionates containing from 3 to 22 sulphur atoms have been isolated chromatographically both from cultures of *Thiobacillus ferroxidans* and from reactions of the type

$$Na_2S_2O_3 + \text{`}S_7Cl_2\text{'} \rightarrow Na_2S_xO_6 + 2NaCl \quad (x = 4\text{--}18)$$

$$K_2S_2O_3 + S_2Cl_2 \xrightarrow[\text{HCl}]{\text{conc}} K_2S_yO_6 + 2KCl \quad (y = 3\text{--}22)$$

It is thought probable that even higher polythionates with as many as 150 sulphur atoms exist, but so far it has not proved possible to isolate them.

HALIDES OF THE GROUP VI ELEMENTS

Table 40 lists the more common halides of this group. An unusual trend is that the iodides only become stable at tellurium and polonium, because normally M—X bonds decrease in strength as the atomic weight of M increases. Similarly, tellurium tetrachloride is the only tetrachloride which is stable in the gaseous state.

The 'monohalides' M_2X_2 can, theoretically, adopt two possible structures, but only the highly unstable sulphur monofluoride S_2F_2 has been found to exist in both the forms shown in Table 40; all the other monohalides adopt the gauche (H_2O_2) conformation. Some of the tetrahalides are associated in the solid state; TeF_4 has a polymeric chain structure whereas tetrameric units occur in $SeCl_4$, $SeBr_4$, $TeCl_4$,

Table 39—Reactions of some Group VI oxo-anions in aqueous solution

Most common sulphates and selenates are soluble in water exceptions being those of Ca^{2+}, Sr^{2+}, Ba^{2+}, Pb^{2+} and Hg^{2+}. Both SO_4^{2-} and SeO_4^{2-} are very stable anions and, apart from precipitation by the above cations, they undergo little aqueous chemistry, merely acting as 'bystander' ions in many reactions; selenate oxidizes hydrochloric acid to chlorine on boiling.

Reaction	Sulphite ion (SO_3^{2-})	Thiosulphate ion (SSO_3^{2-})	Peroxodisulphate ion ($O_3SOOSO_3^{2-}$)
1 Boil solution	Oxidized to SO_4^{2-} by air	No change	Decomposes to SO_4^{2-} and O_2 (plus some O_3)
2 Dilute HCl	Slow release of SO_2; more rapid on warming	Precipitation of sulphur	SO_3^{2-} and Cl_2 formed
3 Dilute alkali	No action	No action	No action
4 Addition of $BaCl_2$	White precipitate, $BaSO_3$	White precipitate BaS_2O_3	No reaction; on boiling, $BaSO_4$ precipitate formed owing to decomposition of $S_2O_8^{2-}$
5 Addition of KI solution	No reaction	No reaction	Iodine formed
6 Addition of I_2 in KI	Decolorized owing to formation of I^-	Decolorized owing to formation of I^- and $S_4O_6^{2-}$	No reaction
7 Addition of $KMnO_4$	Decolorized owing to oxidation of SO_3^{2-} to SO_4^{2-}	Decolorized; $S_2O_3^{2-}$ oxidized to SO_4^{2-}	No reaction
8 Addition of Fe^{2+} or Cr^{3+} ions	No reaction	No reaction	Oxidation to Fe^{3+} or $Cr_2O_7^{2-}$
9 Addition of lead ethanoate	White precipitate, $PbSO_3$	White precipitate, PbS_2O_3, soluble in excess of reagent as complex ion	No reaction in cold; precipitation of $PbSO_4$ on boiling owing to decomposition of $S_2O_8^{2-}$

Table 40—Some common chalcogen halides

	Oxygen	Sulphur	Selenium	Tellurium	Polonium
M_2X_2	O_2F_2	S_2F_2 (two isomers)	—	—	—
	—	S_2Cl_2	Se_2Cl_2	—	—
	—	S_2Br_2	Se_2Br_2	—	—
	—	S_2I_2	—	—	—
M_nX_2	O_4F_2	S_nCl_2 $(n=3\text{–}8)$	—	—	—
		S_nBr_2 $(n=3\text{–}8)$	—	—	—
MX_2	OF_2	SF_2	—	$TeCl_2$	$PoCl_2$
	OCl_2	SCl_2	—	$TeBr_2$	$PoBr_2$
	OBr_2	—	—	$TeBr_2$	PoI_2?
MX_4	—	SF_4	SeF_4	$(TeF_4)_x$	PoF_4
	—	SCl_4	$(SeCl_4)_4$	$(TeCl_4)_4$	$PoCl_4$
	—	—	$(SeBr_4)_4$	$(TeBr_4)_4$	$PoBr_4$
	—	—	—	$(TeI_4)_4$	PoI_4
MX_6	—	SF_6	SeF_6	TeF_6	PoF_6
M_2X_{10}	—	S_2F_{10}	—	Te_2F_{10}	—

Oxygen also forms Cl_2O_6, Cl_2O_7, I_2O_4, I_4O_9 and I_2O_5,

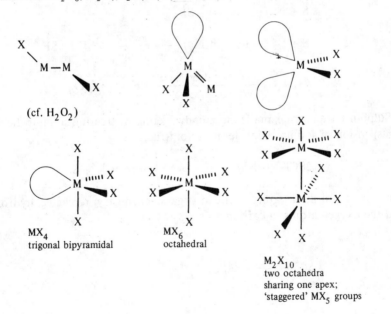

(cf. H_2O_2)

MX_4
trigonal bipyramidal

MX_6
octahedral

M_2X_{10}
two octahedra
sharing one apex;
'staggered' MX_5 groups

$TeBr_4$ and TeI_4:

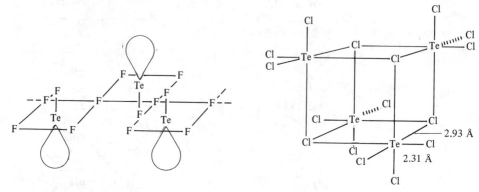

Part of chain in $(TeF_4)_x$

$(TeCl_4)_4$ unit present in solid tellurium tetrachloride; $SeCl_4$, $SeBr_4$ and $TeBr_4$ are similar

In most cases the halides can be synthesized from the elements, using carefully controlled conditions when necessary:

$$S + F_2 \text{ (in excess)} \rightarrow SF_6$$

$$S + F_2 \text{ (diluted in } N_2) \xrightarrow[\text{temperature}]{\text{low}} SF_4 + SF_6$$

$$O_2 + F_2 \xrightarrow[\text{at } -185\,°C]{\text{electrical discharge}} F_2O_2, F_2O_4$$

Sulphur tetrafluoride, made more conveniently by treating sulphur dichloride with a suspension of sodium fluoride in acetonitrile,

$$2SCl_2 + 4NaF \xrightarrow{70\,°C} SF_4 + 4NaCl\downarrow + S$$

is a very useful fluorinating agent in organic chemistry, replacing hydroxyl groups and the oxygen atoms of carbonyl groups by fluorine:

$$RC\overset{\displaystyle O}{\underset{\displaystyle OH}{\big\|}} \xrightarrow{SF_4} RCF_3$$

$$\underset{R}{\overset{R}{>}}C=O \xrightarrow{SF_4} \underset{R}{\overset{R}{>}}CF_2$$

$$ROH \xrightarrow{SF_4} RF$$

Sulphur hexafluoride is normally considered to be a highly stable molecule;

apparently the stability must be purely kinetic in nature, because many reactions which are thermodynamically favoured do not occur. An example is the hydrolysis

$$SF_6(g) + 6H_2O(g) \rightarrow SO_3(g) + 6HF(g), \quad \Delta G^\ominus = -301.2 \text{ kJ}$$

Presumably the six fluorine atoms sterically shield the sulphur atom from the close approach of possible attacking groups. The less sterically crowded sulphur tetrafluoride in contrast, is exceedingly water-sensitive:

$$SF_4 + H_2O \rightarrow SOF_2 + 2HF$$

A developing area of sulphur chemistry is that involving substitution products of sulphur hexafluoride, many of which are made photochemically via the more reactive SF_5Cl:

$$Cl_2 + CsF + SF_4 \xrightarrow{\ 110\,^\circ C\ } SF_5Cl + CsCl$$

$$SF_5Cl + H_2 \xrightarrow{\ h\nu\ } S_2F_{10} + 2HCl$$

$$SF_5Cl + O_2 \xrightarrow{\ h\nu\ } SF_5-O-O-SF_5 + SF_5-O-SF_5$$

$$SF_5Cl + C_2H_2 \rightarrow ClCH{=}CHSF_5$$

$$ClF + (CF_3)_2S \xrightarrow[25\,^\circ C]{\text{pressure}} cis\text{- and } trans\text{-}(CF_3)_2SF_4$$

$$ClF + SF_4 \rightarrow SF_5Cl$$

ORGANO-DERIVATIVES OF THE GROUP VI ELEMENTS

All the elements except polonium form MR_2 derivatives (those of oxygen, the ethers, belong to organic chemistry and will not be discussed here) which are easily prepared either by treating a halide with the corresponding Grignard or lithium reagent in an ether solvent or by direct reaction at elevated temperature:

$$MX_2 + 2LiR \rightarrow MR_2 + 2LiX \quad (M = S, \text{ Se or Te}; X = Cl)$$

$$RI + S \xrightarrow[\text{tube}]{\text{sealed}} SR_2 + RS-SR + RS-S-SR \quad (R = CF_3 \text{ or } C_6F_5)$$

$$R_2Se + X_2 \rightarrow R_2SeX_2 \xrightarrow{\ HgR_2\ } R_3SeX \quad (X = Cl \text{ or } Br)$$

$$R_2M + RX \rightarrow R_3MX \quad (\text{e.g. } X = I)$$

The structure of the MR_2 compounds is similar to that of the water molecule, the two lone pairs of electrons making them useful donor molecules:

boron trifluoride etherate

Chapter 10

Group VII: The Halogens. Fluorine, Chlorine, Bromine, Iodine and Astatine

INTRODUCTION

Except for fluorine, the first ionization energies of the halogens are lower than that of hydrogen (1310 kJ mol^{-1}) but the resulting ions are relatively large and hence cannot be stabilized by either high lattice energies or high solvation energies; as a result no simple ionic salts of the halogens are yet known. However, it is possible to isolate a few compounds of Br$^+$ and I$^+$ in which some of the expended ionization energy has been offset by the formation of strong covalent bonds between the cations and suitable Lewis bases as in [I(pyridine)$_2$]NO$_3$:

$$AgNO_3 + I_2 + 2C_5H_5N \xrightarrow{CHCl_3} AgI + [I(C_5H_5N)_2]^+ NO_3^-$$

$$[Ag(C_5H_5N)_2]SbF_6 + Br_2 \xrightarrow{CH_3CN} AgBr + [Br(C_5H_5N)_2]^+ SbF_6^-$$

(linear ion)

The fact that astatine follows iodine in tracer studies on the [I(pyridine)$_2$]$^+$ ion demonstrates the existence of At$^+$ as would be expected from trends within the group.

The halogens form a few rather unstable compounds containing nitrate, perchlorate, fluorosulphate and trifluoroethanoate groups, typical of which are FClO$_4$, BrNO$_3$, I(NO$_3$)$_3$ and I(ClO$_4$)$_3$. It is clear from spectroscopic studies that these are not ionic species but contain oxo-acid radicals *covalently* bound to the halogen via their oxygen atoms:

(planar)

$$BrF_5 + LiNO_3 \text{(excess)} \xrightarrow{0\,^\circ C} BrONO_2 \quad \text{(yellow solid; m.p. } -28\,^\circ C\text{)}$$

$$NF_4^+ ClO_4^- \xrightarrow{\text{heat}} FOClO_3$$

$$I_2 + 6ClONO_2 \rightarrow 2I(ONO_2)_3; \text{ quantitative yield}$$

$$F_2 + KNO_3 \rightarrow FONO_2$$

In many of their compounds the halogens achieve a rare gas electron configuration on acquiring a further electron by either the formation of X^- ions or M—X covalent bonds. The simplest covalent compounds which the halogens form are the X_2 diatomic molecules whose dissociation energies are shown in Table 41. Reasons for the anomalous value of fluorine's dissociation energy were discussed on page 58, one being the possibility of π bonding between filled p orbitals on one halogen and empty d orbitals on the other. Although the extent to which this occurs in the heavier diatomic halogens is difficult to estimate accurately, a recent theoretical study does suggest that back-bonding between fluorine and chlorine is important in the ClF molecule. Notwithstanding the weakness of the F—F bond in F_2, the high electronegativity of fluorine ensures that covalent fluorides have a substantial electrostatic contribution in their bonding; as a result many M—F bonds rank among the strongest known interactions.

CONSEQUENCES OF FLUORINE'S SMALL SIZE

From the covalent radii given in Table 41 it can be seen that the volume occupied by a covalently bound fluorine atom is only about one-third that of chlorine and less than one-sixth of that occupied by iodine. The small size of the fluorine atom and the low dissociation energy of the fluorine molecule are mainly responsible for the ability of fluorine to expand the covalence of many elements; for example, iodine heptafluoride (IF_7), sulphur hexafluoride (SF_6), uranium hexafluoride (UF_6), tungsten hexafluoride (WF_6) and many other transition metal hexafluorides. The obvious influence of size is to allow a large number of fluorine atoms to pack round the central element whereas the dissociation energy of the fluorine molecule is an

Table 41—The halogens

Element	1st IE (kJ mol^{-1})	Covalent radius^a (Å)	Ionic radius, X$^-$ (Å)	Electron affinity (kJ mol^{-1})	Dissociation energy, X$_2$ (kJ mol^{-1})
9 Fluorine, F $\quad$ 1s^{2}2s^{2}2p^5	1682	0.71	1.19	332.7	157.7
17 Chlorine, Cl $\quad$ [Ne]3s^{2}3p^5	1255	0.99	1.67	348.5	243.5
35 Bromine, Br $\quad$ [Ar]3d^{10}4s^{2}4p^5	1142	1.14	1.82	324.5	192.9
53 Iodine, I $\quad$ [Kr]4d^{10}5s^{2}5p^5	1009	1.33	2.06	295.4	150.6
85 Astatine, At $\quad$ [Xe]4f^{14}4d^{10}6s^{2}6p^5	(912)	—	—	(255.3)	(116)

^{a} Defined as half the bond distance in X$_2$.

important thermodynamic step in the production of a covalent fluoride:

$$
\begin{array}{l}
\left.
\begin{array}{ccccc}
\text{M} & \xrightarrow{\text{atomization}} & \text{M(g)} & \xrightarrow{\text{excitation}} & \text{M} \\
\text{(standard} & & \text{(gaseous} & & \text{(valence} \\
\text{state)} & & \text{atoms)} & & \text{state)} \\
\\
& & \frac{1}{2}nF_2 & \xrightarrow{\frac{1}{2}nD_{F_2}} & \begin{array}{c} + \\ nF \\ \text{(gaseous} \\ \text{atoms)} \end{array}
\end{array}
\right\} \to \text{MF}_n
\end{array}
$$

For a stable fluoride to be formed

$$B_{M-F} > (\text{energy of atomization of M} + \text{excitation energy of M to valence state} + \tfrac{1}{2}nD_{F_2})$$

where B_{M-F} is the *total* M—F bond energy in MF_n. Clearly, formation of the fluoride MF_n is favoured by the small value of D_{F_2}, particularly as several fluorine molecules may be involved.

The heats of formation of halide ions from gaseous X_2 molecules,

$$\tfrac{1}{2}X_2(g) \to X^-(g)$$

are listed in Table 42. They obviously show little variation within the group and hence play only a minor role in determining the differences in chemical behaviour between the halogens. Of more importance is the small size of the fluoride ion relative to the other halides because this results in F^- having the highest hydration energy (see Table 42) and in ionic fluorides having high lattice energies.

The substantial change in size accompanying the formation of a halide ion such as fluoride from a halogen atom is due to the increased mutual shielding of the electrons from the nuclear charge (or put another way, to increased electron–electron repulsions which occur on addition of an extra electron to the system).

COMPARISON OF FLUORIDE AND OXIDE IONS

The radius of the fluoride ion (1.19 Å) is very similar to that of the oxide ion (1.26 Å) and it is often found that fluorides and oxides which have similar stoichiometries

Table 41—Heats of formation and hydration energies of the halide ions

Ion	Heat of formation (kJ mol^{-1}) $\frac{1}{2}X_2(g) \to X^-(g)$	Hydration energy (kJ mol^{-1}) $X^-(g) \to X^-(aq)$
Fluoride, F$^-$	−259.8	−506
Chloride, Cl$^-$	−233.0	−369
Bromide, Br$^-$	−234.3	−335
Iodide, I$^-$	−226.0	−293

have similar structures. An example of this behaviour is shown by calcium oxide and potassium fluoride, both of which adopt the sodium chloride structure. However, the oxide always has the higher melting point and is very much less soluble than the fluoride because the ions in a crystalline oxide necessarily carry a greater charge and this results in extremely high lattice energies.

HYDRATION OF HALIDE IONS

A combination of X-ray and neutron diffraction studies on aqueous solutions of several metal chlorides and bromides has shown that there are six water molecules hydrogen-bonded to the anions in a well-defined hydration shell, although the residence time of individual water molecules within the shell is probably only about 5×10^{-12} s:

2.24 Å

$(OH)_5Cl^- \longrightarrow H\text{------}$

angle θ is usually small,
$0°-10°$; i.e. angle ClHO
is close to $180°$

With some cations at high molarities the hydration number of Cl^- falls below six because of ion pairing; even a univalent cation like Li^+ begins to attach to chloride ions giving $LiCl(H_2O)_5$ and $LiCl_2(H_2O)_4^-$ as lithium chloride concentrations rise above 3M.

However, no ion pairing occurs with long-lived ($\sim 10^{-5}$ s) hydrated cations like $Ni(H_2O)_6^{2+}$ and hence there is competition between anion and cation for the available water. In 4.35M nickel chloride, for example, there are only 12.6 H_2O molecules per $NiCl_2$ unit and under these circumstances it is found that the chloride ions have to hydrogen-bond with water already coordinated to the cations:

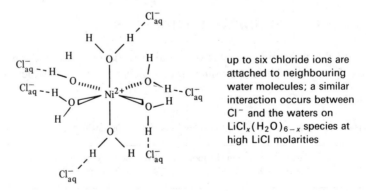

up to six chloride ions are attached to neighbouring water molecules; a similar interaction occurs between Cl^- and the waters on $LiCl_x(H_2O)_{6-x}$ species at high LiCl molarities

In marked contrast to Cl^- it is found that the *polyatomic* anions ClO_4^- and NO_3^- have only a very weak and ill-defined hydration shell. When sufficient water is present to satisfy the hydration requirements of both anion and cation, no ion pairing occurs with perchlorate under conditions where it can be detected for chloride. This is gratifying since chemists often assume that perchlorate ions have little affinity for cations *when in solution* (in the solid state ClO_4^- ions can act as monodentate, bidentate, or tridentate bridging ligands).

HYDROGEN-BONDED COMPLEXES OF HALIDE IONS

Unlike cations, negatively charged halide ions are unable to interact with Lewis bases to form complexes; however, they are able to hydrogen-bond with some protonic species. The simplest such complexes are the hydrogen dihalide ions $[X—H—X]^-$ formed when ionic halides are treated with the appropriate hydrogen halide; stabilities are generally in the order $F \gg Cl > Br > I$ and large cations are required to preserve those anions containing the heavier halogens:

$$KF + HF(l) \rightarrow K^+[F—H—F]^-$$

$$NMe_4X + HX(l) \rightarrow NMe_4^+[X—H—X]^- \quad (X = Cl \text{ or } Br)$$

$$(n–Bu)_4NI + HI(g) \xrightarrow{CH_2Cl_2} (n–Bu)_4N^+[I—H—I]^-$$

$$NMe_4F + HX(l) \rightarrow NMe_4^+[F—H—X]^- \quad (X = Cl, Br \text{ or } I)$$

Halide ions can also be trapped inside protonated macropolycyclic ligands of the type H_4L^{4+} shown in Fig. 95. For example, one of the chloride ions in the salt hydrate $H_4L^{4+}(Cl^-)_4.7H_2O$ is *fully encapsulated* by the ligand where it is assumed to hydrogen-bond with the protons of the four quarternary nitrogen atoms. However, it is not necessary to use crypt-like ligands when hydrogen-bonding an organic molecule to chloride; in the salt $(pyH)_3Cl(AlCl_4)_2$ three protonated pyridine rings

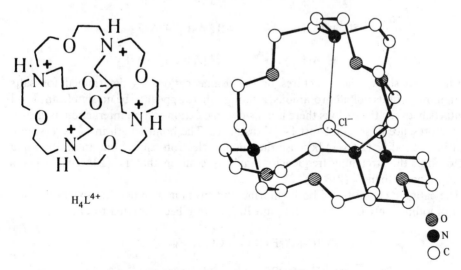

$$H_4L^{4+}$$

◒ O
● N
○ C

Fig. 95—The X-ray structure of the complex cation $[ClLH_4]^{3+}$. [Reproduced with permission from *Journal of the Chemical Society, Chemical Communications* (1976) p. 533.]

hydrogen-bond to the chloride ion giving a pyramidal $[(pyH)_3Cl]^{2+}$ cation which packs into the crystal with two tetrachloroaluminate anions:

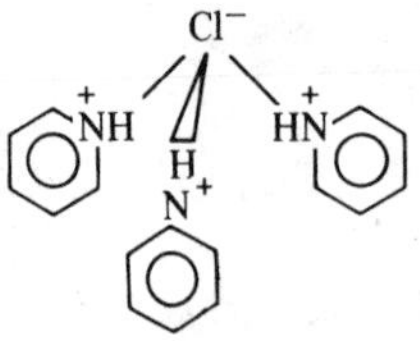

HOMONUCLEAR POLYATOMIC CATIONS

No cations of fluorine have yet been prepared. As might be expected it becomes progressively easier down the group to make cationic species and hence those of iodine have been the most widely studied. There are presently no known compounds containing free I^+ but I_2^+, I_3^+, I_5^+ and I_4^{2+} can be generated by reacting iodine dissolved in fluorosulphonic acid, HSO_3F, with stoichiometric amounts of the extremely powerful oxidant bis(fluorosulphuryl)peroxide, $(FSO_2)OO(SO_2F)$:

$$2I_2 + S_2O_6F_2 \rightarrow I_2^+ \quad (+SO_3F^-)$$

$$3I_2 + S_2O_6F_2 \rightarrow I_3^+ \quad (+SO_3F^-)$$

$$5I_2 + S_2O_6F_2 \rightarrow I_5^+ \quad (+SO_3F^-)$$

Cooling of the deep blue I_2^+ solution to about $-70\,°C$ results in a reversible colour change to red as dimerization of the cation to I_4^{2+} occurs. Subsequently, crystalline salts containing these ions have been isolated from more convenient solvents such as liquid sulphur dioxide:

$$3I_2 + 3AsF_5 \xrightarrow{\text{1.SO}_2} 2I_3^+ AsF_6^- + AsF_3$$

$$5I_2 + 3AsF_5 \xrightarrow{\text{1.SO}_2} 2I_5^+ AsF_6^- + AsF_3$$

$$2I_2 + 3AsF_5 \xrightarrow{\text{1.SO}_2} I_4^{2+}(AsF_6^-)_2 + AsF_3$$

Figure 96 shows the structures of the iodine cations as determined for salts containing various polyfluoro-anions; although there appear to be no significant I---I contacts between the cations there is considerable cation–anion interaction as shown by the presence of several short I---F distances. The bond length in I_2^+ is *less* than that in free iodine (2.66 Å) because the missing electron came from an antibonding π^* orbital; the stretching frequencies similarly confirm that I_2^+ (238 cm^{-1}) has a stronger bond than I_2 (213 cm^{-1}).

The cationic chemistry of both chlorine and bromine is much less extensive than that of iodine, only Cl_3^+, Br_5^+, Br_3^+ and Br_2^+ having been isolated to date:

$$Cl_2 + ClF + AsF_5 \underset{20\,°C}{\overset{-78\,°C}{\rightleftharpoons}} Cl_3^+ AsF_6^-$$

$$Br_2 + BrF_5 + SbF_5 \rightarrow Br_2^+ Sb_3F_{16}^-$$

$$Br_2 + SbF_5 \rightarrow Br_3^+ Sb_3F_{16}^-$$

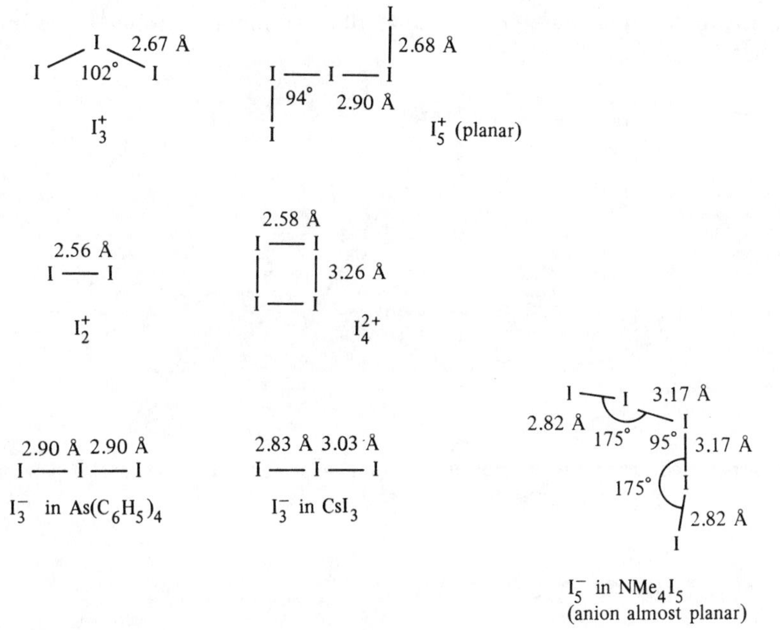

I — 2.67 Å
I 102° I

I_3^+

I
| 2.68 Å
I — I — I
| 94° 2.90 Å
I

I_5^+ (planar)

2.56 Å
I — I

I_2^+

2.58 Å
I — I
| | 3.26 Å
I — I

I_4^{2+}

2.90 Å 2.90 Å
I — I — I

I_3^- in $As(C_6H_5)_4$

2.83 Å 3.03 Å
I — I — I

I_3^- in CsI_3

I — I 3.17 Å
2.82 Å 175° 95° | 3.17 Å
175° | I
| 2.82 Å
I

I_5^- in NMe_4I_5
(anion almost planar)

Fig. 96—Structures of polyiodine cations and anions.

HOMONUCLEAR POLYATOMIC ANIONS

Although Cl_3^- and Br_3^- have been prepared, for example, by

$$As(C_6H_5)_4^+ Cl^- + Cl_2 \xrightarrow[\text{solution}]{\text{aqueous}} As(C_6H_5)_4^+ Cl_3^- \quad \text{(note the large cation)}$$

poly*iodide* anions are far more numerous. The best known is undoubtedly I_3^- which is widely used as a means of holding 'iodine' in aqueous solution because its formation is readily reversible:

$$I^- + I_2 \rightleftharpoons I_3^- \quad \text{equilibrium constant } c. \ 700 \ (25°C)$$

The triiodide ion is linear and, in solution, is probably also symmetrical; however, in the solid state its symmetry depends markedly on the accompanying cation:

$$\left[I \xrightarrow{2.90 \text{ Å}} I \xrightarrow{2.90 \text{ Å}} I \right]^- \quad \text{as in } As(C_6H_5)_4^+ I_3^-$$
(symmetrical)

$$\left[I \xrightarrow{2.83 \text{ Å}} I \xrightarrow{3.03 \text{ Å}} I \right]^- \quad \text{as in } Cs^+ I_3^-$$
(unsymmetrical)

Many higher polyiodides have been isolated, the simplest of which contain the *formal* species I_5^-, I_7^- or I_9^- stabilized in the solid state by large cations such as tetramethylammonium and pyridinium; weak I---I interactions link these anions into polymeric chains as in $C_5H_5NH^+I_5^-$:

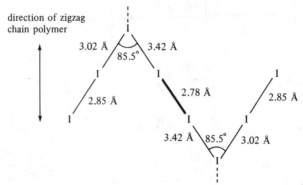

(I_5^- can be regarded as an association of alternating I_2 molecules and I_3^- ions; only *one* of the terminal iodines in I_3^- is involved in this association)

STARCH IODIDE TEST FOR IODINE

The well-known blue-black coloration produced by a mixture of starch and iodide ions in the presence of free iodine has long been used as a sensitive test for I_2. A recent investigation using both Raman and ^{129}I Mössbauer spectroscopy has revealed that the major chromophore having an absorption maximum at about 600 nm is an I_5^- – amylose complex. Amylose, a component of starch, has a helically coiled structure derived from 1,4-linked α-D(+)glucopyranose units,

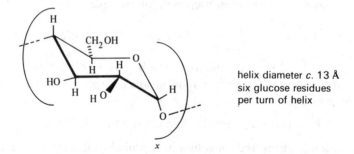

helix diameter *c.* 13 Å
six glucose residues
per turn of helix

and it is within this helix that the pentaiodide ion is trapped. In contrast, the branched-polymeric fraction of starch called amylpectin does not form a similar complex.

CHLORINE ISOTOPE EFFECTS IN NMR AND MASS SPECTROMETRY

The greatly improved resolution now possible with very-high-field nuclear magnetic resonance spectrometers has shown up an unusual type of isotope effect in proton spectra of chlorinated hydrocarbons such as CH_3Cl (but the effect is not necessarily restricted to chlorine isotopes). Methyl chloride has only one type of hydrogen environment and hence should show only a single peak in its proton spectrum; however, at 500 MHz two peaks are observed having relative intensities of about 3:1 which is the approximate abundance ratio of the ^{35}Cl and ^{37}Cl isotopes. Although the chlorine atom is two bonds away from the hydrogens, the NMR spectrometer can just resolve the different peaks arising from the two *isotopomeric* molecules $CH_3{}^{35}Cl$ and $CH_3{}^{37}Cl$:

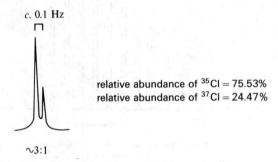

relative abundance of $^{35}Cl = 75.53\%$
relative abundance of $^{37}Cl = 24.47\%$

Since isotopes differ in mass their effect is more commonly observed in mass spectrometry. This can again be illustrated by methyl chloride which has a molecular ion cluster for CH_3Cl^+ in its mass spectrum consisting of four peaks corresponding to the isotopomeric species $^{12}CH_3{}^{35}Cl$, $^{13}CH_3{}^{35}Cl$, $^{12}CH_3{}^{37}Cl$ and $^{13}CH_3{}^{37}Cl$ (neglecting the minute amounts of deuterium present). The intensities of the peaks can be *statistically* calculated knowing the relative abundance of both the carbon and chlorine isotopes and the group would appear as

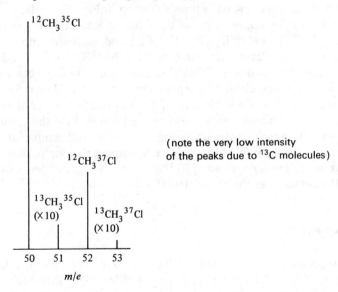

(note the very low intensity of the peaks due to ^{13}C molecules)

When an element has many stable isotopes (e.g. tin has 10) or when a molecule contains several elements with more than one isotope each, many isotopomeric species will be present in the molecular ion and its fragmentation ions. Computers can be

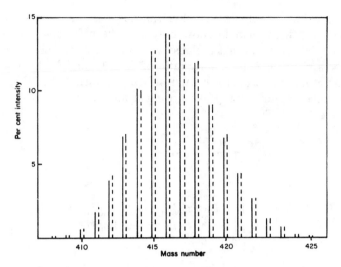

Comparison of the observed (solid line) and theoretical (broken line) intensities
for the molecular ion in B_9Cl_9.

Fig. 97—Comparison of the observed (solid line) and theoretical (broken line) intensities for the molecular ion in B_9Cl_9. [Reprinted with permission from *Journal of Inorganic and Nuclear Chemistry* vol. 32, G. F. Lanthier and A. G. Massey, 'A yellow-orange boron sub-chloride, B_9Cl_9', © 1970, p. 1807, Pergamon Press PLC.]

used to calculate the expected peak intensities in each ion cluster and these may be compared with the experimental values as shown in Fig. 97 for the molecular ion of B_9Cl_9. Under low resolution some of the different isotopomers will have the same mass (e.g. $^{11}B_9{}^{35}Cl_9$ and $^{10}B_2{}^{11}B_7{}^{37}Cl{}^{35}Cl_8$) and each will add to the observed peak intensity in Fig. 97; certain combinations of the ^{10}B, ^{11}B, ^{35}Cl and ^{37}Cl isotopes have such a low statistical probability that their peaks are too weak to be observed. The close fit of the theoretical and experimental intensities shown for B_9Cl_9 proves beyond reasonable doubt the identity of the molecular ion and hence that of the original molecule. A further check can be made by measuring the accurate masses of the main peaks in the ion cluster under high resolution and comparing them to those calculated from the isotopomeric formulae. For example, one peak in the spectrum of B_9Cl_9 is due to the isotopomer $^{11}B_8{}^{10}B_1{}^{35}Cl_8{}^{37}Cl_1$ which has a calculated mass of 414.8041 compared to the experimental value of 414.8039.

ASTATINE CHEMISTRY

All 21 isotopes of astatine are radioactive and have very short half-lives, the most stable being ^{210}At (8.3 h), ^{211}At (7.2 h) and ^{209}At (5.5 h). No suitable polonium isotope exists which could be used in (n,γ) reactions so that the synthesis of astatine normally starts from bismuth. Because two places have to be 'jumped' in the periodic table, the bismuth-209 must be bombarded with α particles ($^4_2He^{2+}$), when three reactions become possible depending on the energy of the α particles:

$$^{209}_{83}Bi + \alpha(c.\ 26\ \text{MeV}) \rightarrow {}^{211}_{85}At + 2n$$

$$^{209}_{83}\text{Bi} + \alpha(>29 \text{ MeV}) \rightarrow ^{210}_{85}\text{At} + 3n$$

$$^{209}_{83}\text{Bi} + \alpha(c.~60 \text{ MeV}) \rightarrow ^{209}_{85}\text{At} + 4n$$

Luckily, α bombardment of bismuth gives virtually pure astatine, but a minimum α energy of about 21 MeV is required to overcome the effects of nuclear repulsion otherwise fusion of target and projectile will not occur.

The above astatine isotopes undergo radioactive decay mainly by electron capture which is accompanied by the release of intense γ radiation. Usually tracer work is carried out using concentrations in the range 10^{-11}–10^{-15} M, the maximum possible concentration being only of the order of 10^{-8} M. As with polonium, radiolysis of the solvent water leads to production of hydrogen peroxide and oxygen which, in some cases, may interfere with the chemical study. Even though such tiny amounts of astatine are available, mass spectrometry has been used successfully to identify species such as HAt, CH_3At, AtBr and AtCl (Fig. 98).

In many ways astatine resembles iodine, even to the extent of being selectively concentrated in thyroid tissue after ingestion. At least four oxidation states (-1; 0; $+1$ (or $+3$?); $+5$) have been recognized by adding tracer amounts of astatine to iodine and noting if the characteristic radioactivity follows iodine through a specific sequence of reactions.

Astatine(0), removed from the bismuth target by heating, is volatile off glass apparatus at room temperature but is retained by a silver surface even at $325\,°C$, presumably owing to the formation of AgAt. Like iodine, At(0) is readily extracted into carbon tetrachloride from aqueous solutions but it has not yet been proved conclusively if At_2 molecules are present in this oxidation state. Reduction of astatine(0) by zinc or sulphur dioxide gives the astatide ion At$^-$ which co-precipitates with silver iodide on the addition of a soluble iodide and silver nitrate. Some of the other typical reactions of astatine are summarized below:

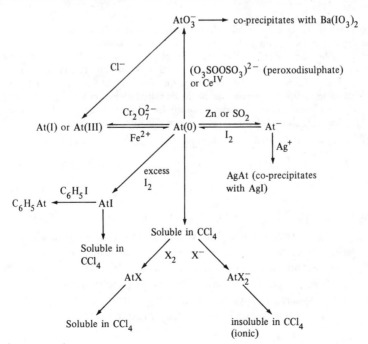

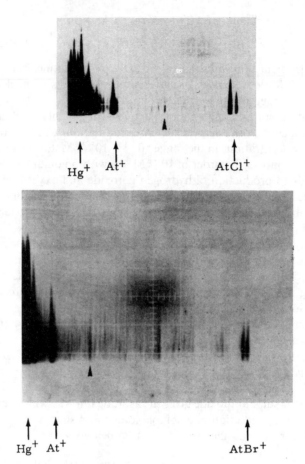

Fig. 98—Mass spectra of astatine chloride and astatine bromide. The two spectra, which have different scale expansions, had to be observed at high gain because of the tiny amounts of material available; note the 'splitting' arising from the presence of two chlorine or two bromine isotopes in the $AtCl^+$ and $AtBr^+$ ions. The arrowed peaks are due to $AtCH_3^+$ which comes from minute traces of hydrocarbons present in the spectrometer; the several isotopes of mercury from the pumping system are responsible for the intense cluster of peaks on the far left of the spectra. [Reprinted with permission from *Inorganic Chemistry* vol. 5, p. 766. Copyright (1966) American Chemical Society. Original photographs kindly made available, by Dr E. H. Appelman.]

Although it cannot be absolutely confirmed that an ion such as AtO_3^- is produced in a vigorous oxidation reaction using, for example, peroxodisulphate, it is nevertheless possible to demonstrate the negative charge using electromigration experiments: on application of an electric field the radioactivity associated with the ion moves towards the anode. Thin layer, paper or gas chromatographic techniques have all been used to separate astatine species, particularly those arising from complex organic syntheses where much unwanted debris is simultaneously produced; see Fig. 99.

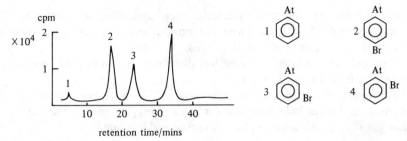

Fig. 99—Gas chromatogram of a mixture arising from a reaction of At$^-$ ions with ortho-, meta- and para-benzene diazonium salts. The invisible amounts of the eluted species were detected by their radioactivity; the retention times of several dihalobenzenes were found to be almost linearly related to their molecular weights, which allowed the astatine compounds to be positively identified. Furthermore, for a range of dihalobenzenes the elution sequence was always in the order para, meta and ortho, the sequence being assumed to hold for astatine as a substituent. [For details of the experiment see *Radiochemical and Radioanalytical Letters* vol. 21 (1975) pp. 247–259.]

PSEUDOHALIDES AND PSEUDOHALOGENS

Some pseudohalogens and pseudohalides are:

Pseudohalogen[a]	Pseudohalide
Cyanogen, $(CN)_2$	Cyanide, CN^-
—	Cyanate, OCN^-
Thiocyanogen, $(SCN)_2$	Thiocyanate, SCN^-
Selenocyanogen, $(SeCN)_2$	Selenocyanate, $SeCN^-$
—	Tellurocyanate, $TeCN^-$
—	Azide, N_3^-
—	Isocyanate, ONC^-

[a] In addition to the X_2 molecule the pseudohalogens can also be obtained as highly polymeric materials, e.g. $(CN)_x$, a black solid called paracyanogen, and $(SCN)_x$, an orange solid called parathiocyanogen or poly-thiocyanogen.

There are a number of uni-negative ions known which have many of the properties expected of halide ions; because of these similarities to the Group VII elements they are often called pseudohalides. Of the pseudohalides listed above, probably the best-known example is the cyanide ion, which has the following properties, among others, in common with chloride, bromide and iodide ions:

(a) It can be oxidized to cyanogen, $N{\equiv}C{-}C{\equiv}N$, which is a volatile and has two cyanide radicals joined in a symmetrical fashion. Cyanogen is therefore a pseudohalogen.

(b) The cyanide group is able to form 'interhalogen' derivatives such as FCN, ClCN, BrCN, ICN and $CN.N_3$, the latter compound being an 'interpseudohalogen' formed between cyanide and azide groups.
(c) Silver cyanide is insoluble in water but dissolves on addition of ammonia (cf. silver chloride and silver bromide).
(d) A wide variety of complex anions containing CN^- have been prepared which have their analogous halo-complexes (e.g. $Cu(CN)_4^{2-}$, $CuCl_4^{2-}$ and $CuBr_4^{2-}$).
(e) Cyanogen reacts with alkali to form cyanide and cyanate:

$$(CN)_2 + 2KOH \rightarrow KCN + KCNO + H_2O$$

$$cf. \ Cl_2 + 2KOH \rightarrow KCl + KClO + H_2O$$

Note, however, that the cyanate ion CNO^- is another pseudohalide and has properties quite unlike the hypochlorite ion ClO^-.
(f) The acid HCN is known but is very weak relative to the halogen acids ($pK = 8.9$).
(g) Copper(II) cyanide is unstable and readily loses cyanogen:

$$Cu^{2+} + 2KCN \rightarrow Cu(CN)_2 \rightarrow CuCN + \tfrac{1}{2}(CN)_2$$

Compare this with the rapid decomposition of copper(II) iodide into copper(I) iodide and iodine.

OCCURRENCE OF THE HALOGENS

Fluorine

270 ppm of the Earth's crust; occurs fairly commonly as fluorspar, CaF_2 and fluorapatite, $CaF_2.3Ca_3(PO_4)_2$.

Chlorine

480 ppm overall abundance in the Earth's crust, but 15 000 ppm in sea-water; main source is sodium chloride, which is either mined (rock salt) or obtained from sea-water.

Bromine

30 ppm; principal source is the bromides present in sea-water.

Iodine

0.3 ppm; occurs as sodium iodate to the extent of about 1 % in Chilean deposits of sodium nitrate. Some natural brines contain about 30 ppm of iodide.

Astatine

Minute amounts of short-lived astatine isotopes occur in uranium ores with ^{219}At ($t_{1/2} = 0.9$ min) being the longest lived. Estimates suggest that in the outermost mile of the Earth's crust there are only about 70 mg of astatine compared with 24.5 g of francium and 4000 tons of polonium. Astatine probably has the dubious distinction of being the rarest element in nature. All its isotopes are made artificially, e.g. by the reaction $^{209}Bi(\alpha,2n)^{211}At$; after irradiation by the α particles, the bismuth target is heated to 430°C in a stream of gaseous nitrogen and the astatine collected (as At_2) on a cold finger held at liquid nitrogen temperatures.

EXTRACTION

Fluorine

Produced by the electrolysis of dry, fused potassium hydrogen fluoride, KHF_2 (m.p. 217°C), carried out in a copper or steel cell. The cell must be provided with a diaphragm (or similar device) to prevent explosive recombination between fluorine and the hydrogen which is liberated simultaneously at the cathode. The main impurity is about 10% of hydrogen fluoride which is removed using sodium fluoride:

$$NaF + HF \rightarrow NaHF_2 \quad \text{(i.e. NaF.HF)}$$

Essentially, this preparation is the electrolysis of hydrogen fluoride which has been made conducting by the presence of potassium fluoride, pure anhydrous hydrogen fluoride being a very poor conductor of electricity.

Chlorine

Since chlorine is not as reactive as fluorine and does not attack water to any meaningful extent it may be prepared by the electrolysis of salt brine, and practically all commercially required chlorine is made in this way; the secondary product of the electrolysis is a solution of sodium hydroxide.

Bromine

Bromine is released from sea-water by chlorinating it at pH 3.5 (mainly to prevent the extensive formation of oxy-salts which would occur at higher pH). Approximately 8000 l of wea-water yield 0.5 kg of bromine.

Iodine

Sodium iodate is reduced in aqueous solution (e.g. with sodium hydrogen sulphite) to iodine which is purified by sublimation. Iodide in brines is oxidized by chlorine and the solutions stripped of iodine in towers using a counter-current of air.

THE ELEMENTS

Under normal conditions the halogens exist mainly as diatomic molecules, although

X_4 species have been detected both in the gas phase and in some solvents for bromine and iodine: e.g.

$$2I_2 \rightleftharpoons I_4 \quad (1.4 \text{ mol}\% \text{ at } 240\,^\circ\text{C and } 2.5 \text{ atm})$$

These dimers are thought to be held together by van der Waals attraction rather than true covalent bonds. There is obviously a weak interaction between the molecules of solid iodine because the distance between non-bonded iodine atoms (3.56 Å) is much shorter than twice the van der Waals radius of iodine (2×2.15 Å) and, on the application of high pressures, the solid begins to develop electrical properties normally associated with metals. A simplified description of the bonding in diatomic halogen molecules is shown in Fig. 100, but some p_π–d_π bonding undoubtedly occurs in Cl_2, Br_2 and I_2 as revealed by theoretical calculations on FCl. Back-bonding involving

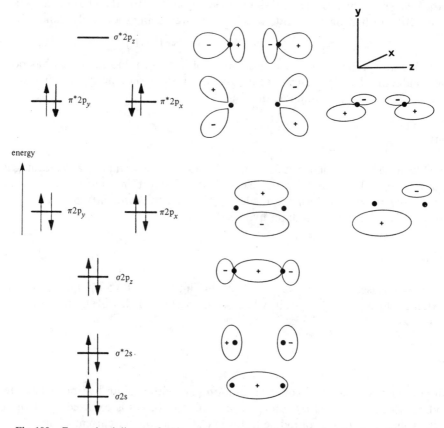

Fig. 100—Energy level diagram for the molecular orbitals in F_2. The bond order in F_2 is one; because of the large 2s–2p energy difference in the fluorine atom, the amount of mixing which occurs between σ2s and σ2p in F_2 is not sufficient to push the σ2p orbital above the degenerate pair π2p$_x$ and π2p$_y$. The other halogen atoms have an ns–np separation which is less than half that of fluorine and consequently their diatomic molecules have the configuration $(\sigma ns)^2(\sigma^* ns)^2(\pi np)^4(\sigma np)^2(\pi^* np)^4$.

chlorine d orbitals had to be included in the calculations in order to obtain an accurate F---Cl bond distance.

A $\sigma^* \leftarrow \pi^*$ electronic transition is responsible for the absorption band which gives rise to the colour of the halogens: fluorine, yellow; chlorine, green; bromine, red-brown; iodine, purple. It has been known for many years that the colour of iodine in solution can vary from violet through red to brown depending on the nature of the solvent used, even though cryoscopic measurements show that 'I_2 molecules' are always present. Some solvents are able to share one of their electron pairs with iodine by interacting with the empty $\sigma^* np$ orbital; overlap of the solvent donor orbital with $\sigma^* np$ creates a bonding (B) and an antibonding (A) orbital as shown in Fig. 101. Electronic transitions in the weakly bonded complex now occur between A and π^* which move the absorption band towards the blue owing to the higher energy involved.

Amines, ethers and alcohols interact with iodine via lone pairs whereas aromatics and alkenes use their filled π orbitals; however, saturated hydrocarbons and CCl_4 possess no suitable donor electrons and iodine in such solvents retains its normal purple colour. Although many of the complexes are weak it is often possible to isolate them as solid species and hence determine their structures using X-ray crystallography. Three representative structures are shown below but, as usual, it must be remembered that the solid phase separating from a solution does not necessarily have the same structure as the dissolved molecules. The X—X bond distance in these halogen complexes is invariably found to be larger than that of the free X_2 molecule because the interaction shown in Fig. 101 places electron density into an orbital ($\sigma^* np$) which

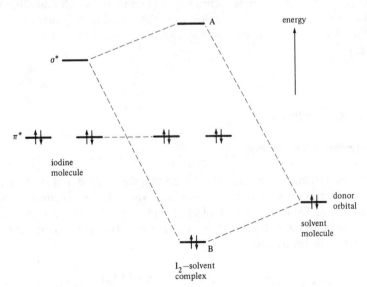

Fig. 101—Interaction between the I_2 σ^* orbital and a donor orbital on the solvent molecule. The colour of an iodine solution is determined by the energy of the $A \leftarrow \pi^*$ electronic transition. The corresponding absorption band will progressively move towards the blue end of the spectrum as the iodine–solvent interaction becomes stronger. (The absorption maximum for violet solutions is in the 520–540 nm range and that of a typical brown solution is about 460–480 nm.)

is antibonding relative to the two halogen atoms:

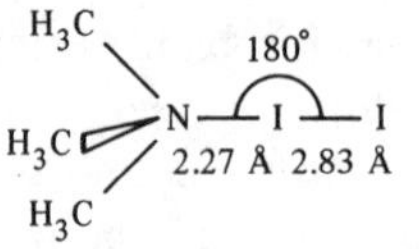

trimethylamine–iodine complex

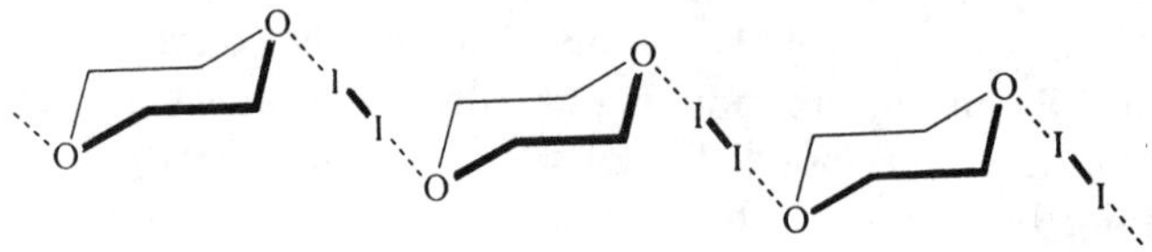

(also for Cl_2 and Br_2)

1,4-dioxan–iodine complex

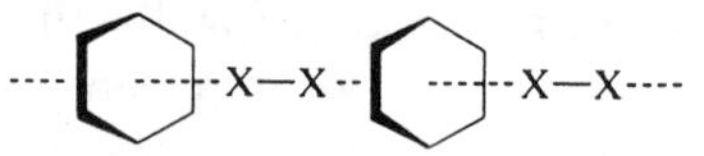

(X = Cl or Br)

benzene–X_2 complex

Many of the complexes are thermally labile and, for example, heating brown solutions of iodine in strong donor solvents can often change the colour to violet as the complexes dissociate. In addition to the absorption which is responsible for the colour of halogen solutions there is another intense band located in the ultraviolet region which is due to the transition $A \leftarrow B$ (Fig. 101). This new transition represents the transfer of an electron from an orbital on the complex which has a high ligand character to one which is largely localized on the halogen molecule (i.e. has high halogen character); quite naturally such a transition is said to give rise to a *charge transfer band*.

INTERHALOGEN SPECIES

Diatomic interhalogen molecules

One of the most fascinating properties of the halogens is their ability to react with each other and form mixed halogen derivatives which can be neutral, cationic or anionic as shown in Table 43. The simplest of these various interhalogen species are the diatomic molecules XY; the small entropy increase (between 4.6 and 5.4 J K^{-1}) associated with their preparation,

$$\tfrac{1}{2}X_2(g) + \tfrac{1}{2}Y_2(g) \xrightarrow[\text{1 atm}]{25°C} XY(g)$$

is close to the theoretical value calculated for homonuclear diatomic molecules changing into heteronuclear diatomic molecules:

$$\Delta S = R \log_e 2 = 5.9 \text{ J K}^{-1} \text{ mol}^{-1}$$

Table 43—Some mixed halogen compounds

Neutral species			
ClF	ClF_3	ClF_5	
BrF	BrF_3	BrF_5	
IF	IF_3	IF_5	IF_7
$BrCl$			
ICl	$(ICl_3)_2$		
IBr			

Cationic species (central atom on left)		
$ClFCl^+$	ClF_4^+	ClF_6^+
ClF_2^+	BrF_4^+	BrF_6^+
ICl_2^+	IF_4^+	IF_6^+

Anionic species (central atom on extreme left)			
$BrCl_2^-$	ClF_4^-		
ICl_2^-	BrF_4^-	BrF_6^-	
IBr_2^-	ICl_4^-	IF_6^-	$IF_8^-(?)$
$IBrCl^-$	ICl_3F^-		
$IICl^-$			
$BrBr_2^-$			
II_2^-			

cationic species

anionic species

YX_2^+

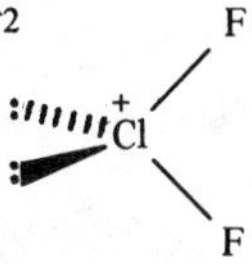

ion is non-linear

trigonal bipyramidal
arrangement of orbitals

YX_2^-

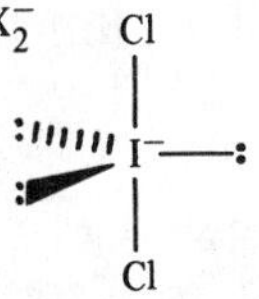

linear ion

trigonal bipyramidal arrangement of orbitals

YX_4^+

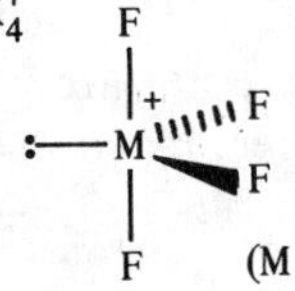

ion has 'distorted
tetrahedral' shape

(M = Cl, Br or I)

trigonal bipyramidal
arrangement of orbitals

YX_4^-

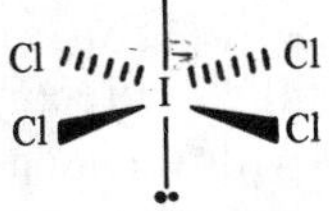

ion is square planar

octahedral arrangement of orbitals

YX_6^+

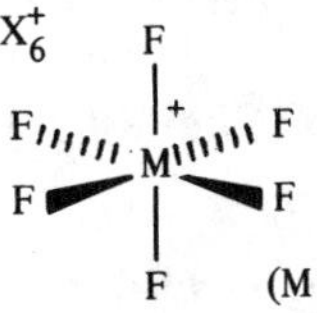

ion is octahedral

(M = Cl, Br or I)

octahedral arrangement
of orbitals

YX_6^-

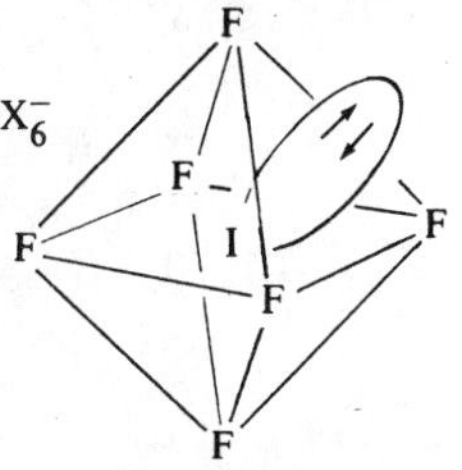

distorted
octahedral ion

(suggested structure; not proven)

Table 44—Some physical data for diatomic interhalogen molecules

Molecule	$H_f^{\ominus}$ (298 K) (kJ mol^{-1})	Bond length (Å)	Dipole moment (gas phase) (Debye)	Dissociation energy (kJ mol^{-1})
IF	-94.5(g)	1.91	—	276.6(?)
BrF	-58.5(g)	1.76	1.29	248.6
ClF	-53.6(g)	1.63	0.88	252.5
ICl	-35.3(s)	2.32	0.65	207.7
IBr	-10.5(s)	2.49	1.21	175.4
BrCl	$+14.6$(g)	2.14	0.57	215.1

Since ΔS is constant for all the diatomic interhalogens, the stability of XY is determined essentially for ΔH measured under the same conditions (gaseous reactants at 25 °C and 1 atm). The values of ΔH_f° are given in Table 44 and show that the stability sequence is IF > BrF > ClF > ICl > IBr > BrCl, which parallels the magnitude of the electronegativity difference between the two halogens in XY. This is due to an electrostatic contribution to the bonding arising from the charge separation $Y^{\delta+}$—$X^{\delta-}$, which occurs when X is more electronegative than Y. For the same reason the bond length X—Y is close to the sum of the covalent radii of X and Y *only* when the electronegativity difference between the two halogens is small.

Although the heat of formation of IF shows that it is the most stable diatomic interhalogen relative to the parent halogen molecules, it is highly unstable towards disproportionation into IF_5 (and a little IF_7),

$$5IF \rightarrow 2I_2 + IF_5$$

for which $\Delta G^{\ominus}$ is -166.5 kJ (mol IF$_5$)$^{-1}$.

Polyatomic interhalogen compounds

From Table 43 it is seen that the polyatomic interhalogens have the general formula YX$_n$ where n is an odd number (3, 5 or 7). Although other routes are possible, the most conceptually simple method of preparation is that of direct reaction carried out under conditions which vary with the identity of both X and Y:

$$Y_2 + 3F_2 \xrightarrow{T\,^{\circ}C} 2YF_3 \quad (Y = Cl,\ T = 200\text{–}300;\ Y = Br,\ T = 20;$$

$$Y = I,\ T = -45 \text{ in } CFCl_3 \text{ solvent}$$

$$I_2 + 3Cl_2(l) \xrightarrow{-80\,^{\circ}C} I_2Cl_6 \quad \text{(readily dissociates to ICl and Cl}_2\text{)}$$

$$Y_2 + 5F_2 \xrightarrow{T\,^{\circ}C} 2YF_5 \quad (Y = Cl,\ T = 350 \ (250 \text{ atm});\ Y = Br,\ T > 150;$$

$$Y = I,\ T = 20)$$

$$I_2 + 7F_2 \xrightarrow{250\text{–}300\,^{\circ}C} 2IF_7$$

Organo-substituted interhalogens can also be prepared:

$$C_2F_5I + ClF_3(\text{or } BrF_5) \rightarrow C_2F_5IF_4$$

$$CH_3I + XeF_2 \rightarrow CH_3IF_2$$

$$C_6H_5I + Cl_2 \xrightarrow[\text{solution}]{CHCl_3} C_6H_5ICl_2$$

Those reactions involving the formation of fluorides must be carried out in apparatus made from copper, nickel or chlorofluorocarbon polymers under strictly anhydrous conditions; cork, rubber, lubricating greases, moisture and glass are all readily attacked. Their very reactivity makes interhalogens useful fluorinating agents, particularly in the nuclear industry for making UF_6 and for separating reactor products: plutonium and most of the fission products form involatile fluorides when treated with ClF_3 or BrF_3 thus allowing UF_6 to be sublimed away.

The interhalogen structures are shown below, distortions from ideal symmetry being discernible owing to the steric effects of lone pairs when they are present:

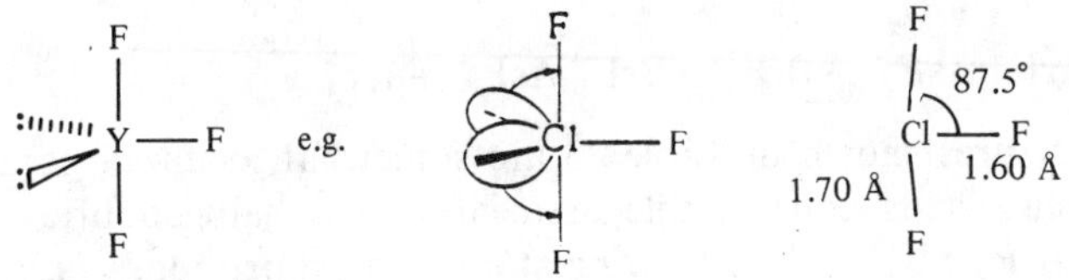

trigonal bipyramidal
arrangement of orbitals

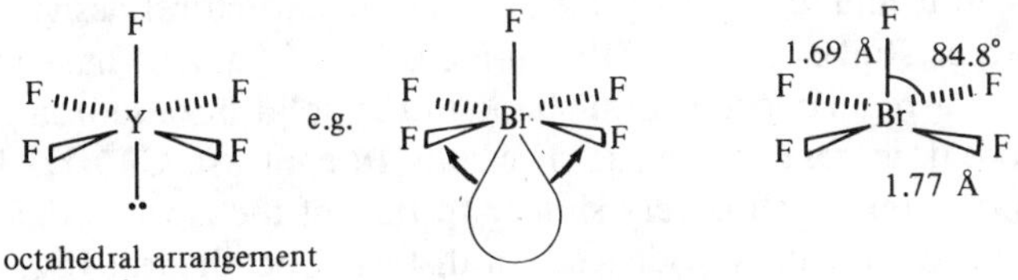

octahedral arrangement
of orbitals

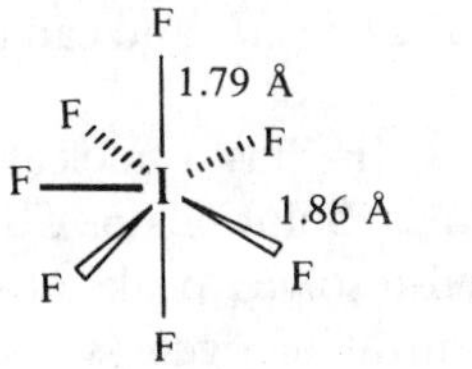

molecule is possibly
pentagonal bipyramidal

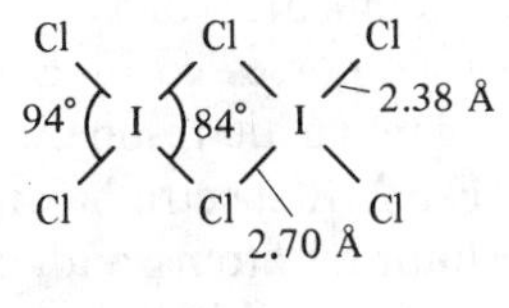

planar dimer

Cationic and anionic interhalogen species

These ions have the general formulae YX_m^+ and YX_m^- where the value of m can be 2, 4, 6 and, possibly, 8; the central halogen Y is usually, but not always, heavier than X. The unique ability of fluorine to force an element into unusually high oxidation states is clearly demonstrated in Table 43 which lists many of the known ions.

Several types of synthesis are available, including *direct halogenation*,

$$NR_4^+ I^- + Cl_2 \rightarrow NR_4^+ ICl_2^- \xrightarrow{Cl_2} NR_4^+ ICl_4^- \quad (e.g.\ R = n\text{-Bu})$$

$$CsF + ClF \rightarrow Cs^+ ClF_2^-$$

$$CsF + ClF_3 \rightarrow Cs^+ ClF_4^-$$

formation of complex fluoro-anions via *fluoride ion transfer*,

$$ClF_3 + AsF_5 \rightarrow ClF_2^+ AsF_6^-$$

$$CsF + IF_5 \rightarrow Cs^+ IF_6^-$$

$$IF_5 + 2SbF_5 \rightarrow IF_4^+ Sb_2 F_{11}^-$$

$$IF_7 + AsF_5 \rightarrow IF_6^+ AsF_6^-$$

$$CsF + IF_7 \rightarrow Cs^+ IF_8^-$$

or extremely *vigorous oxidation* using krypton fluoride derivatives:

$$KrF_2 + ClF_5 + AsF_5 \xrightarrow{1.HF} ClF_6^+ AsF_6^-$$

$$KrF^+ AsF_6^- + BrF_5 \rightarrow BrF_6^+ AsF_6^- + Kr\uparrow$$

The ion structures shown in Table 43 are somewhat idealized because in many of their crystalline salts considerable association via halogen bridging occurs. For example, in $BrF_4^+ Sb_2 F_{11}^-$ four fluorine atoms form short bonds (average 1.81 Å) but the resulting cation is linked via two longer interactions (2.24 and 2.49 Å) to fluorines on the $Sb_2 F_{11}^-$ anion.

Spectroscopic techniques have proved useful for structural assignments in many cases. Mössbauer studies using ^{129}I (half-life, 1.7×10^7 years) have shown that the IF_6^+ cation has a regular octahedral shape in the solid hexafluoroarsenate because only a single peak is observed in the spectrum. In contrast, $Cs^+ IF_6^-$ has a complex Mössbauer spectrum which is very similar to that of the isoelectronic IF_7; this can reasonably be interpreted as evidence of a distorted octahedral IF_6^- ion possessing a sterically active lone pair of electrons.* Raman spectra of YF_6^+ cations show *considerable similarity* to the isoelectronic species SF_6 or SeF_6 and can be rationalized in terms of undistorted octahedral ions.

The single stable isotope possessed by fluorine ($^{19}_9F$) has a nuclear spin of $\frac{1}{2}$ and is widely used in nuclear magnetic resonance studies. Solutions of ClF_6^+, BrF_6^+ and IF_6^+ cations display ^{19}F NMR spectra having well-resolved peaks arising from spin coupling between the fluorine atoms and the central halogen (see Fig. 102). The isotopes of chlorine, bromine and iodine possess nuclear spins greater than $\frac{1}{2}$ (^{35}Cl, ^{37}Cl, ^{79}Br and ^{81}Br have a spin $\frac{3}{2}$, whilst that of ^{127}I is $\frac{5}{2}$); coupling to nuclei with such large spin values can *only* be observed when the molecule or ion has a *high* (e.g. tetrahedral or octahedral) *symmetry*. The ^{19}F NMR spectra thus provide the strongest possible proof that these ions retain their octahedral shape when in solution. The

---

* X-ray and Raman studies on BrF_6^- show the ion has a regular octahedral structure. Bromine, unlike iodine, has a coordination maximum of six and is thus too small to accommodate six fluorines *and* a sterically-active lone pair of electrons. Furthermore, due to the poor shielding characteristics of the 3d shell of bromine, the outer s electrons are unavailable for bonding and would thus tend to remain in the spherically symmetrical 4s orbital. Both BrF_6^- and IF_6^- are fluxional on the nmr time scale.

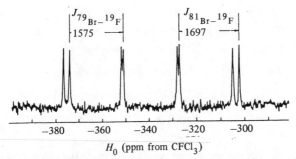

Fig. 102—The ^{19}F NMR spectrum (58.3 mHz, 26°C) of the BrF_6^+ cation in HF solution. [Reprinted with permission from *Inorganic Chemistry* vol. 13, p. 1230. Copyright (1974) American Chemical Society.]

number of peaks arising from spin coupling in YF_6^+ cations is given by the expression $2nI + 1$, where n is the number of chlorine, bromine or iodine atoms and I is the nuclear spin of the isotope coupling to the fluorine atoms. Hence $^{79}BrF_6^+$ and $^{81}BrF_6^+$ will each show a quartet of lines in their ^{19}F spectra, the two quartets being of equal intensities because the relative abundances of ^{79}Br and ^{81}Br are virtually identical (50.57% and 49.43% respectively). A sextet is observed in the ^{19}F spectrum of IF_6^+ since $(2 \times 1 \times \frac{5}{2} + 1) = 6$.

Two equally intense resonances occur in the ^{19}F NMR spectrum of ClF_4^+ but, because of the cation's low symmetry, no fine structure due to fluorine–chlorine coupling is observed; thus there are two different types of fluorine in this cation, as would be expected for the structure shown below:

The hexafluoro-cations are exceptionally strong oxidizing agents even under ambient conditions, as shown by the following examples:

$$BrF_6^+ AsF_6^- + O_2 \rightarrow O_2^+ AsF_6^-$$

$$BrF_6^+ AsF_6^- + Xe \rightarrow XeF^+ AsF_6^-$$

$$IF_6^+ SbF_5^- + NO_2 \rightarrow NO_2^+ SbF_6^-$$

$$IF_6^+ SbF_6^- + SO_2 \rightarrow SO_2F_2$$

The limiting oxidation states among interhalogens containing the heavier halogens appear to be shown by $(ICl_3)_2$ and IBr, hence it seems highly unlikely that many YX_4^+, YX_6^+, YX_6^- or YX_8^- ions will be stable under normal conditions when $X = Cl$, Br or I.

ASPECTS OF HALOGEN BONDING

Although halogens use their half-filled p orbitals when forming normal covalent bonds to a single neighbour atom, in polyatomic species some of the interactions may be of the three-centre, four electron type described on page 69. (Compounds such as IF_7, IF_6^- and IF_8^-, of course, defy a simple bonding description.) A simplified outline of the bonding which occurs in hypervalent molecules of various symmetries is given below:

(trigonal bipyramidal array of·electron pairs)

The equatorial electrons (either bonding pairs or lone pairs) are held in sp^2 hybrid orbitals on Y. The two axial X atoms are bound to Y via 3c–4e bonds.

(octahedral ion)

A 3c–4e interaction along each of the x, y and z axes binds the six X atoms in a regular octahedral array.

(octahedral array of electron pairs)

The four X atoms in the equatorial (or xy) plane are held to Y by two 3c–4e interactions. Mixing of the s and p_z orbitals on Y gives two hybrid orbitals lying along the z axis which can either hold lone pairs or form a covalent bond with an orbital on the axial X atom.

(tetrahedral array of electron pairs)

sp^3 hybrid orbitals on Y overlap with p orbitals on O to give covalent bonds; this initial Y—O interaction is probably augmented by d_π–p_π bonding and/or $Y^{\delta+}$—$O^{\delta-}$ polar contributions (see page 65).

HALOGEN HYDRIDES

All the halogens form a hydride HX, the stability of which falls from HF to HI owing to the increasing size incompatibility between the hydrogen 1s and the halogen np orbitals. Fluorine has the highest electronegativity of all the elements and consequently many of the physical properties of HF are considerably modified by hydrogen bonding. For example, the melting and boiling points of HF are higher than expected by extrapolation from the other hydrogen halides (Fig. 23). Solid hydrogen fluoride has a polymeric structure built up of zigzag chains, and similar chains occur in the liquid and gaseous states as well as in the hydrogen fluoride solvates of metallic fluorides such as KF.HF, KF.2HF, KF.3HF and KF.4HF:

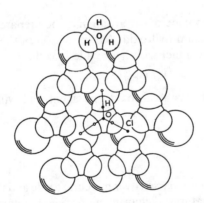

solid HF KF.HF KF.2HF

$K^+ [F--H--F]^-$
symmetrical H--F bonds

In aqueous solution hydrogen chloride, bromide and iodide are completely dissociated and behave as strong acids, whereas hydrogen fluoride is a relatively weak acid in dilute solution:

$$HF \rightleftharpoons H^+(aq) + F^- \quad K_1 = \frac{[H^+][F^-]}{[HF]} = 6.7 \times 10^{-4} \text{ at } 25°C$$

The situation is not as simple as shown by the above equilibrium because even dilute solutions of hydrogen fluoride contain the HF_2^- ion,

$$HF + F^- \rightleftharpoons HF_2^-, \quad K_2 = \frac{[HF_2^-]}{[F^-][HF]} = 3.86 \text{ at } 25°C$$

and as the concentration of hydrogen fluoride rises the formation of other ions such as $H_2F_3^-$ and $H_3F_4^-$ becomes increasingly important. The difference between hydrogen fluoride and the other hydrogen halides in dilute aqueous solution is largely due to the greater bond strength of hydrogen fluoride. However, several other factors contribute to the acidity, and the outcome is so finely balanced that if the dissociation

Fig. 103—Structure of $HCl.H_2O$ showing the presence of oxonium ions. Hydrogen chloride, like the other hydrogen halides, forms several hydrates which have been studied by X-ray crystallography. The monohydrate shown above is actually oxonium chloride, $H_3O^+Cl^-$, which melts at $-15°C$. Other hydrates of hydrogen chloride are

$HCl.2H_2O$ water molecules held together in pairs by short hydrogen bonds to form $H_5O_2^+Cl^-$, m.p. $-25°C$

$HCl.3H_2O$ $[H_5O_2^+Cl^-].H_2O$, m.p. $-70°C$

$HCl.6H_2O$ $[H_9O_4^+Cl^-].2H_2O$, m.p. $-70°C$

[Reproduced from *Acta Crystallographica* vol. 12 (1959) p. 17.]

energy of hydrogen fluoride were 35–40 kJ mol^{-1} lower, then aqueous hydrofluoric acid would also be a strong acid in line with the other hydrogen halides.

Convenient laboratory preparations of gaseous hydrogen halides, which may be dissolved in water to give the aqueous acids, are

$$CaF_2 + H_2SO_4 \rightarrow CaSO_4 + 2HF$$

$$NaCl + H_2SO_4 \rightarrow NaHSO_4 + HCl$$

$$2P(red) + 6H_2O + X_2 \rightarrow 6HX + 2H_3PO_3 \quad (X = Br \text{ or } I)$$

OXIDES AND OXO-ACIDS OF THE HALOGENS

Most of the oxides formed by the halogens are unstable and tend to detonate on shock or exposure to light.

Dihalogen monoxides

These are of general formula X_2O (X = F, Cl or Br):

X = F: $\theta = 103.2°$, $a = 1.41$ Å
X = Cl: $\theta = 110.9°$, $a = 1.69$ Å

Difluorine monoxide

This, the most stable monoxide of the group, may be prepared by bubbling fluorine through 2% aqueous sodium hydroxide (cf. the action of the other halogens with a base), but on prolonged contact with hydroxide ions fluorine produces oxygen:

$$F_2 + 2OH^- \rightarrow 2F^- + H_2O + \tfrac{1}{2}O_2$$

Difluorine monoxide behaves as an oxidizing agent and will oxidize bromide and iodide to the free halogen.

Dichlorine monoxide

When chlorine and dry air are passed over mercuric oxide and the products condensed at liquid nitrogen temperatures, dichlorine monoxide is obtained; it is extraordinary unstable and explodes on shock or impact. A kinetic study has shown the photo-decomposition of dichlorine monoxide to occur via the following steps:

$$Cl_2O + hv \rightarrow ClO + Cl$$

$$Cl_2O + Cl \rightarrow ClO + Cl_2$$

$$ClO + ClO \rightarrow Cl_2 + O_2$$

Both dichlorine and dibromine monoxides react with water to give the corresponding hypohalous acids (thus acting as acid anhydrides) and with alkalis to produce hypohalites—compare this with difluorine monoxide, which is made in the presence of base.

Other fluorine oxides

Two other fluorine oxides are known, F_2O_2 and F_2O_4, of which the former is the more robust but still decomposes at $-57°C$ into oxygen and fluorine. Their synthesis is accomplished by subjecting mixtures of oxygen and fluorine to an electrical discharge, the cell being immersed in liquid oxygen ($c. -185°C$) so that the reaction products may be frozen out of the gas phase before they have time to decompose (Fig. 104). The structure of F_2O_2, determined by continuously pumping the vapour through a Teflon-lined microwave cavity and observing the microwave spectrum, is similar to that of hydrogen peroxide but with a much shorter O—O bond length (1.22 Å); the reason for this short bond is not clearly understood. The compound O_4F_2 appears to be a loosely bound dimer of the radical OOF, i.e. $(OOF)_2$, and its solutions exhibit enormous electron spin resonance signals owing to extensive dissociation into OOF.

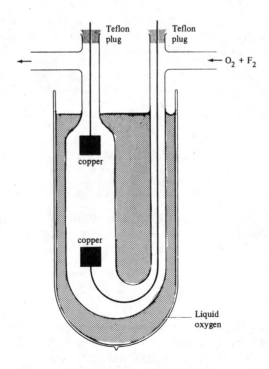

Fig. 104—Discharge apparatus used for the preparation of dioxygen difluoride, O_2F_2.

Chlorine dioxide, ClO_2, and bromine dioxide, BrO_2

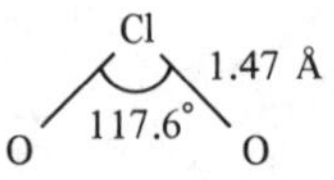

Bromine dioxide is highly unstable and above $-40\,^{\circ}C$ decomposes into bromine, oxygen and dibromine monoxide. Chlorine dioxide is a paramagnetic, yellowish gas which shows little tendency to dimerize owing to the unpaired electron being extensively delocalized over the whole molecule. It is prepared by the action of sulphuric acid on potassium chlorate in the presence of a reducing agent; usually oxalic acid is used so that the dangerously explosive chlorine dioxide is diluted with carbon dioxide, making manipulations somewhat less hazardous. Hydrolysis yields chlorite and chlorate as the final products.

Dichlorine hexoxide, Cl_2O_6

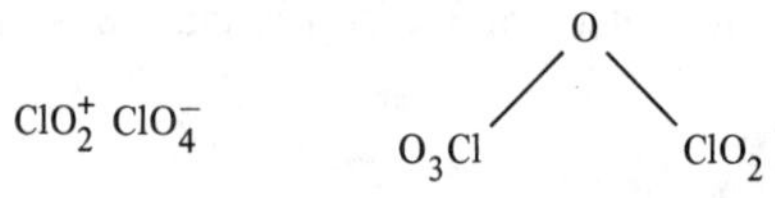

$$ClO_2^+\ ClO_4^-$$

solid phase

$$O_3Cl \qquad ClO_2$$

gas phase

This highly unstable molecule, the mixed anhydride of chloric and perchloric acids, is best made by treating chlorine dioxide with oxone. In the gas phase it decomposes within minutes to form ClO_2, O_2 and $ClOClO_3$.

Dichlorine heptoxide, Cl_2O_7

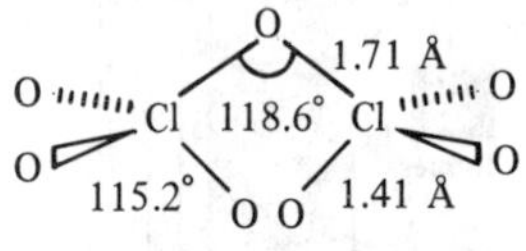

perchloric acid

This oxide is the anhydride of perchloric acid and may be prepared by carefully dehydrating the concentrated acid with phosphorus pentoxide and distilling the product at about $-35\,^{\circ}C$ under 1 mmHg pressure.

Diiodine pentoxide, I_2O_5

This polymeric oxide, the most stable and best studied of the iodine oxides, may be prepared by dehydrating iodic acid at $200\,^{\circ}C$ in a stream of dry air; dissolution in water regenerates HIO_3. It is a powerful oxidizing agent and will quantitatively convert carbon monoxide into CO_2. The iodine liberated simultaneously may be estimated with standard sodium thiosulphate, thus allowing a convenient method for assaying CO:

$$I_2O_5 + 5CO \rightarrow I_2 + 5CO_2$$

Perchloryl fluoride, FClO₃

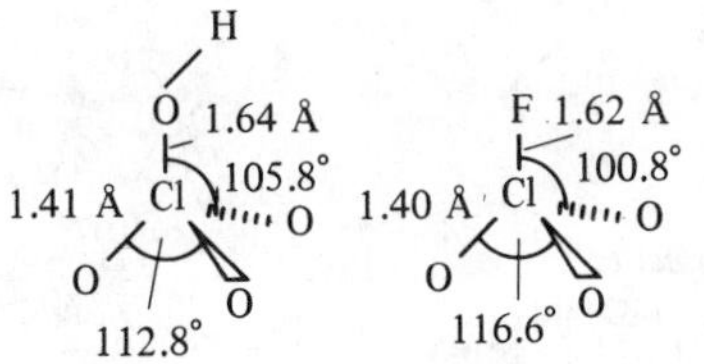

This relatively stable molecule may be considered to arise from perchloric acid via substitution of the —OH group by its isoelectronic partner —F. It is prepared by reacting potassium perchlorate with a mild fluorinating agent:

$$KClO_4 + SbF_5 \text{ (or } FSO_3H) \rightarrow FClO_3 \quad (\text{b.p. } -46.7°C)$$

Perchloryl fluoride is a convenient reagent for the controlled introduction of F or ClO₃ groups into organic molecules:

$$CH_2(CO_2Me)_2 \xrightarrow{\text{FClO}_3} CF_2(CO_2Me)_2 \quad \text{(activated CH}_2 \text{ groups)}$$

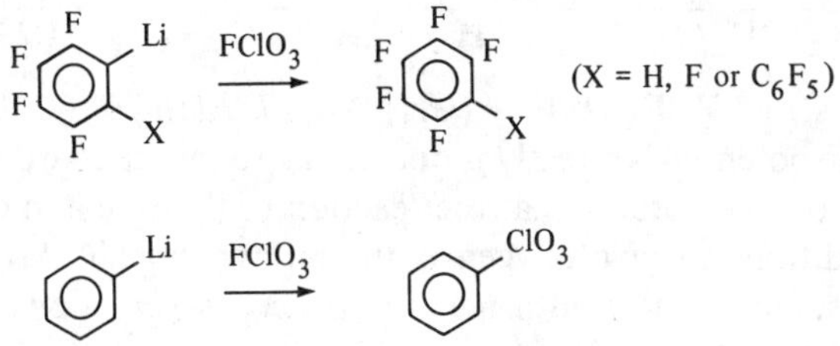

In the last two reactions the introduction of fluorine is favoured by having electronegative R groups coupled to the lithium, but often readily separable mixtures of RClO₃ and RF are obtained.

Oxo-acids of the halogens

Chlorine, bromine and iodine form the four types of oxo-acid shown below; in addition, ortho-periodic acid (H_5IO_6) and its salts are known. By contrast, hypofluorous acid is the only oxo-acid of fluorine which has been isolated to date and is obtained as a thermally unstable species when ice is fluorinated at low temperature:

$$F_2 + H_2O \xrightarrow{-40°C} HOF + HF$$

$$Cl-O\underset{103°}{\overset{1.69 \text{ Å}}{\diagup}}\underset{H}{O}\ 0.97 \text{ Å} \qquad HOXO \qquad HOXO_2 \qquad \underset{O\ \underset{112.8°}{}\ O}{\overset{105.8°\ \overset{H}{\diagup}O}{1.41 \text{ Å}\ \big(Cl\cdots O\ 1.64 \text{ Å}}}$$

hypochlorous acid   halous acid   halic acid   perchloric acid

$$X^- - O \qquad \underset{O}{\overset{O}{\diagdown}}C\overset{\alpha}{\diagup}X^- \qquad \underset{O \cdots \overset{X^-}{\underset{\sim 107°}{|}} \cdots O}{O} \qquad \underset{O}{\overset{O}{\diagdown}}X\overset{\cdots O}{\diagup}O^-$$

hypohalite   halite   halate   perhalate
                 (angular: (pyramidal: (tetrahedral:
                 two lone pairs)   one lone pair)   no lone pairs)
                 $\alpha = 111°$ (Cl)
                 $\alpha = 105°$ (Br)

The complex interrelations between the oxo-acids of a particular halogen are best summed up by a list of the various standard oxidation potentials which operate in aqueous solution (Table 45). Few other non-transition elements exist in a variety of oxidation states as do the halogens in their oxo-acids, so that this is the first time in this book we have been able to use standard potentials as an aid to understanding chemical reactions. From the data given in Table 45,

$$
\begin{array}{ll}
X_2 + 2e \rightarrow 2X^- & E_1 \text{ V} \\
H_2O \rightarrow \tfrac{1}{2}O_2 + 2H^+ + 2e & -1.23 \text{ V} \\
\hline
X_2 + H_2O \rightarrow \tfrac{1}{2}O_2 + 2H^+ + 2X^- & E_2 = (E_1 - 1.23) \text{ V}
\end{array}
$$

E_2 has the values 1.62 V (F_2), 0.13 V (Cl_2), -0.17 V (Br_2), -0.70 V (I_2) and -1.43 V (At_2). Since the potential is directly proportional to the standard free energy change† $\Delta G^\ominus$ for a reaction, we observe that the reaction of fluorine and chlorine with water under acid conditions to give oxygen is thermodynamically favoured. In practice, however, it is usually found that a reaction proceeds very slowly if the potential is less than about half a volt, and consequently only fluorine is found to release oxygen from water (even with fluorine the reaction is fairly sluggish as evidenced by the preparation of difluorine monoxide by passing fluorine through dilute aqueous alkali). Conversely, we note from the high negative potentials of iodine and astatine for the equation as written that oxygen is capable of releasing halogen from acidic solutions of iodide or astatide.

Chlorine, bromine and iodine are slightly soluble in water (saturated solutions at 25°C are about 0.06, 0.21 and 0.0013 molar respectively), the equilibria set up in solution being

$$X_2 + H_2O \rightleftharpoons HOX + HX$$

---

† $-\Delta G^\ominus = zFE$, where F is the faraday (96 500 C), E is the potential in volts and z is the number of electrons associated with the reaction. If E is positive, then the free energy change for the reaction is favourable.

Table 45—Some oxidation–reduction potentials pertinent to the chemistry of the halogens in aqueous solution

Acid solutions
$O_2 + 4H^+ + 4e \rightarrow 2H_2O$,   1.23 V

	F	Cl	Br	I	At
$\frac{1}{2}X_2 + e \rightarrow X^-$	2.85	1.36	1.06	0.53	−0.2
$\frac{1}{2}F_2 + H^+ + e \rightarrow HF$	3.03	—	—	—	—
$HXO + H^+ + e \rightarrow \frac{1}{2}X_2 + H_2O$	—	1.63	1.59	1.45	0.7
$HXO_2 + 3H^+ + 4e \rightarrow X^- + 2H_2O$	—	1.56	—	—	—
$HXO_2 + 3H^+ + 3e \rightarrow \frac{1}{2}X_2 + 2H_2O$	—	1.63	—	—	—
$XO_3^- + 6H^+ + 5e \rightarrow \frac{1}{2}X_2 + 3H_2O$	—	1.47	1.52	1.20	—
$XO_4^- + 8H^+ + 7e \rightarrow \frac{1}{2}X_2 + 4H_2O$	—	1.34	—	—	—

Basic solutions
$O_2 + 2H_2O + 4e \rightarrow 4OH^-$,   0.40 V

	Cl	Br	I	At
$XO^- + H_2O + 2e \rightarrow X^- + 2OH^-$	0.94	0.76	0.49	—
$XO_2^- + 2H_2O + 4e \rightarrow X^- + 4OH^-$	0.76	—	—	—
$XO_3^- + 3H_2O + 6e \rightarrow X^- + 6OH^-$	0.62	0.61	0.26	—
$XO_4^- + 4H_2O + 8e \rightarrow X^- + 8OH^-$	0.56	—	0.39	—
$\frac{1}{2}X_2 + e \rightarrow X^-$	1.36	1.06	0.53	−0.2
$XO_2^- + H_2O + 2e \rightarrow XO^- + 2OH^-$	0.59	—	—	—
$XO_3^- + 2H_2O + 4e \rightarrow XO^- + 4OH^-$	—	0.46	0.56	0.50
$XO_3^- + H_2O + 2e \rightarrow XO_2^- + 2OH^-$	0.35	—	—	—

With chlorine, about half is present as hypochlorous and hydrochloric acids, the rest as free chlorine. This reaction can be made into a preparation of hypochlorous acid by having present in the system an insoluble compound which will remove the chloride ions by forming an insoluble or sparingly ionized chloride (usually mercuric oxide is used). Hypochlorous acid is a far weaker acid than is acetic acid (K_a is about 3.7×10^{-8} at 20°C); hypobromous and hypoiodous acids are even weaker, their dissociation constants being 2×10^{-9} and 4.5×10^{-13} respectively.

The potentials for basic solutions show that bromine is a stronger oxidizing agent than hypobromite:

$$Br_2 + 2e \rightarrow 2Br^- \qquad\qquad 1.06\ V$$
$$Br^- + 2OH^- \rightarrow BrO^- + H_2O + 2e \qquad -0.76\ V$$
$$\overline{Br_2 + 2OH^- \rightarrow BrO^- + H_2O + Br^- \qquad +0.30V}$$

Hence $\Delta G^\ominus$ is negative so that the reaction as written is favoured and the equilibrium constant K may be calculated from the expression

$$\log K = \frac{zFE}{2.303\ RT}$$

$$K = \frac{[BrO^-][Br^-]}{[Br_2][OH^-]^2} \simeq 2 \times 10^5$$

Therefore the reaction of bromine with a base should give a mixture of hypobromite and bromide, the extent of the conversion being dependent on the *square* of the hydroxyl ion concentration.

However, the situation is not quite so simple because hypobromite is thermodynamically unstable towards disproportionation:

$$3BrO^- + 3H_2O + 6e \rightarrow 3Br^- + 6OH^- \qquad 0.76 \text{ V}$$
$$\underline{Br^- + 6OH^- \rightarrow BrO_3^- + 3H_2O + 6e \quad -0.61 \text{ V}}$$
$$3BrO^- \rightarrow 2Br^- + BrO_3^- \qquad +0.15 \text{ V}$$

This disproportionation is slow at $0°C$ so that the reaction of bromine with *ice-cold* base can be used to prepare hypobromite, but at room temperature or above the products are bromide and bromate.

It is obvious from Table 45 that in basic solution all the halogens except astatine are capable of releasing oxygen from water, but this reaction occurs only relatively slowly even with fluorine (which initially forms difluorine monoxide). Hence, in the case of bromine reacting with a base, three reactions are *thermodynamically* favoured, one of which (oxygen evolution) is not *kinetically* important, but the balance between the other two is temperature-sensitive, owing to the rate of the hypobromite disproportionation being markedly temperature-dependent. Similar products are formed at comparable temperatures in the reaction of chlorine with a base, whereas iodine reacts to give iodate and iodide under all conditions because hypoiodite ions disproportionate too rapidly to be isolated.

The disproportionation of aqueous chlorate to perchlorate and chloride,

$$4ClO_3^- \rightarrow 3ClO_4^- + Cl^-$$

is energetically favoured, but it occurs so very slowly that perchlorate is not formed even at the boiling point of water; however, careful heating of *solid* potassium chlorate does give the perchlorate but, of course, this reaction cannot be predicted from the *solution* potentials given in Table 45:

$$4KClO_3 \rightarrow 3KClO_4 + KCl$$

Only recently has it been proved possible to prepare perbromates by the action of the very powerful oxidizing agent xenon difluoride (XeF_2) on bromates; acidification of perbromate solutions gives perbromic acid ($HBrO_4$), which was detected as the parent peak $HBrO_4^+$ in the mass spectrum. No mass spectral evidence for bromine heptoxide (Br_2O_7) was obtained on heating perbromic acid, but the presence of bromine dioxide in the decomposition products was noted. Owing to the 3d inner shell of electrons it is unexpectedly difficult to oxidize As, Sb and Br (e.g. see page 256). The 4d orbitals appear less efficient at back-bonding than the 3d orbitals of the previous period which *also* helps to make arsenates and, especially, perbromates strong oxidizing agents.

Periodates

Although periodic acid (HIO_4) and several salts containing the tetrahedral IO_4^- periodate ions are known, a variety of other periodic acids and their derivatives exist. Iodine resembles tellurium to some extent by forming hydroxo-derivatives in which

Table 46—Reactions of some chlorine-containing anions in aqueous solution

Reaction	Chloride (Cl^-)	Chlorate (ClO_3^-)	Perchlorate (ClO_4^-)
1 Solubility of metal salts	Most chlorides are soluble; common exceptions are AgCl, TlCl, Hg_2Cl_2, $PbCl_2$, BiOCl	Virtually all chlorates are soluble	Most perchlorates are soluble; exceptions are $RbClO_4$, $CsClO_4$; $KClO_4$ is relatively insoluble in the cold
2 Addition of $AgNO_3$	White precipitate, AgCl; soluble in dilute NH_3 owing to ammine formation, $Ag(NH_3)_2Cl$	No reaction	No reaction
3 Concentrated H_2SO_4	Evolution of HCl gas on warming	*Explosive* green gas, ClO_2, formed	No reaction in cold; $HClO_4$ evolved on boiling
4 Addition of $NaNO_2$	No reaction	Reduction to Cl^-	No reaction
5 Addition of $BaCl_2$	No reaction	No reaction	No reaction
6 Acidified I^-	No reaction	I_2 liberated	No reaction
7 Concentrated KCl	No reaction	Cl_2 formed in acidified solution	White precipitate, $KClO_4$; dissolves on heating, reprecipitates on cooling
8 Boil with acidified Fe^{2+}	No reaction	Reduction to Cl^- as Fe^{2+} is oxidized to Fe^{3+}	No reaction

the iodine atoms are octahedrally coordinated, and aqueous solutions of periodate at low pH contain at least four species:

$$IO_4^- \rightleftharpoons HIO_4 + 2H_2O \rightleftharpoons H_5IO_6 \rightleftharpoons H_4IO_6^- \rightleftharpoons H_3IO_6^{2-}$$

Under the correct conditions, salts of these periodate species can be isolated (e.g. $NaIO_4$, $AgIO_4$, NaH_4IO_6, $Na_2H_3IO_6$ and $(NH_4)_2H_3IO_6$). Evaporation of aqueous solutions of periodic acid gives crystals of H_5IO_6 which can be dehydrated by careful heating:

$$H_5IO_6 \text{ (i.e. 'HIO}_4.2H_2O\text{')} \xrightarrow[\text{12 mmHg}]{100\,°C} HIO_4 \xrightarrow{138\,°C} HIO_3 + O_2$$

(octahedral iodine coordination)

$I — O(H) = 1.89$ Å

$I — O = 1.78$ Å

The acid H_5IO_6 has the maximum number of —OH groups surrounding the iodine(VII) atom, hence it can be called *ortho*-periodic acid. It is an essentially dibasic acid with $K_1 = 0.02$ and $K_2 \simeq 10^{-8}$, giving two main series of salts, $M^IH_4IO_6$ and $M_2^IH_3IO_6$, although other salts containing $H_2IO_6^{3-}$ and IO_6^{5-} ions can be prepared: e.g.

$$5Ba(IO_3)_2 \xrightarrow{\text{heat}} Ba_5(IO_6)_2 + 4I_2 + 9O_2$$

At high pH ($\sim 10-11$) dimerization of the periodate anion occurs to give $H_2I_2O_{10}^{4-}$, in which iodine is octahedrally surrounded by six oxygen atoms, the two octahedra having one edge in common. On crystallization from potassium hydroxide solutions the potassium salt $K_4H_2I_2O_{10}.8H_2O$ is obtained, the anion of which has been shown to have the structure

$I—O$ (shared oxygens and hydroxyls) $= 2.00$ Å
$I—O = 1.81$ Å

Spectroscopic evidence suggests that orthoperiodic acid (H_5IO_6) can be protonated under the very acidic conditions existing in 10M perchloric acid to give the cationic species $I(OH)_6^+$.

FLUOROCARBONS

Strictly, organo-derivatives of the halogens belong to the realm of organic chemistry and a discussion of them would be out of place in this book. However, perhaps some mention ought to be made of highly fluorinated carbon derivatives which are now becoming widely used in polymers (e.g. 'Teflon' and 'Fluon'), refrigerant fluids, aerosol propellants and non-inflammable anaesthetics.

Of the groups attached to carbon in organic compounds, fluorine is rapidly becoming established as being second only to hydrogen in importance, and a host of highly fluorinated analogues of the hydrocarbons and their derivatives, especially in the aromatic field, has been synthesized in recent years. Fluorine, like hydrogen, is always monovalent and, owing to shielding effects operative from lithium to fluorine as the $n = 2$ quantum shell is filled, it has an exceptionally small van der Waals radius of only 1.35 Å (that of hydrogen is 1.2 Å). This means that steric crowding round a carbon atom is minimized relative to the other halogen atoms. The C—F bond energy varies between 460 and 500 kJ mol^{-1} (C—H bond energy ~ 414 kJ mol^{-1}), so that polymers containing a —CF$_2$— chain are very stable from both energetic and kinetic viewpoints—the fluorine atoms, being slightly larger than hydrogen atoms, form a more effective shield round the —C—C— backbone of the polymer and inhibit the approach of attacking reagents. Fluorine atoms are not very polarizable, so that intermolecular forces are weaker than between hydrocarbons, resulting in higher volatilities and solubilities of the fluorocarbons for a given number of carbon atoms. Three main methods of forming fluorocarbons are available.

Using fluorides of metals in high oxidation states, e.g. cobalt(III) fluoride

$$2CoF_2 + F_2 \rightarrow 2CoF_3$$

$$R—H + 2CoF_3 \rightarrow R—F + 2CoF_2 + HF \quad (R = \text{alkyl or aryl})$$

When aromatic hydrocarbons are used, the main products of fluorination by cobalt(III) fluoride are the *saturated* alicyclic fluorides: e.g.

The aromatic ring can be regenerated by passing the alicyclic fluorocarbons over heated iron or nickel:

(hexafluorobenzene)

Halogen exchange reactions

A variety of halogen exchange reagents can be used; the two finding the widest application are potassium fluoride and antimony trifluoride:

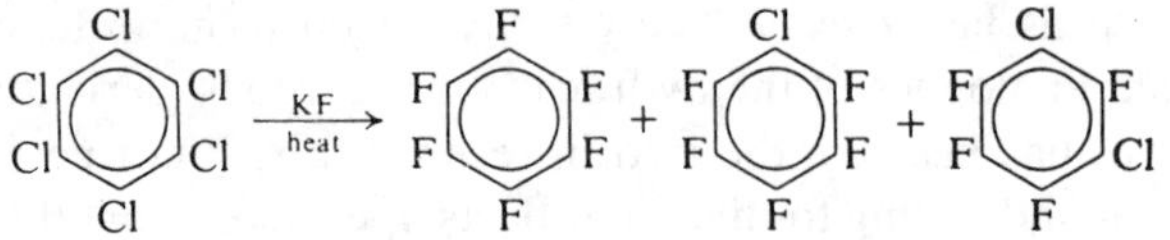

Electrochemical fluorination

The organic compound is dissolved in liquefied anhydrous hydrogen fluoride and the solution is electrolysed at low voltage:

$$\left.\begin{array}{l} R-H \rightarrow R_F-F \\ OR_2 \rightarrow O(R_F)_2 \\ NR_3 \rightarrow N(R_F)_3 \\ NHR_2 \rightarrow FN(R_F)_2 \end{array}\right\} \quad (R_F = C_n F_{2n+1})$$

The highly fluorinated aromatic derivatives are the most amenable to further study. For example, C_6F_5Cl synthesized by halogen exchange methods can be used to form a wide variety of new products, including organometallic derivatives:

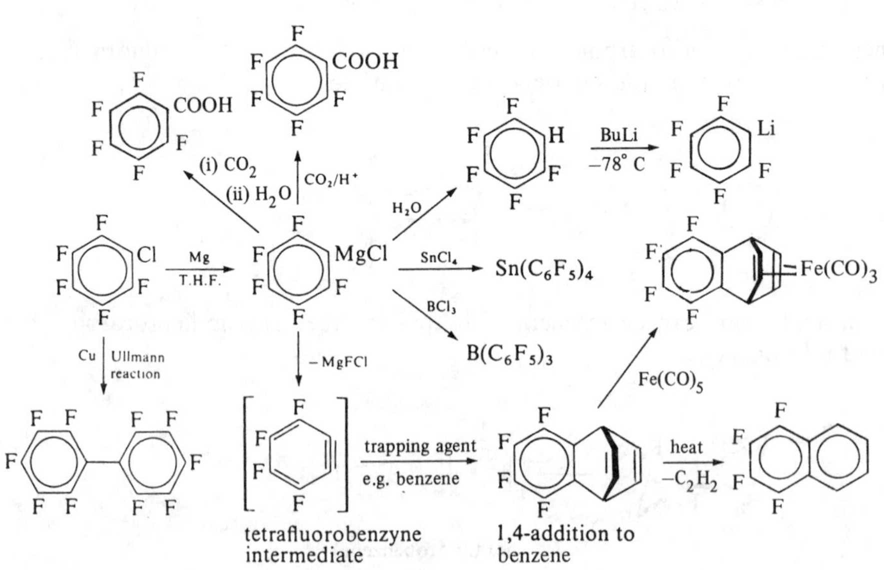

tetrafluorobenzyne
intermediate

1,4-addition to
benzene

CHLOROFLUOROCARBONS AND THE OZONE LAYER

At a height of 25–40 km in the stratosphere, oxygen molecules are continuously photolysed by ultraviolet sunlight of less than 242 nm wavelength:

$$O_2 \xrightarrow{\ hv\ } 2O$$

These highly reactive oxygen atoms can combine with oxygen molecules in the presence of a third body (required to remove excess kinetic energy) to give ozone:

$$O + O_2 + B \rightarrow O_3 + B$$

Ozone itself is photolysed by UV radiation in the 230–290 nm range,

$$O_3 \xrightarrow{\ hv'\ } O + O_2$$

so that ozone is continuously being produced and destroyed by incident sunlight to give a steady state concentration of some 5 billion tons. It is the latter ultraviolet filtering reaction which makes ozone so important because it cannot be accomplished by O_2 *at these wavelengths*.

Nitrogen oxides and chlorofluorocarbons are able to change the position of the mobile ozone equilibrium by removing O_3 in a series of 'catalytic' steps:

(i) $\quad\quad\quad NO + O_3 \rightarrow NO_2 + O_2$
(ii) $\quad\quad\quad NO_2 + O \rightarrow NO + O_2$ $\quad$ net reaction $O_3 + O \rightarrow 2O_2$

(a) $\quad\quad$ chlorofluorocarbons (e.g. $CFCl_3$, CF_2Cl_2) $\xrightarrow{\ hv\ }$ Cl atoms

(b) $\quad\quad\quad Cl + O_3 \rightarrow OCl + O_2$
(c) $\quad\quad\quad ClO + O \rightarrow Cl + O_2$ $\quad$ net reaction $O_3 + O \rightarrow 2O_2$

The ozone layer, which would only be about 3 mm thick if it were compressed worldwide to 1 atm at sea level, protects all life from the hazardous effects of short-wavelength ultraviolet radiation. The ozone holes recently discovered over Antarctica show that the depletion of O_3 by atmospheric impurities is not simply a gradual change in the global equilibrium concentration but is markedly affected by local climatic conditions.

As a result of various pressure groups the use of chlorofluorocarbons as propellants in aerosol sprays is declining, but the major contamination continues to be the refrigeration industry where such compounds are released during the thoughtless destruction of old appliances.

Chapter 11

Group 0: The Rare Gases. Helium, Neon, Argon, Krypton, Xenon and Radon

INTRODUCTION

The rare gases are characterized by both their high ionization energies and the fact that all the available s and p orbitals in the outer n quantum shell are full. Furthermore, the energy required to promote electrons into orbitals of the next (i.e. $n + 1$) shell or into d and f orbitals of the n shell is high, being about 950 kJ mol^{-1} for the promotions

$$5s^2 5p^6 \rightarrow 5s^2 5p^5 6s^1$$

$$5s^2 5p^6 \rightarrow 5s^2 5p^5 5d^1$$

in the case of xenon. Thus covalent bonding involving sp^{3}d or sp^{3}d^2 valence states is highly unfavourable.

For many years the rare gases were considered to be completely unreactive and indeed their inertness led chemists to the concept of the octet theory of chemical bonding. Even van der Waals forces between the rare gas atoms are very weak and lead to the observed low melting and boiling points. The forces of attraction between helium atoms are so slight that the zero-point vibration energy is sufficient to stop the solid from forming. Only by application of at least 25 atm pressure is it possible to cause liquid helium to freeze. The formation of crystalline derivatives of the heavier gases with hydrogen-bonded species such as water and polyphenols caused some interest in the 1920s and 1930s, but these are not real covalently bonded compounds; the gases are trapped mechanically in holes which occur in the open structures of the host solids. A typical 'clathrate' of the above type is $Xe(H_2O)_n$ in which n is about 5 or 6; it has a melting point of 24 °C and a xenon dissociation pressure of 1.3 atm at 0 °C.

A Born–Haber cycle calculation may be used to investigate the feasibility of preparing a rare gas salt such as Xe^+F^-:

$$\text{Xe} + \tfrac{1}{2}\text{F}_2 \xrightarrow[1\ \text{atm}]{25\,^\circ\text{C}} \text{Xe}^+\,\text{F}^-(s), \quad \Delta H_f^{\ominus}$$

$$+\tfrac{1}{2}D \downarrow \qquad\qquad \uparrow -L_{\text{XeF}}$$

$$\text{Xe} + \text{F}(g) \xrightarrow[-EA_F]{I_{\text{Xe}}} \text{Xe}^+(g) + \text{F}^-(g)$$

where D is the dissociation energy of the F_2 molecule, I_{Xe} the first ionization of xenon, EA_F the electron affinity of fluorine and L_{XeF} the lattice energy of ionic xenon monofluoride. Note that a term for the sublimation energy of xenon is not required because its standard state is that of a gas at $25\,^\circ\text{C}$.

All the quantities are known in this cycle with the exception of L_{XeF}; to a reasonably close approximation this may be equated to the lattice energy of caesium fluoride, hence

$$\Delta H_f^{\ominus} = \tfrac{1}{2}D + I_{\text{Xe}} - EA_F - L_{\text{XeF}}$$

$$= \tfrac{1}{2}(157.7) + 1162 - 332.7 - 719.7$$

$$= +188.4 \text{ kJ mol}^{-1}$$

Thus unless for some obscure reason the reaction has a large, positive entropy term, the free energy of formation, $\Delta G_f^{\ominus}$, for Xe^+F^- will be positive and no compound formation will occur under ambient conditions. The chloride, bromide and iodide can be expected to be even less thermodynamically stable because of the high dissociation energy of the halogen molecules and the comparatively low lattice energies of the product salts.

If the electron affinity (oxidizing power) of the species reacting with xenon could be greatly increased then ionic xenon compounds might be anticipated. Such an oxidant appears to be platinum hexafluoride which reacts with xenon in the presence of SF_6 (added to suppress side-reactions) forming the yellow, paramagnetic solid $\text{Xe}^+\text{PtF}_6^-$.

DIATOMIC CATIONS OF THE RARE GASES

Bombardment of a rare gas with electrons in a mass spectrometer causes an electron to be knocked off the atoms: e.g.

$$\text{He} + \text{e}^- \rightarrow \text{He}^+ + 2\text{e}^-$$

Such ions are stable in the isolation of high vacuum but on contact with the container walls of the vacuum system they readily pick up electrons to reform the neutral atoms. In a mass spectrometer such ions are kept away from the vessel walls by magnetic and electric fields until they reach the collector plates of the detector. The reactivity of the ions is so high that if the pressure in the spectrometer is allowed to rise slightly to about 10^{-5} mmHg they will combine with neutral gas atoms to give diatomic cations such as $[\text{He}-\text{He}]^+$; again these species are only stable in a vacuum. Although the bond order in He_2^+ and the other noble gas diatomic ions is only 0.5, their bond energies are surprisingly high as shown in Table 47. Very recently evidence has been

Table 47—The rare gases

Element		1st IE (kJ mol^{-1})	Melting point (K)	Boiling point (K)	Bond energy of X_2^+ (kJ mol^{-1})
2	Helium, He	$1s^2$	2372	1 (25 atm)	4.2	126
10	Neon, Ne	$1s^2 2s^2 2p^6$	2080	24.5	27.2	67
18	Argon, Ar	[Ne]$3s^2 3p^6$	1520	83.8	87.9	104
36	Krypton, Kr	[Ar]$3d^{10} 4s^2 4p^6$	1351	105.9	120.9	96
54	Xenon, Xe	[Kr]$4d^{10} 5s^2 5p^6$	1170	161.3	165.1	88
86	Radon, Rn	[Xe]$4f^{14} 5d^{10} 6s^2 6p^6$	1037	202	211	—

presented for the formation of He_2^{2+}, which is formally isoelectronic with H_2; so far its spectroscopic properties have not been investigated.

In contrast to the gaseous ions described above which are stable only at low pressure, the reduction of $XeF^+Sb_2F_{11}^-$ with xenon in the presence of excess SbF_5 gives a green solution of the paramagnetic Xe_2^+ ion:

$$3Xe + XeF^+ + 2SbF_5 \rightarrow 2Xe_2^+$$

The cation is characterized by a particularly strong band at 123 cm^{-1} in the Raman spectrum and its solutions in SbF_5 are stable indefinitely at room temperature when kept under a pressure of xenon gas; see Fig. 105(a).

KrF₂ AND XeF₂ AS FLUORINATING AGENTS

The rare gas difluorides make very clean fluorinating reagents because in virtually all their reactions the only by-products formed are Kr and Xe which can easily be pumped out of the reaction vessel. Krypton difluoride is much the more reactive of the two and will even fluorinate xenon or a noble metal such as gold:

$$3KrF_2 + Xe \rightarrow XeF_6 + 3Kr$$

$$7KrF_2 + 2Au \xrightarrow{\;-5Kr\;} 2KrF^+AuF_6^- \xrightarrow{\;60-65\,^\circ C\;} AuF_5 + Kr + F_2$$

Owing to its commercial availability, xenon difluoride is becoming quite widely used in synthetic organic chemistry; the stability of many alkyl or aryl groups towards XeF_2 also allows it to be used to oxidize and fluorinate simultaneously the central element in a variety of organometalloid systems: e.g.

$$(C_6H_5)_2S + XeF_2 \rightarrow (C_6H_5)_2SF_2 + Xe$$

$$(CH_3)_3As + XeF_2 \rightarrow (CH_3)_3AsF_2 + Xe$$

$$CH_3I + XeF_2 \rightarrow CH_3IF_2 + Xe$$

$$C_6F_5I + XeF_2 \rightarrow C_6F_5IF_2 \xrightarrow{\;XeF_2\;} C_6F_5IF_4$$

$$(C_6H_5)_2PH + XeF_2 \rightarrow (C_6H_5)_2PHF_2 + Xe$$

$$(C_6H_5)_2PCl_2 + XeF_2 \rightarrow (C_6H_5)_2PF_3 + Xe$$

The reactivity of xenon difluoride can be markedly increased by using it in anhydrous liquid hydrogen fluoride, probably owing to the formation of XeF^+ cations:

$$S_8 + XeF_2 \xrightarrow{\;HF\;} SF_6$$

$$Mo(CO)_4Cl_2 \xrightarrow{\;HF\;} Mo(CO)_4F_2 \xrightarrow[HF]{XeF_2} Mo(CO)_3F_3 \xrightarrow[HF]{XeF_2} Mo(CO)_2F_4 \xrightarrow[HF]{XeF_2} MoF_6$$

$$Ir + XeF_2 \xrightarrow{\;HF\;} IrF_5$$

$$CrF_2 + XeF_2 \xrightarrow{HF} CrF_3 \xrightarrow[HF]{XeF_2} CrF_4$$

$$OsBr_4 + XeF_2 \xrightarrow{HF} OsF_5 \xrightarrow[HF]{XeF_2} OsF_6$$

$$MoO_3 + XeF_2 \xrightarrow{HF} MoF_6$$

THE STRUCTURE OF XENON HEXAFLUORIDE

More effort has been expended on trying to determine and understand the structure of XeF_6 than any other entity of similar size because the molecule 'will just not sit still' except in the crystal. There are 14 electrons present in the outer shell of xenon when it forms XeF_6 and hence there are two theoretically possible structures for the isolated molecule: regular octahedral, with the lone pair of electrons in a spherically symmetrical s orbital, or distorted octahedral, where the lone pair is sterically active.

The wealth of data now available on the behaviour of gaseous XeF_6 suggests that the molecule has no static structure but is continually interchanging between eight possible C_{3v} structures in which the lone pair occupies a triangular face of the (now distorted) octahedron:

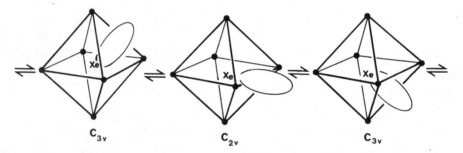

In a variety of non-aqueous solvents xenon hexafluoride changes shape yet again and forms a tetramer Xe_4F_{24} which undergoes a rapid, and complex, intramolecular fluorine interchange even at temperatures as low as $-150\,°C$. Tetramers are also present in each of the four crystal modifications adopted by solid XeF_6, the xenon atoms being arranged in the form of a tetrahedron. Two sketches are given below showing the relative orientations of the four pyramidal XeF_5 groups which are linked by fluorine bridges in the tetramer:

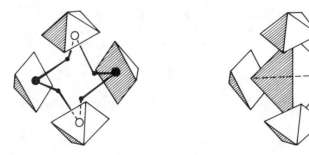

RADON CHEMISTRY

Radon-222, the longest-lived isotope (α; $t_{1/2} = 3.8$ days), is the immediate daughter product of radium-226, itself occurring as a member of the extensive decay series of ^{238}U:

$$^{238}_{92}U \rightarrow \ldots \rightarrow {}^{226}_{88}Ra \xrightarrow{\quad \alpha \quad} {}^{222}_{86}Rn$$
$$t_{\frac{1}{2}} = 1600 \text{ y}$$

One gram of radium in a sealed container generates about 1.2×10^{-6} g of Rn per day, radioactive equilibrium being attained after about a month when barely 0.00066 cm^3 of radon are present at STP ($c.\ 6.5 \times 10^{-6}$ g). To alleviate problems of adsorption onto solid surfaces, aqueous solutions of $RaCl_2$ are normally used in the preparation of radon; however, this causes certain problems in itself because α irradiation of the water generates relatively vast quantities of O_2 and H_2 which can often pollute the radon by a factor $500\,000:1$ or more; from this it is clear that only very tiny amounts of radon are available for tracer studies.

Frequently the position of radon in an experimental rig is deduced, not from its own α activity, but by observing the much more penetrating 1.8 MeV γ radiation arising from one of its daughters, $^{214}_{83}Bi$, which can be detected through glass or metal container walls. Radon is volatile in a vacuum at temperatures above $-80\,^\circ C$ so that any drastic reduction in its mobility after reaction is suggestive of compound formation.

When radon is treated with gaseous fluorine at room temperature, or liquid fluorine at $-196\,^\circ C$ (the boiling point of liquid nitrogen), the γ radiation is found to be 'fixed' in the reaction vessel owing to formation of an involatile fluoride; the α activity provides the driving force for this reaction by producing highly reactive ionic and excited atomic intermediates. By analogy with xenon chemistry the fluoride is usually assumed to be RnF_2, but its properties suggest an ionic structure, perhaps Rn^+F^- or $(RnF)^+F^-$, since it is very considerably less volatile than XeF_2. (Unfortunately the identity of the material cannot be deduced by conventional analyses because the samples are far too small; perhaps significantly, the only ion detectable in the mass spectrum is Rn^+.) The radon is released by reduction of the fluoride with hydrogen gas at $500\,^\circ C$.

Higher fluorides, and oxides, might be expected from radon's position below xenon in the periodic table but none have yet been identified. One problem, noted also with polonium on page 314, is that highly oxidized species may be rendered unstable by the intense α radiation of radon's short-lived isotopes.

Radon is oxidized by several interhalogen cations to give salts which possibly contain RnF^+ or Rn^+. Solutions to the cation in trichlorotrifluoroethane can be collected as a narrow band on a sulphonated fluorocarbon ion exchange resin $(R_FSO_3K)_n$ and *eluted quantitatively* using BrF_3 in sulphuryl chloride, SO_2Cl_2. When similar oxidations are carried out in liquid interhalogen fluorides such as bromine trifluoride the radon is again immobilized in solution and can be shown to carry a positive charge by its movement towards the cathode on electrolysis. If caesium fluoride is added to the interhalogen solution radon is found to move towards both the cathode and the anode, suggesting the formation of complex fluoro-anions; see

Fig. 106(b). It is possible that reactions such as the following occur in these solvents:

$$Rn \xrightarrow{\text{1. BrF}_3} RnF_2 \rightarrow RnF^+ + F^-; \quad RnF_2 \xrightarrow{F^-} RnF_3^-$$

or

$$Rn \xrightarrow{\text{1. BrF}_3} RnF \rightarrow Rn^+ + F^-; \quad RnF \xrightarrow{F^-} RnF_2^-$$

OCCURRENCE OF THE RARE GASES

Helium

5 ppm by volume of the atmosphere. Owing to its extreme lightness helium is able to exceed the escape velocity of 11.2 km s^{-1} and leave the Earth's atmosphere; the half-life for escape has been calculated to be about 5×10^5 years. Hence virtually all helium now present on Earth has come from the radioactive decay of other elements, the alpha particles they eject being doubly charged helium nuclei. The principal source of helium is natural gas which can contain between 0.1% and 2% helium; the hydrocarbons are removed by liquifying them at about $-150\,°C$ and pumping off the helium. A typical US plant has the capacity to produce up to $3.6 \times 10^6 \text{ m}^3$ of helium per day.

Neon, argon, krypton and xenon

These four gases are present in the atmosphere to be extent of 0.002%, 0.93%, 0.0001% and 0.00001% by volume respectively. They are separated from liquid air by distillation; put in perspective, a separation unit processing about $25\,000 \text{ m}^3$ of air per hour will generate only 200 m^3 of krypton per year.

Radon

Radon-222, the longest-lived isotope with a half-life of 3.8 days, is the immediate decay product of radium, approximately 1.2×10^{-6} g being produced by each gram of radium-226. Estimates suggest that every square mile of the Earth's surface to a depth of six inches contains about 1 g of radium, hence there is always a very tiny background concentration of radon being discharged into the atmosphere. Radon and its short-lived daughter elements are the principal cause of lung cancer in uranium miners; to reduce such exposure the mines have to be continuously ventilated at high rates.

THE ELEMENTS

These elements are monatomic in all phases and, as gases, have specific heat ratios C_p/C_v which are close to the theoretical value of 1.67 for ideal monatomic gases. In the solid state the rare gas atoms are held together by non-directional London forces which vary according to the inverse sixth power of the distance between atoms. For this reason the atoms of Ne, Ar, Kr and Xe take up the space-economizing cubic close-packed structure. When the mass of an atom is small, its zero-point energy also exerts an influence on the crystal structure: thus at 1 K and 25 atm helium-4 has a hexagonal close-packed arrangement whereas the lighter helium-3 adopts the body-centred cubic structure. Above 1100 atm helium-4 can be forced to assume the cubic close-packed structure like the other rare gases.

Under ambient conditions radon is a colourless gas, but when cooled below the melting point of $-71\,°C$ it emits a brilliant phosphorescence which changes from yellow through to orange-red as the temperature is lowered slowly to that of liquid nitrogen $(-196\,°C)$.

THE RARE GAS FLUORIDES

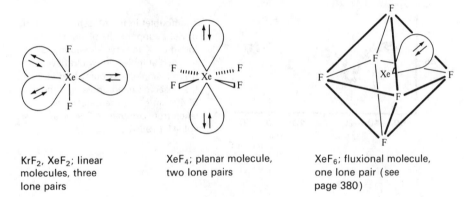

KrF$_2$, XeF$_2$; linear molecules, three lone pairs

XeF$_4$; planar molecule, two lone pairs

XeF$_6$; fluxional molecule, one lone pair (see page 380)

The only known halide of krypton is KrF_2, made by passing an electrical discharge through a gaseous Kr–F_2 mixture. Owing to the thermal instability of KrF_2, the discharge cell, similar in design to that shown in Fig. 104, is immersed in a bath of liquid oxygen $(c.\ -185\,°C)$ so that the fluoride is rapidly frozen onto the cold cell walls as it forms. The three xenon fluorides can be formed by heating differing ratios of xenon and fluorine in sealed nickel vessels:

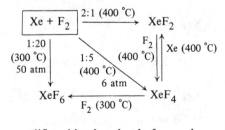

Colourless crystals of xenon difluoride also slowly form when xenon and fluorine are exposed to sunlight in glass flasks, the reaction probably occurring via the photochemical production of fluorine atoms:

$$F_2 \xrightarrow{\ hv\ } 2F + Xe \rightarrow XeF_2$$

The simplicity of this synthesis is remarkable when it is remembered that xenon was long considered to be an inert gas incapable of forming any compounds. Possibly the scarcity of xenon before the Second World War had much to do with this situation, since it made a careful study of the chemistry of xenon very difficult. In 1933 Yost and Kaye must have been very close to synthesizing the xenon fluorides when they passed xenon and fluorine through an electrical discharge, but, no doubt somewhat polarized in their chemical thinking, they came to the conclusion that no compounds were formed. Only slight modifications to their procedure are required to make both xenon and krypton difluorides.

All three xenon fluorides are colourless, volatile solids which readily grow large crystals when allowed to stand in closed vessels; only the hexafluoride attacks silica in the absence of moisture, so that xenon difluoride and tetrafluoride can be conveniently handled in a glass vacuum system.

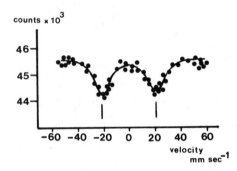

The doublet in the Mössbauer spectrum of XeF_4 supports the planar structure found by X-ray crystallography; a tetrahedral molecule would have given a single peak compared to the observed quadrupole splitting of 42 mm s^{-1}. [The data were taken from *Journal of Chemical Physics* vol. 41 (1964) p. 1157.]

CHEMICAL PROPERTIES OF XENON FLUORIDES

Some of the earliest reactions to be studied involved reduction of the fluorides with simple species like hydrogen or mercury in order to provide suitable methods for analysis and characterization:

$$XeF_2 + H_2 \xrightarrow{\ 400\,°C\ } Xe + 2HF$$

$$XeF_4 + 2H_2 \xrightarrow{\ 130\,°C\ } Xe + 4HF$$

$$XeF_6 + 3Hg \longrightarrow Xe + 3HgF_2$$

Subsequently it was found that xenon tetrafluoride can be readily estimated by using potassium iodide,

$$XeF_4 + 4KI \rightarrow Xe + 2I_2 + 4KF$$

and titrating the liberated iodine in the usual manner with standard sodium thiosulphate solution.

COMPLEXES WITH FLUORIDE ION ACCEPTOR MOLECULES

Apart from studies on their use as fluorinating agents (page 379), much recent research on xenon fluorides has concentrated on their reactions with a wide range of pentafluorides which give rise to complexes with the stoichiometries $(XeF_n)_2MF_5$, XeF_nMF_5 and $XeF_n(MF_5)_2$, where M can be As, Sb, Bi, Nb, Ta, Ru, Os, Ir or Pt. These compounds may be prepared either by fusing together the neat component fluorides or by working in a solvent such as HF or BrF_5. When defined as the number of such complexes formed, the reactivity of the xenon fluorides is in the order $XeF_2 > XeF_6 \gg XeF_4$.

The pentafluoride molecules act essentially as fluoride ion acceptors so that these complexes can be formulated as containing cationic xenon species; however, X-ray structural data show that there is very considerable fluorine bridging between the anions and cations. This is illustrated below for the $XeF^+M_2F_{11}^-$, $XeF^+MF_6^-$ and $Xe_2F_3^+MF_6^-$ complexes formed by xenon difluoride, but the situation is similar for those compounds containing XeF_3^+ and XeF_5^+ 'ions':

$F_5AsF \xrightarrow{2.21 \text{ Å}} Xe \xrightarrow{1.87 \text{ Å}} F$

179°

$(XeF)AsF_6$

$F_4SIFSI(F)_4F \xrightarrow{2.35 \text{ Å}} Xe \xrightarrow{1.84 \text{ Å}} F$

$(XeF)SI_2F_{11}$

1.90 Å   2.14 Å   180°

$F \text{—} Xe \text{—} F \text{—} Xe \text{—} F$

150°

$Xe_2F_3^+$ ion in AsF_6^- salt
(cation–cation and cation–
anion contacts present in the range 3.0–3.4 Å)

KrF_2 forms similar
complexes containing
KrF^+ and $Kr_2F_3^+$ cations
fluorine-bridged to the
anions

The xenon atom of XeF^+ is able to accept an electron pair from Lewis bases such as acetonitrile and pentafluoropyridine to form $[BXeF]^+$ cations:

$$\left[CH_3CN \rightarrow Xe - F \right]^+ \qquad \left[F \underset{F\ F}{\overset{F\ F}{\bigcirc}} N \rightarrow Xe - F \right]^+$$

XeF$_5^+$ SALTS IN NON-AQUEOUS SOLVENTS

Although the essentially square pyramidal structure of the XeF_5^+ ion is plainly visible in X-ray structural analyses of salts such as $XeF_5^+AsF_6^-$, weak fluorine bridging between cation and anion does occur in the solid state:

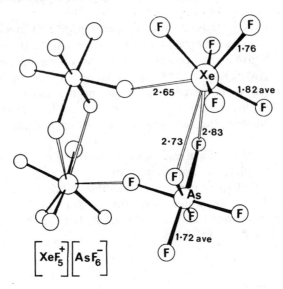

$$\left[XeF_5^+\right]\left[AsF_6^-\right]$$

In liquid hydrogen fluoride and several other fluorinated solvents, these salts display well-resolved ^{19}F NMR spectra characteristic of the (solvated) pentafluoroxenonium(VI) cation: a low-field *quintet* of relative intensity one due to the single axial fluorine, and a *doublet* representing the four equivalent equatorial fluorines (Fig. 105). The multiplets, which arise from F—F coupling, and their 1:4 intensity ratio immediately prove that XeF_5^+ retains its square pyramidal structure in solution; the presence of xenon in the species under study is established by the observation of *twin* satellites on each multiplet due to $^{129}Xe-^{19}F$ coupling (^{129}Xe, relative abundance 26.4%, has a nuclear spin of $\frac{1}{2}$).

COMPLEXES WITH FLUORIDE ION DONOR MOLECULES

Xenon hexafluoride reacts with NOF and the heavier alkali metal fluorides to form complexes such as $MXeF_7$ and M_2XeF_8 which probably all contain either XeF_7^- or XeF_8^{2-} ions, but only the nitrosyl salt $(NO)_2XeF_8$ has been fully characterized by X-ray crystallography. The xenon atom in XeF_8^{2-} has square antiprismatic coordination with no clearly defined position for the expected lone pair of electrons.

HYDROLYSIS OF XeF$_2$ AND XeF$_4$

Xenon difluoride can be dissolved unchanged in water, but a slow reaction does occur on standing, with the liberation of xenon gas:

$$XeF_2 + H_2O \rightarrow Xe + \tfrac{1}{2}O_2 + 2HF \quad (t_{1/2} \sim 7 \text{ h at } 0°C)$$

On the other hand, only a partial evolution of xenon is observed when the tetrafluoride hydrolyses because part of the xenon remains in solution as xenon trioxide:

$$XeF_4 + 6H_2O \rightarrow 2Xe + 1\tfrac{1}{2}O_2 + XeO_3 + 12HF$$

HYDROLYSIS OF XeF$_6$

During experiments on xenon hexafluoride, several explosions have occurred as a result of accidental hydrolysis of the hexafluoride to xenon trioxide, a highly explosive white solid:

$$XeF_6 + 3H_2O \rightarrow XeO_3 + 6HF$$

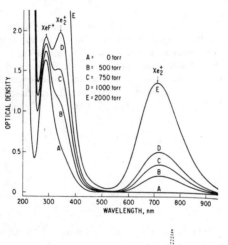

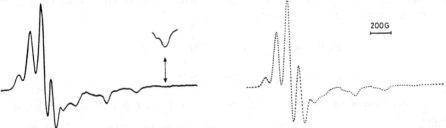

Fig. 105(a)—A spectral investigation of the XeF$^+$ + Xe⇌Xe$_2^+$ equilibrium. *Top.* The effect of xenon pressure upon the UV–visible spectrum of a solution of XeF$^+$Sb$_2$F$_{11}^-$ in antimony pentafluoride at 25°C: curve A, 0.404M XeF$^+$Sb$_2$F$_{11}^-$ in the absence of xenon gas (yellow); curves B–D, the same solution in equilibrium with xenon at partial pressures shown in the figure (green). The bands at 287 and 335 nm are off scale in curve E. Upon removal of the xenon, the bands of Xe$_2^+$ dissapear to leave spectrum A. [Reproduced with permission from *Journal of the American Chemical Society* vol. 102 (1980) p. 2856.] *Bottom.* The electron spin resonance (ESR) spectrum of Xe$_2^+$. ESR spectra are drawn as the first derivative of the absorption peak, nuclei having spin split an electron resonance peak into $(2nI + 1)$ hyperfine lines, where n is the number of nuclei with spin I. The complex nature of this spectrum arises from the large number of xenon isotopes which exist, some of them having nuclear spin. The spectrum on the right is computed by assuming the green compound is Xe$_2^+$ and, as can be seen, the comparison with the experimental curve is excellent. [Reproduced with permission from *Chemical Communications* (1978) p. 502.]

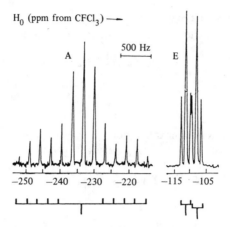

Fig. 105(b)—^{19}F NMR spectrum of the XeF_5^+ cation in $XeF_5^+ SbF_6^-$ dissolved in liquid hydrogen fluoride. The relative intensity of the axial fluorine resonance A has been artificially increased to show up the two weakest lines in the xenon satellites. The lines due to ^{129}Xe—^{19}F coupling are indicated below the spectrum and it can be seen that overlap occurs between the outer lines of the main A resonance and its xenon satellites. Partial overlap of the xenon satellite peaks occurs in the resonance of the equatorial fluorines, E. [Reprinted with permission from *Inorganic Chemistry* vol. 13, p. 765. Copyright (1974) American Chemical Society.]

With small quantities of water, or when reacted with silica, xenon hexafluoride gives the colourless liquid xenon oxide-tetrafluoride,

$$2XeF_6 + SiO_2 \xrightarrow{50°C} 2XeOF_4 + SiF_4$$

$$XeF_6 + H_2O \rightarrow XeOF_4 + 2HF$$

but this latter hydrolysis is further complicated by two side-reactions involving xenon trioxide:

$$XeO_3 + 2XeF_6 \rightarrow XeOF_4$$

$$XeO_3 + XeOF_4 \rightarrow 2XeO_2F_2$$

Rather than using the somewhat hazardous hydrolysis of XeF_6 to make xenon oxide-tetrafluoride, it has been found much more convenient to treat dry sodium nitrate with a slight excess of xenon hexafluoride using stainless steel apparatus:

$$NaNO_3 + XeF_6 \xrightarrow{70°C} NaF + XeOF_4 + FNO_2 \quad (80\% \text{ yield})$$

If the sodium nitrate is in excess, a secondary process occurs:

$$NaNO_3 + XeOF_4 \rightarrow NaF + XeO_2F_2 + FNO_2$$

XENON TRIOXIDE, XENATES AND PERXENATES

XeO₃   perxenate

Xenon trioxide is not ionized in solution and concentrations as high as 4M in XeO_3 can be achieved, but in strongly alkaline solution above pH = 10.5 the oxide behaves as a weak acid and exists as the xenate ion $HXe^{VI}O_4^-$,

$$HXeO_4^- \rightleftharpoons XeO_3 + OH^-, \quad K \sim 7 \times 10^{-4}$$

from which salts such as sodium xenate, $NaHXeO_4.1.5H_2O$, can be derived. The aqueous $HXeO_4^-$ ion slowly disproportionates to perxenate and xenon gas under ambient conditions:

$$2HXeO_4^- + 2OH^- \rightarrow Xe^{VIII}O_6^{4-} + Xe + O_2 + 2H_2O$$

Xenon trioxide is one of the strongest oxidants known in aqueous media; its solutions should (thermodynamically) oxidize water; the non-occurrence of this reaction indicates a high activation barrier which probably arises from the absence of any stable intermediate oxidation states of xenon between free xenon and xenon trioxide:

Oxidation potentials for xenon in aqueous solution

Acid solution

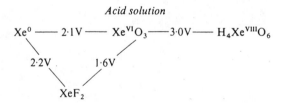

$$Xe^0 \; \text{—— 2·1V ——} \; Xe^{VI}O_3 \text{——3·0V——} H_4Xe^{VIII}O_6$$

2·2V   1·6V

XeF_2

Alkaline solution

$$Xe^0 \text{——1·3V——} XeO(?) \text{——0·7V——} HXeO_4^- \text{——0·9V——} HXeO_6^{3-}$$

$$\text{—— 1·2V ——}$$

When treated with potassium iodide, XeO_3 liberates exactly $3I_2$ per mole and thus the reaction can be used to estimate the oxide quantitatively using standard thiosulphate. In its oxidizing reactions xenon trioxide releases xenon gas,

$$XeO_3 + 6H^+ + 6e^- \rightarrow Xe + 3H_2O, \quad E^\ominus = 2.10 \text{ V}$$

and hence a novel way of detecting the endpoint of an oxidation is to monitor the pressure change in a closed system using a pressure transducer. Thus a *pressuremetric titration* of an iron(II) solution with xenon trioxide showed that the pressure increased linearly up to an Xe:Fe ratio very close to 1:6 and then remained constant (Fig. 106(a)):

$$XeO_3 + 6H^+ + 6Fe^{2+} \rightarrow Xe + 3H_2O + 6Fe^{3+}$$

The perxenate ion is also an extremely powerful oxidizing agent and will oxidize Fe^{2+} to Fe^{3+}, Cr(III) to Cr(VI), Ag(I) to Ag(II), iodate to periodate, Mn(II) to MnO_4^-, water to oxygen (although only slowly in alkaline solution) and concentrated hydrochloric acid to chlorine.

With concentrated sulphuric acid, sodium and barium perxenates given xenon tetroxide (XeO_4), a yellow, volatile, highly unstable solid which decomposes explosively in the solid state even at temperatures as low as $-40°C$:

$$XeO_4 \rightarrow Xe + 2O_2$$

OTHER COVALENT COMPOUNDS OF XENON

Rather few groups have been successfully bonded to xenon and in all cases they have proved to be only those which are highly electronegative. Even chlorine is apparently not able to form derivatives which are stable under ambient conditions; however, when xenon and chlorine mixtures are passed through a microwave discharge and rapidly quenched to about 20 K, infrared and Raman spectra of the resulting matrix show bands which are probably due to $XeCl_2$. The bands are not present when chlorine alone is passed through the discharge and cooled rapidly or when other rare gases replace the xenon.

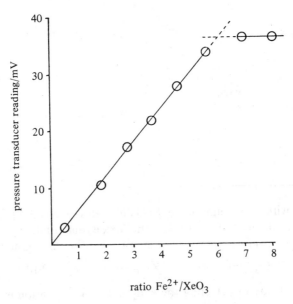

Fig. 106(a)—Pressuremetric titration of Fe^{2+} ions with xenon trioxide.

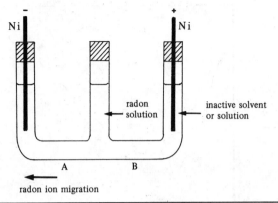

			Radioactivity (mCi cm^{+3})		
Radon added (mCi)	Voltage (V)	Current (mA)	Cathode solution	Centre solution	Anode solution
4.6	50	~2	0.00	1.55	0.00
5.6	32	50	0.31	0.68	0.00
19.5^a	18	50	0.98	1.11	0.11

^{a} Solution contained added KF.

Fig. 106(b)—Electrolysis cell for electromigration studies on radon. Owing to the corrosive solvents used, the apparatus is made of chlorofluorocarbon polymer and fitted together with brass flare connectors; the nickel electrodes are 15 cm apart. The cell is filled with about 20 ml of inactive solution (BrF_3 or $HF-BrF_3$ as solvent) and 2 ml of radon in the same solvent added to the centre leg without stirring. After the solution has been electrolysed for 1 h at 18–50 V, the solution in each of the three legs is monitored for radioactivity, the intensity being measured in millicuries (mCi), to give the results shown above. Normally electrolysis cells of this design have the three legs separated by porous discs at A and B, but these were dispensed with here to stop any possibility of the radon sticking to the large surface area of the pores. Surface adsorption is always a major problem when handling ultra-trace quantities of material. [Data taken from *Science* vol. 168 (1970) p. 362.]

The v_3 band in the infrared spectrum of $XeCl_2$, shown in Fig. 107, has considerable fine structure associated with it. This arises from the various combinations of xenon and chlorine isotopes, each isotopomer having its characteristic vibration frequency. By assuming that the peak at 314.1 cm^{-1} is due to $^{129}Xe^{35}Cl_2$, the theoretical contour of fine structure can be computed from the relative abundances of the various isotopes and, as can be seen, the agreement with the experimental curve leaves little doubt that $XeCl_2$ is present in the matrix. Three closely spaced peaks of intensities 9:6:1 which occur at 254 cm^{-1} in the Raman spectrum of the matrix are indicative of a linear $XeCl_2$ molecule because the observed peak ratios are those statistically expected from the 3:1 abundance of ^{35}Cl and ^{37}Cl. (The symmetric vibration giving rise to the Raman band does not involve movement of the xenon atom and hence there will be no observable effect from the various xenon isotopes.)

Several, often highly explosive, species are known in which xenon binds to NO_3, ClO_4, SO_3F, $OSeF_5$ and $OTeF_5$ substituents via Xe—O bonds. The most stable of

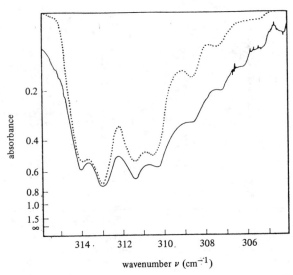

Fig. 107—Infrared spectrum obtained on condensing Xe/Cl$_2$ discharge at 20 K. The dotted line shows the band shape calculated from the known masses and abundances of Xe and Cl isotopes. [Reproduced with permission from *Chemistry in Britain* (1971) p. 189.]

these contain the latter pentafluorotellurate group and can be formed by treating B(OTeF$_5$)$_3$ with xenon fluorides:

$$XeF_4 + B(OTeF_5)_3 \rightarrow Xe(OTeF_5)_4$$

$$XeF_6 + B(OTeF_5)_3 \rightarrow Xe(OTeF_5)_6$$

$$Xe(OTeF_5)_6 \xrightarrow{-10\,°C} Xe(OTeF_5)_4 \xrightarrow[\text{(slow)}]{20\,°C} Xe(OTeF_5)_2$$

$$OXeF_4 + B(OTeF_5)_3 \rightarrow OXe(OTeF_5)_4$$

The highly unusual chromium complex XeCr(CO)$_5$ has been recognized spectroscopically as a short-lived, soluble product arising from ultraviolet photolysis of chromium hexacarbonyl, Cr(CO)$_6$, dissolved in liquid xenon; it has a half-life of about 2 s at $-98\,°C$.

Organo-derivatives of xenon

Many attempts to make compounds containing xenon–carbon bonds have failed and to date the only known examples are those containing fully fluorinated methyl and phenyl groups:

$$Xe + CF_3 \text{ radicals} \rightarrow Xe(CF_3)_2$$

$$XeF_2 + B(C_6F_5)_3 \rightarrow [C_6F_5Xe]^+[C_6F_5BF_3]^- + [C_6F_5Xe]^+[(C_6F_5)_2BF_2]^-$$

Bis(trifluoromethyl)xenon is a waxy, thermally unstable solid which has a half-life of about half an hour at 20°C. In contrast, the (pentafluorophenyl)xenonium cation was stable enough in CH_3CN solvent for it to be characterized by ^{19}F NMR spectroscopy; it is a strong oxidizing agent which readily transfers the fluoro-aromatic groups to other elements

$$Te(C_6F_5)_2 + [FXe(C_6F_5)]^+ \rightarrow Te(C_6F_5)_3^+$$
$$C_6F_5I + [FXe(C_6F_5)]^+ \rightarrow I(C_6F_5)_2^+.$$

An X-ray structure determination on $[CH_3CNXeC_6F_5]^+[(C_6F_5)_2BF_2]^-$ recrystallized from acetonitrile confirmed both the presence of a xenon-carbon bond and a weakly-coordinating CH_3CN molecule:

Chapter 12

Solution Chemistry

INTRODUCTION

Notwithstanding the alchemists' fruitless but extensive search for the perfect solvent, water probably remains our best all-round liquid for dissolving many inorganic materials. However, since there is an immense variation in both the chemical and physical properties of solutes, chemists require access to a wide range of different solvents and there are currently few liquids which have not been studied with this in mind. Liquefied gases such as ammonia and sulphur dioxide were among the first non-aqueous solvents to be studied scientifically during the last century, and to this list we can now add anhydrous acids, fused salts, molten metals and, very recently, even liquefied xenon.

There are three main reasons why solvents are used in chemistry: (1) to facilitate the intimate mixing of reactants; (2) to allow control of reactions which might otherwise be violent; and (3) to aid separation of pure products using the techniques of crystallization and precipitation. Since relatively large amounts of a solvent are required during reactions, its cost, availability and ease of purification may sometimes have to be weighed against its chemical suitability.

The *melting point* and *boiling point* of a liquid largely determine the method by which it can be studied as a solvent. When the boiling point is below room temperature, precautions must be taken to prevent excessive build-up of pressure otherwise explosions may result. This is conveniently done by cooling the reaction tube below the solvent's boiling point using a refrigerant contained in a vacuum flask as shown in Fig. 108. For rough work a low-boiling solvent may be contained in an open vacuum flask, when only very slow evaporation takes place owing to the highly insulating properties of the flask. However, the cold nature of the liquid (held at its normal boiling point under these conditions) means that gross contamination inevitably occurs owing to moisture condensation from the air.

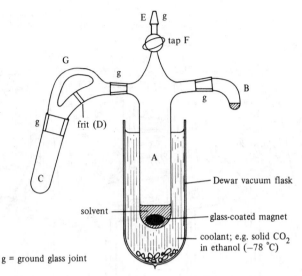

Fig. 108—Schematic diagram of a typical reaction vessel used to study very volatile solvents. The reaction vessel A is evacuated on a vacuum system before the solvent is distilled in so that any gaseous products can be identified and measured without contamination by air or moisture. Solutes are added as required using one or more of the rotatable side-arms B and the resulting solution is stirred magnetically. If an insoluble product is obtained it can be isolated by removing tube A from the Dewar flask, cooling tube C in liquid nitrogen (b.p. $-196\,^\circ$C) and carefully tilting the apparatus to allow the solvent to pass through the glass filter-frit D. The solvent is distilled away by attaching the ground glass joint E to the vacuum system and slowly warming up C with tap F open; any involatile but soluble material will be left in C. Depending on the properties of the collected products, the apparatus can be dismantled either in the open air or in a glove box flushed with dry nitrogen. The bypass tube G allows for rapid pumping of air or solvent vapour past the glass frit.

Solvents which are liquid at room temperature can be handled in apparatus fitted with ground glass joints and taps similar to that used in conventional organic preparations, taking the precaution, however, to keep the whole system under a 'blanket' of *dry* nitrogen gas. Many of these solvents (e.g. ethanoic acid and pure sulphuric acid) are extremely hygroscopic and moisture must be rigorously excluded from them. Complex manipulations are difficult to carry out with this type of experimental set-up and some chemists prefer to use a glove box; this is, in effect, a miniature laboratory to which the worker has access via a pair of long gauntlets sealed into the front panel, the whole box being flushed continuously with dry, purified nitrogen (Fig. 109).

Corrosion is often the biggest problem when dealing with very-high-melting solvents hence this may severely limit the choice of materials from which apparatus is fabricated; below about 600 °C silica, alumina or refractory metals are usually suitable. The 'frozen wall' technique can sometimes be employed industrially to protect the container from attack by molten salts; a high current is passed through the solid salt so that the heat generated melts only the material contained between the electrodes to leave a coating of frozen solvent on the vessel walls. For small-scale work a crucible is often suspended in a tube furnace flushed continuously with a dry,

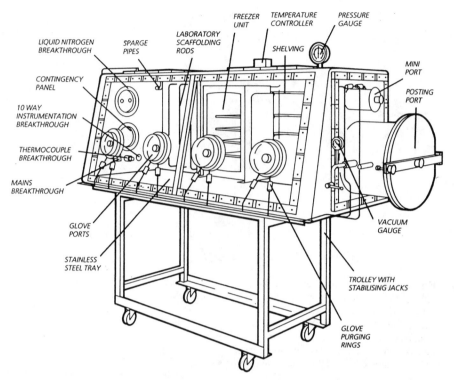

Fig. 109—A general-purpose glove box. Long gauntlets (not shown) are fastened with C-clips onto the round glove ports and flushed with dry, inert gas using the glove-purging rings before the circular doors sealing off the glove ports are opened. The posting port is provided with double doors so that equipment and chemicals can be placed inside the first door and the port flushed with gas before being opened to the interior of the box; to assist in this purging process, the posting port can be evacuated to remove most of the air. Large windows sealed into the front panel provide visibility and illumination; electrical cables and other services enter via the gas-tight 'breakthroughs'. [The diagram is reproduced by courtesy of Faircrest Industrial and Scientific Equipment, Blaina, Gwent NP3 3JW, UK.]

inert gas and the system arranged so that the solution can be examined using a suitable spectroscopic technique.

Solubilities of solutes are virtually impossible to predict even for a much studied solvent like water. The two general factors affecting the solubility of an ionic compound are its lattice energy and the interaction which occurs between the solvent molecules and the dissolved ions. The electrostatic attraction between two ions of opposite charge, $Z(+)$ and $Z(-)$, placed a distance r cm apart in a medium of relative permittivity (dielectric constant) k is $Z(+)Z(-)e^2/4\pi k_0 r^2 k$; hence a solvent with a high relative permittivity should be capable of holding the ions in solution where their mutual attraction is low, approximately $1/k$ that experienced in the crystal. The solvent must, of course, be capable of *solvating* the ions in the first place in order to separate them in solution. Some of the major interactions which can occur between ions and solvent are ion–dipole attractions, covalent bonding and hydrogen bonding (the latter presumably accounts in a large part for the water solubility of many ammonium salts).

For covalent solutes the mechanisms available to achieve solubility are more limited in variety. Such compounds are usually insoluble in solvents like water, which have a highly ordered structure due to hydrogen bonding, unless they can form covalent bonds with the solvent using any available lone pairs of electrons on the latter, or take part in mutual hydrogen bonding (for example, as occurs with the hydroxy compounds ethanol and sucrose so widely enjoyed in aqueous solutions). Non-polar, covalent solutes can be dissolved in a non-polar solvent by virtue of van der Waals attractions, which are usually sufficient to overcome the very low lattice energies and weak solvent–solvent interactions involved. Thermodynamically, the gain in entropy of the solute on dissolution is the driving force of the system (because the enthalpy of solution is usually close to zero and hence ΔG for the dissolution process is negative by virtue of the negative value of $-T\Delta S$).

It was seen above that the attraction between two dissolved ions is inversely proportional to the relative permittivity k of the solvent. Hence when k is small there is an increased probability of *ion pair formation* in solution; a typical solvent with a low relative permittivity is anhydrous ethanoic acid ($k = 6.2$ at $25\,^{\circ}C$), in which it is found that 'free' ions are rather rare and ion pairs, or larger clusters, predominate:

$$K^{+}_{solv} \qquad Br^{-}_{solv} \qquad\qquad (K^{+}Br^{-})_{solv} \qquad\qquad\qquad (Br^{-}K^{+}Br^{-})_{solv}$$

free, but solvated, ions   solvated ion pair   solvated ion triplet

Solutions of similar concentration in water ($k = 78.5$ at $25\,^{\circ}C$) still contain a small proportion of ion pairs, but in pure sulphuric acid ($k = 100$ at $25\,^{\circ}C$) virtually all the ions are free. From this it might be expected that ionic precipitations (double decomposition, or metathetic, reactions) would occur more rapidly in sulphuric acid than in ethanoic acid; in fact, the opposite is found to be true. Within ethanoic acid the ion pairs (and ion clusters) are in mobile equilibrium and hence a metathetic reaction can still occur readily:

$$KBr + AgNO_3 \rightarrow AgBr\downarrow + KNO_3$$

i.e.

$$\left. \begin{array}{l} (K^{+}Br^{-}) \rightleftharpoons K^{+} + Br^{-} \\ (Ag^{+}NO_3^{-}) \rightleftharpoons NO_3^{-} + Ag^{+} \end{array} \right\} \rightarrow AgBr\downarrow$$

Such precipitation reactions do not occur in sulphuric acid because this solvent is about 25 times more viscous than water which results in only slow aggregation of the ions to form solid particles; such particles, if and when formed, tend to remain suspended in the liquid rather than settling out to the bottom of the container.

This short introduction shows why the physical properties of a liquid must be carefully studied when considering it as a possible reaction medium. However, the chemical phenomenon of solvolysis can also be very troublesome but may usually be overcome by simply changing the solvent: typical of many covalent halides, boron trichloride hydrolyses with explosive violence in water but can be studied quite safely in phosphoryl chloride, $OPCl_3$. Although sulphuric acid has a high relative permittivity, and hence supports extensive ionization of solutes, its usefulness as a solvent in which to study anions is considerably limited by the widespread occurrence of solvolysis even under ambient conditions.

SOLVATION OF IONS IN SOLUTION

Although we tend to take hydration of cations in aqueous solution for granted, its proof was not easily accomplished in the earlier days of chemistry. However, we can now show by X-ray crystallography that all six water molecules in the *solid* hydrate $[Cr(H_2O)_6]Cl_3$ are octahedrally coordinated to the chromium(III) ion. If this hydrate, which contains normal water (mainly $H_2{}^{16}O$), is dissolved in $H_2{}^{18}O$ and the solution evaporated, it is found that $[Cr(H_2{}^{16}O)_6]Cl_3$ crystallizes out, demonstrating that the same six water molecules must remain with chromium throughout the dissolution process (the half-life for water exchange is about 40 h). Unfortunately, Cr^{3+} is the only common ion for which this type of direct experiment works; in all other cases cations are found to exchange water with the $H_2{}^{18}O$ solvent so rapidly (residence times at $20°C$ are usually in the range $10^{-9}–10^{-11}$ s) that no $H_2{}^{16}O$ molecules are left in the hydrated crystals which separate out at the end of the experiment.

It is interesting to compare the results of the above $Cr(H_2{}^{16}O)_6^{3+}$ experiment with those obtained when $[Cr(D_2O)_6]Cl_3$ is dissolved in H_2O and then rapidly crystallized out again; all the hydrate water in the crystals is found to be H_2O and not D_2O as might have been expected. This result is still compatible with the $^{16}O/^{18}O$ tracer experiments because the mechanism involves an H/D exchange which occurs via a series of rapid ionic reactions:

$$(D_2O)_5CrOD_2^{3+} \rightleftharpoons (D_2O)_5CrOD^{2+} + D_{aq}^+ \quad \text{(polarization of } D_2O \text{ by } Cr^{3+}\text{)}$$

$$H_2O \rightleftharpoons OH^- + H_{aq}^+ \quad \text{(self-ionization of solvent)}$$

$$D_{aq}^+ + OH^- \rightleftharpoons DHO \quad \text{(loss of } D^+ \text{ into bulk solvent)}$$

$$(D_2O)_5CrOD^{2+} + H_{aq}^+ \rightleftharpoons (D_2O)_5CrOHD^{3+} \rightleftharpoons \text{(repeated for all D atoms)}$$

When a 'tracer' isotope is used to explore a reaction mechanism it must be placed within the actual bond system (here the Cr^{3+}—O bond) which is being investigated so that misleading results from secondary reactions are avoided. During dissolution of $Cr(H_2{}^{16}O)_6^{3+}$ ions in the $H_2{}^{18}O$ solvent, rapid H/H exchange was occurring but the all-important Cr^{3+}—^{16}O bonds remained intact. An interesting effect noted in these exchange reactions was that cations have a slightly greater affinity for $H_2{}^{18}O$ than for $H_2{}^{16}O$.

For the coloured transition ions, spectroscopy can be used to demonstrate cation coordination by solvent water. As an example, X-ray crystallography shows that six water molecules octahedrally coordinate the nickel ion in crystalline $NiSO_4.6H_2O$; since the ultraviolet–visible spectrum of both the crystals and their aqueous solution are virtually identical, it proves conclusively that the nickel coordination remains unchanged in solution, because the d–d spectral bands arise as a direct consequence of *octahedral* ligand field splitting of the metal d orbitals.

During recent years it has become possible to determine the hydration numbers of both cations and anions using either X-ray or neutron diffraction studies on aqueous solutions of salts. In this way it has been shown that the small Be^{2+} ion is tetrahedrally linked to four water molecules, whilst the larger Na^+ and Ca^{2+} ions have 6- and ~ 10-coordination respectively; similarly the presence of the H_3O^+ ion, closely associated with four water molecules, has been firmly established in aqueous solutions of strong acids. The six water molecules which are found to surround a chloride or bromide ion each have one of their hydrogen atoms directed towards the anion:

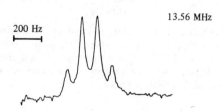

(lifetime c. 10^{-11} s)

One oxygen isotope, the rather rare ^{17}O, has a nuclear spin and hence can be studied by nuclear magnetic resonance (Fig. 110). Thus by placing an ion like Al^{3+} in water enriched in $H_2{}^{17}O$ two resonances can be observed at 20°C, one due to the free solvent and the other to $Al(^{17}OH_2)_n^{3+}$. The area under each resonance peak is proportional to the number of water molecules in that environment and so, by knowing the initial concentration of aluminium ions, the value of n can be calculated. The proton NMR spectrum of aluminium ions dissolved in water at room temperature shows only one peak because very rapid exchange occurs between protons on $Al(H_2O)_n^{3+}$ and those in the solvent via the mechanism described above for chromium. This exchange has to be slowed down by cooling the sample to about -40°C using acetone as a co-solvent (to stop the solution freezing), when the NMR spectrometer has time to 'recognize' each component in the mixture; no acetone appears to coordinate to the aluminium so that the hydration number n can be readily calculated from the peak areas (see Fig. 111(a)).

The isotopes ^{27}Al, ^{69}Ga and ^{71}Ga each have a nuclear spin, and hydration, or complexation in general, can be demonstrated directly for aluminium and gallium ions in a variety of environments. When $AlCl_3$ is studied in acetonitrile, CH_3CN, there is competition between Cl^- and acetonitrile for the Al^{3+} ion, and nine out of the possible ten species (including *cis* and *trans* isomers in some cases) having the

200 Hz

13.56 MHz

Fig. 110—Oxygen-17 NMR spectrum of the oxonium ion in $H_3O^+SbF_6^-$. The spectrum was recorded of an approximately 1:1:1 mixture of H_2O, HF and SbF_5 dissolved in liquid sulphur dioxide at -15°C. The 1:4:4:1 quartet arises from coupling between the nuclear spins of three equivalent hydrogen atoms and the oxygen-17 nucleus. This was proved by proton decoupling in a double-irradiation experiment, when the quartet collapsed into a singlet. [Reproduced with permission from G. D. Mateescu and G. M. Benedikt, *Journal of the American Chemical Society* vol. 101 (1979) p. 3959.]

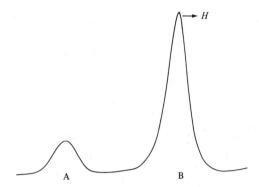

Fig. 111(a)—Diagrammatic representation of a proton NMR spectrum used to investigate solvation numbers of ions in water and liquid ammonia. The peak A arises from the protons located on H_2O or NH_3 molecules coordinated to a metal ion such as aluminium; peak B is due to the uncoordinated solvent molecules. If the area under each peak (proportional to the number of protons in that particular environment) is measured and the metal concentration known, then it is possible to calculate the solvation number of the metal ion. The technique cannot be used for the univalent alkali metal cations because the weakly coordinated solvent molecules undergo such a rapid exchange with the bulk solvent that the NMR spectrometer is unable to detect any magnetic distinction between the two types of proton.

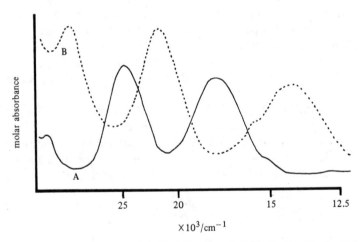

Fig. 111(b)—UV–visible spectra of $Cr(H_2O)_6^+$ (A) and Cr^{3+} dissolved in liquid hydrogen fluoride (B). The similarity of the spectra shows that the chromium ion is probably hexacoordinated in hydrogen fluoride to give $Cr(FH)_6^{3+}$. (It is the number of bands and their relative shapes which are the important features, not their position on the wavenumber scale.] [Data taken from *Inorganic Chemistry* vol. 27 (1988) p. 4504.]

general formula $AlCl_n(CH_3CN)_{6-n}$, $n = 0$–6, can be detected. Gallium trichloride, on the other hand, gives evidence only of the species Ga_2Cl_6, $Ga(CH_3CN)_6^{3+}$ and $GaCl_4^-$. Fig. 112 shows the complexity of products which are obtained when aluminium tribromide is dissolved in a mixture of two complexing solvents, water and acetonitrile.

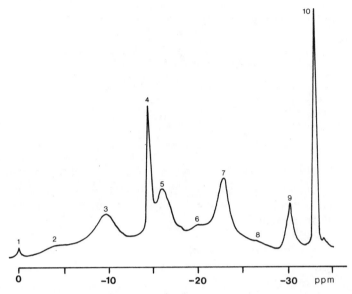

Fig. 112—A ^{27}Al NMR spectrum, recorded at 93.8 MHz, showing the species present in a solution of aluminium tribromide dissolved in an acetonitrile–water solvent mixture (S = CH$_3$CN): (1) Al(H$_2$O)$_6^{3+}$; (2) AlS(H$_2$O)$_5^{3+}$; (3) AlS$_2$(H$_2$O)$_4^{3+}$ (*cis* and *trans*); (4) *cis*-AlS$_3$(H$_2$O)$_3^{3+}$; (5) *trans*-AlS$_3$(H$_2$O)$_3^{3+}$; (6) *trans*-AlS$_4$(H$_2$O)$_2^{3+}$; (7) *cis*-AlS$_4$(H$_2$O)$_2^{3+}$; (8) AlS$_5$H$_2$O^{3+}; (9) AlS$_5$Br^{2+}; (10) AlS$_6^{3+}$. [Redrawn from *Bruker Reports*; paper presented by F. W. Wehrli and S. L. Wehrli at the 22nd Experimental NMR Conference, Asilomar CA, 5–10 April 1981.]

If metal fluorides are involved in coordination studies it is possible to investigate the fluoro-complexes formed using ^{19}F NMR spectrometry. For example, dissolving aluminium in aqueous hydrofluoric acid gives a solution which exhibits a large number of ^{19}F resonances owing to the presence of the following species: AlF$_6^{3-}$, AlF$_5$H$_2$O^{2-}, AlF$_4$(H$_2$O)$_2^-$, AlF$_4^-$, AlF$_3$(H$_2$O)$_3$, AlF$_2$(H$_2$O)$_4^+$ and AlF(H$_2$O)$_5^{2+}$. The hexa-aquo cation cannot be observed using this technique but has been shown to be present by ^{1}H NMR studies on the same solutions. These NMR experiments demonstrate very convincingly how complicated systems can become when a metal ion is placed in an environment containing more than one type of potential ligand molecule.

The number of water molecules associated with a cation can also be calculated from measurements on the amount of water migrating with the ion when it moves towards a cathode under the influence of an electric field. Hydration numbers which arise from such experiments are much larger than those obtained by the methods outlined above because further water molecules are quite strongly hydrogen-bonded to those directly coordinating the cation. The polarizing effect of the cationic charge makes the hydrogen atoms on the inner water molecules somewhat positively charged, thus enhancing their hydrogen-bonding potential:

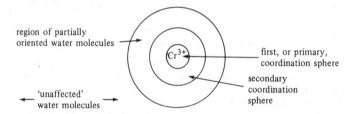

Such a solvation model, proposing the presence of two or more 'coordination spheres' around dissolved ions, is further supported by diffraction studies on aqueous solutions of the Cr^{3+} ion which show that the hexahydrated cation is hydrogen-bonded to twelve more water molecules:

region of partially
oriented water molecules —→

Cr^{3+}

— first, or primary,
   coordination sphere

secondary
coordination
sphere

←— 'unaffected'
     water molecules —→

What happens in very concentrated aqueous solutions of salts when there are not enough water molecules to satisfy the coordination requirements of the cations and anions? In dilute solutions of lithium chloride the predominant cationic species is $Li(OH_2)_6^+$, but when the concentration rises to 10M or higher, chloride ions begin to enter the first coordination sphere; at 14.9M each Li^+ cation, on average, is surrounded by three water molecules and three chloride ions. In the case of $Ni(H_2O)_6^{2+}$, which has an exceptionally long lifetime of about 30 ms, there is no evidence of Cl^- ions entering the first coordination sphere even at high salt concentrations; under these conditions the chlorides presumably have to attach themselves to protons of water molecules retained in the first coordination sphere of the nickel cations.

In 8M zinc bromide there are only 3.8 available water molecules per $ZnBr_2$ unit and the tetrahedrally coordinated cation is found to have an averaged set of 2.4 Br^- and 1.2 H_2O neighbours; since the number of coordinated bromide ions is greater than two per zinc atom, some polymerization via $Zn—Br—Zn$ bridges must occur under these conditions.

Although research on the solvation of ions in non-aqueous solvents is not yet particularly extensive, similar features to the aqueous system are beginning to emerge. For example, when lithium nitrate is dissolved in liquid ammonia the cation has four ammonias in the first coordination sphere which is surrounded by as many as twenty further solvent molecules. The fact that ammonia has three hydrogen atoms to water's two is probably the reason why only four, and not six, NH_3 molecules can fit into the first coordination sphere of the small lithium ion; the diameter of the $Li(NH_3)_4^+$ ion was deduced to be 5.5 Å from X-ray and neutron diffraction data. Many of the other larger cations are found to have six ammonia molecules in their first coordination sphere when in solution.

One has always to be cautious in assuming that cation coordination found in crystals is identical to that in the mother liquor. However, it seems reasonable to accept that the solvates $Fe(NCH)_6(FeCl_4)_2$ and $M(NCH)_6(InCl_4)_2$ emerging from liquid hydrogen cyanide solutions represent good evidence for the occurrence of octahedrally coordinated cations in this solvent when $M = Mg^{2+}, Mn^{2+}, Fe^{2+}, Co^{2+}$, Ni^{2+}, Cu^{2+} and Zn^{2+}. In the crystals, and presumably also in solution, the copper(II) ion has an irregular octahedral coordination owing to Jahn–Teller distortion.

As was found in aqueous solutions, the very characteristic UV–visible spectra arising from d–d electronic transitions within octahedrally coordinated metal ions can be used to investigate solvation in other solvents. The number and relative intensities of the d–d spectral bands depend largely on the coordination geometry, all the bands simply being shifted when the ligands on the metal are changed. Thus the fact that the spectrum of V^{2+} ions dissolved in liquid hydrogen fluoride is very similar to that of $V(H_2O)_6^{2+}$ cations present in solid alums strongly suggests that the octahedral ion $V(FH)_6^{2+}$ occurs in this solvent; similarly the spectrum of Cr^{2+} in HF closely resembles that of aqueous solutions containing $Cr(H_2O)_6^{2+}$, from which we deduce that similar coordination of solvent HF occurs for many other metal ions; see Fig. 111(b). Because of the different chemical properties of liquid HF it is possible to produce stable solutions of ions such as Ti^{2+}, U^{3+}, Sm^{2+} and Yb^{2+} which are oxidized by water, again illustrating the need for a range of solvents.

SELF-IONIZATION OF SOLVENTS

Many solvents, including water, have a very small, though finite, electrical conductivity which can be attributed to a low concentration of ions produced in self-ionization reactions such as

$$2H_2O \rightleftharpoons H_3O^+ + OH^-, \quad K_w = 10^{-14} \text{ at } 25°C$$

Other self-ionization processes which have been suggested for several of the more common solvents are shown in Table 48. In most cases the extent of ionization is so minute that direct evidence for the schemes proposed in the table is very difficult to obtain and is not always unequivocal. An obvious starting point is to check if the postulated ions are stable in the solvent. In the aqueous system a number of compounds analysing as 'solvates' are actually found to contain either oxonium or hydroxide ions (e.g. $HClO_4.H_2O \equiv H_3O^+ClO_4^-$ and $K_2O.H_2O \equiv 2K^+OH^-$); hence solvates emerging from other solvents can be similarly checked for their ionic make-up. If the species in Tables 48 and 49 are compared it will be seen that sulphite ions SO_3^{2-} are not stable in liquid sulphur dioxide, showing that the suggested self-ionization scheme must be incorrect. This is borne out when isotopically labelled SO_3 is dissolved in sulphur dioxide; sulphur exchange with the solvent *only* occurs when sulphite ions are added to the system, showing that the concentration of SO_3^{2-} (or $O_3S—SO_2^{2-}$) ions arising from the self-ionization of sulphur dioxide must be vanishingly small. It is currently assumed that sulphur dioxide, in common with several other solvents, including $OPCl_3$, does not undergo any self-ionization.

Isotopic tracer experiments can also be used to investigate the postulated self-ionization schemes of other solvents. For example, $^{15}NH_4Cl$ rapidly exchanges

Table 48—Proposed self-ionization reactions in some non-aqueous solvents

Solvent	Suggested self-ionization	Remarks
NH_3	$2NH_3 \rightleftharpoons NH_4^+ + NH_2^{2-}$	Ionic product 1.9×10^{-33} at $-50\,°C$
HF	$2HF \rightleftharpoons H_2F^+ + F^-$	Complicated by strong solvation of F^- to give $F(HF)_n^-$
BrF_3	$2BrF_3 \rightleftharpoons BrF_2^+ + BrF_4^-$	
N_2O_4	$N_2O_4 \rightleftharpoons NO^+ + NO_3^-$	Further complicated by the dissociation $N_2O_4 \rightleftharpoons 2NO_2$
H_2SO_4	$2H_2SO_4 \rightleftharpoons H_3SO_4^+ + HSO_4^-$ and $2H_2SO_4 \rightleftharpoons H_3O^+ + HS_2O_7^-$	Two simultaneous reactions
$OPCl_3$	$OPCl_3 \rightleftharpoons OPCl_2^+ + Cl^-$ or $2OPCl_3 \rightleftharpoons OPCl_2^+ + OPCl_4^-$	Both apparently somewhat doubtful
SO_2	$2SO_2 \rightleftharpoons SO^{2+} + SO_3^{2-}$	Now considered unlikely

Table 49—The structures of some solvates

Solvent	Solvate	Structure
NH_3	$NH_3.HCl$	$NH_4^+ Cl^-$
HF	$H_2O.HF$	$H_3O^+ F^-$
	$KF.HF$	$K^+ [F{-}H{-}F]^-$
	$KF.2HF$	
BrF_3	$KF.BrF_3$	$K^+ BrF_4^-$
	$SbF_5.BrF_3$	$BrF_2^+ SbF_6^-$
$OPCl_3$	$BCl_3.OPCl_3$	$Cl_3B \leftarrow OPCl_3$
	$AsCl_3.OPCl_3$	Held by dipole–dipole forces
	$(TiCl_4.OPCl_3)_2$	
SO_2	$K_2SO_3.SO_2$	$K_2[O_3S{-}SO_2]$ (i.e. SO_3^{2-} *reacts* with solvent)

nitrogen atoms with liquid $^{14}NH_3$ so that on recrystallization only $^{14}NH_4Cl$ is recovered. This is compatible with the following sequence of reactions:

$$2^{14}NH_3 \rightleftharpoons {}^{14}NH_4^+ + {}^{14}NH_2^- \quad \text{(self-ionization of solvent)}$$

$$^{15}NH_4Cl \rightarrow {}^{15}NH_4^+ + Cl^- \quad \text{(dissolution of labelled solute)}$$

$$^{15}NH_4^+ + {}^{14}NH_2^- \rightarrow {}^{15}NH_3 + {}^{14}NH_3 \quad \text{(participation of } {}^{15}NH_4^+ \text{ ions in the self-ionization reaction)}$$

By virtue of the last reaction, the ^{15}N atoms are lost to the bulk solvent and, because only small amounts of solute are used relative to solvent, the ammonium ions remaining in solution will be almost wholly labelled with ^{14}N atoms. Whilst this exchange mechanism supports the suggested self-ionization scheme for liquid ammonia, a simple proton transfer reaction such as

$$^{15}NH_4^+ + {}^{14}NH_3 \rightarrow {}^{14}NH_4^+ + {}^{15}NH_3$$

cannot be ruled out.

Just over 3% of pure nitric acid is ionized, giving it one of the highest electrical conductivities for a 'molecular' solvent $(3.8 \times 10^{-2}\ \Omega^{-1}\ cm^{-1})$,

$$2HNO_3 \rightleftharpoons H_2O + NO_2^+ + NO_3^-$$

and at this concentration, 0.25M, it is possible to detect the various species directly using Raman spectroscopy; estimates have suggested that a further 10% of the solvent is required to solvate these dissociation products.

An interesting variation in extent of self-ionization with relative permittivity has been demonstrated for dinitrogen tetroxide, N_2O_4, which has particularly low values for both its electrical conductivity $(c.\ 10^{-12}\ \Omega^{-1}\ cm^{-1})$ and permittivity (2.42); correspondingly, the self-ionization

$$N_2O_4 \rightleftharpoons NO^+ + NO_3^-$$

is not very extensive and no metals dissolve in the pure liquid. The electrical conductivity rises markedly as inert nitromethane $(k = 37)$ is added, reaching a maximum at about 90% nitromethane. Presumably the ionic dissociation of N_2O_4 increases with the permittivity of the mixed system, the fall in conductivity after the maximum being simply due to dilution of the ions as more nitromethane is added. Not surprisingly, metals now react with N_2O_4 and the dissolution rate of copper, for example, almost exactly parallels the enhanced solvent ionization as shown in Fig. 113.

ACID–BASE BEHAVIOUR IN NON-AQUEOUS SOLVENTS

Essentially, the classification of substances into acids and bases gives us two types of active compounds such that members of one type have a greater tendency to react with members of the other type than among themselves. Such a classification thus allows the correlation of an immense amount of data and also assists in the planning of new research.

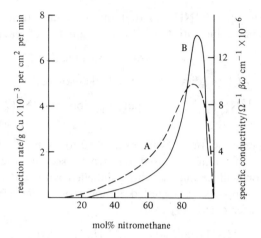

Fig. 113—Variation of specific conductivity of N_2O_4 with added nitromethane. The conductivity of dinitrogen tetroxide reaches a maximum (dashed line, curve A) at about 90 mol% of nitromethane. The enhanced ionization of N_2O_4 makes the mixed solvent increasingly more reactive as shown by the change in rate of dissolution of copper (solid line, curve B) with added nitromethane. [Redrawn from *Chemical Reviews* vol. 80 (1980) p. 21.]

In the aqueous system we know that acids are defined as substances which form H_3O^+ ions in solution whilst bases dissolve to give OH^- ions. Oxonium and hydroxide ions are exactly the same ions postulated to arise from the self-ionization of water,

$$2H_2O \rightleftharpoons H_3O^+ + OH^-$$

Hence we have a simple way of defining acids and bases in other solvents: *acids* are those compounds which produce cations characteristic of the chosen solvent's self-ionization,

$$NH_4Cl \xrightarrow{\text{l.NH}_3} NH_4^+ + Cl^- \quad \text{(via direct dissociation)}$$

$$HCl(gas) \xrightarrow{\text{l.NH}_3} NH_4^+ + Cl^- \quad \text{(via reaction with solvent)}$$

whilst *bases* give the solvent's characteristic anion,

$$NaOH \rightarrow Na^+ + OH^- \quad \text{(in water)}$$

$$NaNH_2 \rightarrow Na^+ + NH_2^- \quad \text{(in liquid ammonia)}$$

$$KF \rightarrow K^+ + BrF_4^- \quad \text{(in liquid } BrF_3 \text{ by reaction with the solvent)}$$

For solvents like liquid sulphur dioxide which undergo no self-ionization we obviously cannot define acids and bases in this manner.

In strongly proton-donating solvents like sulphuric acid and liquid hydrogen fluoride some of the compounds we associate with acidic behaviour in the aqueous system are protonated and thus act as bases by increasing the concentration of anions characteristic of these solvents' self-ionization: e.g.

$$HNO_3 + HF(l) \rightarrow H_2NO_3^+ + F^-$$

base in liquid HF

$$CH_3COOH + H_2SO_4(l) \rightarrow CH_3C(OH)_2^+ + HSO_4^-$$

base in sulphuric acid

As in water, it is possible to carry out acid–base *neutralization reactions* in many non-aqueous solvents by observing the endpoint either potentiometrically or conductimetrically: e.g.

$$KNH_2 + NH_4Cl \xrightarrow{l.NH_3} KCl + 2NH_3$$

base + acid = salt + solvent

Zinc displays *amphoteric behaviour* in liquid ammonia by dissolving in solutions of either ammonium salts or alkali metal amides with the evolution of hydrogen. For the same reason, when potassium amide is added to zinc iodide solution a white precipitate of zinc amide forms which dissolves in the presence of excess amide owing to the production of soluble tetra-amidozincate ions:

$$2KNH_2 + ZnI_2 \rightarrow Zn(NH_2)_2\downarrow \xrightarrow{2KNH_2} K_2Zn(NH_2)_4$$

LEVELLING ACTION OF PROTONIC SOLVENTS

In solvents which contain ionizable protons, the solvated proton is responsible for the acidity whereas the base changes with each solvent. When a strong proton donor HA (i.e. a strong acid) dissolves in a solvent HB which is a proton acceptor (i.e. is basic), *complete proton exchange* between solute and solvent occurs and cations characteristic of the solvent's self-ionization are formed:

$$2HB \rightleftharpoons H_2B^+ + B^-$$

$$HA + HB \rightarrow H_2B^+ + A^-$$

The acid strength of HA (i.e. its proton-donating power when dissolved in HB) is therefore equal to that of the cation H_2B^+ and it follows that all strong acids will display this same acidity: in other words, they are all *levelled* to the proton-donating ability of the H_2B^+ ion. For example, hydrochloric, nitric and sulphuric acids are all levelled to the acidity of the oxonium ion in water,

$$HCl + H_2O \rightarrow H_3O^+ + Cl^-$$

$$H_2SO_4 + 2H_2O \rightarrow 2H_3O^+ + SO_4^{2-}$$

$$HNO_3 + H_2O \rightarrow H_3O^+ + NO_3^-$$

and to the acidity of the ammonium ion in liquid ammonia: e.g.

$$HCl + NH_3 \rightarrow NH_4^+ + Cl^-$$

If, on the other hand, the solvent HB is also a proton donor (i.e. is acidic), an *equilibrium* will be set up when the acid HA dissolves,

$$HA + HB \rightleftharpoons H_2B^+ + A^-$$

and fewer H_2B^+ ions will be formed. The acid strength of the solution will depend on the number of such ions and will be a function of the proton-donating power of HA. Thus acids, when dissolved in an acidic medium, are *differentiated* according to their proton-donating ability. Conversely, bases are differentiated in basic solvents and are levelled in acidic solvents.

This levelling action has several far-reaching consequences regarding the usefulness of a solvent for a particular reaction. Thus it would obviously be impossible to use a comparatively acidic solvent such as water for dissolving very strong bases like the hydride and ethoxide ions because they are immediately levelled, by protonation, to the basicity of the hydroxide ion:

$$H^- + H_2O \rightarrow H_2 + OH^-$$

$$OEt^- + H_2O \rightarrow EtOH + OH^-$$

Such compounds must be handled in either a basic or an inert solvent.

In a basic solvent all acids, including those classed as 'weak' in the aqueous system, will be levelled to the acid strength of the solvent cation. Analysts have exploited this fact by titrating weak acids like carboxylic acids and phenols in liquid ammonia or anhydrous primary amine solvents, for example butylamine and ethylene diamine:

$$R'COOH + RNH_2 \rightarrow RNH_3^+ + R'COO^- \quad \text{(levelling reaction)}$$

$$RNH_3^+ + (OR'')^- \rightarrow RNH_2 + R''OH \quad \text{(neutralization reaction)}$$
$$\uparrow$$
standard base (ethoxide)

The endpoint of such a titration is normally detected potentiometrically by placing an indicator electrode in the solution and comparing the potential between this and some standard electrode; at the endpoint there is an abrupt change in the observed potential as shown schematically in Fig. 115.

SOLUTIONS OF METALS IN NON-AQUEOUS SOLVENTS

The fact that several metals dissolve in liquid ammonia to form deep blue or bronze-coloured solutions has been known for many years and the great reducing power of such solutions has been exploited in both preparative inorganic and organic chemistry. The phenomenon is not restricted to liquid ammonia, and coloured metal solutions are now known in ethers, amines, amides and molten metallic salts.

Dilute solutions of alkali metals in liquid ammonia consist primarily of solvated cations and *solvated electrons* which further interact to give short-lived ($c.\ 10^{-12}$ s) ion pairs:

$$M \xrightarrow{\ 1.NH_3\ } M_{am}^+ + e_{am}^-$$

$$M_{am}^+ + e_{am}^- \rightleftharpoons (M_{am}^+ e_{am}^-)$$

The electrons are present in relatively large cavities of radius 3.2–3.4 Å formed by the orientation of protons from eight or nine solvent molecules towards the negative charge; a further sheath of coordinating ammonia molecules surrounds this inner solvation shell:

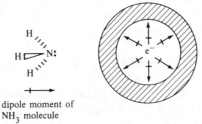

dipole moment of
NH_3 molecule

Two types of ion pair probably co-exist in ammonia, one in which intervening ammonia molecules are shared and the other where both the first coordination spheres remain intact:

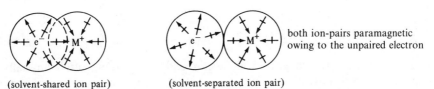

both ion-pairs paramagnetic owing to the unpaired electron

(solvent-shared ion pair)   (solvent-separated ion pair)

Up to about 10^{-2} mol% of metal (mpm), the concentration of unpaired electrons released into solution is equal to the concentration of dissolved metal, but as more metal is added some of the electrons form species in which their spins become 'paired'. For example, about 90% of the electrons in a 0.1 mpm sodium solution are spin-paired, but the phenomenon remains far from being clearly understood. It is usually assumed that diamagnetic 'ion triplets' $e_{am}^- M_{am}^+ e_{am}^-$ are present, but their estimated lifetime $(>10^{-6}$ s) is nearly a million times too long for the binding to be purely Coulombic. There is no evidence to suggest that the diamagnetic species are simple anions, M^-, in which two electrons reside in the outer ns orbital of the metal, although such anions are now known to occur in some amines and ethers from which they can be isolated in crystalline solids $[M(crypt)]^+ M^-$ by adding 'crypt' ligands (page 117).

As the metal concentration is increased still further, the solutions begin to exhibit metallic character, the electrical properties, for example, changing from those typical of an electrolyte to those more typical of a metal; this is generally referred to as the non-metal–metal (or NM–M) transition. The metallic state consists of solvated cations and a 'Fermi sea' of conducting electrons (i.e. the electrons are present in a conduction band similar to those found in pure metals). Finally, cooling such concentrated solutions for Li, Ca, Sr, Ba, Eu and Yb gives rise to solid metal ammines of approximate formulae $Li(NH_3)_4$ and $M(NH_3)_6$. In these 'expanded metal' species the metal atoms are almost completely ionized, with the electrons contained in a conduction band; the ammonia acts essentially as a diluent by solvating the cations and so increasing the M—M distance relative to that in the pure metal. Neutron diffraction studies of these ammines show the ammonia molecules have an unusual geometry in that they are almost planar with one short N—H bond and two extremely long bonds.

An intriguing phenomenon occurs on cooling a sodium–ammonia solution in the intermediate concentration range (*ca.* 4 mpm) to below −42 °C because two immiscible liquid phases separate, giving a denser, but less concentrated, dark blue solution beneath a bronze-coloured solution. The former contains the electron-in-a-cavity species discussed above while the bronze-looking system is typical of the metallic phase. Other metal solutions show the same property but the critical temperature changes in each case.

Coloured solutions are also formed when alkali metals dissolve in their corresponding molten halides. At low metal concentrations the dissociated electrons are 'solvated' in the sense that they lie in vacant anion sites present within the body of the melt; short-range order persists in the liquid salts so that ions have, on average, mainly oppositely charged species as neighbours. (These electron sites are often referred to as 'liquid F-centres'.) As more metal dissolves, spin-paired systems begin to form, and eventually at high concentration a non-metal–metal transition occurs in which metal clusters are thought to develop.

CHEMISTRY IN ETHANOIC ACID

Large quantities of ethanoic acid (acetic acid) are manufactured annually, giving ethanoic acid the advantages of ready availability and low cost, two points of prime importance when considering the potentialities of a solvent. The 99.7% commercial acid has water as its main impurity and this can be removed by refluxing with ethanoic anhydride for about 3 h:

$$(CH_3CO)_2O + H_2O \rightarrow 2CH_3COOH$$

A fractional distillation from chromium(VI) oxide (added to destroy any oxidizable organic impurities) then gives the anhydrous solvent, the purity of which may be checked by its melting point.

As can be seen from Table 50, the melting and boiling points allow ethanoic acid to be handled in ordinary glassware at room temperature and pressure. For accurate work, strict precautions must be taken to exclude the entry of atmospheric moisture, for example by working in a glovebox under a slight positive pressure of dry nitrogen.

From the low relative permittivity one would predict most metal salts to be insoluble in ethanoic acid, but in practice it is found that a surprising number do dissolve, particularly those containing large ions (i.e. salts with comparatively low lattice energies). These include nitrates, thiocyanates, cyanides and iodides; many ethanoates, the base analogues in this solvent, are also soluble, but owing to the low permittivity

Table 50—Physical properties of ethanoic acid

Melting point	16.6 °C
Boiling point	118.1 °C
Relative permittivity	6.2 (25 °C)
Specific conductivity	$(2.4–3.0) \times 10^{-9} \ \Omega^{-1} \ cm^{-1}$
Viscosity	1.15 cP (25 °C)

they, like other salts, form ion clusters, resulting in ionic dissociation constants of less than 10^{-5} at $25\,^{\circ}$C.

On the other hand, the low permittivity suggests that the solvent molecules are only slightly polar and have relatively little interaction with each other; not surprisingly then, many covalent compounds, such as $SnCl_4$, $AsCl_3$, $GeCl_4$ and H_2S, are freely soluble. Water, ammonia and the common (anhydrous) acids dissolve by interaction with the solvent either by forming hydrogen bonds (as in the case of water) or by undergoing protonation reactions:

$$NH_3 + CH_3COOH \rightarrow NH_4^+ + CH_3COO^- \quad \text{(protonation by the solvent)}$$

$$HClO_4 + CH_3COOH \rightarrow CH_3C(OH)_2^+ + ClO_4^- \quad \text{(protonation of the solvent)}$$

A useful solubility guide is that compounds insoluble in water are usually insoluble in ethanoic acid. Realizing this, it is possible to predict a number of ionic precipitation reactions from a knowledge of aqueous chemistry:

$$AgNO_3 + KI \rightarrow AgI\downarrow + KNO_3$$

$$AgNO_3 + KSCN \rightarrow AgSCN\downarrow + KNO_3$$

$$ZnI_2 + Na_2C_2O_4(\text{sodium oxalate}) \rightarrow ZnC_2O_4\downarrow + 2NaI$$

The specific conductivity of pure ethanoic acid, though very low, is assumed to be due to a slight self-ionization:

$$2CH_3COOH \rightleftharpoons \left[CH_3C \underset{OH}{\overset{OH}{<}} \right]^+ + CH_3COO^-$$

Such an equilibrium is compatible with the rapid exchange which occurs between potassium ethanoate, labelled with radioactive carbon-14, and ethanoic acid. However, the exchange could also take place by direct proton transfer between an ethanoate ion and a solvent molecule.

Ethanoic acid is a protonic (i.e. acidic) solvent and reactions can thus be expected to occur between it and solutes capable of attacking protons. For example, sodium carbonate evolves carbon dioxide when added to ethanoic acid,

$$Na_2CO_3 + 2CH_3COOH \rightarrow 2NaOOCCH_3 + CO_2 + H_2O$$

sodium hydroxide dissolves to give sodium ethanoate,

$$NaOH + CH_3COOH \rightarrow NaOOCCH_3 + H_2O$$

and electropositive metals react producing hydrogen:

$$2Na + 2CH_3COOH \rightarrow 2NaOOCCH_3 + H_2$$

In the case of zinc only slight reaction occurs to give hydrogen and a coating of insoluble zinc ethanoate on the metal surface, but when sodium ethanoate (a base in this solvent) is added the zinc dissolves readily owing to the formation of a soluble complex, sodium tetra-ethanoatozincate, which cleans up the metal surface:

$$2NaOOCCH_3 + Zn(OOCCH_3)_2 \rightarrow Na_2Zn(OOCCH_3)_4$$

This demonstrates the amphoteric nature of zinc and is analogous to the dissolution of zinc hydroxide in aqueous NaOH which gives sodium zincate. Neodymium tri-ethanoate is also soluble in the presence of potassium ethanoate to give a bluish-coloured complex; under the influence of an electric field the colour migrates towards the *anode* as expected for an ethanoatoneodymiate ion:

$$Nd(OOCCH_3)_3 + nKOOCCH_3 \rightarrow K_n Nd(OOCCH_3)_{3+n}$$

From the self-ionization of the solvent, ethanoates can be recognized as bases in ethanoic acid whilst compounds producing solvated protons will behave as acids. The number of solvated protons formed by a proton donor HA will be proportional to the proton donating power of HA owing to the *differentiating* effect of the 'acidic' solvent (see page 408):

$$HA + CH_3COOH \rightleftharpoons CH_3C(OH)_2^+ + A^-$$

It is therefore possible to compare the strength of several acids by determining the proportion of solvated protons formed on their dissolution. Such experiments have shown that the strengths of several inorganic acids are in the order $HClO_4 > HBr > H_2SO_4 > HCl > HNO_3$.

Conversely, when bases which are classed as 'weak' in the aqueous system are dissolved in ethanoic acid they are levelled to the base strength of the ethanoate ion by protonation, e.g.

$$NH_3 + CH_3COOH \rightarrow NH_4^+ + CH_3COO^-$$

and, since the ethanoate ion can be titrated readily against a strong acid such as perchloric acid, these weak bases (weak, that is, in aqueous media) can be easily estimated in ethanoic acid. The acid–base neutralization reaction

$$CH_3COO^- + CH_3C(OH)_2^+ \rightarrow 2CH_3COOH$$

can be followed by noting changes either in the electrical conductivity of the solution or in the potential between an indicator electrode placed in the solution and a standard reference electrode. At the end point, when equivalent amounts of acid and base have been mixed, there is an abrupt change in both the conductivity and the potential (Fig. 114).

By a suitable procedure, the constituents in a mixture of primary, secondary and tertiary aliphatic amines (all of which are weak bases in water) can be determined by potentiometric titration against perchloric acid in ethanoic acid solvent. The mixture is divided into three equal portions (aliquots). One aliquot is titrated directly against the perchloric acid to give the total amine content. The second portion is treated with ethanoic anhydride for 3 h at room temperature before being titrated; in this way the primary and secondary amines are acylated to non-basic products, when titration thus gives the tertiary amine content. Salicylaldehyde is added to the final aliquot to form a non-basic derivative with the primary amine so that the secondary and tertiary amines can be determined. From these three titrations it is obviously possible to calculate the concentration of each amine in the original mixture.

These few reactions give an indication of the chemistry which can occur in ethanoic acid. As previously pointed out, this solvent shows certain similarities to water, which makes its study slightly less strange to those unfamiliar with non-aqueous solvents.

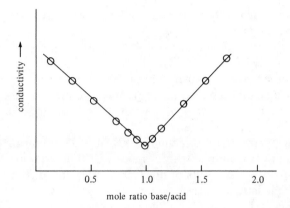

Fig. 114—Conductivity curve for the titration of perchloric acid with sodium acetate in ethanoic acid.

Most solvents, however, bear little or no resemblance to water; some are exceedingly reactive, such as bromine trifluoride in which even asbestos dissolves with incandescence, or the liquid structure may be composed completely of ions, as in the case of molten salts.

CHEMISTRY IN LIQUID AMMONIA

Ammonia (Table 51) was one of the first non-aqueous solvents to be studied scientifically and its chemistry dates back to around the middle of the last century. Huge quantities of ammonia are produced industrially by the direct reaction of nitrogen and hydrogen in the Haber process:

$$N_2 + 3H_2 \xrightarrow[\text{pressure}]{\text{catalyst}} 2NH_3$$

The main impurity in the liquefied ammonia sold commercially in pressurized cylinders is water, which can be removed readily by adding sodium to the ammonia until the characteristic 'inky blue' colour of the dissolved metal is produced:

$$H_2O + Na \rightarrow NaOH + \tfrac{1}{2}H_2$$

$$Na \text{ (excess)} \xrightarrow[\text{NH}_3 \text{ is dry}]{\text{dissolves when}} \text{blue solution}$$

Table 51—Physical properties of liquid ammonia

Melting point	$-77.7\,°C$
Boiling point	$-33.4\,°C$
Density	0.68 gm ml^{-1} ($-33\,°C$)
Viscosity	0.135 cP ($25\,°C$)
Relative permittivity	22 ($-33\,°C$)
Specific conductivity	$4 \times 10^{-10}\ \Omega^{-1}$ cm^{-1} ($-15\,°C$)

The anhydrous ammonia is then vacuum-distilled directly from the sodium solution into a cooled reaction flask or storage vessel.

The low boiling point of ammonia ($-33.4\,°C$) means that rather special apparatus has to be used to study its chemistry. Normally liquid ammonia is handled in a glass vacuum apparatus where the temperature of the reaction vessel can be held below $-33\,°C$ by use of a suitable cold bath; a typical refrigerant for the cold bath is solid CO_2 suspended in ethanol, which gives a temperature close to $-78\,°C$ when held in a vacuum flask.

The low relative permittivity of ammonia (22 at $-33\,°C$) means that fewer salts will be soluble in liquid ammonia than in water, and those that are tend to form ion pairs and ion triplets. Indeed salts of divalent or trivalent ions are seldom soluble, owing to their high lattice energies, unless the accompanying ions are large and polarizable as occurs, for example, in the case of thiocyanates or iodides. For the same reason even chlorides and fluorides of the alkali metals do not have appreciable solubility in ammonia; this is often used to advantage in synthetic chemistry by arranging that an alkali chloride is precipitated from solution in 'double-decomposition' (metathetic) reactions. The lone pair on the ammonia molecule make it an excellent donor, hence solvates are formed with many metallic salts, including those of the alkali metals (e.g. $NaI.4NH_3$) and the transition metals (normally giving octahedrally coordinated $[M(NH_3)_6]^{n+}$ cations). NMR and isotopic tracer studies have shown that cations such as Mg^{2+}, Al^{3+} and Cr^{3+} are solvated by six ammonia molecules in solution (see Fig. 111). The exchange of labelled NH_3 with the hexammine ion $Cr(NH_3)_6^{3+}$ is thought to involve an amido intermediate whose formation is induced by the polarizing effect of the trivalent cation,

$$Cr(NH_3)_6^{3+} \rightleftharpoons Cr(NH_3)_5NH_2^{2+} + H_{am}^{+} \quad \text{(compare page 398)}$$

and in agreement with this, the exchange rate decreases in the presence of ammonium ions since addition of acid will force the ammonolysis reaction to the left. There is no decrease in exchange rate between solvent and $Mg(NH_3)_6^{2+}$ under acid conditions, showing that ammonolysis does not occur with a divalent cation.

The slight electrical conductivity of purified ammonia is explained by assuming the occurrence of a self-ionization reaction,

$$2NH_3 \rightleftharpoons NH_4^{+} + NH_2^{-}$$

so that ammonium salts behave as acids and amides as bases. Hence when a reactive metal is added to an ammonium salt dissolved in liquid ammonia, hydrogen is evolved,

$$NH_4Cl + Na \rightarrow NaCl\downarrow + \tfrac{1}{2}H_2 + NH_3$$

$$2NH_4BF_4 + Mg \rightarrow Mg(BF_4)_2 + H_2 + 2NH_3$$

and the reaction between a metal amide and an ammonium salt is an acid–base neutralization. Not surprisingly it has been found, calorimetrically, that the heats of neutralization reactions are constant at about 109 kJ mol^{-1} irrespective of the amide or ammonium salt used, because the same reaction is occurring in all cases:

$$NH_4^{+} + NH_2^{-} \rightarrow 2NH_3$$

Titrations can be carried out in ammonia, as in water, but the endpoints are usually detected either conductimetrically or potentiometrically (Fig. 115) because most

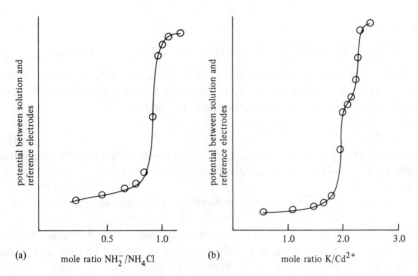

Fig. 115—Potentiometric titrations in liquid ammonia. (a) Simple acid–base titration between an alkali metal amide and ammonium chloride. (b) Reduction of a cadmium salt using a solution of potassium in ammonia. The first endpoint at a K/Cd^{2+} ratio of 2.0 corresponds to the reduction of the salt to cadmium metal:

$$Cd^{2+} + 2K \rightarrow 2K^+ + Cd\downarrow$$

The second endpoint at a ratio of about 2.3 corresponds to the formation of an intermetallic compound, 'KCd_3'.

indicators do not function satisfactorily in liquid ammonia (many indicators used in water are weak organic acids and thus are fully ionized in ammonia owing to the solvent's levelling action).

Solutions of alkali metal amides are convenient reagents for synthesizing other amides,

$$IrBr_3 + 3KNH_2 \rightarrow 3KBr + Ir(NH_2)_3$$

$$AgCl + NaNH_2 \rightarrow NaCl + AgNH_2$$

$$ZnI_2 + 2KNH_2 \rightarrow 2KI + Zn(NH_2)_2$$

and in several cases it is found that these precipitated amides will dissolve on the addition of an excess of NH_2^- ions owing to the formation of soluble amidometallates: e.g.

$$Zn(NH_2)_2 + 2KNH_2 \rightarrow K_2Zn(NH_2)_4 \quad \text{potassium tetra-amidozincate}$$

This amphoteric nature of zinc amide in liquid ammonia parallels the behaviour of zinc hydroxide in aqueous bases:

$$Zn(OH)_2 + 2KOH \rightarrow K_2Zn(OH)_4$$

Like water, liquid ammonia solvolyses many covalent chlorides, bromides and iodides, the initial step in the reaction sequence probably involving coordination of NH_3 to the central atom followed by rapid loss of hydrogen halide from the unstable solvate: e.g.

$$BCl_3 + NH_3 \rightarrow [Cl_3B{:}NH_3] \rightarrow HCl + [Cl_2BNH_2] \rightarrow \rightarrow B(NH_2)_3$$

As described on page 408, the alkali metals dissolve in liquid ammonia to give solutions of great reducing power which are widely used in both organic and inorganic chemistry. Simple salts are normally reduced to the finely divided metal,

$$CuBr + K \rightarrow KBr + Cu{\downarrow}$$

but the reductions are often complicated by a side-reaction in which the precipitated metal catalyses the solvolysis of the alkali metal:

$$K + NH_3 \xrightarrow[\text{catalyst}]{\text{metal}} KNH_2 + \tfrac{1}{2}H_2 \quad \text{(rapid)}$$

(The alkali metal solutions are thermodynamically unstable with respect to amide formation but, unless a catalyst is present, the kinetic barrier to decomposition is sufficiently high for the solutions to be kept for many months at $-78\,^\circ$C.) In more complex salts the anions are often attacked and, for example, iodates are reduced to iodides and manganates(VII) to manganates(VI).

Unusually low oxidation states can be produced in a number of elements by reaction with metal–ammonia solutions, cyanometallates and carbonyl derivatives of the transition metals being reduced so far that the metal actually attains a zero, or even a *negative*, oxidation state:

$$K_2Ni^{(II)}(CN)_4 + 2K \rightarrow K_4Ni^{(0)}(CN)_4$$

$$K_4Mn^{(II)}(CN)_6 + 2K \rightarrow K_6Mn^{(0)}(CN)_6$$

$$Mn_2(CO)_{10} + 2Na \rightarrow 2NaMn^{(-I)}(CO)_5$$

$$Fe_3(CO)_{12} + 6Na \rightarrow 3Na_2Fe^{(-II)}(CO)_4$$

Among the non-transition elements, highly coloured anions such as Sb_7^{3-}, Pb_5^{2-}, Sn_9^{4-}, Te_4^{2-} and Bi_5^{3-} are formed by reducing the free element. Removal of the solvent yields rather ill-defined and amorphous ammoniates such as $[Na(NH_3)_n]_3Sb_7$ which decompose if the adhering ammonia is removed in a vacuum. However, by working in 1,2-diaminoethane (ethylene diamine) and solvating the cation with crypt ligands it is possible to isolate crystalline derivatives of several such anions, thus allowing their structures to be determined by X-ray diffraction.

Owing to the different solvating properties of water and ammonia, some distinctly unfamiliar metathetic reactions can occur in the latter solvent: e.g.

$$AgCl + KNO_3 \rightarrow KCl{\downarrow} + AgNO_3$$

Silver ions have a strong affinity for ammonia ligands (recall the solubilities of AgCl and AgBr in *aqueous* ammonia owing to the formation of ammine complexes) and hence they are strongly solvated in liquid ammonia, thus leading to a range of soluble silver salts. As mentioned earlier, the low solubility of alkali metal chlorides in ammonia can be used to advantage in preparative chemistry by removing unwanted cations and anions from solution:

$$[Cr(NH_3)_6]Cl_3 + 3KBH_4 \rightarrow [Cr(NH_3)_6][BH_4]_3 + 3KCl{\downarrow}$$

$$[Mg(NH_3)_6]Cl_2 + 2KBH_4 \rightarrow [Mg(NH_3)_6][BH_4]_2 + 2KCl\downarrow$$

After filtration to remove the precipitated potassium chloride, the pure tetrahydroborates are obtained by simple crystallization. Although KBH_4 slowly hydrolyses in water, it may be dissolved unchanged in ammonia which thus acts as a convenient solvent for such syntheses.

Reduction of Group IV and Group V hydrides by alkali metal solutions, followed by suitable metathetic reactions, has allowed a variety of element-to-element coupling reactions to be achieved:

$$R_3GeH + K \rightarrow \tfrac{1}{2}H_2 + KGeR_3 \begin{cases} \xrightarrow{R_3GeCl} R'_3Ge\!-\!GeR'_3 + KCl\downarrow \\ \xrightarrow{R_3SnCl} R_3Ge\!-\!SnR'_3 + KCl\downarrow \\ \xrightarrow{ClMn(CO)_5} R_3Ge\!-\!Mn(CO)_5 + KCl\downarrow \end{cases}$$

$$PH_3 + K \rightarrow \tfrac{1}{2}H_2 + KPH_2 \xrightarrow{EtCl} EtPH_2 + KCl\downarrow$$

From this short discussion we note that liquid ammonia is readily available and easily purified (two essential points if a solvent is to have wide applicability), that many compounds which are hydrolytically unstable in aqueous media can be studied in ammonia solution (which opens up new areas of synthetic chemistry) and that, since ammonia is 'basic', the effect of its levelling action on weak acids is opposite to that of an acidic medium like ethanoic acid (which levels bases).

CHEMISTRY IN PURE SULPHURIC ACID

As required of a useful solvent, sulphuric acid is readily accessible, cheap and may easily be obtained in a state of high purity by adding aqueous sulphuric acid to oleum until the melting point is 10.371 °C. It has a wide and convenient liquid range (Table 52) and hence can be studied in conventional apparatus provided that atmospheric moisture is excluded.

The high viscosity of sulphuric acid makes precipitation reactions and crystallization procedures very difficult to carry out, whilst at the same time its low volatility severely hinders the thorough drying of any solid products arising from such manipulations. As would be expected for a viscous solvent, mobilities of all ions except HSO_4^- and $H_3SO_4^+$ are low, being less than one-tenth of those measured in aqueous solutions. The species HSO_4^- and $H_3SO_4^+$ do not need to move physically through the solution

Table 52—Physical properties of sulphuric acid

Melting point	10.37 °C
Boiling point	290–317 °C
Viscosity	24.5 cP (25 °C)
Relative permittivity	120 (10 °C), 100 (25 °C)
Specific conductivity	$1.04 \times 10^{-2}\ \Omega^{-1}\ cm^{-1}$ (25 °C)

on the application of an electric field since they take part in a rapid proton transfer mechanism involving the extensively hydrogen-bonded structure of the solvent; their apparent mobility is thus not affected by viscosity:

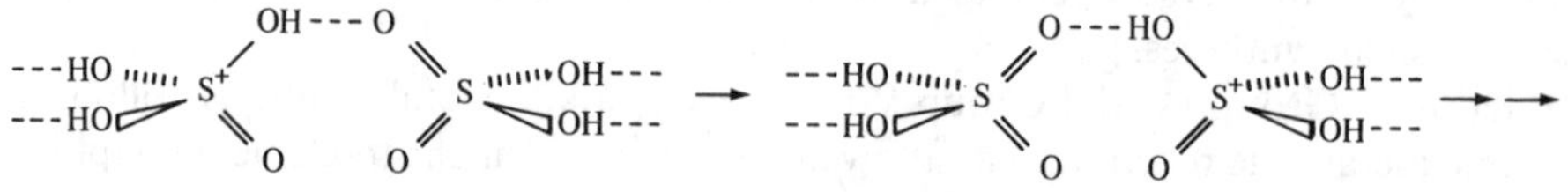

Therefore solutions containing either HSO_4^- or $H_3SO_4^+$ ions have high electrical conductivities whereas solutions of other ions conduct electricity only very feebly. Gillespie has shown that it is possible to determine the number of hydrogen sulphate (HSO_4^-) ions released into sulphuric acid per mole of solute by assuming the molar conductivities are determined almost wholly by these ions. Thus the molar conductivity of a solution of $KHSO_4$ is found to be about half that of dissolved HNO_3:

$$KHSO_4 \rightarrow K^+ + HSO_4^-$$

$$HNO_3 + 2H_2SO_4 \rightarrow NO_2^+ + H_3O^+ + 2HSO_4^-$$

The high molar freezing point depression of sulphuric acid ($6.12°C$) allows an accurate cryoscopic determination of the number of species released into solution to be made. When this information is coupled to the conductimetric estimation of the HSO_4^- concentration, it is normally possible to deduce the stoichiometry and products formed in a reaction. For example, when N_2O_4 dissolves in sulphuric acid, six species are released into solution, three of which conductimetric measurements show are hydrogen sulphate ions; this is consistent with the reaction

$$N_2O_4 + 3H_2SO_4 \rightarrow NO^+ + NO_2^+ + H_3O^+ + 3HSO_4^-$$

Further confirmatory evidence can be obtained from the Raman spectrum of the solution which shows characteristic peaks at 2320 cm^{-1} (NO^+), 1400 cm^{-1} (NO_2^+) and 1050 cm^{-1} (HSO_4^-).

Sulphuric acid has a high electrical conductivity (1.04×10^{-2}) $\Omega^{-1} \text{ cm}^{-1}$ at $25°C$) corresponding to extensive self-ionization:

$$2H_2SO_4 \rightleftharpoons H_3SO_4^+ + HSO_4^-; \quad \text{ionic product,} \quad [H_3SO_4^+][HSO_4^-] = 1.7 \times 10^{-4}(10°)$$

However, the system is more complex than this because there is a secondary self-dehydration reaction occurring simultaneously:

$$2H_2SO_4 \rightleftharpoons H_3O^+ + HS_2O_7^-$$

Oxonium and hydrogen disulphate ($HS_2O_7^-$) ions have only low mobilities in sulphuric acid so that the conductivity of the solvent arises almost totally from the ions produced in the self-ionization process. Very few solutes produce *sulphuric acidium* ions ($H_3SO_4^+$) in this highly acidic solvent; most of the 'aqueous acids' are protonated and thus act as bases owing to the concomitant formation of hydrogen sulphate ions:

$$RCOOH + H_2SO_4 \rightarrow RC(OH)_2^+ + HSO_4^-$$

$$H_3PO_4 + H_2SO_4 \rightarrow P(OH)_4^+ + HSO_4^-$$

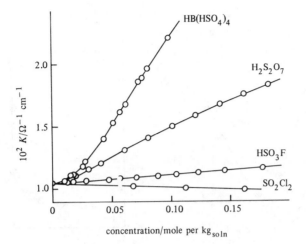

Fig. 116—Conductivities of the non-electrolyte SO_2Cl_2 and three acids in sulphuric acid solution. The virtually constant conductivity of SO_2Cl_2 (sulphuryl chloride) is typical of a non-electrolyte. The curves for the three acids reflect their acid strengths because the conductivity of each solution is determined by the concentration of $H_3SO_4^+$ ions they each produce. [Reproduced with permission from *Advances in Inorganic Chemistry and Radiochemistry* vol. 1 (1959) p. 403.]

$$HNO_3 + 2H_2SO_4 \rightarrow NO_2^+ + H_3O^+ + 2HSO_4^-$$

$$B(OH)_3 + 6H_2SO_4 \rightarrow 3H_3O^+ + B(HSO_4)_4^- + 2HSO_4^-$$

Hydrogen chloride is undissociated and perchloric acid, usually considered to be the strongest acid in aqueous media, is only slightly ionized. Disulphuric acid, present in dilute oleum, is about 30% ionized in 0.1M solution:

$$H_2S_2O_7 + H_2SO_4 \rightarrow H_3SO_4^+ + HS_2O_7^-$$

One of the strongest known acids in the sulphuric acid system, $HB(HSO_4)_4$, can be formed by dissolving boric oxide in oleum (see its high conductivity in Fig. 116):

$$B_2O_3 + 3H_2S_2O_7 + 4H_2SO_4 \rightarrow 2H_3SO_4^+ + 2B(HSO_4)_4^-$$

Alkali metal sulphates and hydrogen sulphates behave as typical bases,

$$Na_2SO_4 + H_2SO_4 \rightarrow 2Na^+ + 2HSO_4^- \quad \text{(protonation)}$$

$$KHSO_4 \rightarrow K^+ + HSO_4^- \quad \text{(simple ionization)}$$

acting in an analogous fashion to their oxides and hydroxides in water. As in other solvents, acid–base neutralization reactions in sulphuric acid can be followed conductimetrically, the conductivity falling virtually to zero at the endpoint owing to the low ionic mobilities of the solute ions remaining in solution; as the endpoint is passed, the conductivity rises sharply again owing to the added HSO_4^- (or $H_3SO_4^+$) ions; see Fig. 117:

$$H_3SO_4^+[B(HSO_4)_4]^- + K^+HSO_4^- \rightarrow K^+[B(HSO_4)_4]^- + 2H_2SO_4$$

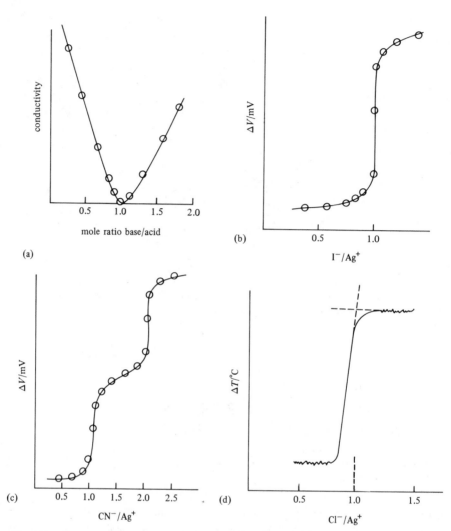

Fig. 117—Titrations in non-aqueous solvents. (a) Sketch of a conductimetric titration curve for an acid–base neutralization carried out in sulphuric acid. (b,c) Potentiometric and (d) thermometric titrations in molten $LiNO_3$–KNO_3.

Since the neutralization corresponds to the reaction

$$H_3SO_4^+ + HSO_4^- \rightarrow 2H_2SO_4$$

there is virtually no change in the melting point of the solvent as the titration proceeds because the same number of species is present in solution the whole time; after the endpoint, excess base (or acid) is being added and the melting point correspondingly falls in a linear manner. Of course, similar behaviour should occur generally but little or no accurate cryoscopic work has been reported on neutralizations carried out in other solvent systems.

While many of the common hydrogen sulphates follow a similar trend in solubilities to their hydroxides in water (e.g. alkali metals > alkaline earths $\gg$ Mg), virtually all other salts suffer extensive solvolysis in sulphuric acid. This occurs primarily because the parent 'acids' of the salts are very weak in this solvent and thus their anions are readily protonated; the similar hydrolysis of weak acid salts is a well-known phenomenon in water. Therefore sulphuric acid is not normally a useful solvent in which to study the chemistry of anions.

Hydrogen sulphates are often isolated as solvates from their sulphuric acid solutions, the composition of the crystals ranging quite widely from metal to metal: $LiHSO_4.0.5H_2SO_4$; $KHSO_4.H_2SO_4$; $Ba(HSO_4)_2.2H_2SO_4$ and $LiHSO_4.4H_2SO_4$. Gillespie has deduced from a variety of techniques that the extent of cation solvation is in the order $Sr^{2+} > Ba^{2+} > Li^+ > Na^+ > K^+ > NH_4^+$; furthermore, by assuming a solvation number of unity for the ammonium ion, he was able to calculate the approximate solvation numbers as Sr, Ba = 8, Li, Na = 3 and K = 2.

Although many anions are solvolytically unstable in sulphuric acid, it makes a valuable medium in which to study a range of quite unique cations. Probably one of the first cations to be used widely in sulphuric acid was the nitronium ion, NO_2^+. Over a period of many years organic chemists had used a mixture of nitric and sulphuric acids for nitrations before it was appreciated why the reactions occurred more readily in the presence of H_2SO_4; the active nitrating species is the NO_2^+ ion:

$$HNO_3 + 2H_2SO_4 \rightarrow NO_2^+ + H_3O^+ + 2HSO_4^-$$

Nitronium ions are present in pure nitric acid owing to the occurrence of self-ionization, but their concentration is greatly increased by the solvent action of sulphuric acid.

It has been pointed out by O'Donnell that as the acidity of a solvent increases, the possibility of forming cations which are unstable in water becomes more favourable. This can be illustrated by the various cationic iodine species which have been prepared in recent years; I_3^+ and I_5^+ are produced by oxidizing iodine in H_2SO_4, but blue I_2^+ ions, which are stable in the more acid solvent mixture AsF_5–HSO_3F, disproportionate in sulphuric acid:

$$8I_2^+ + 3HSO_4^- \rightarrow 5I_3^+ + I(HSO_4)_3$$

The AsF_5 is added to the fluorosulphuric acid (which can be regarded as sulphuric acid with one OH group replaced by F) to react with, and thus reduce the concentration of, FSO_3^-, the base analogue in HSO_3F; even when prepared in fluorosulphuric acid, I_2^+ will slowly disproportionate in the presence of added FSO_3^- ions. Similarly, brown Br_3^+ is unstable in sulphuric acid but can be produced in the 'super-acid' system SbF_5–SO_3–HSO_3F by oxidizing bromine with $S_2O_6F_2$; further amounts of oxidant give red Br_2^+.

Selenium reacts with sulphuric acid at about $50\,^\circ$C to form a solution of Se_8^{2+} ions, which are oxidized further to Se_4^{2+} on addition of SeO_2. Tellurium gives red Te_4^{2+}

directly in H_2SO_4 whilst in oleum (a mixture of polysulphuric acids formed by adding SO_3 to H_2SO_4) the more oxidized species Te_6^{4+} results. No sulphur ions appear to be stable in pure sulphuric acid but S_4^{2+}, S_8^{2+} and S_{19}^{2+} can be prepared in either oleum or HSO_3F.

Sulphuryl chloride (Fig. 116) is one of the very few non-electrolytes to be soluble in sulphuric acid and was thus used to determine the freezing point depression constant for the solvent. Presumably it is able to dissolve by entering into the hydrogen-bonded network of the solvent.

A surprisingly wide range of organic compounds dissolve in pure sulphuric acid and can be recovered unchanged by careful addition of water. Those which contain atoms with lone pairs of electrons (O, N, S) tend to be protonated and hence behave as bases:

$$R_2C{=}O + H_2SO_4 \rightarrow R_2C^+{-}OH + HSO_4^-$$

$$RNH_2 + H_2SO_4 \rightarrow RNH_3^+ + HSO_4^-$$

$$R_2NH + H_2SO_4 \rightarrow R_2NH_2^+ + HSO_4^-$$

$$R_3N + H_2SO_4 \rightarrow R_3NH^+ + HSO_4^-$$

$$RC{\overset{O}{\underset{OR'}{\big<}}} + H_2SO_4 \rightarrow RC^+{\overset{OH}{\underset{OR'}{\big<}}} + HSO_4^-$$

$$RCONH_2 + H_2SO_4 \rightarrow RC{\overset{O}{\underset{NH_3^+}{\big<}}} + HSO_4^-$$

Thus sulphuric acid, which might have been considered as being far too corrosive a liquid to be a useful solvent, has an extensive and varied chemistry. The fact that it is a strong proton donor means that most anions are solvolysed (i.e. protonated), but this very acidity allows the formation and study of some unusual cations. In extremely acidic conditions provided by the super-acids (e.g. $SbF_5{-}HSO_3F{-}SO_3$ mixtures) it is possible to stabilize a wide variety of carbonium ions (R_3C^+), and even 'hyper-valent' carbon species such as

$$R_3C^+{-}{-}{-}{\overset{H}{\underset{H}{\big<}}}$$

are considered possible products in some reactions.

CHEMISTRY IN MOLTEN SALTS

The use of molten salts as solvents stretches far back into history because the ancients developed such species as fluxes and slag-formers in metallurgy. However, it is only comparatively recently that we have had the techniques available to make a serious scientific study of them. A few salts are liquid below $100\,^{\circ}C$ but the majority melt only at high temperatures and under these extreme conditions corrosion is often a

Table 53—Melting points (°C) of some salts and salt mixtures

Pure salts		Eutectic mixtures	
LiCl	605	$LiNO_3$–KNO_3	125
NaCl	801	$NaNO_3$–KNO_3	218
KCl	770	LiCl–KCl	350
$LiNO_3$	264	$LiNO_3$–$NaNO_3$–KNO_3	120
KNO_3	334		
$AlCl_3$	190 (2.5 atm)		

major problem; lower-melting eutectic mixtures can be used to reduce the operating temperature but this necessarily introduces 'foreign' ions into the system and may not always be acceptable (Table 53).

Borosilicate glasses (e.g. Pyrex) have been used to make container vessels for work up about 400°C, but there is always a danger of contamination by ion-leaching from the glass and, furthermore, extensive chemical attack occurs with many fluorides, oxides, hydroxides and carbonates. The transparency, relative inertness and high melting point of silica make it useful for the construction of optical cells working up to 600°C or even, in some cases, to 1000°C. In Raman studies a sealed silica cell can be suspended in a tube furnace provided with three side-arms in the form of a letter T; this allows the exciting laser light to pass through the molten sample whilst the scattered Raman radiation is observed at right angles as shown schematically in plan below:

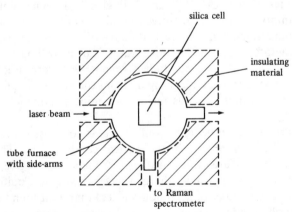

The Raman side-arm can be dispensed with if research demands only a spectral study of the ultraviolet–visible region. By using very efficient insulating materials it is possible to hold the temperature of the sample constant to ± 1 °C or better even at 1000°C. The lack of space between the magnet pole faces of NMR spectrometers rather limits the degree of insulation which can be achieved around a tube furnace, so that small thermal gradients (5–10 °C along the sample lengths) inevitably occur; however, successful NMR studies have been carried out up to 1000°C using borosilicate glass, silica or alumina sample tubes, depending on the temperature range and salt identity.

Synthetic sapphire (i.e. crystalline alumina) and diamond are among the more exotic materials which have been used to construct the windows of optical cells. Unfortunately, diamond tends to darken slowly when used above $600\,^{\circ}C$ owing to partial transformation into graphite, but it has the double advantage of being resistant to attack by the extremely aggressive molten fluorides and transparent to infrared radiation. This spectral region is a difficult one for study because of the lack of suitable window materials. One simple way of avoiding the problem is to use surface tension effects to suspend the molten salt on a fine platinum gauze heated in a horizontal tube furnace; by passing the infrared beam along the open furnace tube and through the sample it is possible to record the spectrum without the need of any cell windows.

For non-optical studies, noble metals can be used if the more easily fabricated silica is not convenient. For example, in studies on the solubility of metals in their molten salts each of the chosen mixtures was brought to equilibrium in a tantalum crucible placed inside a furnace-heated silica tube purged with an inert gas. The apparatus was so arranged that the tube, and crucible, could be dropped into a bath of hot water to quench the solution *rapidly*; it was then a simple matter to determine the solubility of the metal, at the furnace temperature, by analysing the composition of the resulting solid matrix.

Although the operations are more difficult to perform with molten salts because of the high temperatures involved, it is possible to use many familiar techniques employed in the study of aqueous media: decantation is possible when two phases are formed, silica or metal frits allow the filtration of non-viscous melts, cryoscopy provides information about the number of species released into solution and even chromatography can be achieved on heated alumina columns. In an example of the latter, the chlorides of Fe^{III}, Co^{II}, Ni^{II} and Cu^{II} give well-defined coloured bands on elution with ammonium chloride dissolved in an $LiNO_3$–KNO_3 eutectic.

Perhaps the most important physical property of molten salts is the low vapour pressure they generate even at high temperatures, thus providing a wide liquid range within which each solvent may be studied in relative safety. Furthermore, many different solvent types are available, ranging from the simple alkali halides through oxo-anion salts to the more viscous melts containing polymeric borate or silicate anions.

Molten salts are unique in being the only solvents composed entirely of ions and, correspondingly, they exhibit very high electrical conductivities. The liquid state is not simply a random mixture of ions: rather there is a high degree of order in that each ion is surrounded by oppositely charged species owing to Coulombic attraction. However, the average number of neighbours is less than that found in the solids, as, for example, in molten sodium chloride where the coordination number is about four compared to six in the crystalline state.

Over the years there has been considerable interest shown in the dissolution of metals in molten halides. In many cases it appears that the process is one of essentially passive solubility such as occurs when nickel dissolves to the extent of 9% in fused $NiCl_2$; however, if a lower oxidation state is available or a halide of a different metal is used then reduction may occur:

$$M + MCl_3 \rightarrow M^{2+} \quad (M = Nd, Sm \text{ or } Eu)$$

$$Bi + BiCl_3 \rightarrow Bi_9^{5+} \quad (\text{structure shown below})$$

$$Na + LiCl \rightleftharpoons NaCl + Li, \quad K_{equil} = 0.45 \text{ at } 900\,^\circ C$$

$$Cs_2UCl_6 + 2Mg \xrightarrow[550\,^\circ C]{CsCl-MgCl_2 \text{ melt}} U\downarrow + 2MgCl_2 + 2CsCl$$

Although divalent species of titanium, vanadium and chromium are unstable in the oxygenated aqueous media normally encountered in the laboratory, they can be prepared and studied in a fused LiCl–KCl eutectic, e.g.

$$M + LiCl(41\%)-KCl(59\%) \xrightarrow[450\,^\circ C]{anodic\ oxidation} M^{II} \quad (M = V, Cr \text{ or } Fe)$$

where they are probably present as octahedrally coordinated MCl_6^{4-} ions. The concentration of the V^{II} and Cr^{II} solutions can be determined by potentiometric titrations using Fe^{III} as oxidant: e.g.

$$Cr^{II} + Fe^{III} \rightarrow Fe^{II} + Cr^{III}$$

Fused aluminium trichloride has proved to be a useful medium in which to prepare a variety of unusual cations. An early example was the linear Hg_3^{2+} ion formed by dissolving mercury in an $HgCl_2/AlCl_3$ melt; evaporation of excess mercury and solvent left yellow crystals of the tetrachloroaluminate $Hg_3(AlCl_4)_2$. (The same ion is detectable during polarographic studies of $HgCl_2$ dissolved in $NaCl–AlCl_3$ mixtures where three waves represent the reduction sequence $Hg^{2+} \rightarrow Hg_2^{2+} \rightarrow Hg_3^{2+} \rightarrow Hg$.) Reduction of aluminium chloride solutions of $CdCl_2$ or $BiCl_3$ with the respective metals similarly gives polyatomic cations:

$$CdCl_2 + Cd \xrightarrow{AlCl_3} Cd_2(AlCl_4)_2$$

$$BiCl_3 + Bi \xrightarrow{AlCl_3} Bi_5(AlCl_4)_3 + Bi_8(AlCl_4)_2$$

Bi_5^{3+}
trigonal bipyramid

Bi_8^{2+}
square antiprism

Bi_9^{5+}
tri-capped trigonal prism

Cations of non-metals are also attainable in molten salts, an example being I_2^+ which is formed when solutions of iodine in $AlCl_3$-rich $NaCl/AlCl_3$ melts are oxidized by chlorine.

Silver chemistry of a more classical kind has been carried out in the low-melting $LiNO_3/KNO_3$ eutectic. Silver nitrate is very soluble in this solvent system and the addition to it of potassium halides, chromate or cyanide results in the immediate and quantitative precipitation of the corresponding silver salts:

$$AgNO_3 + KX \rightarrow AgX\downarrow \quad (X = Cl, Br \text{ or } I)$$

$$AgNO_3 + K_2CrO_4 \rightarrow Ag_2CrO_4\downarrow$$

$$AgNO_3 + KCN \rightarrow AgCN\downarrow$$

When these reactions are carried out as titrations, their endpoints can be determined either thermometrically (i.e. by monitoring the rise in temperature of the solution) or potentiometrically; Fig. 117. The two endpoints indicated in the cyanide titration correspond to the consecutive reactions

$$Ag^+ + CN^- \rightarrow AgCN\downarrow \quad \text{(light brown precipitate)}$$

$$AgCN + CN^- \rightarrow Ag(CN)_2^- \quad \text{(soluble; colourless solution)}$$

The infrequent occurrence of solvolytic reactions in molten salts makes them particularly valuable for the large-scale electrochemical production of highly electropositive elements like sodium or aluminium (however, some metals such as barium are so soluble in their fused halides that they cannot easily be isolated by this method). Similarly, fluorine is made commercially by electrolysis of KHF_2, a medium which can be considered either as a concentrated solution of potassium fluoride in liquid HF or as a molten salt consisting of K^+ and HF_2^- ions. When alkali or alkaline earth fluorides are electrolysed at 600°C or above using carbon electrodes, the generated fluorine attacks the anode forming a mixture of fluorocarbons, mainly CF_4 and C_2F_6; by adding chlorides to the melt it is possible to isolate the industrially important chlorofluorocarbons. A less vigorous fluorinating system can be made by dissolving potassium fluoride in molten $KCl/ZnCl_2$ mixtures:

$$SOCl_2 \xrightarrow{\text{KF}} SOF_2$$

$$(CH_3)_3SiCl \xrightarrow{\text{KF}} (CH_3)_3SiF$$

$$CCl_4 \xrightarrow{\text{KF}} CCl_{4-n}F_n$$

Gases are not very soluble in fused salts but they often undergo rapid, high-yield reactions because of the high temperatures which are employed: e.g.

$$SiCl_4(g) + LiH \text{ (dissolved in } LiCl/KCl) \xrightarrow{400°C} SiH_4(100\%)$$

As the lithium hydride is used up it can be recharged by first electrolysing the melt to give lithium metal and then passing hydrogen through the resulting solution:

$$2Li + H_2 \xrightarrow{LiCl/KCl} 2LiH$$

On a large scale (*c.* 30000 tons per year), methane is blown through a melt of $CuCl/CuCl_2/KCl$ held at 370–450°C to form chlorinated methanes; the melt is periodically regenerated by oxidizing the CuCl by-product with chlorine or hydrogen chloride. If ethane is used as the feedstock a variety of C_2 derivatives are produced, including chloroethene (vinyl chloride monomer).

Molten sulphates are used in sulphuric acid manufacture to dissolve the vanadium oxides required to catalyse the oxidation of sulphur dioxide to SO_3. To increase the surface area of the catalyst and so make the process more efficient, the vanadium oxide solution is dispersed on kieselguhr (a form of silica), about 80 tons being packed into the huge oxidation towers.

The high temperatures achieved with fused salts promote elimination reactions involving organic compounds, particularly alcohols. For example, when ethanol is passed through $NaCl/ZnCl_2$ melts up to 98% conversion to ethene is possible; other alcohols similarly give propene and butene.

The coverage of this solvent type is obviously less cohesive than that given for other solvents because so many different fused salts or salt mixtures are available; a wide variety of examples has been given, including some commercial systems, to illustrate the scope of possible reactions. It is not possible to apply our definition of acids and bases in fused salts; for mixtures such as $KCl/AlCl_3$, some workers use the Lewis concept of acid and bases; acids are electron pair acceptors and bases are electron pair donors. In this way it is possible to talk of acidic $KCl/AlCl$ melts as those containing more than 50 mol% of $AlCl_3$ whereas basic melts have more than 50 mol% of KCl; when the stoichiometry is exactly $KAlCl_4$ the melt can be considered as neutral.

CHEMISTRY IN LIQUID ALKALI METALS

The impetus to a study of these rather exotic solvents was the decision, taken in the 1950s, to cool fast nuclear reactors with liquid sodium. In the sense that lithium, sodium and potassium are cheap and have a wide liquid range (Table 54), they make ideal additions to our list of available solvents. However, a big drawback to their usefulness is their opacity to electromagnetic radiation which makes a spectroscopic study of solutes impossible. The identity of material held in solution has to be deduced from less direct techniques such as cryoscopy, changes in solvent resistivity with solute concentration and the composition of precipitated products.

Structurally, the liquid alkali metals behave as though the outer s electrons are present in a vast number of molecular orbitals delocalized over the whole liquid, like the conduction (i.e. energy) bands of the solid metals; in the very simplest terms the cations can be thought of as being buried in a sea of electrons. Chemical reactions

Table 54—Physical properties of molten alkali metals

	Li	Na	K	Rb	Cs
Melting point (°C)	180.5	97.8	63.2	39.0	28.5
Boiling point (°C)	1317	883	754	688	671
Viscosity (cP) (200°C)	0.57	0.45	0.30	0.35	0.35
Density (g ml^{-1}) (200°C)	0.51	0.90	0.80	1.4	1.7

involve addition or removal of electrons from this conduction band:

$$Ba \xrightarrow{1.Na} Ba^{2+} + 2e^- \quad \text{(addition of electrons to conduction band)}$$

$$H_2 + 2e^- \rightarrow 2H^- \quad \text{(removal of electrons from conduction band)}$$

As in other liquids, solvation appears to be very important in molten metals as can be judged from the substantial value of the enthalpy of solvation of chloride ions in molten sodium (-305 kJ mol^{-1}). Although the nature of the solvation process is unknown, it is probable that each Cl$^-$ is surrounded by several sodium ions held in close proximity by Coulombic attraction. Not unexpectedly, as the charge on the anion increases so does the enthalpy of solvation: Cl$^-$ = -343, O^{2-} = -1960, N^{3-} = -3473 kJ mol^{-1} in liquid lithium. Sonic absorption measurements on solutions of sodium in caesium show that a weak 'complex' of composition Na$_3$Cs is formed (Fig. 118); unfortunately the nature of the interaction is unknown.

At relatively low temperatures Pyrex glass can be used for handling all the alkali metals except lithium, but for high-temperature work manipulation must be carried out in stainless steel apparatus because the metals rapidly corrode glasses by leaching out the oxygen atoms. Precipitates may be removed from the solvents by filtration through glass-wool plugs, glass frits or sintered steel discs, depending on particle size and temperature; evaporation of excess solvent by vacuum distillation allows the isolation of any soluble materials as crystals. Because of the very high reactivity of

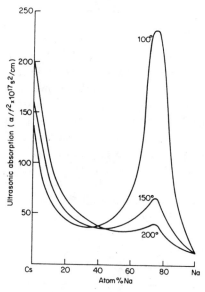

Fig. 118—Influence of temperature and composition on sound absorption in liquid sodium–caesium mixtures. Although metals are opaque to large parts of the electromagnetic spectrum it is sometimes possible to detect compound formation using sonic absorption techniques instead. Here the sonic absorption peak arising at a ratio of 3Na:1Cs shows a 'compound' Na$_3$Cs is formed in the liquid. The sensitivity of the peak towards temperature increases suggests only a weak association occurs between the sodium and caesium. [Reproduced with permission from *The Chemistry of the Liquid Alkali Metals* by C. C. Addison (Wiley, 1984).]

the alkali metals, all manipulations must be carried out under an atmosphere of a scrupulously dry and pure inert gas, preferably argon.

Cryoscopic measurements have revealed that oxygen dissolves in the alkali metals as a monatomic species which all available evidence suggests is the oxide ion O^{2-}. When oxygen is passed into liquid sodium containing dissolved calcium a precipitate of CaO is formed; the same product results if sodium oxide is added to the calcium solution, thus illustrating the feasibility of typical metathetic reactions in these solvents:

$$O_2 \xrightarrow{1.\,Na} 2O^{2-} \xrightarrow{Ca} 2CaO\downarrow$$

Lithium is the only alkali metal which reacts with nitrogen to give nitride ions, N^{3-}, the probable mechanism being the initial adsorption of nitrogen atoms onto the surface of the metal followed by their reduction via electrons taken from the conduction band:

$$N_2(gas) \xrightarrow{dissociation} 2N(adsorbed) \xrightarrow{6e^-} 2N^{3-}$$

The precipitated products formed from the reaction of graphite with liquid alkali metals give ethyne, C_2H_2, on hydrolysis, suggesting that C_2^{2-} ions are present in the mother liquor. By working in molten lithium it is possible to react these C_2^{2-} ions with N^{3-} and so prepare lithium cyanamide:

$$graphite + nitrogen\ gas \xrightarrow{1.\,Li} Li_2(N=C=N)$$

This is a rather unusual reaction because many polyatomic anions are cleaved, rather than synthesized, in liquid lithium:

$$OH^- \rightarrow O^{2-} + H^-$$
$$2CO_3^{2-} \rightarrow 6O^{2-} + C_2^{2-}$$
$$NO_3^- \rightarrow 3O^{2-} + N^{3-}$$
$$2CN^- \rightarrow C_2^{2-} + 2N^{3-}$$

Such reactions are obviously very solvent-dependent because, for example, sodium cyanide can be recovered unchanged from liquid sodium and OH^- ions are completely stable in K, Rb or Cs.

We are all accustomed to the extremely vigorous reaction which occurs when small pieces of sodium are thrown into water. However, when approached from the sodium-rich side and small amounts of water are added to liquid sodium there is little visible sign of reaction:

$$H_2O \rightarrow OH^- + H^-$$
$$OH^- \rightleftharpoons H^- + O^{2-} \quad \text{(note this equilibrium in liquid Na)}$$

This unexpectedly quiet reaction leaves us with a very dramatic demonstration of the fact that we cannot always extrapolate our observations accurately from one solvent to another; each solvent system has its own rich chemistry.

FURTHER READING

Audrieth, L. F. & Kleinberg, J. (1953) *Non-aqueous Solvents*. Wiley

Gillespie, R. J. & Robinson, E. A. (1959) The sulphuric acid solvent system. *Adv. Inorg. Chem. Radiochem.* **1** 385

Blander, M. (ed.) (1964) *Molten Salt Chemistry*. Wiley

Sundheim, R. S. (ed.) (1964) *Fused Salts*. McGraw-Hill

Holliday, A. K. & Massey, A. G. (1965) *Inorganic Chemistry in Non-aqueous Solvents*. Pergamon

Waddington, T. C. (1969) *Non-aqueous Solvents*. Nelson

Lagowski, J. J. (ed.) (1978) *The Chemistry of Non-aqueous Solvents*. Academic Press

Burgess, J. (1978) *Metal Ions in Solution*. Ellis Horwood

Hatt, B. W. & Kerridge, D. H. (1979) Industrial applications of molten salts. *Chem. Brit.* p. 78

Addison, C. C. (1980) Dinitrogen tetroxide, nitric acid and their mixtures as media for inorganic reactions. *Chem. Rev.* **80** 21

Lovering, D. G. & Gale, R. J. (eds) (1983–) *Molten Salt Techniques*. vol. 1 *et seq.* Plenum

Addison, C. C. (1984) *The Chemistry of the Liquid Alkali Metals*. Wiley

Edwards, P. P. (1985) From solvated electrons to metal anions. *J. Solution Chem.* **14** 187

Enderby, J. E. *et al.* (1987) Diffraction and the study of aqua ions. *J. Phys. Chem.* **91** 5851

Chapter 13

Bioinorganic Chemistry

INTRODUCTION

We are accustomed to accepting that life is chemically based on carbon, but in fact hydrogen is by far the most abundant biological element and constitutes 63% of the atoms in a human body. Carbon is only third on the list at 9.5% coming after oxygen (25.5%) and before nitrogen (1.4%). More than twenty other essential elements make up the remaining 0.6% of which Ca, P, Na, K, S, Cl and Mg are, in order, the seven most plentiful in humans. Although only minute amounts of them are required, life also depends on the presence of all the first row transition metals except Sc and Ti; the most common of these trace elements, iron, is located mainly in the oxygen-carrying molecules myoglobin and haemoglobin. Transition metals are usually present in the prosthetic groups (i.e. the active sites) of biomolecules where they are complexed to oxygen, nitrogen and sulphur ligands.

Many elements are required in such tiny amounts (see Fig. 120) that it is sometimes difficult to be completely sure that they are not present in an animal or plant simply as indigenous, but benign, 'impurities'. One way in which this question can sometimes be answered for animals is to keep them for several generations in plastic cages, provide them with food lacking a specific element and note any abnormalities. Such experiments are extraordinarily difficult to perform because of the exacting purification and analytical procedures required in food preparation—typical 'analytical quality' chemicals would be considered highly impure! Even the air has to be rigorously filtered because of trace impurities which could be carried into the cages by dust particles.

Obviously similar studies cannot be carried out on humans or large animals but, by observing an unusually high incidence of a specific disease in certain areas, it is sometimes possible to link the disease to the lack of an essential element in the diet. For example, in some provinces of China it was noted that heart problems were

IA	IIA	IIIA	IVA	VA	VIA	VIAA	VIII	VIII	VIII	IB	IIB	IIIB	IVB	VB	VIB	VIIB	0
(H)																	He
Li	Be											[B]	(C)	(N)	(O)	[F]	Ne
(Na)	(Mg)											Al	[Si]	(P)	(S)	[Cl]	Ar
(K)	(Ca)	Sc	Ti	[V]	[Cr]	[Mn]	[Fe]	[Co]	[Ni]	[Cu]	[Zn]	Ga	Ge	[As]	[Se]	[Br]	Kr
Rb	Sr	Y	Zr	Nb	[Mo]	Tc	Ru	Rh	Pd	Ag	Cd	In	[Sn]	Sb	Te	[I]	Xe
Cs	Ba	La	Hf	Ta	W	Re	Os	Ir	Pt	Au	Hg	Tl	Pb	Bi	Po	At	Rn
Fr	Ra	Ac	Th	Pa	U												

◯ Bulk biological elements.   ☐ Trace elements believed to be essential for plants or animals.
⌐⌐⌐ Possibly essential trace elements.

Fig. 119—Elements essential to the life of animals and plants.

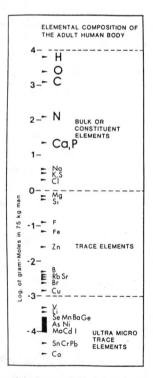

Fig. 120—Elemental composition of the human body on a logarithmic scale. [Reproduced with permission from Chapter 2 of *Biochemistry of the Essential Ultratrace Elements* by E. Freiden (ed.) (Plenum Press).]

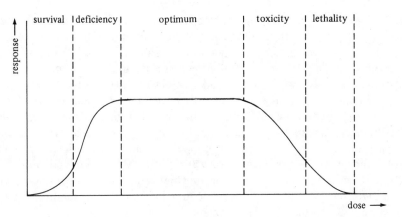

Fig. 121—Idealized curve showing the response of an organism to increasing doses of an essential element.

particularly common within the human population and this was finally shown to be due to the very low levels of selenium present in the local soil from which virtually all the food was derived. Conversely, 'locoweed' growing on soil containing relatively large amounts of selenium causes abnormal neurological behaviour in animals which eat it owing to the high selenium content. From this it is apparent that even essential elements can prove toxic if taken in excess.

Fig. 121 shows schematically the health response of an organism to an essential element. Death or, at best, bare survival results if the element is withdrawn from the diet; as the organism is exposed to increasing amounts of the element, response improves until a plateau of full health is achieved. In excess the element produces undesirable side-effects and gradually results in a deterioration of health until death finally occurs. The quantitative characteristics of the curve vary with the particular element under study and the plateau can be remarkably narrow: a range of only 50–200 μgm per day separates deficiency and toxicity for selenium in man (Table 55).

Table 55—Some recommended dietary intakes of elements for adults

Element	Intake (mg per day)[a]
Iron (males)	10
Iron (females)	18
Zinc	15
Manganese	2.5–5.0
Fluorine	1.5–4.0
Copper	2.0–3.0
Molybdenum	0.15–0.5
Chromium	0.05–0.2
Selenium	0.05–0.2
Iodine	0.15

[a] Figures taken from E. Frieden, *Journal of Chemical Education* vol. 62 (1985) p. 917.

The study of essentiality is further complicated by the phenomenon of *antagonism* in which one element interferes with the uptake of another. Herbivores foraging on pasture with a very low copper content suffer from anaemia and loss of pigmentation, but in extreme circumstances copper deficiency results in death from gross lesions of main arteries, possibly owing to depleted reserves of the copper-containing enzyme monoamine oxidase. (Enzymes are naturally occurring molecular catalysts, often of great complexity, and this particular one is thought to be involved in the cross-linking of polypeptide chains required for structural strength of normal arteries.) Even when the pasture has a plentiful supply of copper, animals may still suffer deficiency symptoms owing to the presence of very high soil levels of molybdenum, a metal which inhibits the uptake of copper by forming highly insoluble copper thiomolybdates in the gut. However, low levels of molybdenum, like copper, are required for healthy growth since it is present in several enzymes.

THE IMPORTANCE OF NON-COVALENT INTERACTIONS IN BIOLOGY

A number of weak intermolecular or, sometimes, intramolecular forces play an important supporting role to the strong covalent bonds responsible for the primary structure of biomolecules. Although each interaction amounts to only a few kilojoules per mole, when they occur in large numbers their total contribution becomes significant: as an example, the haem group of myoglobin is held in place by a coordinate bond (between the iron atom and an imidazole nitrogen) and about 60 van der Waals contacts.

Some molecules possess ionic groups such as COO^- or NR_3^+ and these can add substantial energy (up to 40 kJ mol^{-1}) to molecular attractions when they are in close proximity to appropriately charged species. Eight such 'salt links' are present in haemoglobin and are partially responsible for the proper functioning of this molecule.

Perhaps the most important of the non-covalent interactions is that of hydrogen bonding (page 98) because without it water would boil at about $-100°C$ and hence life, having an absolute dependence on liquid water, could not have evolved on our warm planet. Hydrogen bonding is not a particularly strong interaction (10–40 kJ mol^{-1}) but most importantly the 'bonds' are often directional and, being relatively weak, they can be readily broken and remade; these properties are utilized to maximum advantage in the DNA helix which, though held together by hydrogen bonds, must be capable of unwinding during replication (page 442).

Many biomolecules with a large number of purely organic substituents have to function in aqueous media where their non-polar groups, having no mechanism for breaking open the hydrogen-bonded network of water molecules, are incompatible with the surroundings. To minimize the repulsion energy, these hydrophobic groups are 'pressed' together by the polymeric water structure and adhere to each other by van der Waals-like forces which contribute about 2–4 kJ mol^{-1} for each pair of CH_2 groups in contact; the molecule is thus forced into a conformation where the non-polar groups are on the inside of the structure and any hydrophilic polar parts are presented to the aqueous medium on the outside. Thus in a subtle way hydrogen bonding in water is responsible for the intricate folding of the protein chains in myoglobin and haemoglobin where it results in a protective 'pocket' being formed for the haem groups (page 470). As described on page 446, the hydrophobic pushing together of non-polar groups also results in the spontaneous alignment of lipids into the double-layer structure found in cell membranes.

HYDROGEN

Nature presumably picked hydrogen for a major involvement in life-processes because it can satisfy all the residual substitution positions of C, N and O without conferring any of the structural complexities which would be demanded by polyvalent elements. However, when coupled to nitrogen and oxygen, this element also has the added ability of forming strong hydrogen bonds (page 98). These unique bonds are responsible for the life-supporting properties of water and are also intimately involved in many other functions from the operation of several enzymes to the stabilization of the double helix in deoxyribonucleic acid (DNA) molecules. An idea of the biological importance of water can be gained from the fact that a 70 kg man contains about 40 l, of which 28 l are on the inside of the body cells and 12 l are outside (of the latter, some 2–3 l are to be found in the blood plasma).

CARBON†, NITROGEN AND OXYGEN

Carbon is unique among the elements in having the dual ability to catenate, producing chains and rings, and to form strong p_π–p_π bonds with itself, nitrogen and oxygen. All three elements are involved in the formation of the 20 amino-acids required for protein synthesis (Tables 56 and 57), the purine and pyrimidine bases used to produce the nucleic acids, and multidentate ligands such as the porphyrin and corrin rings of haem and chlorophyll.

The naturally occurring amino acids all have the general structure

$$R - \overset{H}{\underset{\alpha}{C}} \begin{smallmatrix} \text{COOH} \\ \text{NH}_2 \end{smallmatrix}$$

and, with the exception of glycine where R = H, they are optically active owing to the asymmetry of their α carbon atoms; normally only the L forms are found in biological systems. Proteins are derived from the L form of these amino acids via condensation reactions in which a molecule of water is lost when an amine group on one acid reacts with the carboxylic acid of its neighbour:

$$\text{H}_2\text{N}-\underset{\text{H}}{\overset{\text{R}}{\text{C}}}-\overset{\text{O}}{\overset{\|}{\text{C}}}-\text{OH} \quad + \quad \text{H}-\underset{\text{H}}{\overset{\text{R}'}{\underset{\text{H}}{\text{N}}}}-\text{C}-\text{COOH} \rightleftharpoons \text{H}_2\text{N}-\underset{\text{H}}{\overset{\text{R}}{\text{C}}}-\overset{\text{O}}{\overset{\|}{\text{C}}}-\underset{\text{H}}{\text{N}}-\overset{\text{R}'}{\text{C}}-\text{COOH} \quad + \quad \text{H}_2\text{O}$$

---

† An estimated 73% of the carbon in the Earth's crust is locked up as calcium carbonate and more than 50 times the amount of carbon in the biosphere is fixed as coal.

Table 56—The amino acids

$$H-\underset{\underset{R}{|}}{\overset{\overset{NH_3^+}{|}}{C}}-COOH \; \underset{}{\overset{+H^+}{\rightleftharpoons}} \; H-\underset{\underset{R}{|}}{\overset{\overset{NH_3^+}{|}}{C}}-COO^- \; \underset{}{\overset{+H^+}{\rightleftharpoons}} \; H-\underset{\underset{R}{|}}{\overset{\overset{NH_2}{|}}{C}}-COO^-$$

| predominant form at pH 1 | predominant form at pH 7 | predominant form at pH 11 |

Ionization states of an amino monocarboxylic acid as a function of pH

Amino acid	Three-letter abbreviation
Alanine	Ala
Arginine	Arg
Asparagine	Asn
Aspartic acid	Asp
Cysteine	Cys
Glutamine	Gln
Glutamic acid	Glu
Glycine	Gly
Histidine	His
Isoleucine	Ile
Leucine	Leu
Lysine	Lys
Methionine	Met
Phenylalanine	Phe
Proline	Pro
Serine	Ser
Threonine	Thr
Tryptophan	Trp
Tyrosine	Tyr
Valine	Val

Although the shaded *peptide group* (Fig. 122) is rigidly planar, the bonds around the α carbons of the amino acid *residues* retain their rotational freedom. Proteins often contain a hundred or more acid residues whose sequence is written using the three-letter code shown in Table 56, the beginning of the chain being taken as the amine end (see Fig. 139 which illustrates this for the protein of whale myoglobin).

Four nitrogen heterocycles are responsible for holding the DNA molecule in its familar double-helix form. These are the pyrimidine bases thymine and cytosine, and the purines adenine and guanine:

thymine (T)   cytosine (C)   adenine (A)   guanine (G)

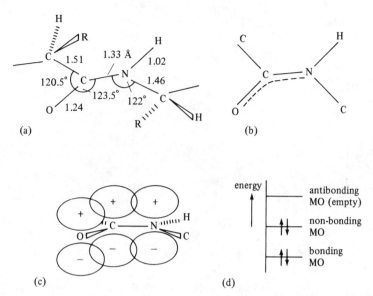

Fig. 122—The peptide bond. (a) Bond lengths (in Å) and angles in the peptide bond; for comparison, the lengths of C—N single and double bonds are 1.47 and 1.26 Å respectively. (b) Schematic representation of the delocalized π bonding in the O—C—N region which holds the peptide bond planar. (c) The three 2p orbitals involved in the O—C—N π system. (d) Simple energy level diagram showing the relative energies of the three molecular orbitals arising from the interaction shown in (c).

These amines link to a cyclic sugar, deoxyribose, via intermolecular loss of water involving the ringed hydrogen atoms:

The backbone of DNA is constructed of phosphate bridges which link successive deoxyribose molecules via the 3′ and 5′ oxygen atoms into long chains as shown in Fig. 123. Two such chains, which can be several million bases long, are held in a double-helix conformation by hydrogen bonds between opposite bases, but, owing to steric restrictions, adenine can only hydrogen-bond with thymine and, similarly, guanine only with cytosine. Note from Fig. 123 that in the double helix the two different but complementary chains run in *opposite directions*.

Hence hydrogen bonding, made possible by the high electronegativities of nitrogen and oxygen, is seen to be responsible for controlling genetic codes: as the two chains unwind during replication (under the influence of an enzyme) only one specific

Table 57 — Names, formulae and structure of the amino acids

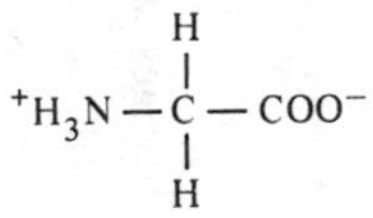

glycine
(Gly)

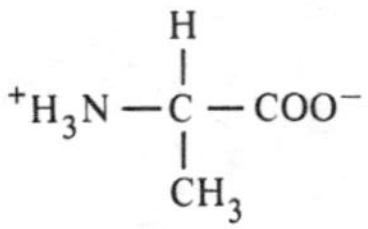

alanine
(Ala)

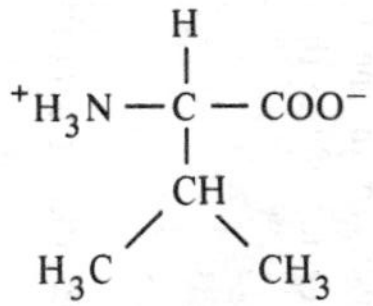

valine
(Val)

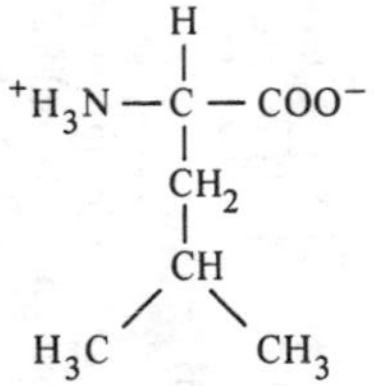

leucine
(Leu)

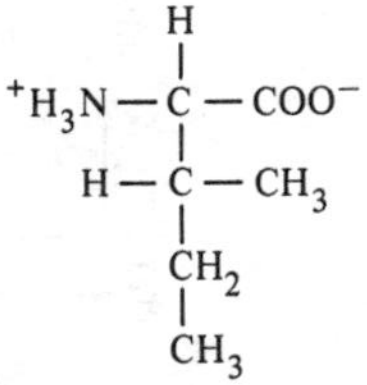

isoleucine
(Ile)

Amino acids having aliphatic side chains

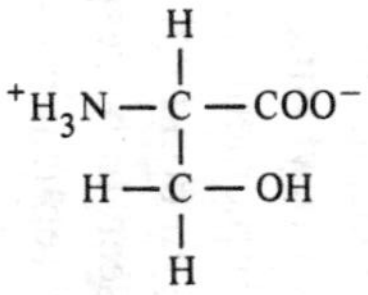

serine
(Ser)

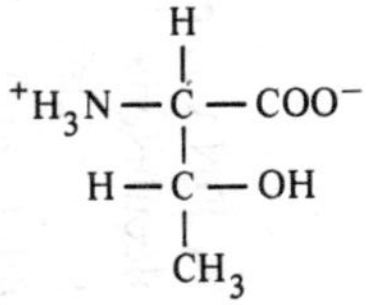

threonine
(Thr)

Serine and threonine have
aliphatic hydroxyl side
chains

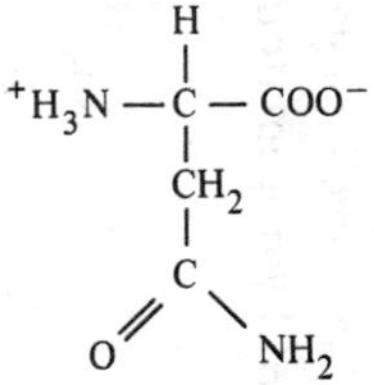

asparagine
(Asn)

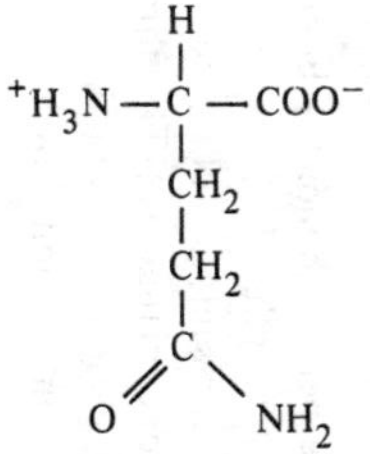

glutamine
(Gln)

Asparagine and glutamine
have amide side chains

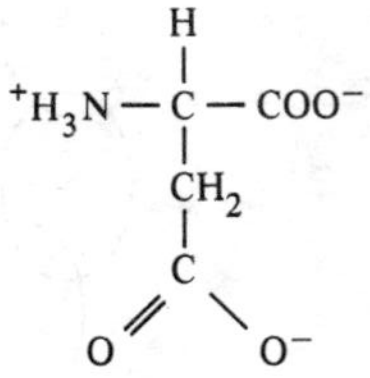

aspartate
(Asp)

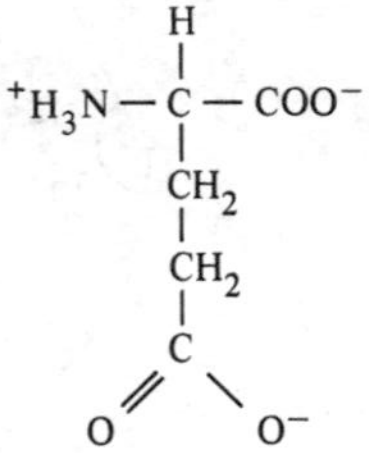

glutamate
(Glu)

Aspartate and glutamate
have acidic side chains

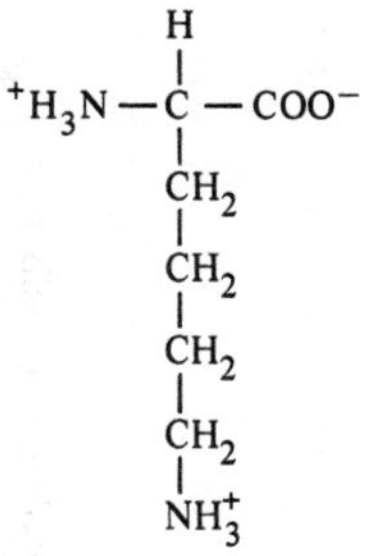

lysine
(Lys)

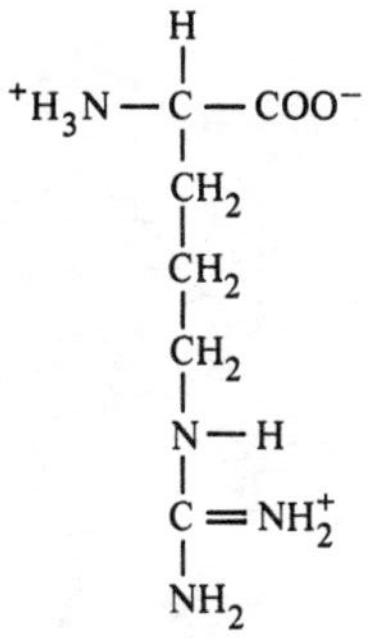

arginine
(Arg)

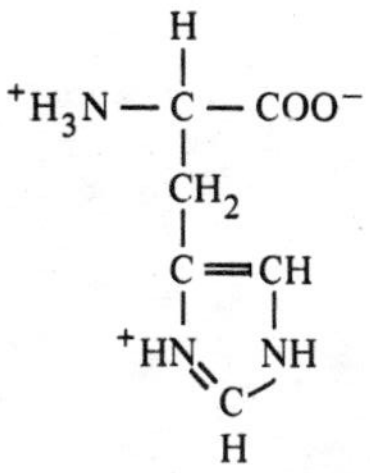

histidine
(His)

Lysine, arginine and histidine have basic
side chains

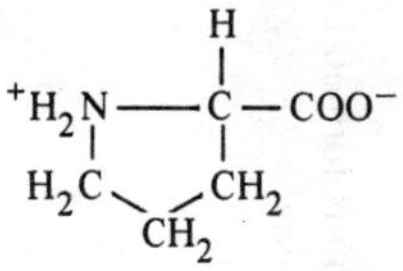

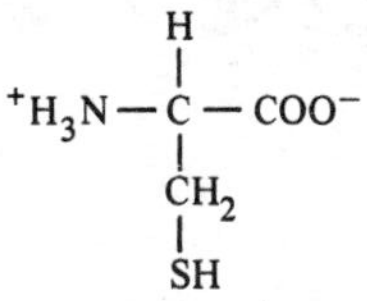

cysteine
(Cys)

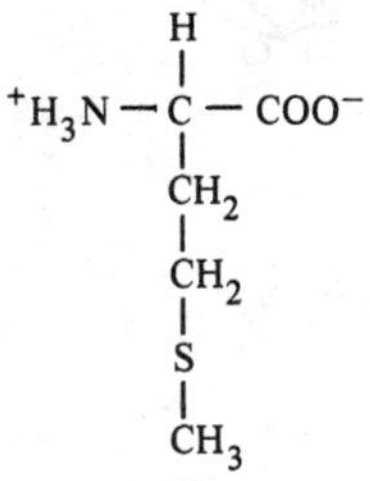

methionine
(Met)

Cysteine and methionine have sulphur-
containing side-chains

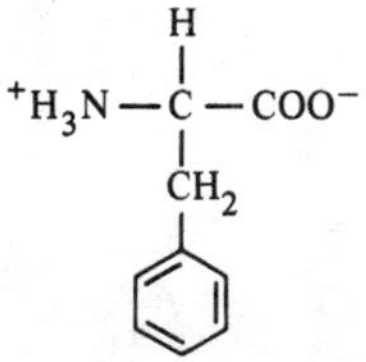

proline
(Pro)

Proline differs from the other common amino
acids in having a secondary amino group

phenylalanine
(Phe)

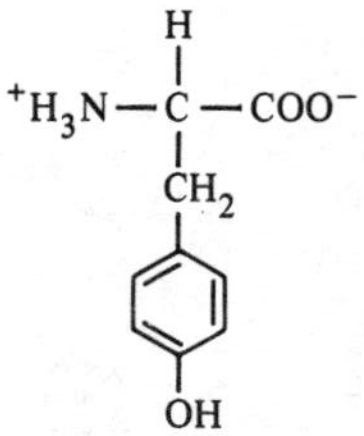

tyrosine
(Tyr)

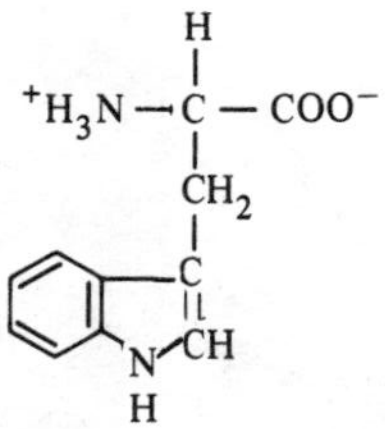

tryptophan
(Trp)

Phenylalanine, tyrosine and tryptophan
have aromatic side-chains

Part of one chain of DNA
showing phosphate bridges
linking the deoxyribose units
each with its specific base

Fig. 123(a)

Two hydrogen bonds formed between
thymine and adenine

Three hydrogen bonds formed between
cytosine and guanine

Fig. 123(b)

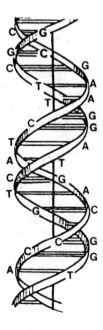

DNA double helix held
together by hydrogen bonding
between adenine and thymine
and between cytosine and
guanine

Fig. 123(c)

sequence of bases on a neighbouring chain will satisfy the steric constraints imposed by hydrogen bonding. Thus each chain acts as a template upon which is constructed an exact facsimile of the previous partner.

The basic structure of ribonucleic acids (RNA) is very similar to that of DNA shown in Fig. 123 except the molecules normally have only *single chains* and contain ribose sugar units:

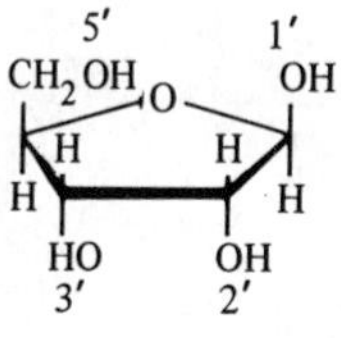

ribose
(contains a 2′
hydroxyl group)

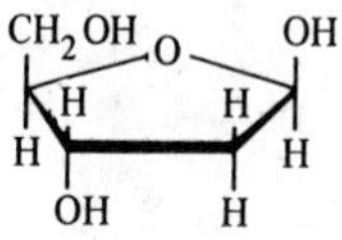

deoxyribose
lacks a 2′ hydroxyl
group)

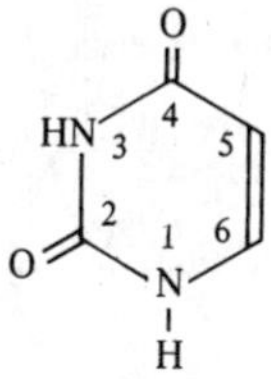

uracil (U)
(lacks the 5–CH_3
group of thymine)

In all three known types of RNA the base thymine is replaced by uracil which, like thymine, is able to hydrogen-bond effectively with adenine; the number of bases in RNA ranges from about 75 to a few thousand, hence the molecules are very much smaller than those of DNA. Replication of RNA usually occurs on partially unwound DNA chains but, owing to the dissimilarity in chain length, certain 'signal' sequences of bases on the DNA template are present to initiate, and then terminate, the process of synthesis.

PHOSPHORUS

We saw in the last section that PO_4 groups play a crucial role in biology by acting as bridges between the sugar units of DNA and RNA. During replication this phosphate is incorporated into the nucleic acids via an enzyme-catalysed reaction involving *nucleoside 5′-triphosphates*:

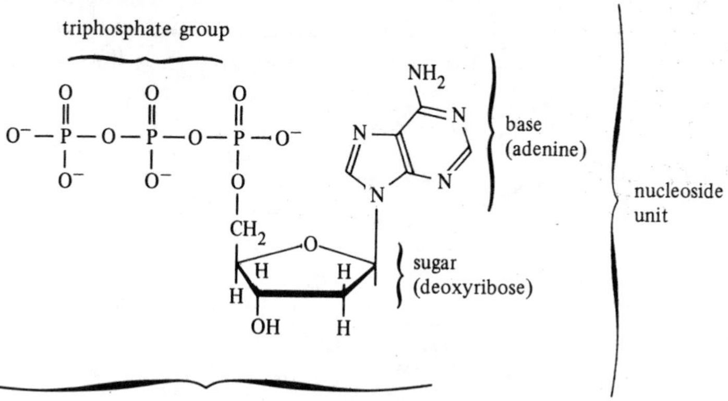

nucleotide: symbol dATP (for deoxy-adenosine triphosphate)
(d = deoxyribose; A = adenine; TP = triphosphate)

The whole nucleoside triphosphate unit is called a *nucleotide* and five such nucleotides are used in nucleic acid synthesis, each containing one of the heterocyclic bases adenine, cytosine, guanine, thymine and uracil. About 15 separate enzymes also assist in DNA replication, one of which, DNA polymerase (containing Mg and Zn), is required to catalyse the stepwise addition of each deoxyribonucleotide unit to the chain growing on its template in the 5′–3′ direction:

growing chain

5′ _____⟨3′OH⟩ + dATP ⟶   GTTCA   + $P_2O_7^{4-}$ + H_{aq}^{+}
GTTC   CAAGTG
CAAGTG
3′ _____ 5′

template chain

The mechanism at the molecular level involves the nucleophilic attack by the 3′-OH group on the α phosphorus atom of dATP, possibly assisted by coordination of the magnesium in polymerase to the α PO_4 group†:

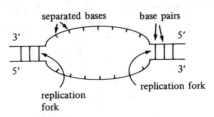

Although a single-chain template is required for this replication process, the parent DNA does not completely unwind and separate into two free chains. Only partial separation occurs and, if this does not begin at an end, then 'loops' are produced to give two replication forks:

separated bases   base pairs

3′   5′

5′   3′

replication   replication fork
fork

† For convenience, the *tetrahedral* phosphate groups in this and other formulae are drawn as though they are planar.

A protein stabilizes the loop by temporarily binding to the separated chains, and then an *RNA* enzyme, primase, actually initiates the growth of a new chain by starting a short RNA chain (later hydrolysed off by DNA polymerase) which generates the free 3'-OH group required for DNA synthesis:

DNA template chain

3' ↓   5'

G C A A G T A C

C G U U C A T G
                      ↑   3'
   5' ↑
short RNA chain   growing DNA chain

The other chain in the replication fork is unable to grow directly in the 3'–5' direction; it forms short pieces of chain in the normal 5'–3' direction, initiated each time by primase, and these pieces are then joined up by the enzyme ligase:

new chain   3'

5'   5'

            3'
         5'   joined here by ligase
3' parent
  DNA
                              3'

        short lengths
        of new chain   5'

   Species containing phosphate and polyphosphate groups occur extensively in many other biological systems, but space limits brief discussion to only two of them: adenosine triphosphate and the phospholipids involved in membrane construction. Those readers interested in the wider aspects of DNA, RNA and biophosphorus chemistry are referred to the relevant sections of Stryer's stimulating book on biochemistry listed at the end of this chapter.
   Adenosine triphosphate (ATP) has a very similar structure to that given above for dATP, the only difference being that ribose rather than deoxyribose is the constituent sugar:

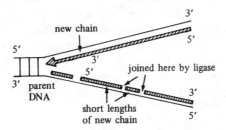

adenosine triphosphate (ATP)

adenosine diphosphate (ADP)

adenosine monophosphate (AMP)

ATP is the major energy source for all biological processes and organisms require enormous amounts of it for molecular synthesis, muscle movement and ion transport against concentration gradients. Stryer estimates that a resting human consumes about 40 kg of ATP per 24 h period but, during strenuous exertion, this may rise to a staggering 500 g per minute (30 kg per hour). Since muscle reserves would last for less than one second under these circumstances, ATP needs to be continuously generated by the glycolytic and oxidative metabolism of food, about 12 molecules being produced for each two-carbon fragment oxidized to carbon dioxide and water.

The cleavage of ATP into orthophosphate groups and adenosine diphosphate releases considerable energy, as shown, for example, in simple hydrolysis,

$$ATP^{4-} + H_2O \rightarrow ADP^{3-} + H^+ + HPO_4^{2-}, \quad \Delta G^{\ominus\prime} = -41 \text{ kJ mol}^{-1}$$

where $\Delta G^{\ominus\prime}$ refers to the standard free energy change when the reaction occurs at 25°C and pH 7.4 (the biological standard state). Electrostatic repulsion between the negative charges on neighbouring oxygen atoms may be the major driving force for these cleavage reactions, further assisted by extra π bonding made possible by the new terminal P—O bonds formed; bridging oxygen atoms, judged by the relatively long P—O bonds found in P—O—P systems, do not appear to be extensively involved in π bonding. Since the same considerations apply to the hydrolysis of the pyrophosphate group in ADP it is not surprising to find that the free energy change is similar:

$$ADP^{3-} + H_2O \rightarrow AMP^{2-} + H^+ + HPO_4^{2-}, \quad \Delta G^{\ominus\prime} = -43.5 \text{ kJ mol}^{-1}$$

The simple hydrolysis of ATP, liberating the bond energy as heat, is of no use to an organism and represents a pure waste of the ATP built up at some expense via the oxidation of foodstuff or absorption of light (as in plants). More usefully, ATP is able to transfer phosphate groups enzymatically, usually the terminal phosphate, to other molecules which then become *activated* towards further transformations. For example, glucose is phosphorylated by ATP under the influence of the enzyme hexokinase:

A good deal of the energy required for this and similar processes arises from the thermodynamically favoured splitting of ATP into ADP and monophosphate groups; since the hydrolysis products of ATP are very similar, chemists tend to talk loosely about the hydrolysis of ATP providing the energy for many biological reactions. In fact this is really jargon for a statement such as 'the reactions involve a sequence of events in which a certain quantity of ATP disappears to be replaced by equivalent amounts of ADP and monophosphate species'.

Since ATP is constantly reformed in metabolism, it functions in a catalytic capacity, acting as a link between the reactions that serve as sources of energy for the organism and those that lead to biosynthesis and growth[†].

Cell membranes contain yet another series of organophosphates, called phospho-lipids, most of which are derived from glycerol:

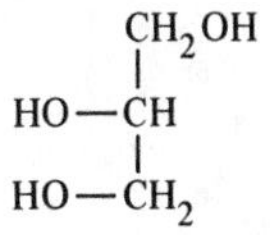

$$
\begin{array}{l}
CH_2OH \\
| \\
HO-CH \\
| \\
HO-CH_2
\end{array}
$$

Since glycerol has three hydroxy groups it may be tri-esterified, and in phospholipids two of the esterifying agents are long-chain fatty acids and the third is phosphoric acid:

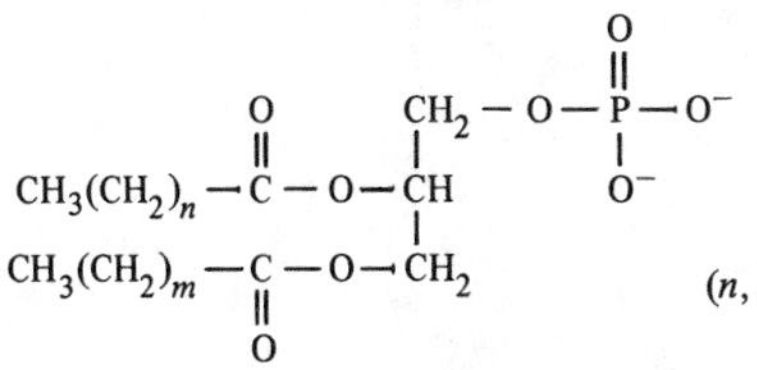

$(n, m = 14-24 \text{ (even numbers)})$

(a phosphatidate)

Although the carboxylic acid residues are shown fully saturated in this general formula, in fact these acids sometimes have a double bond part way down the chain; to complete the membrane phospholipid, the phosphate group forms an ester with a further alcohol such as ethanolamine, $HOCH_2CH_2NH_2$:

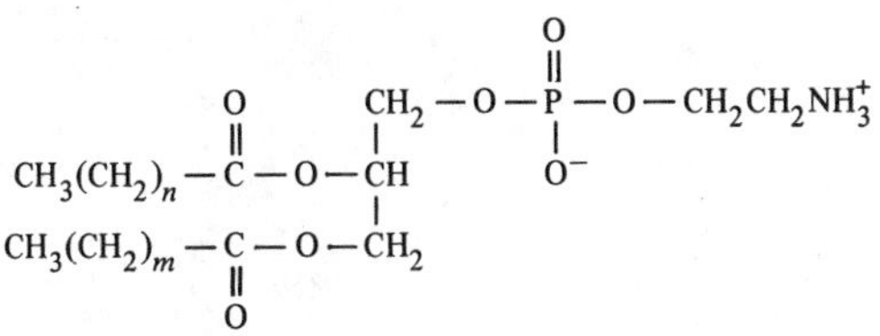

(a phosphatidyl ethanolamine)

Some of the common membrane phosphoglycerides are shown in Table 58 from which it will be noticed that they all have two properties in common—a 'tail' with two

† *General Microbiology* 2nd edn, by R. Y. Stanier, M. Duodoroff and E. A. Adelberg (MacMillan, 1968) p. 231.

long hydrophobic carbon chains and a hydrophilic 'head':

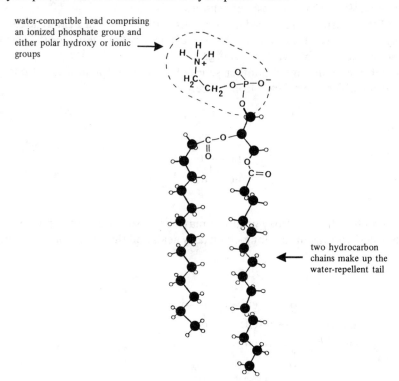

water-compatible head comprising
an ionized phosphate group and
either polar hydroxy or ionic
groups

two hydrocarbon
chains make up the
water-repellent tail

Phospholipids and the similarly shaped glycolipids, together with cholestrol and proteins, make up the major components of most cell membranes (Figs 124 and 125). Since the membranes are in an aqueous environment, mutual van der Waals attraction between the tails and hydrogen bonding of the heads with the surrounding water cause the ligands to spontaneously take up a double-layer structure about 45 Å thick:

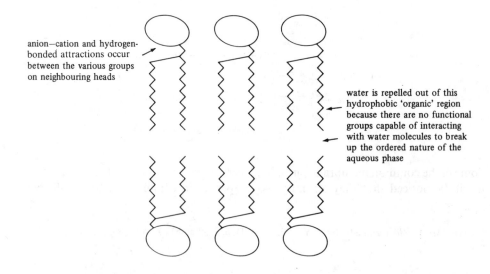

anion–cation and hydrogen-
bonded attractions occur
between the various groups
on neighbouring heads

water is repelled out of this
hydrophobic 'organic' region
because there are no functional
groups capable of interacting
with water molecules to break
up the ordered nature of the
aqueous phase

Table 58—Common phosphoglycerides

$$\left.\begin{array}{l} CH_3(CH_2)_n - \overset{\displaystyle O}{\overset{\|}{C}} - O - \overset{\displaystyle CH_2-}{\overset{|}{CH}} \\ CH_3(CH_2)_m - \overset{}{C} - O - CH_2 \\ \overset{\|}{O} \end{array}\right\} = R -$$

$$R - O - \overset{\displaystyle O}{\overset{\|}{\underset{\underset{O_-}{|}}{P}}} - O - CH_2CH_2 - NH_3^+ \qquad \text{phosphatidyl ethanolamine}$$

$$R - O - \overset{\displaystyle O}{\overset{\|}{\underset{\underset{O_-}{|}}{P}}} - O - CH_2CH_2 - N^+(CH_3)_3 \qquad \text{phosphatidyl choline}$$

$$R - O - \overset{\displaystyle O}{\overset{\|}{\underset{\underset{O_-}{|}}{P}}} - O - CH_2\overset{\displaystyle H}{\underset{\underset{OH}{|}}{\overset{|}{C}}}CH_2 - O - \overset{\displaystyle O}{\overset{\|}{\underset{\underset{O_-}{|}}{P}}} - O - R \qquad \text{diphosphatidyl glycerol}$$

$$R - O - \overset{\displaystyle O}{\overset{\|}{\underset{\underset{O_-}{|}}{P}}} - CH_2\overset{\displaystyle N^+H_3}{\underset{\underset{H}{|}}{\overset{|}{C}}} - COO^- \qquad \text{phosphatidyl serine}$$

$$R - O - \overset{\displaystyle O}{\overset{\|}{\underset{\underset{O_-}{|}}{P}}} - O \qquad \text{phosphatidyl inositol}$$

Although the resulting membrane is extremely stable, the lipid molecules are free to *diffuse sideways* in the plane of the bilayer. Embedded within the cell membranes are a variety of proteins which, usually by acting as enzymes, perform many vital functions, including the selective transport of ions across the inhospitable organic barrier and localizing chemical reactions in the proximity of the membrane; not just the whole cell but also its various organelles (e.g. the nucleus, mitochondia, endoplasmic reticulum, lysosome) are surrounded by membranes.

The membrane phospholipids are not passive but are continuously being broken down by phospholipases and replaced; as an example, the half-life of phospholipids in liver mitochondrial membranes is only about 10 days. A few unfortunate individuals lack these enzymes so that undegraded phospholipids tend to accumulate in nerve tissue and result in mental retardation.

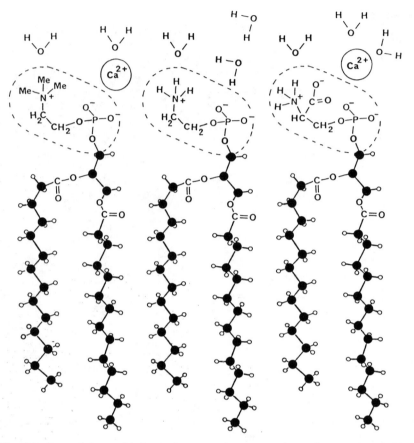

Fig. 124—Some of the possible interactions occurring at a water–membrane surface. Calcium ions appear to be essential for the integrity of membranes, but other cations such as sodium and potassium undoubtedly interact with some of the ionized phosphate groups. In some membranes there is a difference in the phospholipid distribution between the two layers. A typical example is the membrane of red blood cells where there is more phosphatidyl choline in the external layer than in the internal layer; the reverse is true for phosphatidyl ethanolamine and phosphatidyl serine. This is not fully understood but may relate to both the geometric factors affecting overall stability of the bilayer configuration and specific enzyme–membrane– surface interactions.

Orthophosphate ions, PO_4^{3-}, are triply charged and when combined with calcium result in ionic solids possessing very high lattice energies and hence low solubilities. Not unreasonably, then, it is found that the bony skeleton of vertebrates is mainly made up of a fibrous protein (collagen) and crystals of hydroxyapatite, a calcium phosphate having the formula $Ca_5(PO_4)_3OH$; bone accounts for 85% of the 700 g of phosphorus present in our bodies. Tooth enamel is almost wholly hydroxyapatite; the fluoride ions used in the prevention of dental caries possibly act by replacing OH groups in hydroxyapatite to give the much less soluble $Ca_5(PO_4)_3F$.

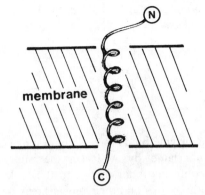

Fig. 125—Many large ('globular') proteins protrude right through the cell membrane, their amino acid chains being folded so that the hydrophobic portions are buried within the hydrocarbon core of the bilayer; some smaller globular proteins are only partially embedded within one layer of the membrane.

SODIUM AND POTASSIUM

Between them, the four cations Na^+, K^+, Mg^{2+} and Ca^{2+} make up slightly more than 1% of the body weight in man or 99% of the metal content; a typical 70 kg man contains about 90 g Na, 170 g K, 25 g Mg and 1200 g Ca compared with only 5 g of iron and 0.06 g of copper.

Sodium ions are found primarily on the outside of cells, being located in blood plasma and in the interstitial fluid which surrounds the cells; the ions participate in the transmission of nerve signals, in regulating the flow of water across cell membranes and in the transport of sugars and amino acids into cells. Sodium and potassium, although so similar chemically, differ quantitatively in their ability to penetrate cell membranes (both at rest and during excitation), in their transport mechanisms and their efficiency to activate enzymes. Thus potassium ions are the most abundant cations within cell fluids, where they activate many enzymes (probably via conformational control), participate in the oxidation of glucose to produce ATP and, with sodium, are responsible for the transmission of nerve signals.

There is a very considerable variation in the concentrations of sodium and potassium ions found on the opposite sides of cell membranes. As a typical example, in blood plasma sodium is present to the extent of 143 mmol l^{-1} whereas the potassium level is only 5 mmol l^{-1}; within the red blood cells these concentrations change to 10 mmol l^{-1} (Na^+) and 105 mmol l^{-1} (K^+). These ionic gradients demonstrate that a discriminatory mechanism, called the sodium–potassium pump, operates across the cell membrane which Stryer estimates to consume more than one-third of the ATP used by a resting animal—about 15 kg per 24 h in a resting human.

An enzyme, Na^+–K^+ ATPase, acts as the carrier of alkali cations across the membrane (which is an 'organic' region of low dielectric constant and hence ionophobic). This enzyme is probably a 'tetrameric' protein, $\alpha_2\beta_2$, having a combined molecular weight of about 270 000. The α units (molecular weight 95 000) are in contact with each other and are large enough to protrude right through the membrane;

each of the smaller β units adheres to an α chain but they are not in mutual contact as shown diagrammatically below:

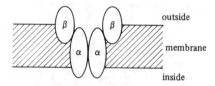

On the inside of the cell the enzyme binds three sodium ions before one of its α unit aspartyl groups is phosphorylated by ATP (in the presence of Mg^{2+} ions); phosphorylation causes a conformational change in the molecule which results in weakened binding of the sodium ions and their replacement by two potassium ions. Under the influence of the potassium, hydrolysis of the phosphoryl group to free phosphate occurs and consequent reversion of the enzyme to its original conformation in which potassium is more weakly bound than sodium takes place, and hence the cycle repeats continuously; this is shown schematically in Fig. 126.

Nerve cells, like other cells, have a higher concentration of potassium ions on the inside of the membrane than in the bathing fluid, the reverse being true of sodium ions. A resting nerve cell membrane is slightly permeable to potassium ions but not to either sodium ions or the negatively charged protein molecules which accompany potassium in the cell; thus leakage of potassium ions occurs, leaving the inside of the cell negatively charged, until an equilibrium is reached where the protein charge resists further loss of cations. The negative membrane potential thus created under these resting conditions is about 60–95 mV, the potential outside the cell being defined as zero. Removal of this potential by an outside stimulus results in the membrane becoming highly permeable to sodium ions, which rapidly move into the cell and

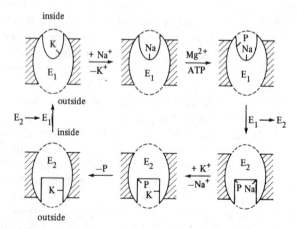

Fig. 126—Schematic representation of the Na^+–K^+ pump mechanism. Only one α unit at a time appears to be active in transport; two conformations of the enzyme, E_1 and E_2, are involved in the mechanism. Conformation E_1 has a high affinity for sodium and E_2 a high affinity for potassium ions. The hydrophilic parts for the α unit are shown as broken lines while the lipophilic region, buried in the membrane, is drawn with a solid line. It is likely that the conformation differences in E_1 and E_2 are quite small.

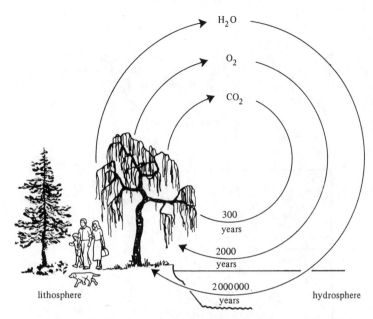

H₂O

O₂

CO₂

300
years

2000
years

2 000 000
years

lithosphere

hydrosphere

Fig. 127—The biosphere exchanges water vapour, oxygen and carbon dioxide with the atmosphere and hydrosphere in a continuing cycle. Water is split by plant cells and reconstituted by animals and plants about every two million years. The oxygen generated by splitting of the water enters the atmosphere and recycles about every 2000 years. Carbon dioxide has an average residence time in the atmosphere of 300 years before it is fixed by plant cells; respiration by plants and animals regenerates the CO_2. [Redrawn from 'The oxygen cycle' by P. Cloud and A. Gibor, *Scientific American* (September 1970).]

reverse the potential so that the inside of the cell is then positively charged; this occurs within about 1 ms but involves only tiny amounts of sodium, perhaps as little as one-millionth part of the cations available. For a few milliseconds after this potential reversal, the membrane becomes very permeable to potassium ions, which flood out to re-establish the resting potential; finally, the sodium–potassium pump restores the original cation ratios in the cell.

A nerve is a roughly spherical cell which in some places is drawn out into fibres, one (the axon) being much longer than the others—some axons can be about 1 m long in large animals; Fig. 128. The reversal of the membrane potential and its attendant cation movement occur under stimulus at one end of the axon and are confined to a very small area of axon membrane, but these voltage changes excite the neighbouring parts of the membrane to undergo the same cycle of events and so the original stimulus, starting in the nervous system or at a sense organ site, is passed down the fibre. Thus nerve impulses which pass along an axon are electrical signals induced by the flow of sodium or potassium across the axon membrane.

An active area of research at the present time is the study of possible mechanisms by which ions can passively diffuse through cell membranes. Fenton has described five possible ways by which such an ion diffusion process could occur and these are shown diagrammatically in Fig. 129. Movement of a bare (or hydrated) ion through a carbon-rich region of low dielectric constant is energetically very unfavourable and

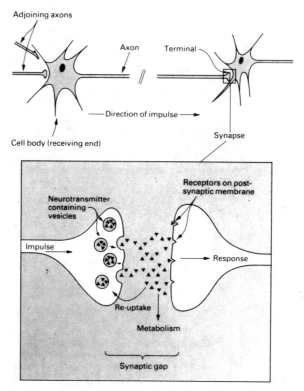

Fig. 128—Diagrammatic representation of the junction (or synapse) between two nerve cells. The axon of one nerve cell is not actually joined to the next cell but ends in a synaptic knob; the space between the two cells is called the synaptic gap or cleft. The arrival of an electrical nerve impulse at the knob causes release of specific transmitter chemicals which cross the synaptic gap and interact with receptor sites on the neighbouring cell thus passing on the impulse. A motor nerve axon terminates in an 'end-plate' which is attached to a muscle fibre, the synaptic gap between them spanning about 200 Å. [Reproduced with permission from *Chemistry in Britain* (September 1984) p. 798.]

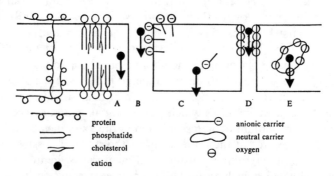

Fig. 129—Some modes of ion transport: A, bare cation permeation; B, negatively charged pore; C. anionic carrier molecule; D, neutral pore; E, neutral carrier molecule. [Reproduced with permission from *Chemical Society Reviews* vol. 6 (1977) p. 328.]

this mode of transport can presumably be dismissed as being highly unlikely. Several hundred molecules are now known which will either form ion-permeable channels extending through the membranes or complex with cations and actively transport them across the membrane; in both cases the molecules have a non-polar outer surface which is presented to the membrane and polar groups (usually oxygen) inside which can bind to the metal ion. Some of these molecules show antibacterial action and are collectively known as transport antibiotics, although currently they appear to have no useful clinical function. Their bacterial toxicity appears to be connected with loss of cell potassium which inhibits protein synthesis and ultimately leads to cell death.

A typical ion-carrying ligand (an ionophore) is valinomycin which is a cyclic molecule containing alternate ester and peptide bonds:

----L-valine;   D-hydroxy-; D-valine;   L-lactate----
isovalerate

When this ligand wraps itself around a potassium ion the conformation is such that the six ester carbonyls (marked *) point inwards and interact equally with the metal. An infrared spectrum of the sodium complex shows asymmetry in the ester carbonyl-stretching region, indicating that this cation does not bind equally well to all six oxygens and presumably 'rattles around' inside a cavity which is slightly too large for it. Such a situation should lead to weakened ligand–sodium binding and, in agreement with this, it is found that the potassium complex is about 1000 times more stable than that of sodium. Thus a *precisely tailored cavity* within the ligand is essential for selectivity, in this case for potassium ions.

The hydration sheath around the metal ion must be removed before satisfactory coordination of the encompassing ligand is possible; although this requires the input of a considerable amount of energy (322 kJ mol^{-1} in the case of potassium) it is more than compensated for by the formation of six strong ion-ligand interactions. However, there remains the problem as to how the water is stripped from the cations. In free valinomycin four oxygen atoms lie on the outside of the molecule and two or more of these could initially displace some water off the ion and, to quote Fenton, 'entice it into the hydrophilic interior where stepwise removal of the remaining hydration sheath could occur as the sinuous ligand enfolds the metal'. In the case of flexible ligands capable, initially, of forming cavities sufficiently large to encapsulate the hydrated cation, stepwise removal of the water would be made possible by the ligand undergoing a number of conformational changes.

Gramicidin A, a linear polypeptide containing 15 amino acid residues, is a channel-forming antibiotic. It dimerizes end-to-end as a β helix and spans the membrane (Fig. 130) allowing as many as 10^7 cations to pass through its tubular, and polar, cavity each second. The channel is about 4 Å wide in the inactive state but, when passing a cation, a molecular conformational change opens this to about 6.8 Å.

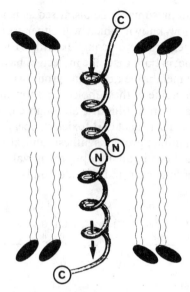

Fig. 130—Gramicidin A channel passing through a membrane; the channel is formed by the association of two polypeptides. When the 'dimer' dissociates, the channel, or pore, closes; the channel stays open for about 1 s when 10^7 cations are transported.

CALCIUM

Although 99% of body calcium is present in bones and teeth, this metal also plays extremely important roles in neuromuscular function, interneuronal transmission, cell membrane integrity and blood coagulation; to stay healthy, an adult requires a calcium intake of between 400 and 600 mg per day.

The bony skeleton of an animal has two basic functions in that it gives mechanical strength (required both for locomotion and prevention of damage to a number of sensitive organs) and also acts as a calcium store for the body; a human skeleton contains 1100–1500 g of calcium present mainly as hydroxyapatite, $Ca_5(PO_4)_3OH$, but also containing appreciable amounts of carbonate, sodium, magnesium, citrate and fluoride. In plasma, the calcium concentration is closely regulated at about 100 mg l^{-1} and any adjustments required in maintaining this level are made by two hormones, calcitonin and parathyroid hormone, and closely involve the store in bone tissue. Bone is not an inert and unchanging substance but is continuously being solubilized and redeposited to the extent of 400 mg per day in man; all this calcium passes through the plasma. When plasma levels fall, the parathyroid glands secrete parathyroid hormone, a polypeptide of some 85 amino acid residues, which releases bone calcium. Thyroid secretion of calcitonin occurs when plasma calcium levels rise and arrests the resorption of bone. Thus the cellular activity of bone tissue, and with it the calcium equilibrium in body tissue, are controlled by two partly synergistic and partly antagonistic processes. Parathyroid hormone has secondary functions in that it enhances gastrointestinal absorption of calcium from food and reduces calcium losses via excretion through the kidneys.

Remodelling of bone in man is most vigorous during the years of active growth when deposition must necessarily predominate over resorption. However, in middle age resorption exceeds deposition and results in 5%–10% loss of bone mass every 10 years, or 15–30 mg per day. In old age, or sometimes after the menopause, this can lead to osteoporosis, which is particularly manifested by brittle bones. Vitamin D, synthesized by the action of ultraviolet light on 7-dehydrocholesterol, is important in promoting resorption of calcium and deposition of bone so that insufficient intake, or inadequate exposure to sunlight, can result in bone diseases such as rickets in childhood or osteomalacia (softening of bones) in later life.

Many invertebrates have skeletal material composed of either the calcite or aragonite form of calcium carbonate, often containing small amounts of the neighbouring Group II elements, magnesium and strontium. Marine sediments are rich in calcium carbonate skeletons and over millenia may be turned into rocks and mountain ranges, thus locking up huge reserves of calcium (and 73% of the available carbon in the Earth's crust) for very long periods of time.

Skeletal muscle is made up of long cells composed of bundles of fibres (myofibrils); each myofibril itself is a collection of thick and thin filaments containing myosin and actin respectively (Fig. 132). The space between the myofibrils contains two membranes, one of which is a system of transverse channels (tubules) running at right angles to the myofibrils and the other is a series of sac-like membranes called the sarcoplasmic reticulum; the sarcoplasm is a special name given to the cytoplasm of muscle fibres. The sarcoplasmic reticulum controls the level of calcium ions in the sarcoplasm.

At rest, calcium is pumped from the sarcoplasm into the sarcoplasmic reticulum by an enzyme called Ca^{2+}, Mg^{2+}-ATPase and stored at 10^{-3}M as calcium–protein complexes (e.g. calsequestrin, molecular weight 44 000, which can bind about 40 calcium ions); this leaves the calcium concentration in the sarcoplasm as low as 10^{-6}M. A nerve impulse (page 450) that leads to muscle contraction results in an

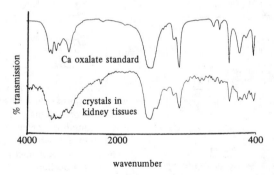

wavenumber

Fig. 131—Use of infrared microscopy to identify deposits in human tissue. Insoluble deposits of calcium salts sometimes form in the kidneys and bladder. It is now possible, using an infrared/microscope combination, to identify very tiny samples of such deposits taken in tissue biopsy; the infrared spectrum of surrounding tissue is subtracted from that of an area pin-pointed by the microscope to contain inorganic material. The spectrum of crystals found in a kidney shows how positive the identification can be even when samples measure only a few micrometres. [D. A. Levison, P. R. Crocker and S. D. Allen, *European Spectroscopy News* vol. 62 (1985) p. 18. Reprinted by permission of John Wiley and Sons, Ltd.]

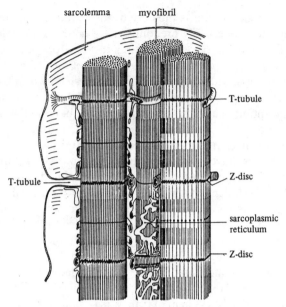

The internal cellular membranes of cardiac muscle

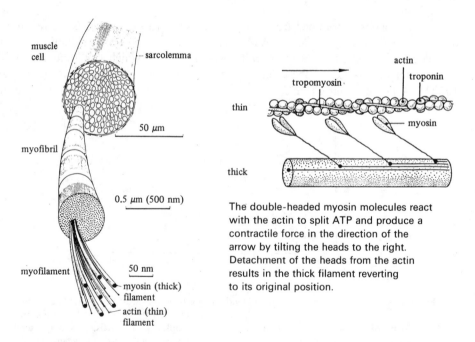

The double-headed myosin molecules react with the actin to split ATP and produce a contractile force in the direction of the arrow by tilting the heads to the right. Detachment of the heads from the actin results in the thick filament reverting to its original position.

Fig. 132—The suggested mode of action and structure of muscle. [Reproduced with permission from Chapter 16 of *A Companion to Medical Studies* vol. 1, 2nd edn, by R. Passmore and J. S. Robson (eds) (Blackwell Scientific Publications, 1976).]

electrical potential at the excitable membrane surrounding the muscle fibre (the sarcolemma) which is communicated to the transverse tubules and causes a rapid release of relatively large amounts of calcium ions from the neighbouring sarcoplasmic reticulum into the sarcoplasm, where the calcium concentration rises about 100-fold.

The thick and thin myofilaments contain the proteins actin, myosin, tropomyosin and troponin. The latter (molecular weight 50 000) binds calcium ions very strongly and in doing so undergoes a conformational change which makes actin and myosin react together causing a contraction as the two types of myofilaments slide over each other; the required energy is provided by ATP.

The mechanism by which blood clots after injury is exceedingly complex and involves a 'cascade' of enzyme-controlled reactions resulting finally in the production of an insoluble matrix of fibrin. Blood contains a soluble protein called prothrombin which, by the action of prothrombin activator in the presence of calcium ions, is converted into the enzyme thrombin; this enzyme then converts soluble fibrinogen into fibrin:

$$\text{prothrombin} \xrightarrow[\text{Ca}^{2+}]{\substack{\text{prothrombin} \\ \text{activator}}} \text{thrombin} \xrightarrow[\text{Ca}^{2+}]{+\text{fibrinogen}} \text{fibrin}$$

MAGNESIUM

An adult body contains about 25 g of magnesium, 60% of which is present in the skeleton while most of the remainder resides inside cells where it is the next most important cation after potassium. The daily requirement in the human diet is not known with certainty but has been estimated to be 200–300 mg.

All enzymes which utilize ATP in phosphate transfer, and many which aid either transfer of other groups or hydrolysis, require magnesium as the cofactor. For example, in phosphorylation reactions the Mg^{2+} ion is thought to complex with oxygens on the various phosphate groups; since it is the smallest, doubly charged cation which is biologically available, magnesium will normally form the strongest complexes of this type. Figures 133 and 134 illustrate postulated mechanisms which involve magnesium–phosphate coordination; the former reversible reaction involving creatine kinase is extremely important since it provides both a means of storing energy (as phosphocreatine) and a rapid synthesis for ATP (to use this stored energy when required).

A further crucial function performed by intracellular magnesium ions is to keep ribosomes intact. (A ribosome, the site of protein synthesis, is a complex particle of about 200 Å diameter found in the cytoplasm and comprising about 40% proteins and 60% RNA molecules.) The extent of bacterial cell growth is directly proportional to the concentration of magnesium ions, largely owing to the inhibition of protein synthesis caused by a limited number of intact ribosomes available in the cells.

Chlorophyll, the green colouring matter in plants, is a macrocyclic ligand which chelates a central magnesium ion as shown in Fig. 135. It is involved in photosynthesis where radiant energy from the Sun is absorbed and used to generate carbohydrates from water and carbon dioxide:

$$xH_2O + xCO_2 \rightarrow (CH_2O)_x + xO_2$$

Fig. 133(a)—A possible mechanism for the phosphorylation of creatine by creatine kinase (creatine phosphotransferase). Creatine phosphate is used as an energy store in muscle since the amount of ATP present is sufficient for less than 1s of vigorous activity; extra ATP, enough for up to 100 further muscle contractions, is formed from creatine phosphate via the reverse reaction. [Reproduced with permission from *Advances in Enzymology* vol. 33 (1970).]

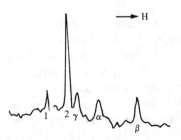

Fig. 133(b)—Phosphorus-31 NMR spectrum of a human forearm with the subject at rest. Resonances of the α, β and γ phosphorus atoms in ATP are clearly visible. Resonance 1 is due to inorganic phosphate ions and 2 arises from the store of creatine phosphate. [Redrawn from *Nature* vol. 295 (1982) p. 608.]

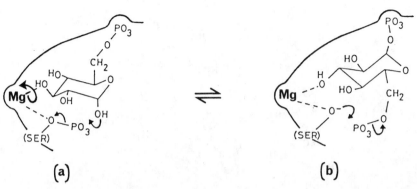

Fig. 134—A possible mechanism for the phosphoglucomutase-assisted isomerization of glucose-6-phosphate to glucose-1-phosphate. (a) A phosphorylated serine group close to the magnesium atom of the enzyme transfers its phosphate group to the 1-OH position of a glucose-6-phosphate substrate molecule which is already coordinated to Mg via its 3-hydroxyl oxygen atom; two binding sites about 5.5 and 10 Å away from the magnesium hold the glucose-phosphate unit in the correct position. (b) The right-hand diagram shows the intermediate glucose-1,6-diphosphate after it has undergone rotation about the Mg—O bond to present the 6-phosphate group for nucleophilic attack by the serine OH and so release glucose-1-phosphate. Glucose, an important energy source for organisms, is phosphorylated by ATP to glucose-6-phosphate, a reaction catalysed by the magnesium-containing enzyme hexokinase. Glucose-1-phosphate is required as an intermediate in the formation of the energy storage material glycogen, which is a high polymer made up of glucose residues. [Reprinted with permission from *Metals in Biochemistry* by P. M. Harrison and R. J. Hoare (Chapman and Hall, 1980).]

Perhaps unexpectedly from this simplified equation, tracer experiments using $H_2{}^{18}O$ have shown that all the oxygen is derived from water:

$$2xH_2{}^{18}O + xCO_2 \rightarrow (CH_2O)_x + x{}^{18}O_2 + xH_2O$$

The end-products within green plants are energy-rich carbohydrates which become food for herbivorous organisms and a succession of animal-eating organisms; thus, directly or indirectly, magnesium provides the key to nearly all living processes. Only about 1% of incident sunlight ends up as plant material, itself undergoing only 10% conversion into grazing animal tissue; since humans change a mere 1% of animal meat into human tissue, this results in a tiny 0.001% utilization of radiant energy. Notwithstanding the small fraction of sunlight used in photosynthesis, it has been estimated that about 10^{10} tons of carbon are converted into carbohydrates annually.

Photosynthesis occurs in the chloroplasts of plants where the chlorophyll molecules take an active part in the photosynthesis; all the chlorophyll b and most chlorophyll a molecules act as *antennae*, collecting a light photon and passing the acquired energy over several similar molecules to an 'active' chlorophyll a residing at a special site on the protein, the whole process taking about 10^{-12} s. (The active chlorophyll a is sometimes called P-700 since it is a pigment and one of its absorption bonds occurs at 700 nm.) The mechanism of photosynthesis is exceedingly complex and in fact involves absorption of light at two different wavelengths giving rise to two different

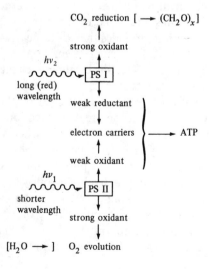

Fig. 135—The structure of a chlorophyll molecule. Chlorophyll a and chlorophyll b are the two most common types found in land plants. Both contain a magnesium cation located about 0.4 Å above the centre of a puckered macrocyclic ring and linked to four pyrrole-like nitrogen atoms; *in vivo* other ligands complete the coordination sphere about magnesium because a square planar array is energetically unfavourable.

photosystems, PS I and PS II:

The strong reductant in photosystem I is NADPH formed by the photoexcitation of P-700 (via the antenna chlorophylls) which transfers its energy to a membrane-bound ferredoxin then to a soluble ferredoxin which is finally used to transform $NADP^+$ to NADPH:

NAD$^+$, R = H
NADP$^+$, R = PO$_3^{2-}$

NADH, R = H
NADPH, R = PO$_3^{2-}$

The identity of the strong oxidant in photosystem II is not yet known but it is able, with the assistance of a manganese(II) enzyme, to oxidize water:

$$2H_2O \xrightarrow[\text{strong oxidant}]{(Mn^{2+})} O_2 + 4e^- + 4H^+ \qquad \text{(photosystem II)}$$

$$2NADP + 4e^- + 2H^+ \xrightarrow[\text{ferredoxin}]{\text{soluble}} 2NADPH \qquad \text{(photosystem I)}$$

$$\underline{2H^+ + 2NADPH + CO_2 \rightarrow 2NADP + H_2O + (CH_2O)} \qquad \text{(dark reaction)}$$

$$H_2O + CO_2 \rightarrow O_2 + (CH_2O) \qquad \text{(net reaction)}$$

ZINC

An adult human contains about 2 g of zinc which is an active component of many enzymes, including the much studied carbonic anhydrase so essential to respiration; the metal is stored in the kidneys and liver on metallothionein, a small protein of about 60 amino acid residues, one-third of which are sulphur-containing cysteines. Dietary zinc deficiency, manifested by poor growth, impaired bone development and hindered function of the reproductive organs, can occur in humans if there is a cultural dependence on cereals containing a high proportion of phytic acid (inositol hexaphosphate) which strongly complexes with zinc and inhibits its uptake in the intestine:

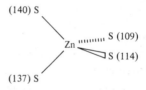

phytic acid

Nature presumably chose zinc, copper and, to some extent, iron because of the strong preference they show in Lewis acid–base complexes for the (biochemically important) tetrahedral conformation; unlike copper and iron, however, zinc cannot undergo redox reactions and its biological role depends wholly on its Lewis acidity.

When bound to some proteins, the function of zinc seems simply to stabilize the three-dimensional nature of the molecule and not to function as the substrate-binding site. An example of this is to be found in the enzyme aspartate transcarbamylase which is involved in pyrimidine biosynthesis and has twelve polypeptide chains, six being catalytic and six regulatory. It is only these latter regulatory chains, each possessing specific binding sites for ATP and cytidine triphosphate, which contain a zinc atom tetrahedrally coordinated to four cysteine residues:

$$(140)\ S \qquad\qquad\qquad\qquad S\ (109)$$
$$Zn$$
$$(137)\ S \qquad\qquad\qquad\qquad S\ (114)$$

In a metalloenzyme the stereochemistry of the metal is strongly dictated by the ligand atom geometry, a fact which has a profound effect on the affinity of the protein for a particular metal ion such as zinc. In many enzymes the metal has a highly distorted coordination geometry and there is little doubt that the strain imposed by such distortions is intimately connected with the metal's catalytic activity. Thus zinc acting as a structure former has essentially regular tetrahedral geometry, but when it is at the active site the 'tetrahedral' coordination is much less regular. Although the reason for strained, or 'entatic', four-coordination is not fully understood, it could mechanistically aid approach of a fifth ligand by forcing the metal ion to adopt a shape part way towards five-coordination.

Such a strained 'tetrahedral' conformation occurs around the zinc ion of carbonic anhydrase. This enzyme, present in red blood cells, catalyses the reversible hydration of carbon dioxide,

$$CO_2 + H_2O \rightleftharpoons HCO_3^- + H^+$$

allowing both the take-up of carbon dioxide from tissue such as active muscle and its release in the alveoli of the lungs. This normally sluggish reaction is so efficiently speeded up by the enzyme that each molecule of carbonic anhydrase is capable of hydrating about one million molecules of carbon dioxide every second at around 37 °C.

Three closely related forms of carbonic anhydrase (A, B and C) are found in mammals, each having a molecular weight of approximately 30 000 and a roughly ellipsoidal shape; the zinc atom is located in a small cleft about 12 Å from the surface (Fig. 136). Although precise details of the enzyme's mechanisms are still open to debate, it seems clear that reaction occurs at, or near, the zinc ion: careful removal of the zinc results in complete loss of activity whereas reintroduction of zinc to this 'apo-enzyme' seems to cause little conformational change in the molecule (thus the zinc is presumably not simply acting as a structure-forming metal ion).

Isotopic tracer studies have shown that dehydration of $HC^{18}O_3^-$ leaves an oxygen-18 atom on the zinc, proving that at some stage during the reaction sequence the bicarbonate ion must be directly coordinated to the metal. The most favoured mechanism for the hydration reaction involves nucleophilic attack of a zinc-bound OH^- on carbon dioxide and displacement of the resulting bicarbonate ion by a water molecule (probably via a transient five coordinate zinc intermediate):

$$\text{—Zn—O}\overset{H}{\cdot}\ \underset{O}{\overset{O}{C}}\rightleftharpoons \text{—Zn—O—C}\overset{\overset{H}{|}}{\underset{O}{\diagup}} \rightleftharpoons \text{—Zn—O—C}\overset{\overset{H}{\diagup}O}{\underset{O}{}} \rightleftharpoons \text{—Zn}\overset{OCO_2H}{\underset{OH_2}{}} \rightleftharpoons \text{—Zn—OH}_2 + HCO_3^-$$

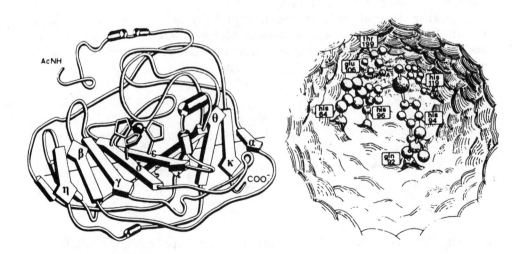

Fig. 136—(a) Secondary and tertiary structure of human carbonic anhydrase C. The cylinders represent the α helixes and the arrows the strands of the β structure. The zinc atom together with its histidine ligands is also shown as a black sphere in the centre of the structure. (b) Pictorial representation of the active site cavity showing the residues in close proximity to the zinc atom. [Reproduced with permission from *Structure and Bonding* vol. 48 (1982) p. 45.]

The carbon dioxide molecule could be suitably oriented for the nucleophilic attack by hydrogen bonding of its solvating water to the OH group, by hydrogen bonding of one of its oxygens to the OH group or, much less probably, by direct coordination to the zinc. One major objection which might be levelled at this simple mechanism is that a water molecule coordinated to zinc would not normally lose a proton at physiological pH. However, 'assisted proton loss' from the coordinated water could be possible via a sequence of hydrogen bonds involving the Glu 106 and Thr 199 residues of the protein chain:

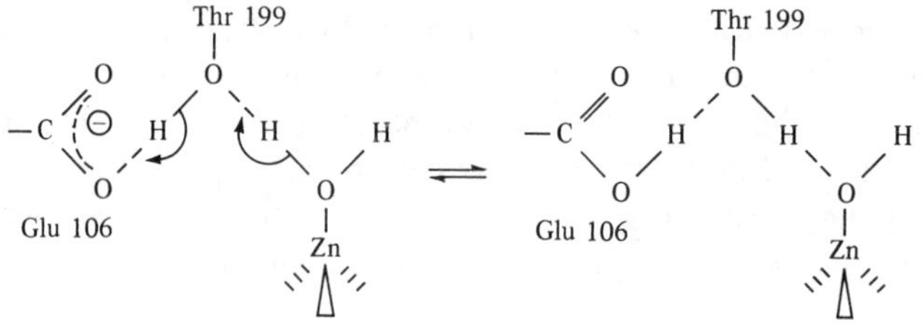

The dimeric enzyme liver alcohol dehydrogenase (LAD) is unusual in that two of its four zinc atoms lie at the catalytic sites whereas the others appear to take on a purely structure-maintaining role. LAD, which oxidizes ethanol in the liver to produce ethanal, can be inherited as either a slow- or a rapid-acting form; people with the latter are sometimes embarassed by prickling sensations and flushing caused by sudden releases of ethanal on drinking alcohol.

The inactive, structure-forming zinc ions in LAD lie in a relatively regular tetrahedral array comprised of four cysteine sulphur atoms, whereas the catalytic zincs have a much more distorted tetrahedral coordination derived from two cysteine sulphurs, a histidine nitrogen and a water molecule,

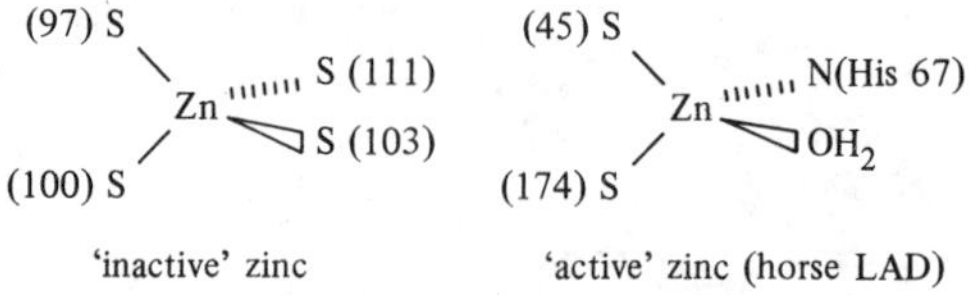

'inactive' zinc   'active' zinc (horse LAD)

and are in a 'pocket' some 20 Å from the surface. The cramped conditions within this very narrow hydrophobic pocket make it unlikely that the zinc atoms can form pentacoordinate intermediates during the reaction sequence. A co-enzyme NAD^+ (page 461) binds to the protein with C-4 of its positively charged pyridine ring lying about 4.5 Å from the active zinc and assists in removal of a 'hydride ion' from the α CH_2 group of the metal-bound alcohol substrate by forming NADH:

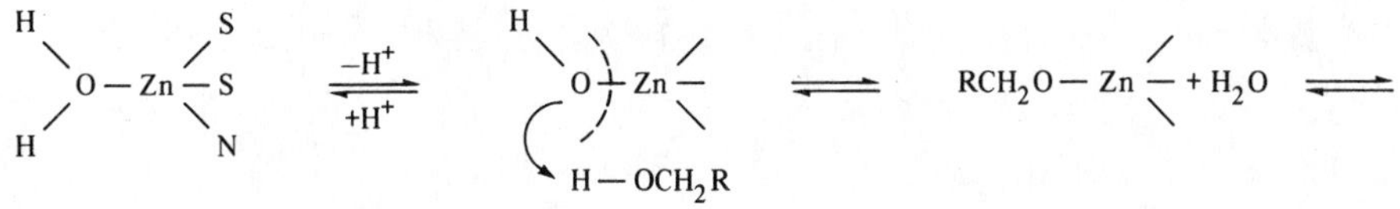

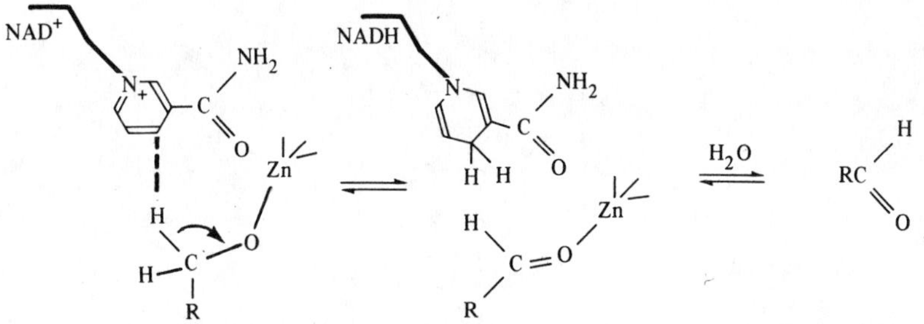

$$-H^+ \quad +H^+$$

H—OCH$_2$R

RCH$_2$O—Zn + H$_2$O

NAD$^+$

NH$_2$

Zn

NADH

NH$_2$

H H

Zn

H$_2$O

RC

R

R

SULPHUR

Sulphur is an important constituent of many protein molecules since it is present in the amino acids cysteine and methionine:

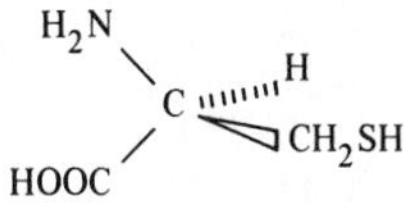

H$_2$N

C

H

HOOC

CH$_2$SH

cysteine (Cys)

H$_2$N

C

H

HOOC

CH$_2$CH$_2$SCH$_3$

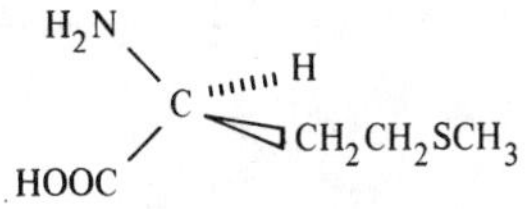

methionine (Met)

The relative ease with which sulphur catenates (i.e. forms chains with itself) makes it unique in the whole periodic table and as a result nature frequently uses S—S bridges between cysteine residues to link either two separate protein chains or two parts of the same chain. Examples of both types of disulphide bridge are found in the hormone insulin:

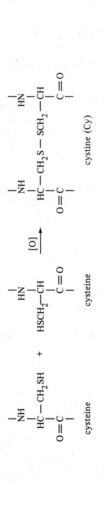

$$HC-CH_2SH \quad + \quad HSCH_2-CH \quad \xrightarrow{[O]} \quad HC-CH_2S-SCH_2-CH$$

cysteine + cysteine → cystine (Cy)

Phe—Val—Asn—Gln—His—Leu—Cy—Gly—Ser—His—Leu—Val—Glu—Ala—Leu—Tyr—Leu—Val—Cy—Gly—Glu—Arg—Gly—Phe—Phe—Tyr—Thr—Pro—Lys—Ala

B chain

Gly—Ile—Val—Glu—Gln—Cy—Cy—Ala—Ser—Val—Cy—Ser—Leu—Tyr—Gln—Leu—Glu—Asn—Tyr—Cy—Asn

A chain

bovine insulin

Between 1 and 2 g of the sulphur ingested daily by humans as animal protein is excreted in the urine, mainly as SO_4^{2-} although organic sulphate esters and thiocyanate are also present in small amounts. Biological decay of organic matter under the reducing conditions prevailing, for example, in marshes and estuaries, leads to the production of hydrogen sulphide which on escaping to the atmosphere or oxygenated water has a half-life of only a few hours, being converted ultimately into sulphate.

Sulphur, either as the free sulphide ion or when covalently bound in amino acids, behaves as a very effective Lewis base towards many transition metals; for example, a wide variety of metallo-enzymes have one or more cysteine sulphur atoms coordinated to iron, copper, zinc or molybdenum. Some iron–sulphur redox proteins contain Fe_2S_2, Fe_3S_3 and Fe_4S_4 clusters in which sulphide ions bridge the iron atoms, the clusters themselves being attached to the protein chain by cysteine sulphur–iron bonding; see Fig. 137. Treatment of these proteins with dilute acid results in the conversion of the S^{2-} bridging groups into hydrogen sulphide.

Iron–sulphur proteins are involved in many important biological processes, including mitochondrial respiration (page 475), nitrogen fixation (page 485) and photosynthesis (page 459). The Fe_4S_4 clusters have been the most extensively studied, both in normal redox proteins and in model compounds used to mimic the natural processes; each behaves as a 'delocalized' unit within which it is not meaningful to assign a specific oxidation state to each iron atom. Although three overall oxidation states, $Fe_4S_4^{3+,2+,1+}$, are potentially accessible it is found that proteins from different sources can only cycle between either the $Fe_4S_4^{2+,1+}$ or $Fe_4^{3+,2+}$ states, probably

(a) This, the simplest of the iron–sulphur systems, is found in rubredoxins. The tetrahedrally coordinated iron atom cycles between the high-spin +II and +III oxidation states.

(b) Probable arrangement of atoms in the Fe_2S_2 clusters of '2-iron ferredoxins'. The oxidized form has two high-spin Fe^{III} atoms but is diamagnetic at low temperatures owing to superexchange through the sulphide bridges; in the reduced form the formal oxidation state of the iron atoms is 2.5.

Fig. 137—Iron–sulphur clusters found in some redox proteins. In (d) the positions of the 54 amino acid α carbon atoms are shown as black circles and those of the Fe_4S_4 sulphurs as open circles; the cysteine sulphur atoms at position γ in the residues are represented by S. Tyrosines 2 and 28 each have their aromatic ring parallel to one face of their neighbouring Fe_4S_4 cube; both rings also present an exposed edge to the solvent. It is possible that electron transfer is mediated by these unique tyrosines, although some ferredoxins are known which lack such residues. [Reproduced with permission from *Journal of Biological Chemistry* vol. 248 (1973) p. 3987.]

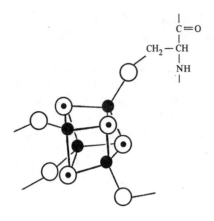

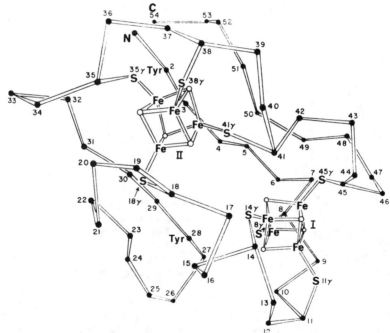

(c) Distorted Fe_4S_4 cube which is bound to the protein chain by four cysteine sulphur atoms; one or more of these clusters can be present in the '4-iron ferredoxins'.

Only one electron is transferred during a redox reaction involving *each unit* shown in (b) and (c).

● Fe

⊙ S^{2-}

○ S (Cys)

(d) Schematic diagram of ferredoxin from *Peptococcus aerogenes* showing two Fe_4S_4 clusters bound to the protein by a total of eight cysteine sulphur ligands

Fig. 137—*continued*

owing, in part, to the rigidity of the protein chain; very considerable flexibility would be required to accommodate the large variation in iron–sulphur bond lengths resulting from the transfer of more than one electron in a redox reaction.

Even with the above restriction imposed on the oxidation state change, nature manages to achieve an astonishingly wide range of reduction potentials in the various ferredoxins: possibly about 650 mV for those using the $Fe_4S_4^{2+,1+}$ couple and over 750 mV for others restricted to the $Fe_4S_4^{3+,2+}$ couple. This range is achieved by a combination of factors such as steric influences of the protein chain, the relative accessibility of the Fe_4S_4 cluster to the surrounding aqueous medium and peptide NH---S hydrogen bonding. Although the Fe_4S_4 groups are stable in water it appears that a further function of the protein chain is to provide them with protection against oxygen degradation.

Mammalian metallothionein is a protein having an unusually high sulphur content in that one-third of its amino acid residues are cysteines. It has a strong affinity for a variety of metals, particularly copper, zinc and cadmium, all of which bind only to the cysteine residues; the metal-free protein will take up seven atoms of Zn and Cd or twelve atoms of copper(I). Liver metallothionein recovered from human biopsies contains zinc almost exclusively whereas that from kidney also has substantial levels of cadmium and copper; presumably this reflects both the variation in metal exposure of the two organs and the changes in metal affinity of different forms of metallothionein. The protein acts as an efficient body store for zinc whereas its function with the highly toxic cadmium, which binds 1000 times more effectively than zinc, may be to hold it tenaciously in an inactive form.

IRON

It has been suggested that life in any form would probably be impossible without the presence of iron because of its wide spectrum of biological activity. However, owing to its ready oxidation, iron is available in nature only as highly insoluble derivatives of iron(III) and organisms have been forced to evolve special iron-sequestering molecules which are able to hold the element in a soluble form for both transport and biosynthesis of essential iron proteins and enzymes.

About 75% of the 5 g of iron present in the adult human body is present as the oxygen-carrying protein haemoglobin in red blood cells and 20% is stored as ferritin and haemosiderin; myoglobin (3%–4%) and cytochromes make up most of the remainder. Since the body cannot excrete absorbed iron through either the urine or faeces, there are dangers associated with excessive intake either long-term, e.g. via the use of iron cooking pots, or short-term e.g. when children eat their mothers' iron sulphate tablets as 'sweets'; the former can result in cirrhosis of the liver whereas death occurs in the children unless prompt action is taken (e.g. removal of the stomach contents by pumping followed by treatment with chelating ligands which bind to the iron and allow its excretion).

During menstruation or pregnancy, women require about twice the dietary amount of iron as men to prevent anaemia: 20–25 mg per day of which only 10% is actually absorbed. Most of this iron is used to make haemoglobin since red blood cells have a relatively short life span of 100–120 days. There are a quarter of a million haemoglobin molecules in each blood cell and 5×10^9 cells are present in each millilitre of an adult's 5 l blood supply; about 1 % of all this haemoglobin is replaced daily! Copper is required during certain stages of haem synthesis and hence symptoms of anaemia can occur in animals and man, even when iron is readily available in the diet, owing to copper deficiency.

The liver and bone marrow contain most of the body's ferritin, which is a water-soluble, high-molecular-weight protein having 24 identical sub-units. These 'cylindrical' sub-units pack together to form a roughly spherical structure having a central cavity with a diameter of about 75 Å. The iron is stored within this cavity as microcrystals of hydrated iron(III) oxide (ferrihydrite, $5Fe_2O_3.9H_2O$) with variable amounts of phosphate; the presence of the hydrophilic protein sheath enables these tiny, highly insoluble grains containing about 4500 iron atoms to move through the plasma and cytoplasm. Relatively little is known about the water-soluble haemosiderin protein; it is probably a breakdown product of ferritin.

The site of *reversible* oxygen binding in both haemoglobin and myoglobin is haem, an iron(II) complex of the macrocyclic ligand protoporphyrin IX shown in Fig. 138. Myoglobin, which is located in muscle tissue, has a long protein chain of 153 amino acid residues intricately folded in such a way that a hydrophobic pocket is created for the single haem group. The four protein chains in haemoglobin, each with a haem pocket, are packed in the form of a roughly tetrahedral $\alpha_2\beta_2$ cluster, the α chains having 141 amino acid residues and the β chains 146. Hence myoglobin is able to bind only one oxygen molecule whereas haemoglobin can carry one, two, three or four oxygens; on the other hand, myoglobin has the higher oxygen affinity and therefore facile oxygen transfer occurs between the blood capillaries and muscle tissue

Fig. 138—Protoporphyrin IX and its iron(II) complex haem. Protoporphyrin IX is a macrocyclic ligand derived from four pyrrole rings linked by —CH— bridges; on being bound to iron(II) its two NH protons are lost to give an uncharged complex called haem. Fifteen isomers are possible for the protoporphyrin molecule depending on the relative positions of the four methyl, two vinyl and two propanoic acid groups. Only the one isomer shown is found in vertebrate and many invertebrate myoglobins and haemoglobins, although the globin part of the molecule differs in amino acid sequence and conformation from species to species. In myoglobin and haemoglobin the iron atom is substantially below the plane of the four nitrogens and the whole haem is dome-shaped. On oxygenation the iron moves towards, but not completely into, the nitrogen plane and the haem flattens appreciably.

where myoglobin stores the oxygen until it is required for oxidative phosphorylation of ADP in the mitochondria:

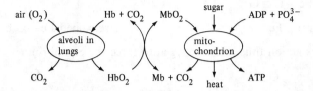

The haem group of myoglobin is held in its pocket by a coordinate bond from an imidazole nitrogen atom of histidine F8 and about 60 van der Waals contacts to various parts of the protein chain; Fig. 139. Before oxygenation the iron(II) atom is in the high-spin state, making the myoglobin molecule paramagnetic owing to the four unpaired electrons. When an oxygen molecule binds to the free site on the iron to complete octahedral coordination, the six iron 3d electrons are forced into the t_{2g} orbitals (i.e. $3d_{xy}$, $3d_{yz}$ and $3d_{xz}$) by the increased ligand field, resulting in a change from a high-spin (paramagnetic) state to a low-spin (diamagnetic) state. The oxygen too becomes diamagnetic on coordination and this probably occurs by a change in electron configuration,

$$(\sigma_{2p_z})^2(\pi_{2p_x})^2(\pi_{2p_y})^2(\pi_{2p_x}^*)^1(\pi_{2p_y}^*)^1 \longrightarrow (\sigma_{2p_z})^2(\pi_{2p_x})^2(\pi_{2p_y})^2(\pi_{2p_x}^*)^2(\pi_{2p_y}^*)^0$$

the filled antibonding π orbital then being able to donate two electrons to the iron atom; overlap of the filled metal $3d_{xy}$ orbital with the now empty $\pi_{2p_y}^*$ orbital could augment the initial bonding interaction:

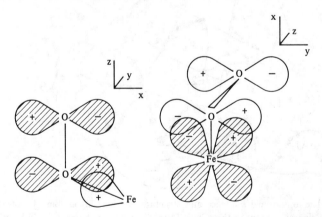

The main function of the haem pocket is probably to prevent oxidation of the iron(II) atom (which occurs rapidly when a free haem group is exposed to *both* water and oxygen); iron(III) haem groups do not reversibly carry oxygen and the life-supporting role of myoglobin and haemoglobin would thus cease if such oxidation were to occur. A secondary function is to sterically exclude all but the smallest molecules from approaching the iron atom's sixth coordination site; the site is so restricted, in fact, that even diatomic molecules are forced to bond to the iron at an angle. This is fortunate because carbon monoxide binds 25 000 times more effectively than oxygen to a *free* haem (where no steric restriction is possible) but only about

100 times more when haem is present in myoglobin or haemoglobin:

$$Fe-C{\equiv}O \qquad\qquad\qquad Fe-C{\nwarrow^{\!\!\!\!O}}$$

(very strong Fe—C bond)   (weakened Fe—C bond)

The law of mass action thus allows oxygen to force the equilibrium

$$Fe-C{\nwarrow^{\!\!\!\!O}} \quad +O_2 \quad \rightleftharpoons \qquad Fe-O{\nwarrow^{\!\!\!\!O}} \quad +CO$$

(in haemoglobin)   (in haemoglobin)

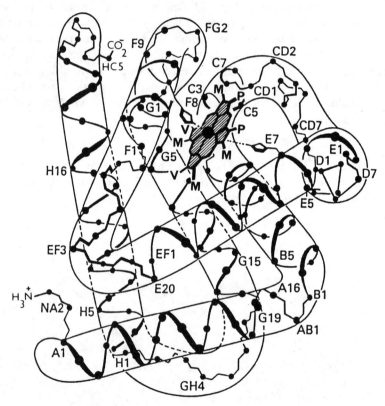

Fig. 139—The amino acid sequence and structure of whale myoglobin. About 75% of the polypeptide chain is present in the α helix form. Since globin chains differ in length but have similar conformations it is not convenient to simply number the residues in sequence. More usefully the eight helical regions are assigned letters A to H; A12 is then the twelfth residue in the A helix and CD3 is the third residue in the non-helical region between helix C and helix D. The amino end of the chain is labelled N and the carboxylic end C. The haem group, shaded grey, is located in a crevice created by protein folding (only the α carbon atoms of the amino acids are shown as black circles). Invariably, residue F8 is found to be the histidine bonded to the iron while E7 is another histidine in van der Waals contact with the iron on the other side of haem; because of their spatial positions relative to the iron atom, F8 and E7 are often called the proximal and distal histidines respectively. The methyl, vinyl and propionic acid groups on haem are labelled M, V and P. [Sequence data are from *Progress in Stereochemistry* vol. 4 (1969) p. 299; diagram adapted from *British Medical Bulletin* vol. 32 (1976) p. 195.]

well over to the right at the normally low ambient concentrations of carbon monoxide. One of the breakdown products of haem is carbon monoxide and hence without the sterically restrictive haem pocket we would slowly asphyxiate as carboxy-haemoglobin built up in the blood stream.

The four haem groups react *cooperatively* in haemoglobin: as first the α and then the β haems become oxygenated, conformational changes occur in the molecule which make it easier to add further oxygen molecules the more oxygen there is already bound. This property gives rise to the S-shaped oxygenation curve shown in Fig. 142 and ensures that maximum amounts of oxygen are transferred to myoglobin. The

1	Val	NA1	41	Glu	C6	81	His	EF4	121	Gly	GH3
2	Leu	NA2	42	Lys	C7	82	His	EF5	122	Asn	GH4
3	Ser	A1	43	Phe	CD1	83	Glu	EF6	123	Phe	GH5
4	Glu	A2	44	Asp	CD2	84	Ala	EF7	124	Gly	GH6
5	Gly	A3	45	Arg	CD3	85	Glu	EF8	125	Ala	H1
6	Glu	A4	46	Phe	CD4	86	Leu	F1	126	Asp	H2
7	Trp	A5	47	Lys	CD5	87	Lys	F2	127	Ala	H3
8	Gln	A6	48	His	CD6	88	Pro	F3	128	Gln	H4
9	Leu	A7	49	Leu	CD7	89	Leu	F4	129	Gly	H5
10	Val	A8	50	Lys	CD8	90	Ala	F5	130	Ala	H6
11	Leu	A9	51	Thr	D1	91	Gln	F6	131	Met	H7
12	His	A10	52	Glu	D2	92	Ser	F7	132	Asn	H8
13	Val	A11	53	Ala	D3	93	His	F8	133	Lys	H9
14	Trp	A12	54	Glu	D4	94	Ala	F9	134	Ala	H10
15	Ala	A13	55	Met	D5	95	Thr	FG1	135	Leu	H11
16	Lys	A14	56	Lys	D6	96	Lys	FG2	136	Glu	H12
17	Val	A15	57	Ala	D7	97	His	FG3	137	Leu	H13
18	Glu	A16	58	Ser	E1	98	Lys	FG4	138	Phe	H14
19	Ala	AB1	59	Glu	E2	99	Ile	FG5	139	Arg	H15
20	Asp	B1	60	Asp	E3	100	Pro	G1	140	Lys	H16
21	Val	B2	61	Leu	E4	101	Ile	G2	141	Asp	H17
22	Ala	B3	62	Lys	E5	102	Lys	G3	142	Ile	H18
23	Gly	B4	63	Lys	E6	103	Tyr	G4	143	Ala	H19
24	His	B5	64	His	E7	104	Leu	G5	144	Ala	H20
25	Gly	B6	65	Gly	E8	105	Glu	G6	145	Lys	H21
26	Gln	B7	66	Val	E9	106	Phe	G7	146	Tyr	H22
27	Asp	B8	67	Thr	E10	107	Ile	G8	147	Lys	H23
28	Ile	B9	68	Val	E11	108	Ser	G9	148	Glu	H24
29	Leu	B10	69	Leu	E12	109	Glu	G10	149	Leu	HC1
30	Ile	B11	70	Thr	E13	110	Ala	G11	150	Gly	HC2
31	Arg	B12	71	Ala	E14	111	Ile	G12	151	Tyr	HC3
32	Leu	B13	72	Leu	E15	112	Ile	G13	152	Gln	HC4
33	Phe	B14	73	Gly	E16	113	His	G14	153	Gly	HC5
34	Lys	B15	74	Ala	E17	114	Val	G15			
35	Ser	B15	75	Ile	E18	115	Leu	G16			
36	His	C1	76	Leu	E19	116	His	G17			
37	Pro	C2	77	Lys	E20	117	Ser	G18			
38	Glu	C3	78	Lys	EF1	118	Arg	G19			
39	Thr	C4	79	Lys	EF2	119	His	GH1			
40	Leu	C5	80	Gly	EF3	120	Pro	GH2			

Fig. 139—*continued*

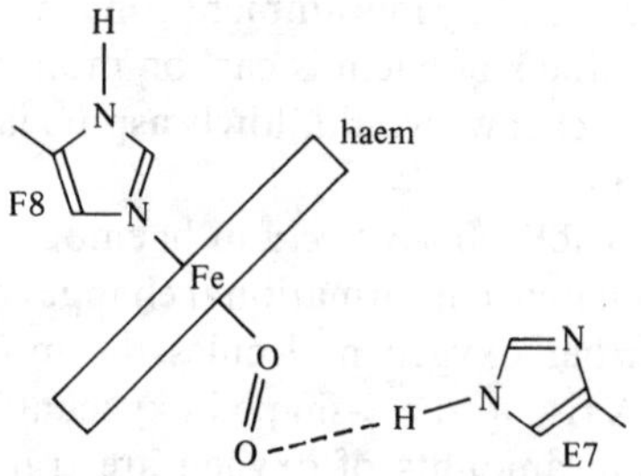

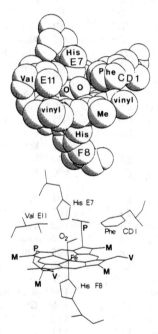

Fig. 140—Schematic representation of haem binding to oxygen. The oxygen molecule is in a tight hydrophobic pocket formed by Phe CD1, Val, E 11 and the side-chain of the distal histidine E7. This distal histidine, which is almost invariably present in myoglobins and haemoglobins from different organisms, hydrogen-bonds to the O_2 and also sterically prevents linear bonding of diatomic molecules to the iron [see *Nature* vol. 292 (1981) p. 81]. In myoglobin and the α and β chains of haemoglobin the haem pockets are so small that the distal histidines must swing out of the way before oxygen can enter.

Fig. 141—The O_2–haem complex in oxymyoglobin (with atoms represented by spheres having van der Waals radii of 1.4 Å for O_2 and 1.7 Å for all other atoms). [Reprinted with permission from S. E. V. Phillips, *Journal of J. Molecular Biology* vol. 142 (1980) p. 531.]

oxygenation curve moves to the right (oxygen released more readily) as the pH of the surrounding medium decreases, a condition which occurs in active muscle owing to carbon dioxide formation:

$$CO_2 + H_2O \underset{\text{anhydrase}}{\overset{\text{carbonic}}{\rightleftharpoons}} HCO_3^- + H^+$$

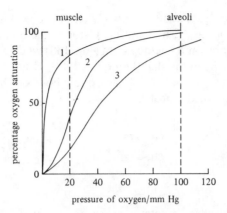

Fig. 142—Oxygenation curves for myoglobin and haemoglobin: 1, myoglobin; 2, haemoglobin at pH 7.5; 3, haemoglobin at pH 6.8. The S-shaped curves for haemoglobin show that it becomes easier to add further oxygen once some is already bound. There is about 70 times more chance of $Hb(O_2)_3$ picking up a final oxygen molecule than for deoxy-Hb to bind its first molecule.

This phenomenon, called the Bohr effect, results in even more oxygen being made available to myoglobin in active muscle than would otherwise be the case.

Iron-containing proteins play an active redox role in *respiration*, the oxidation of organic molecules such as carbohydrates and fats which occurs in the many mitochondria of cells. Mitochondria have an inner and an outer phospholipid membrane, the former being highly folded into cristae; the iron proteins are integral parts of this inner membrane. Aerobic oxidation of glucose, a typical substrate derived from starch or glycogen, involves three phases—glycolysis, oxidative decarboxylation and oxidative phosphorylation:

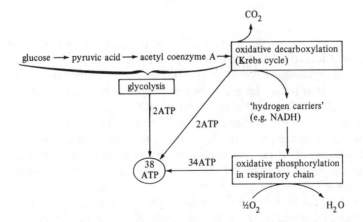

The prosthetic groups of the redox proteins required for the numerous electron transfers occurring in this chain of reactions are either ion–sulphur clusters (page 467) or haems in which the iron atoms undergo cyclical changes between the $+II$ and $+III$ oxidation states.

Five different haem proteins, collectively called *cytochromes*, are utilized in respiration: cytochromes a, a_3, b, c and c_1. Cytochrome b contains the same haem group found in myoglobin and haemoglobin (Fig. 138) whereas in the c cytochromes the protoporphyrin IX ligand is slightly modified; S—H groups of cysteine residues on the protein chain have added across the double bonds of both vinyl substituents to link haem covalently to the protein:

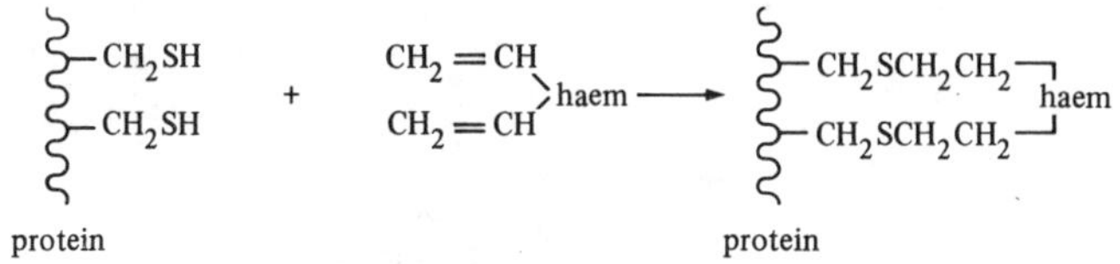

$$\begin{array}{c} \text{—CH}_2\text{SH} \\ \text{—CH}_2\text{SH} \end{array} \quad + \quad \begin{array}{c} \text{CH}_2\!=\!\text{CH} \\ \text{CH}_2\!=\!\text{CH} \end{array}\!\!\!>\text{haem} \quad \longrightarrow \quad \begin{array}{c} \text{—CH}_2\text{SCH}_2\text{CH}_2 \\ \text{—CH}_2\text{SCH}_2\text{CH}_2 \end{array}\!\!\!\bigg]\text{haem}$$

protein   protein

Cytochromes a and a_3 possess a different haem in which one methyl is replaced by a formyl group and a vinyl by a long hydrocarbon chain:

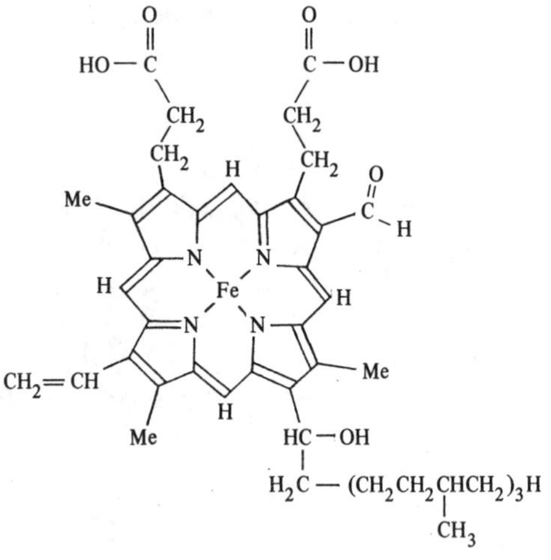

Cytochrome c is by far the most studied of these redox systems (Fig. 143). Its protein chain is folded so as to create a haem pocket and also to present the iron with additional, strong-field fifth and sixth ligands, an imidazole nitrogen atom from His 18 and a sulphur from Met 80. Although the propionic acid groups are enclosed within the protein (unlike myoglobin and haemoglobin), one haem edge is left exposed on the surface. In both the oxidized (Fe^{III}) and reduced (Fe^{II}) forms of cytochrome c the iron–ligand distances are virtually identical, the iron itself adopting the low-spin state in each case; these two properties ensure that the activation energy of any ensuing redox process is minimized, the electron exchange probably occurring via an outer-sphere mechanism closely associated with the exposed haem edge.

The final steps in the respiratory chain involve an enzyme, cytochrome oxidase, which is a large protein with 7–13 sub-units spanning the inner mitochondrial membrane. Incorporated in the molecule are four different metals—iron, copper, zinc and magnesium—the role of the latter two being far from clear. The iron atoms are present as cytochromes a and a_3, both having a closely associated copper Cu_a and Cu_{a_3}. Reduced cytochrome c releases its electrons to the Fe_a–Cu_a site while the

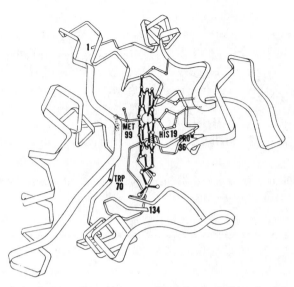

Fig. 143—Schematic diagram of cytochrome c from the denitrifying bacterium *Paracoccus denitrificans*. The —NH$_2$ and —COOH termini of the protein chain are at positions 1 and 134 respectively. The haem, with its exposed edge clearly visible, remains planar in both the (singly charged) oxidized and (uncharged) reduced states. [Reproduced with permission from *Annual Review of Biochemistry* vol. 46. © 1977 by Annual Reviews Inc.]

Fe$_{a_3}$–Cu$_{a_3}$ unit binds an oxygen molecule and reduces it to water:

How is the electron transferred from cytochrome c to cyrochrome a of cytochrome oxidase? Clearly, since the haem iron atoms are buried within the protein chain of both systems an inner-sphere mechanism can be considered highly unlikely, especially as there is no evidence of group transfer. An attractive suggestion is that an encounter complex forms by attraction of positive surface charge near the exposed haem edge of cytochrome c to a similar, but negative, area on the oxidase; in this way the c and a haems are thought to be held parallel and in close proximity. During thermally induced vibrations and/or distortions, the π systems of the interacting haems would be able to approach closely enough for 'overlap' to occur, so allowing electron transfer via an outer-sphere mechanism; Fig. 144.

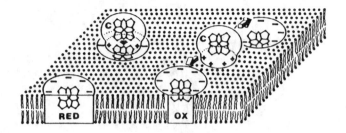

Fig. 144—Schematic diagram of possible interactions between cytochrome c and the membrane-bound oxidase (OX) and reductase (RED). The cytochrome c is shown weakly bound, electrostatically, to negative charges on the membrane surface and is thus able to move over the membrane in two dimensions, as shown by the arrows. Complementary ionic interactions orient the cytochrome c molecule so that it is oxidized (shown top left) or reduced by a mechanism involving direct interaction of its haem with prosthetic groups on the oxidoreductases. [Reproduced with permission from *Annual Review of Biochemistry* vol. 46. © 1977 by Annual Reviews Inc.]

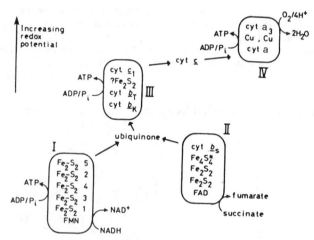

Fig. 145—Components of the mitochondrial respiratory chain. Complex I contains five centres which are thought to be of the Fe_2S_2 type. Complex II is tentatively described as containing two Fe_2S_2 centres and one of the $Fe_4S_4^*$ type. Complex III contains two cyt b and one cyt c as well as an Fe_2S_2 centre. Complex IV, cytochrome oxidase, contains cyt a and cyt a_3, each with a closely associated copper. P_i is inorganic phosphate; the quinone acts as a mobile electron carrier. The electron flow is shown only schematically and is undoubtedly much more complex, but the diagram illustrates how the process is accomplished by a series of graded redox processes. A more comprehensive diagram and discussion is given in *Annual Review of Biochemistry* vol. 54 (1985) p. 1015. [Reproduced with permission from *Metals in Biochemistry* by P. M. Harrison and R. J. Hoare (Chapman and Hall), 1980).]

The rapid lethal action of HCN is due to the formation of a complex, protein–Cu–CN, with copper(I) atoms present in cytochrome c oxidase; the latter, considered to be the single most important metabolic enzyme in mammalian cells, is thus rendered inactive and mitochondrial respiration ceases.

Table 59—Some common copper enzymes

Enzyme	Molecular mass	Metal	Function
Caeruloplasmin	132 000	8Cu	Oxidase; iron release from liver
Tyrosinase	4 × 32 000	4Cu	Hydroxylation of tyrosine (see text)
Dopamine-β-hydroxylase	8 × 36 000	4–7Cu	Hydroxylation of dopamine:
Superoxide dismutase	2 × 16 500	2Cu, 2Zn	Disproportionation of superoxide ion (see text); metal storage (?)
Amine oxidases	2 × 90 000	2Cu	Oxidation of primary amines to aldehydes $-CH_2NH_2 \rightarrow -CHO$
Cytochrome oxidases	c. 150 000	2Cu + 2Fe-haem	Reduction of oxygen to water in mitochondria (see page 475)
Plastocyanin	10 500	1Cu	Electron transfer in plants, algae and bacteria
Laccase	120 000	4Cu	p-diphenols $+ O_2 \rightarrow p$-quinones $+ H_2O$
Lysyl oxidase	120 000	4Cu	Cross-linking of collagen and elastin in connective tissue

COPPER

There are about 60 mg of copper in an adult human, mainly present in a range of metalloproteins distributed throughout the body. The average daily diet provides 0.5–1.0 mg of copper which, after absorption through the intestine, is carried to the liver by two transport proteins, serum albumin and transcuprein, present in the plasma. In the liver most of the copper is incorporated into caeruloplasmin which re-enters the bloodstream to function as a long-term copper transporter and an antioxidant. The inherited Wilson's disease, in which a build-up of high and eventually lethal concentrations of copper occurs in many organs of the body, may be related to a defect in caeruloplasmin biosynthesis because sufferers often show a very marked deficiency of this protein (see page 495 for treatment).

A further important function of caeruloplasmin is to release, and oxidize, iron from liver storage sites prior to its complexation by trans-ferrin and subsequent delivery to developing red blood cells; hence there is a possibility that a low dietary intake of copper can result in anaemia. The serum concentration of caeruloplasmin must fall to an extremely low level ($< 10^{-9}$ M) before iron metabolism is seriously disturbed, and anaemia arising from copper deficiency is not normally encountered in humans, even those suffering from Wilson's disease. However, copper-associated anaemia can occur in farm animals; copper deficiency also causes many other symptoms, including heart failure, loss of wool crimp in sheep, demyelination of the spinal cord and loss of pigmentation. The latter probably results from an insufficient supply of tyrosinase, a copper enzyme required for the production of the pigment melanin from tyrosine; albino mammals are unable to synthesize any tyrosinase and

Table 60—Some functions of caeruloplasmin

Caeruloplasmin	Molecular weight 132 000; 1064 amino acid residues with seven or, more likely, eight copper atoms per molecule. Accounts for 70%–80% of total copper in normal human plasma
Functions	Copper transporter in plasma, the ultimate recipients being several vital copper enzymes
	Mobilization of iron to transferrin
	Major serum antioxidant; the antioxidant activity possibly increases the life span of circulating red blood cells
	Regulation of serum biogenic amines; possible role in inflammatory response; growth promoter for some cells
	Stimulation of angiogenesis in cornea and in cell culture
Deficiency	Results in anaemia owing to reduction of mobilization of iron from liver storage sites (general copper deficiency aggravates anaemia owing to impairment of mitochondrial iron metabolism resulting in reduced haem biosynthesis)

[Data taken from E. Frieden, *Clinical Physiology and Biochemistry* vol. 4 (1986) pp. 11–19.]

are consequently devoid of all pigmentation:

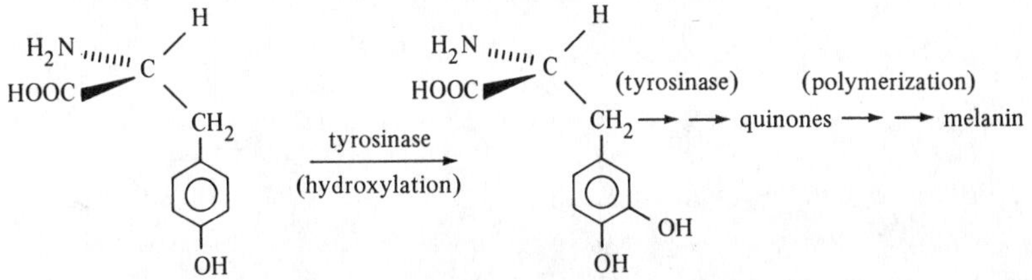

tyrosine, one of the
essential amino acids

dihydroxyphenylalanine
(DOPA)

It appears likely that in tyrosinase the copper atoms operate in pairs to first bind an O_2 molecule, reduce this to peroxide and utilize one oxygen atom to generate the DOPA, which, after further oxidation, is released as a quinone:

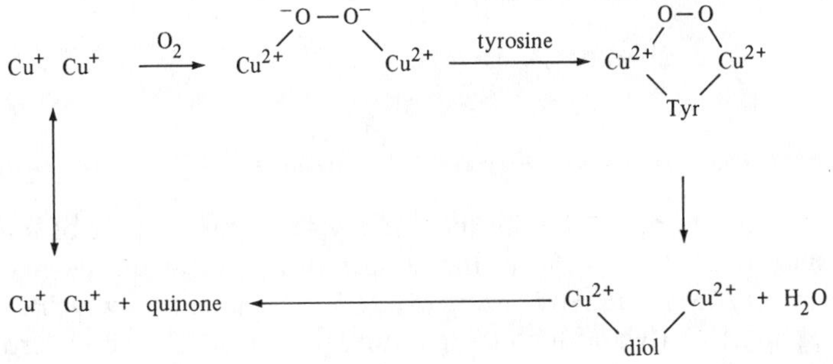

A similar binuclear active site occurs in the oxygen-carrying protein haemocyanin found in the blood of molluscs and arthropods; this protein is colourless in the deoxygenated, copper(I) state but assumes a blue colour on exposure to oxygen owing to oxidation of the copper atoms. In the oxygenated form the Cu(II) atoms are thought to have square pyramidal coordination, but on deoxygenation a bridging hydroxyl (or water) oxygen probably moves away from the active site to leave the copper(I) atoms three-coordinated to imidazole groups of histidine residues:

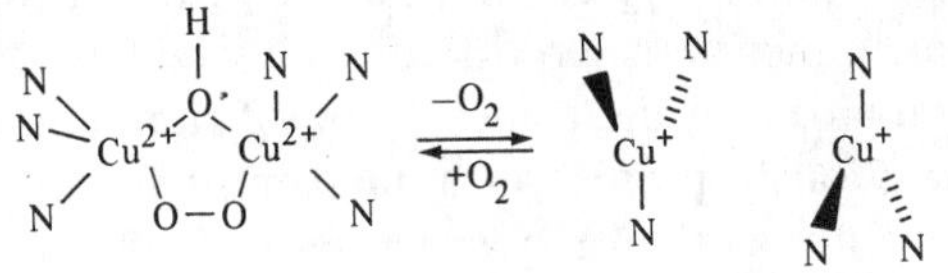

Cu - - - Cu distance 3.67 Å

Cu - - - Cu distance 3.39 Å

The O---O stretching frequency occurs at about 750 cm^{-1}, a value characteristic of peroxides and hence in agreement with the essentially complete transfer of charge shown in the structure above. Although two copper(II) atoms are present in oxygenated haemocyanin they are antiferromagnetically coupled to give a diamagnetic complex.

Copper–zinc superoxide dismutase consists of two identical sub-units, each having a binuclear active site in which the copper and zinc are coordinated to the separate nitrogen atoms of a common imidazole group of histidine 61:

[Reproduced with permission from *Chemistry in Britain* (1986) p. 839.]

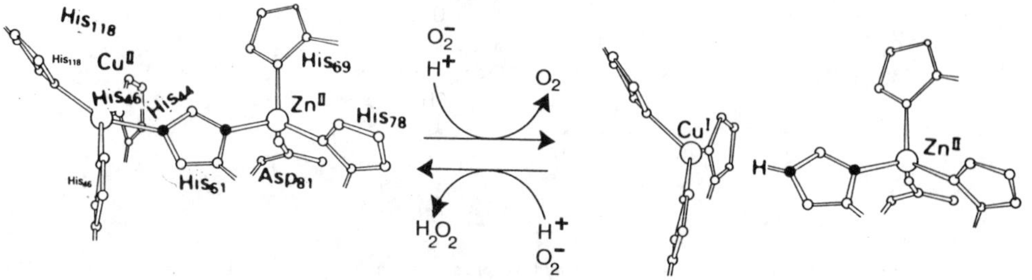

Surprisingly, the enzyme remains almost fully active even when the zinc atom has been removed, from which it has been deduced that this metal adopts only a structure-forming role. The copper oscillates between the +I and +II oxidation states in a cycle of reactions such as

$$\sim Cu^{II} + O_2^- \longrightarrow \sim Cu^{I} + O_2$$

$$\sim Cu^{I} + O_2^- + H^+ \longrightarrow \sim Cu^{II} + H_2O_2 \xrightarrow{\text{catalase}} H_2O + \tfrac{1}{2}O_2$$

Microbiological mining is now an industrially important aspect of 'bioinorganic chemistry' since more than 10% of the copper produced in the USA is currently obtained from very-low-grade ores using extraordinary microbes which can thrive in the prevailing high acidity and high temperature (above 50°C). The microbes obtain their energy by oxidizing iron(II) or sulphide ions and under these conditions copper is leached from billions of tons of rock to give a dilute solution of copper(II) sulphate; treatment of this solution with scrap iron precipitates out the copper.

COBALT

The cobalt-containing vitamin B_{12} co-enzyme is considered to be essential for all growing tissue in man. Since it is produced exclusively by micro-organisms (for example, some food moulds and the rumen fauna of ruminant animals) we have to rely on the traces fortuitously present in animal meat for our daily requirements. Dietary deficiency due to stomach surgery, strict vegetarianism or an inherited inability to absorb the vitamin from the gut results in anaemia owing to decreased production of red blood cells. An injection of only one-millionth of a gram of vitamin B_{12} to such patients can result in a detectable increase in their blood cell count.

Vitamin B_{12} co-enzyme has a cobalt(III) atom placed at the centre of a macrocyclic ligand derived from corrin and is coordinated to four ring nitrogen atoms and a fifth nitrogen from a benzimidazole group attached to one of the corrin substituents. The sixth position is normally occupied by the 5′ carbon atom of deoxyadenosyl (Fig. 146) but this carbon can be replaced by other groups to give a range of compounds collectively known as cobalamins.

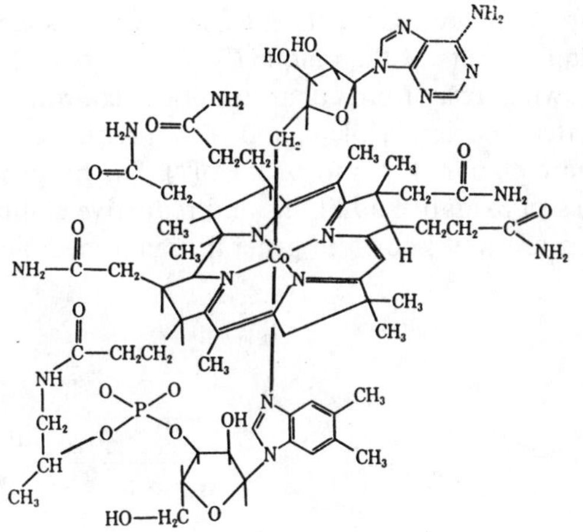

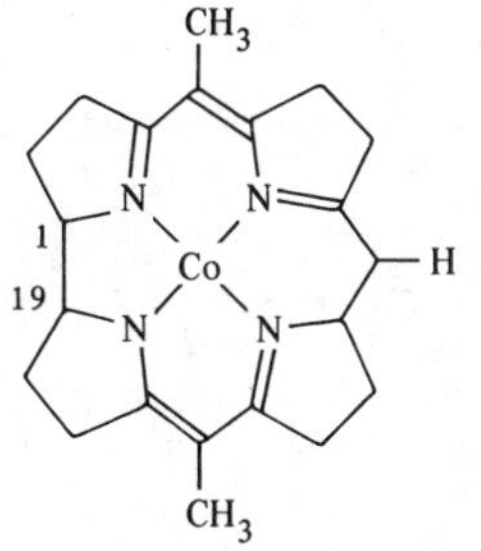

Fig. 146—Vitamin B_{12} co-enzyme, and corrin ring system showing the unique C(1)—C(19) bond.

Cobalamins mediate in two specific reaction types: methyl transfer reactions and stereospecific vicinal interchanges. As an example of the former, methylcobalamin acts as the co-enzyme of homocysteine transmethylase in the production of methionine:

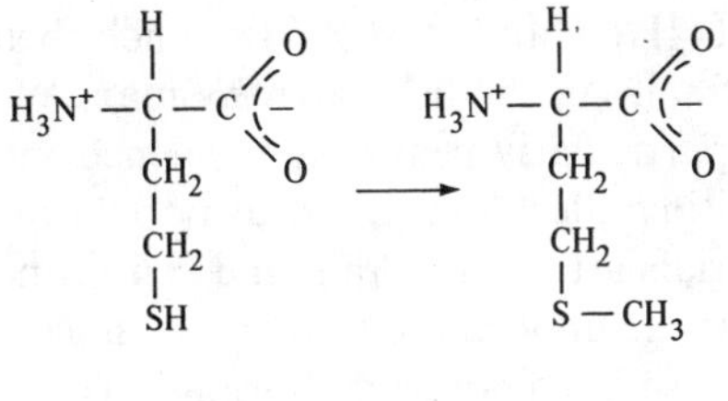

homocysteine methionine

An intriguing mechanism for the B_{12}-catalysed vicinal rearrangement

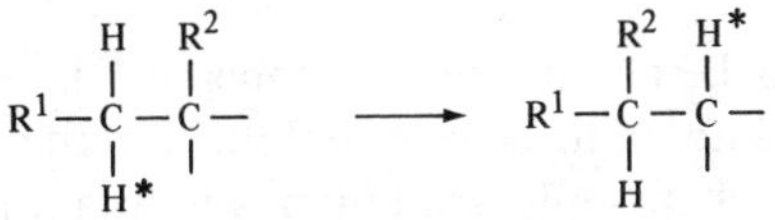

has been put forward by Corey, Cooper and Green which suggests that cleavage of the corrin ring initially occurs at the unique C_1—C_{19} bond. This allows one of the corrin nitrogens to swing away from cobalt leaving a coordination position vacant; the weakly coordinated benzimidazole ligand is also required to dissociate from the metal to leave a square planar, d^8 complex of Co(I). The proposed reaction sequence then involves a series of oxidative additions and reductive eliminations at the cobalt atom (such reactions are well known in general organometallic chemistry):

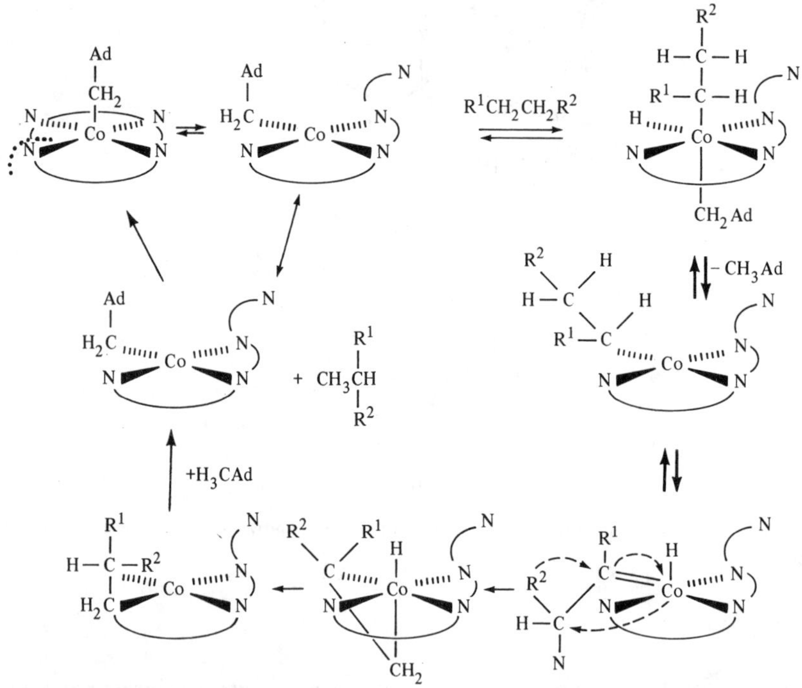

There are some circumstances in which B_{12} derivatives can pose a health hazard. Under the anaerobic conditions prevailing in the mud of polluted rivers and estuaries the methylcobalamin of some micro-organisms is able to methylate traces of inorganic mercury compounds to CH_3Hg^+ and $(CH_3)_2Hg$ which, being lipid-soluble, enter the food chain to end up finally in the fish eaten by man. When mercury pollution from industrial sources is particularly heavy, as happened in Minimata Bay, Japan during the 1950s, the resulting effects on the local population can be catastrophic. The ingested organomercurials attack the brain and central nervous system causing, ultimately, coma and death; the unborn are also at risk since the mercurials are able to pass the placenta and cause horrendous malformations.

MOLYBDENUM

Molybdenum is unique in being the only member of the second- and third-row transition metals which is known to be essential for life; this is in direct contrast to the case of the first row where only scandium and titanium appear to have no demonstrated biological role. The plant kingdom, and hence indirectly all animals, has an absolute requirement for two molybdenum-containing enzymes involved in nitrogen utilization, *nitrate reductase* (which reduces NO_3^- to NO_2^- as the first step in supplying nitrogen to non-leguminous plants) and *nitrogenase* (found in blue-green algae, some free-living bacteria and the bacteria present in legume root nodules); so important is molybdenum that the application of as little as 5 g of it per hectare of land has a very pronounced beneficial effect on the plant life.

Probably more research effort and money have been expended in recent years on the study of nitrogen 'fixation' than on all other topics in bioinorganic chemistry combined. Man now fixes more nitrogen annually than that fixed in the biosphere but the industrial processes involved are very energy-intensive. When we have a full understanding of the nitrogenase enzyme, which converts atmospheric nitrogen directly into ammonia, it may prove possible to make efficient model compounds which will fully mimic its reactions under ambient conditions.

Nitrogenase is an extremely complex molecule composed of two oxygen-sensitive proteins called Component I (or the MoFe protein) and Component II (or the Fe protein). The brown *MoFe protein* is an '$\alpha_2\beta_2$' tetramer with a molecular weight in the range 220 000–245 000 depending on source; it contains two molybdenums and about 30 iron atoms associated with an approximately equal number of acid-labile sulphur groups (these groups release H_2S on treatment with dilute acid, showing that they are free 'sulphide' atoms, and are shown in the formulae below as S*). Spectroscopic studies suggest that the Mo and Fe atoms are present as four Fe_4S_4 cubes (unusual in that their reduced form contains only Fe(II)) and two FeMo-cofactors with a stoichiometry of $MoFe_{6-8}S_{4-9}$. The yellow *Fe protein* is dimeric with a molecular weight of 55 000–75 000 and it probably has an $Fe_4S_4^*$ cube linked to the protein chains by two cysteine groups from each monomeric unit. As might be expected from the presence of these metal–sulphur clusters, both the MoFe protein and Fe protein are capable of donating and accepting electrons (see page 467).

The detailed mechanism by which nitrogenase produces ammonia is not known but probably involves initial coordination of the nitrogen molecule to molybdenum by displacement of H_2 from a reduced metallohydride; for every N_2 reduced to $2NH_3$ at least one equivalent of H_2 is evolved. Kinetic studies with nitrogenase, and the chemistry of known Mo complexes, suggest that reduction of coordinated N_2 proceeds via a hydrazido(2−) intermediate: $Mo\!=\!\underset{\alpha}{N}\!-\!\underset{\beta}{N}H_2$. The strong triple bond of N_2 is weakened by progressive build-up of multiple-bond character between Mo and α nitrogen atom with concomitant protonation of the β nitrogen. Like the industrial formation of ammonia, this biological reaction is very energy-demanding in that 12 moles of ATP are required to reduce each mole of N_2 (Fig. 147).

A number of other molybdenum enzymes are essential to animals, including sulphide oxidase and xanthine oxidase, but unfortunately none of their intimate mechanisms is fully understood at the present time, mainly owing to an incomplete knowledge of their structure; however, it appears likely that the molybdenum atoms cycle between the $+IV$, $+V$ and $+VI$ oxidation states. Sulphite oxidase is a dimeric enzyme of molecular weight 55 000 having two molybdenum atoms and two b cytochromes; it oxidizes toxic sulphite ions into sulphate which is then excreted in the urine. Water has been shown to supply the required oxygen atom with cytochrome c acting as a final acceptor of the electrons transferred to it from the active site via the b

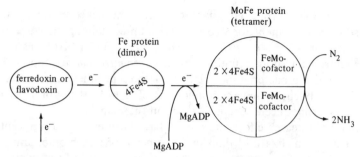

MoFe protein
(tetramer)

Fe protein
(dimer)

ferredoxin or
flavodoxin

e^-

4Fe4S

e^-

2 × 4Fe4S | FeMo-cofactor

2 × 4Fe4S | FeMo-cofactor

N_2

$2NH_3$

MgADP

MgADP

e^-

MgATP and electrons (e^-) arise from normal metabolic processes. Eight electrons transferred and 16 ATP molecules hydrolysed to release $2NH_3$. [See *New Scientist* (18 June 1987) p. 59.]

N_2 probably binds to a single FeMo-cofactor by hydride displacement releasing H_2. Net reaction is

$$N_2 + 8H^+ + 8e^- \longrightarrow 2NH_3 + H_2$$

16ATP

16ADP

Some nitrogenases contain vanadium in place of molybdenum

Fig. 147—Simplified diagram showing the action of nitrogenase.

cytochromes:

$$SO_3^{2-} + H_2O \rightarrow SO_4^{2-} + 2H^+ + 2e^-$$

Xanthine oxidase contains four Fe_2S_2 clusters in addition to two molybdenum atoms and it catalyses the conversion of hypoxanthine to uric acid during degradation of purines:

NH_2

ribose —

ribose —

H

hypoxanthine

xanthine
oxidase

xanthine
oxidase

OH

H H

uric acid

H H

ULTRA-TRACE ELEMENTS WHICH OCCUR IN PLANTS AND ANIMALS†

Several elements which occur at very low concentrations in living organisms are currently being studied for both essentiallity and toxicity.

Lithium

Lithium salts are used to relieve symptoms of manic-depressive psychosis, possibly by altering an abnormal electrolyte distribution in the brain. Goats fed on a lithium-free diet showed 28 % less weight gain than control animals and their females had reduced fertility; hence lithium may prove to be an essential element although it occurs only to the extent of 0.03 ppm in man.

Rubidium

Adults contain up to 300 mg of rubidium cations widely distributed throughout the body; like caesium, rubidium can interchange with potassium but shows no toxic behaviour. At present it is not thought to be an essential element but an apparently close biological control of rubidium levels in the blood of children (at $c.$ 12 ppm) may be indicative of some, as yet unknown, activity.

Strontium and barium

These cations, present to the extent of 4 and 0.3 ppm respectively in man, are relatively non-toxic and in small amounts can replace calcium, for example, in bone. This is particularly dangerous in the case of ^{90}Sr, a highly radioactive isotope produced in nuclear explosions, because its radiation can damage the bone marrow, causing leukaemia or bone cancer. The amount of ^{90}Sr released from nuclear tests is not inconsiderable; during the relatively dirty tests of the 1950s an estimated 63 kg of strontium-90 (3.4×10^{17} Bq) were released into the environment together with 5.6×10^{18} Bq of ^{137}Cs. Since caesium can duplicate some of potassium's biochemistry it concentrates in body cells and becomes a particular hazard to the genes.

Most surprisingly, marine creatures called radioralians use strontium sulphate as their outer skeletal material.

Boron

Boron is essential for healthy plants, the early symptoms of deprivation being detectable in growing root tips as little as 6 h after boron removal. The biochemical roles of boron are not fully understood even 60 years after its recognition as an essential element, although it is known to be involved in nucleic acid synthesis, possibly linked to adequate provision of the pyrimidine nucleotides; boron also plays a part

† For a comprehensive coverage of this topic see *Biochemistry of the Essential Ultratrace Elements* by E. Frieden (ed.) (Plenum Press), from which much of the following data were derived with permission.

in carbohydrate metabolism, hormon action and membrane formation. There appears to be no essential function for boron in animals.

Boron has two isotopes, ^{10}B and ^{11}B, of which the former makes up about 20% of natural boron and has a high affinity for low-energy neutrons:

$$^{10}_{5}B + ^{1}_{0}n \rightarrow ^{7}_{3}Li + ^{4}_{2}\alpha$$

For a short while after injection into human patients many boron compounds preferentially collect in growing tumours, especially those of the brain. By irradiating the tumour area with low-energy neutrons, which do relatively little biological damage, the boron-10 isotopes are transformed into lithium atoms and at the same time cause the release of high-energy α particles actually within the tumour itself; because the α particles travel only short distances in tissue the damage is very localized and does not affect healthy cells.

Aluminium

Although aluminium is the third most abundant element in the Earth's crust after oxygen and silicon it does not seem to be essential for any known life-process; one possible reason for this is that at physiological pH it is only sparingly soluble (Fig. 148). However, there is increasing concern that it is a highly toxic element, particularly towards humans suffering from kidney failure. Trace amounts of aluminium in water used to dialyse the blood of such patients appear to cause *dialysis dementia* (by destroying brain function) which can finally end in convulsions and death. The symptoms are similar to Alzheimer's disease (senile dementia) and both result, for example, in tangled and twisted patches within the nerve cells of the cortex—the part of the brain concerned with memory and thinking. Some, or most, of the aluminium in the water supply comes from aluminium salts used to clarify

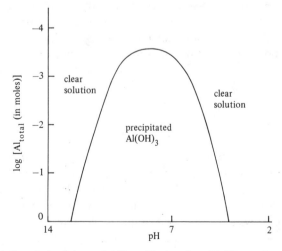

Fig. 148—Variation of aluminium solubility with changing pH. [Data taken from *Journal of Inorganic and Nuclear Chemistry* vol. 33 (1971) p. 791.]

water in surface reservoirs and is able to pass through dialysis membranes directly into the patients' blood plasma. It is recommended that water used for dialysis should contain less than 50 μg of aluminium per litre (c. 50 ppb). *In vitro* experiments show that one effect of the highly charged aluminium ion is to compete with magnesium for coordination to ATP during phosphate transfer reactions, but it is not known if such enzyme inhibition is responsible for brain damage.

It appears to be an extraordinary fact that patients with kidney failure do not suffer from dementia as a result of the huge amounts (3–4 g per day) of orally ingested aluminium hydroxide given to maintain serum phosphate levels within normal limits. However, continued high doses of aluminium can result in osteomalacia, manifested by brittle bones; such patients also have high plasma levels of calcium (hypercalcaemia). Both of these symptoms would occur if aluminium accumulated at the calcification front of bone where it could then inhibit the take-up of available calcium from the serum.

Silicon

Silicon is the second most abundant element after oxygen but many of its compounds are so insoluble that it would appear to be unavailable to life-forms. For this reason it was long considered that silicon was not an essential element, although photosynthesizing diatoms were known to form their shells from silica, perhaps to allow more light through than the more usual, but opaque, calcium carbonate. Relatively recently, silicon has been shown to be intimately involved in the structural framework of connective tissue such as bone, cartilage and skin. Typically, silicon-deficient laboratory animals and birds have stunted growth, retarted skeletal development and many atrophied organs; silicon supplementation results in dramatic increases in growth rates of 25%–50%. Connective tissue in animals contains between 1 and 5 g per kilogram of proteoglycans and it is in this material that the silicon is concentrated. Proteoglycans are complexes containing proteins and acidic chains of repeating monosaccharide residues, ABAB.. etc.; the role of silicon may be to act as a cross-linking agent in the monosaccharide region. During bone growth, silicon is found at the active calcification sites but its concentration falls off as the bone mineral 'matures' and reaches the composition of apatite until, finally, no silicon can be detected.

Tin

This element is widely spread throughout the organs of the human body, except the brain, at the low levels of 0.5–1.0 ppm; there appears to be little or no sound evidence for considering tin to be essential. Inorganic tin has low oral toxicity but organo-tin derivatives, used industrially as stabilizers in the polymer industry and for making anti-fouling paints, are often very toxic. They attack the central nervous system of animals causing, ultimately, paralysis and death; trialkyltins are useful bactericides and fungicides.

Lead

It has been stated that lead compounds adversely affect any life-procceses one cares to study; however, work in the early 1980s on animals fed on a lead-free diet apparently shows that a very low lead level is required for full health. In its toxic role lead seems to affect the brain most rapidly, and especially so in young children. Currently strenuous efforts are being made to keep the lead content of drinking water below 50 ppb, although about 80% of ingested lead is taken in the diet; luckily less than 10% of lead is absorbed from this source. Lead accumulates in the bones where Pb(II) replaces calcium, the amount increasing with age; although such deposited lead is not harmful the danger is that it can be released in old age or during high fever.

Arsenic

The popular notoriety afforded to arsenic as the ultimate poison makes its newly discovered role as an essential element that much more surprising. In laboratory tests, animals fed on an essentially arsenic-free diet showed rough fur, poor growth and low fertility; lactating goats tended to die suddenly of heart failure. No specific function of arsenic is yet known but it appears to be involved in arginine and zinc metabolism in mammals; arseno-lipids have been found in some marine organisms. Several micro-organisms methylate inorganic arsenic compounds to trimethylarsine, $(CH_3)_3As$, possibly via a number of arsenic–oxygen intermediates:

$$AsO_3^{3-} \rightarrow CH_3AsO_3^{2-} \rightarrow (CH_3)_2AsO^{2-} \rightarrow (CH_3)_3AsO \rightarrow (CH_3)_3As$$

Methyl-arsenicals, particularly those containing $(CH_3)_3As^+$ groups, also occur in several compounds of biological origin.

Selenium

This was acknowledged to be an essential element in the 1950s when selenium-deficient rats were shown to suffer from dietary liver necrosis. At about the same period it was found that the widespread disease in chickens, exudative diathesis, could be prevented by supplementation of the diet with as little as 0.1 ppm of selenium. Deficiency problems have even been recognized in man; a common and often fatal heart condition in some regions of China was eventually traced to a selenium deficiency in the local soil and, hence, crops. After selenium had been artificially added to the diet from 1974, the disease gradually decreased in scale until by 1980 it had essentially been eliminated.

Selenium, on the other hand, can be extremely toxic in only relatively small amounts. Fodder containing less than 50 ppm can initially cause loss of appetite, malformation and emaciation, but in the long term more severe symptoms such as liver cirrhosis, atropy of the heart and anaemia become evident. Higher selenium levels (> 100 ppm) lead to a condition in cattle called blind staggers in which the affected animals stumble over obstacles owing to impaired vision and weak forelegs; respiratory failure, paralysis and death finally occur if the animals continue to graze in the same area.

The breath of such animals smells like garlic, probably owing to traces of dimethyl selenide in the exhaled air.

Biochemically, selenium is often found to replace its congener sulphur in the amino acid residues of proteins. An example is the enzyme glutathione peroxidase which has a molecular weight of 84 000 and one selenium atom in each of its four sub-units; the selenium is present as selenocysteine residues in the polypeptide chains:

$$\underset{\text{HOOC}}{\overset{\text{H}}{\diagdown}}\text{C}\overset{\diagup\text{CH}_2\text{SeH}}{\underset{\diagdown\text{NH}_2}{}}\qquad\text{(selenocysteine)}$$

This enzyme is thought to protect the lipids in cell membranes, and particularly those in the liver, from attack by peroxides and oxygen radicals. The recommended human intake for selenium is 60 μgm per day, but levels only 10 times larger are considered to be toxic.

Fluorine

Although small quantities of fluorine occur in animals, it is not known with certainty if it is an essential element, though in experimental animals having a low iron intake, fluoride ions enhance absorption of iron; the effect disappears when the amount of dietary iron is increased. In man, about 99% of body fluoride occurs in bone and teeth where it is able to substitute some OH anions in hydroxyapatite because their ionic radii are so similar ($r_F = 1.29$ Å; $r_{OH} = 1.33$ Å). Fluorapatite, $Ca_5(PO_4)_3F$, contains 37 700 ppm of F^- whereas adult bone contains only 1000–5000 ppm, the level depending on age and available fluoride in food or water. Most of the substituted hydroxy groups in apatite are replaced by the fluoride ions after the bone or tooth enamel has been formed and this had led to the suggestion that, although the total fluoride content of tooth enamel is low, the incorporation of relatively large amounts in the surface layers may reduce the enamel solubility and help prevent dental caries. However, it appears likely that the actual role of fluoride in controlling caries is more complex than just reducing the solubility of tooth enamel by substitution. Excess fluoride inhibits the action of enzymes such as enolase and pyrophosphatase, probably owing to substituting of F for an OH group at, or near, the active sites; continued ingestion of fluoride over many years (c. 50 mg per day) leads to the painful bone condition osteosclerosis in which nearly 50% of the hydroxy groups in apatite may have been replaced by fluoride.

Bromine

Bromide ions, which occur to the extent of 1–9 ppm in human tissues, are not considered to be essential for bacteria, plants or animals. However, there appears to be a definite correlation between poor sleep pattern and lowered serum-bromide concentration in renal patients undergoing haemodialysis. The vanadium-containing enzyme bromoperoxidase, present in some seaweeds, is able to brominate organic compounds.

Iodine

Most of the 20 mg of iodine in humans is bound to a protein, thyroglobin (molecular weight 660 000), in the thyroid gland. Tyrosine residues on thyroglobin are mono- and di-iodinated by iodide ions under the influence of the enzyme thyroid peroxidase (TP) in the presence of hydrogen peroxide; coupling of these iodo-aromatic residues, followed by cleavage from the protein, gives three thyroid hormones:

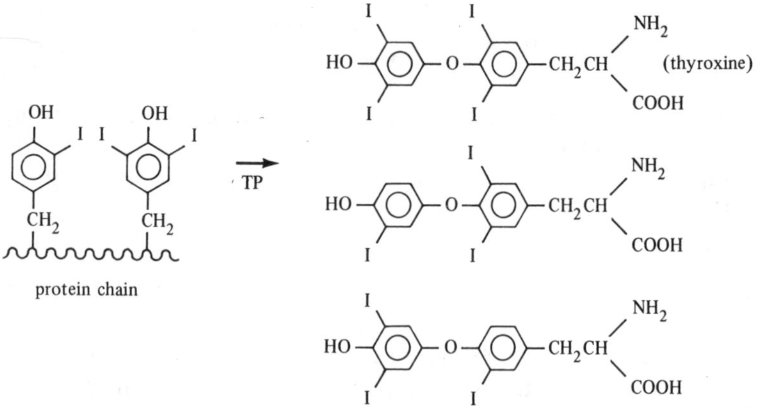

The hormones are released into the bloodstream where they are non-covalently bound to a number of proteins, mainly globulin and albumin; one of their main functions is to control body temperature. Iodine deficiency, which results in a lack of energy, mental sluggishness and a feeling of cold, can usually be avoided by the ingestion of about 200 μgm of iodide per day. Continued absence of iodine in the diet can lead to unsightly swellings of the neck (goitre) as the thyroid gland enlarges; estimates have suggested that at least one million people suffer from this disease, particularly in the undeveloped countries. Disturbance of thyroid activity impairs reproduction in mammals and can lead to reduced growth and cretinism in newborn children.

Iodine-131, produced in atomic tests or accidents at nuclear power stations, is particularly dangerous to man because it becomes localized in the thyroid and can produce cancers. One recommendation to combat the effects of radio-iodine has been to eat potassium iodide tables, using the inactive iodine-127 isotope as a diluent; however, iodide is toxic when ingested over long periods.

Cadmium

After intensive study, no specific role has been found for cadmium although it occurs to a small extent in many animals, including man, where it is found in the kidneys (30–40 ppm) and liver. Like its congener zinc, cadmium has a strong affinity for sulphur ligands and hence is bound strongly in the body to a protein called metallothionein. As its name suggests, this protein contains high levels (c. 30%) of cysteine residues and virtually all the SH groups are bound to tetrahedrally coordinated cadmium (or other metals). Ligand-free cadmium is extremely toxic to the kidneys and it has been speculated that metallothioneins may be produced by an animal's defence system to detoxify the metal; however, in long-term cadmium exposure, cadmium–metallothonein itself causes kidney damage so it is possible that binding of cadmium is fortuitously carried out by a ligand meant for other, more essential, metals such as zinc or copper. The general toxicity of cadmium appears to be related to disturbances it causes in the metabolism of several essential elements, including Zn, Cu, Fe, Se, Ca and Mn.

Transition metals

Vanadium is essential for healthy growth of chicks and rats; although its essentiality has not yet been demonstrated in man, a number of workers argue that it probably is. The human daily intake is thought to be of the order of 20 μgm per day. Some marine species called tunicates contain protein-bound vanadium in their blood at a concentration factor of 10^6 relative to the surrounding water. Some nitrogenases contain vanadium in place of the more usual molybdenum.

Chromium is essential to animals, including man, its primary role being connected with insulin action during glucose metabolism; thus chromium deficiency, like diabetes, results in poor glucose tolerance. In excess, chromium compounds are toxic, several being carcinogenic.

Of the 10–20 mg of manganese present in man, about 25% is to be found in the skeleton; 2 mg per day is considered a sufficient dietary intake. It is an essential part of a few metallo-enzymes such as arginase, pyruvate carboxylase and superoxide dismutase. The latter has a molecular weight of around 80 000 and consists of four sub-units, each of which binds one manganese atom; this enzyme's function is to destroy O_2^- ions which would otherwise oxidize membrane lipids. Manganese also loosely associates with, and activates, several other enzymes, but during manganese deficiency, another metal, thought to be magnesium, seems able to take over the activation role.

Nickel has been found to be essential for plants, micro-organisms and a number of animals. Plant urease, for example, contains strongly bound nickel and is responsible for removing urea via the reaction

$$(NH_2)_2CO + H_2O \rightarrow CO_2 + 2NH_3$$

Abnormal uptake of nickel by plants has been used in the technique of biogeochemical prospecting when the metal is not too deeply buried; one Australian shrub, *Hybanthus floribundus*, can sometimes have as much as 20% nickel in its ash when growing in metal-rich areas.

CHELATION THERAPY

As shown in Table 61, the ingestion of metals causes a number of severe diseases in man and, although relatively large amounts are required to cause clinical symptoms, even the essential elements can become extremely toxic when present in excess. In some cases it is possible to use chelating ligands to bind strongly with the rogue metal and allow its excretion before too much damage is done. Among the first ligands to

Table 61—Some metal-associated diseases in man

Metal	Deficiency	Toxicity
Al	Non-essential element	Neurological disturbance; osteomalacia
Cd	Non-essential element	Kidney and liver necrosis
Co	Anaemia; growth retardation	Polycythemia
Cr	Glucose intolerance	Gastroenteritis; dermatitis
Cu	Menkes' disease (early death)[a]	Neurological disturbance; liver necrosis; Wilson's disease[a]
Fe	Anaemia	Liver necrosis; heart failure
Hg	Non-essential element	Kidney failure; mental disturbance; coma; death
Mn	Skeletal abnormalities; connective tissue defects	Neurological and mental disturbance; ataxia
Pb	Non-essential element(?)	Kidney failure; encephalopathy; neurological disturbance
Zn	Gonadal failure; stunted growth	Metal fume fever; kidney failure

[a] Inherited disease.

be specifically synthesized for medical use was 2,3-dimercaptopropanol developed during the Second World War as an antidote for lewisite:

$$
\begin{array}{c}
\text{H} \quad \text{H} \quad \text{H} \\
| \quad\; | \quad\; | \\
\text{H}-\text{C}-\text{C}-\text{C}-\text{H} \\
| \quad\; | \quad\; | \\
\text{SH} \;\; \text{SH} \;\; \text{OH}
\end{array}
\qquad\qquad
\begin{array}{c}
\text{Cl} \qquad\qquad\quad \text{H} \\
\diagdown \qquad\qquad \diagup \\
\text{As}-\text{CH}=\text{C} \\
\diagup \qquad\qquad \diagdown \\
\text{Cl} \qquad\qquad\quad \text{Cl}
\end{array}
$$

2,3-dimercaptopropanol   lewisite
(British Anti-Lewisite; BAL)   (poison gas)

As is to be expected from its dithiol structure, BAL has a most unpleasant and penetrating smell; on the other hand it is these very —SH groups which allow it to sequester a range of metals in the body, including Hg, Pb, Au, Sb and Bi. Cadmium might be considered a surprising omission from this list considering the avidity with which it binds to the polythiol protein metallothionein (page 469); indeed, the cadmium–BAL complex is formed but it causes massive damage on passing through the kidneys.

Once in the body, the toxic elements rapidly disperse to a variety of sites, including cells and bone. Although BAL itself is exceedingly toxic and has to be administered with care, its great advantage is that, by being uncharged, it is able to pass through membranes and complex with the target metal inside the cells.

Ethylenediamine tetraacetic acid (EDTA), a common chelating agent in analytical chemistry, was first used in 1952 to save a child from death by lead poisoning; since that time amine-polyacids have been employed regularly against a wide spectrum of metals. The very range of metals which EDTA will chelate with raises problems for the clinician because essential elements like calcium and zinc can be leached from

$$\begin{array}{c}
\text{HOOCCH}_2 \diagdown \qquad \qquad \diagup \text{CH}_2\text{COOH} \\
\text{NCH}_2\text{CH}_2\text{N} \\
\text{HOOCCH}_2 \diagup \qquad \qquad \diagdown \text{CH}_2\text{COOH}
\end{array}$$

ethylenediamine tetraacetic acid (EDTA)

body stores during prolonged administration. One way to avoid this is to use the calcium chelate $Na_2CaEDTA$ and rely on the higher stability of the lead complex, for example, to drive the reaction

$$Na_2CaEDTA + Pb^{2+} \rightarrow Na_2PbEDTA + Ca^{2+}$$

to the right; the majority of the calcium drug remaining unaffected is excreted in the urine along with the lead chelate. Lead poisoning is surprisingly common in children, especially those from the poorer areas of large cities. Young children are renowned for picking up tiny items and eating them, a habit known medically as 'pica' from the Greek word for jackdaw; in slum housing the old flaking paint often contains lead (now banned in domestic paints), sufficient being present in a finger-nail-sized flake to kill a small child should all the metal enter the bloodstream.

Young children are also at risk from eating their mothers' iron(II) sulphate tablets which, tragically, are often highly coloured like chocolate beans. The body has no general mechanism for excreting absorbed iron, a legacy remaining from nature's attempt to conserve this metal in all organisms because its highly insoluble sources make it relatively unavailable. The iron absorbed before stomach washing can be completed may be removed using the highly iron-specific desferrioxamine ligands synthesized by micro-organisms:

$$\begin{array}{c}
\text{HO} \diagdown \quad \text{O} \qquad\qquad \text{HO} \diagdown \quad \text{O} \\
\text{N--C} \qquad\qquad\qquad \text{N--C} \\
\text{H}_2\text{N--(CH}_2)_5 \quad (\text{CH}_2)_2 \quad (\text{CH}_2)_5 \quad (\text{CH}_2)_2 \quad (\text{CH}_2)_5 \quad \text{CH}_3 \\
\text{N--C} \qquad\qquad \text{N--C} \qquad\qquad \text{N--C} \\
\text{HO} \quad\quad \text{O} \qquad\quad \text{HO} \quad\quad \text{O} \qquad\quad \text{HO} \quad\quad \text{O}
\end{array}$$

desferrioxamine

Some patients with prolonged anaemias require regular blood transfusions, causing a gradual build-up of body iron (about 3 g per year) which has to be controlled with chelation therapy using desferrioxamine.

Wilson's disease is an inherited disorder in which high concentrations of copper build up in several organs, particularly the liver and brain; if uncontrolled, death usually occurs from liver failure during the early teens. Originally treatment involved painful and nauseating injections of BAL, but happily this has now been superseded by lifelong oral doses of either penicillamine or trien, both of which promote excretion of copper in the urine:

$$\begin{array}{c}
\text{CH}_3 \ \text{H} \\
| \quad | \\
\text{H}_3\text{C--C--C--COOH} \qquad\qquad \text{H}_2\text{NCH}_2\text{CH}_2\text{NCH}_2\text{CH}_2\text{NCH}_2\text{CH}_2\text{NH}_2 \\
| \quad | \\
\text{SH} \ \text{NH}_2
\end{array}$$

penicillamine
(2-mercaptovaline)

triethylenetetramine
(trien)

SELECTED READING

General books

Metals in Biochemistry. P. M. Harrison & R. J. Hoare (Chapman and Hall, 1980)

The Inorganic Chemistry of Biological Processes. 2nd edn. M. N. Hughes (Wiley, 1981)

Biochemistry. 2nd edn. L. Stryer (W. H. Freeman Company, 1981); strongly recommended for a stimulating introduction to biochemistry

Bio-inorganic Chemistry. R. W. Hay (Ellis Horwood, 1984)

Selective Toxicity. 7th edn. A. Albert (Chapman and Hall, 1985)

Articles, reviews and monographs (in approximate order of coverage)

Trace elements in Human and Animal Nutrition. 4th edn. E. J. Underwood (Academic Press, 1977)

Mineral cycles. E. S. Deevey. *Sci. Amer.* (September 1970) 149

New perspective on the essential trace elements. E. Frieden. *J. Chem. Educ.* **62** (1985) 917

The chemical elements of life. E. Frieden. *Sci. Amer.* (June 1972) 52

Non-covalent interactions. E. Frieden. *J. Chem. Educ.* **52** (1975) 754

The water cycle. H. L. Penman. *Sci. Amer.* (September 1970) 99

The carbon cycle. B. Bolin. *Sci. Amer.* (September 1970) 124

The nitrogen cycle. C. C. Delwiche. *Sci. Amer.* (September 1970) 136

The oxygen cycle. P. Cloud & A. Gibor. *Sci. Amer.* (September 1970) 110

How cells make ATP. P. C. Hinkle & R. E. McCarty. *Sci. Amer.* (March 1978) 104

The molecules of the cell membrane. M. S. Bretscher. *Sci. Amer.* (October 1985) 86

The structure of proteins in biological membranes. N. Unwin & R. Henderson. *Sci. Amer.* (February 1984) 56

How receptors bring proteins and particles into cells. A. Doutry-Varsat & H. F. Lodish. *Sci. Amer.* (May 1984) 48

Phospholipids and proteins in biological membranes. J. Seelig & P. M. Macdonald. *Acc. Chem. Res.* **20** (1987) 221

The role of divalent cations in the mechanism of enzyme-catalysed phosphoryl and nucleotidyl transfer reactions (includes a discussion of Na^+-K^+ ATPase). A. S. Mildvan & C. M. Grisham. *Struct. Bonding* **20** (1974) 1

Across the living barrier. D. E. Fenton. *Chem. Soc. Rev.* **6** (1977) 325

Coordination chemistry of alkali and alkaline-earth cations with macrocyclic ligands. B. Dietrich. *J. Chem. Educ.* **62** (1985) 955

The calcium signal. E. Carafoli & J. T. Penniston. *Sci. Amer.* (November 1985) 50

Blood coagulation. C. M. Jackson & Y. Nemerson. *Ann. Rev. Biochem.* **49** (1980) 765

The proton switch of muscle contraction. C. Cohen. *Sci. Amer.* (November 1975) 36

Structure and catalysis of enzymes. W. N. Lipscomb. *Ann. Rev. Biochem.* **52** (1983) 17

Zinc enzymes. I. Bertini, C. Luchinat, & R. Monnanni. *J. Chem. Educ.* **62** (1985) 924

Carbonic anhydrase. I. Bertini, C. Luchinat, & A. Scozzafava. *Struct. Bonding* **48** (1982) 45

Biochemistry of sulphur-containing amino acids. A. J. L. Cooper. *Ann. Rev. Biochem.* **52** (1983) 187

Proteins containing 4Fe–4S clusters: an overview. W. V. Sweeney & J. C. Rabinowitz. *Ann. Rev. Biochem.* **49** (1980) 139

Ferritin. E. C. Theil. *Ann. Rev. Biochem.* **56** (1987) 289

Haemoglobin, R. E. Dickerson & I. Geis (Benjamin Cummings, 1983)

Structure and refinement of oxymyoglobin at 1.6 Å resolution. S. E. V. Phillips. *J. Mol. Biol.* **142** (1980) 531

The biochemistry of some iron porphyrin complexes. C. J. Rix. *J. Chem. Educ.* **59** (1982) 389

Key elements of the chemistry of cytochrome P-450. J. T. Groves. *J. Chem. Educ.* **62** (1985) 928

The mitochondrial electron transport and oxidative phosphorylation system. Y. Hatefi. *Ann. Rev. Biochem.* **54** (1985) 1015

Biochemistry of Nonheme Iron. A. Bezkorovainy (Plenum Press)

Biochemical catalysis involving coenzyme B_{12}. E. J. Corey, N. J. Cooper, & M. L. H. Green. *Proc. Natl Acad. Sci., USA* **74** (1977) 811

Mechanism for the biomethylation of metals and metalloids. J. M. Wood *et al. Fed. Proc.* **37** (1978) 16

Copper in Biology and Medicine (series in five volumes). C. A. Owen (Noyes Publications)

Bioinorganic chemical modelling of dioxygen-activating copper proteins. K. D. Karlin & Y. Gultneh. *J. Chem. Educ.* **62** (1985) 983

Stuctural studies of the hemocyanin active site. T. G. Spiro *et al. J. Amer. Chem. Soc.* **102** (1980) 4210, 4217

Molybdenum enzymes, cofactors and model systems. S. J. N. Burgmayer & E. I. Stiefel. *J. Chem. Educ.* **62** (1985) 943

Structure and function of nitrogenase. L. E. Mortensen & R. N. F. Thorneley. *Ann. Rev. Biochem.* **48** (1979) 387; see also *Phil. Trans. R. Soc., London* **317B** (1987) 131

Molybdenum in nitrogenase. V. K. Shah, R. A. Ugalde, J. Imperial, & W. J. Brill. *Ann. Rev. Biochem.* **53** (1984) 231

Biochemistry of the Essential Ultratrace Elements. E. Frieden (ed.) (Plenum Press) N.Y. 1984

Appendix

Crystal Structures

Simple compounds of stoichiometries MX, MX_2 and M_2X normally adopt one, or more, of a small number of crystal types. In this appendix each of the more common crystal structures is described very briefly. Various important aspects of the same structure can be emphasized using different projections and several of these are given in each case to aid a full appreciation, particularly of the atomic coordination. If, like myself, the reader is unable to 'think in three dimensions', the model kit described in the latter half of the appendix should prove invaluable. Professional chemists constantly use models of various types in their research; students, on the other hand, are often denied the help afforded by such aids owing to either high cost or, worse still, examination regulations.

Table 62—Compounds adopting the more simple crystal structures

Stoichiometry and crystal type	Coordination	Examples
MX: sodium chloride	6:6	Under ambient conditions, all the alkali metal halides *except* CsCl, CsBr and CsI; aobve 445 °C, CsCl; above their transition temperatures NH_4Cl (184.3 °C), NH_4Br (137.8 °C) and NH_4I (-17.6 °C); alkali metal hydrides, MH; oxides, sulphides, selenides and tellurides of Ca, Sr and Ba; MgO, MgS and MgSe; AgF, AgCl and AgBr; MnS, PbS, EuS and CdO; oxides of divalent first-row transition metals

Table 62—*continued*

Stoichiometry and crystal type	Coordination	Examples
MX: caesium chloride	8:8	CsCl, CsBr and CsI; below their transition temperatures: NH$_4$Cl (184.7 °C), NH$_4$Br (137.8 °C) and NH$_4$I (−17.6 °C); TlCl, TlBr and TlI (red form); for a given halide the caesium chloride structure is more dense than that of sodium chloride: hence on the application of about 5000 atm pressure RbCl, RbBr and RbI change to the CsCl structure
MX: zinc blende (very similar to wurtzite structure; often both forms adopted by the same compound)	4:4	ZnS, ZnSe and ZnTe; CdS, CdSe and CdTe; HgS, HgSe and HgTe; BeS, BeSe and BeTe; AgI, MnS and SiC; below their transition temperatures: CuCl (407 °C) CuBr (385 °C) and CuI (369 °C)
MX: wurtzite	4:4	BeO, ZnO, ZnS, ZnTe, CdS, CdSe, MnS, AgI, NH$_4$F and MgTe; above their transition temperatures: CuCl (407 °C), CuBr (385 °C) and CuI (369 °C)
MX$_2$: fluorite*	8:4	CaF$_2$, SrF$_2$, SrCl$_2$, BaF$_2$ and BaCl$_2$; HgF$_2$, PbF$_2$ and PdF$_2$; many dioxides: ThO$_2$, CeO$_2$, HfO$_2$, NpO$_2$, PuO$_2$, ThO$_2$, ZrO$_2$, AmO$_2$, PoO$_2$, CmO$_2$, PrO$_2$ and UO$_2$
M$_2$X: antifluorite	4:8	Li$_2$S, Na$_2$S, K$_2$S and Rb$_2$S; Li$_2$Se, Na$_2$Se and K$_2$Se; Li$_2$Te, Na$_2$Te and K$_2$Te; Li$_2$O, Na$_2$O, K$_2$O and Rb$_2$O
MX$_2$: rutile^a	6:3	MgF$_2$, ZnF$_2$, NiF$_2$, CoF$_2$, FeF$_2$ and MnF$_2$; many dioxides: TiO$_2$, GeO$_2$, SnO$_2$, MnO$_2$, RuO$_2$, OsO$_2$, IrO$_2$, CrO$_2$ and TeO$_2$
MX$_2$: cadmium iodide	6:3	MgBr$_2$, MgI$_2$, CaI$_2$, ZnI$_2$, CdBr$_2$, CdI$_2$ and PbI$_2$; TiI$_2$, MnI$_2$, FeI$_2$, CoI$_2$ and YbI$_2$; FeBr$_2$, CoBr$_2$ and NiBr$_2$; TiCl$_2$ and VCl$_2$; SnS$_2$, TiS$_2$, ZrS$_2$, PtS$_2$ and TaS$_2$
MX$_2$: cadmium chloride	6:3	MgCl$_2$, CdCl$_2$, MnCl$_2$, FeCl$_2$, CoCl$_2$ and NiCl$_2$; ZnBr$_2$, NiI$_2$ and PbI$_2$; NbS$_2$ and TaS$_2$

^{a} Rather few halides other than fluorides have the fluorite or rutile structures; usually those fluorides with small cations adopt the rutile structure.

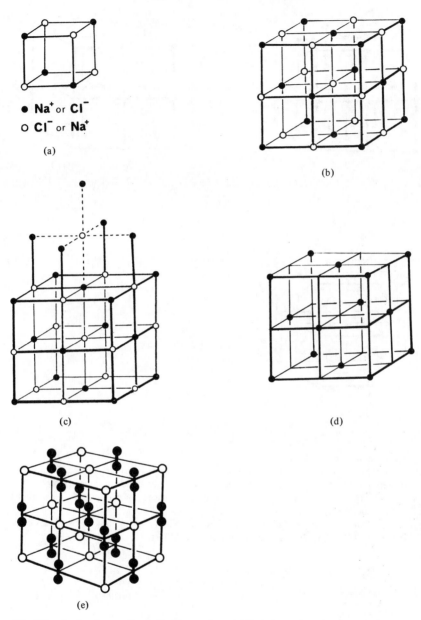

$\bullet$ **Na$^+$ or Cl$^-$**

$\circ$ **Cl$^-$ or Na$^+$**

(a)

(b)

(c)

(d)

(e)

Fig. 149—Rock salt or sodium chloride structure. (a) Cubelet to show the alternate arrangement of ions at the cube corners. (b) Eight cubelets stacked to give the unit cell of sodium chloride. (c) Extension of (b) to show the octahedral coordination of the ions; thus the sodium chloride structure is said to have 6:6 coordination. (d) Face-centred cubic arrangement of *identical* ions. The NaCl structure is made up of two interpenetrating face-centred cubic lattices, one of cations, the other of anions. The structure can also be described as a cubic close-packed array of anions with the sodium cations occupying all the octahedral holes. (Owing to the rather large size of the Na$^+$ ions the anions are not in mutual contact as are the spheres of a normal close-packed structure.) (e) Calcium carbide, CaC$_2$, has an elongated version of the sodium chloride structure in which the C$_2^{2-}$ ions are aligned parallel to each other and take the place of the Cl$^-$ ions. The alkali superoxides MO$_2$ have a similar structure.

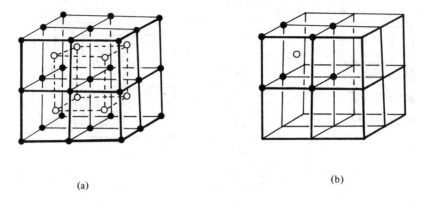

(a)

(b)

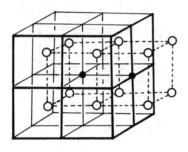

(c)

Fig. 150—Caesium chloride structure.

(a), (b) A cubelet with identical ions (cations or anions) at each corner is the basic entity in the caesium chloride structure. Eight such cubelets, with an interpentrating cubelet of oppositely charged ions, is shown in (a). Each cubelet has a centrally placed ion of opposite charge, giving both ions eigh-coordination.

(c) An extension of the array of open circles shows the eight-coordination of two black ions. The structure has 8:8 coordination giving the required MX stoichiometry.

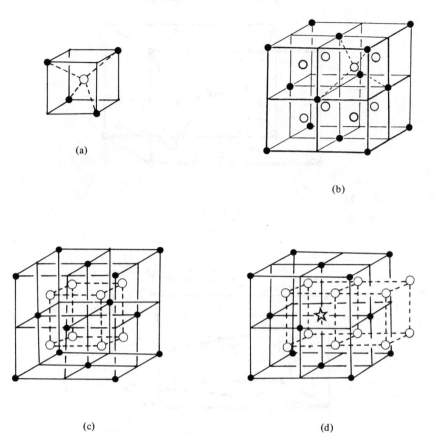

(a)

(b)

(c)

(d)

Fig. 151—Calcium fluoride (fluorite) structure.

(a) This structure is closely related to that of caesium chloride. If half the cations in the CsCl lattice are removed to leave a face-centred cubic array of Ca^{2+} ions, the required MX_2 stoichiometry is achieved if F^- ions replace all the chlorides.

(b) Because there is no cation at the body centre of the unit cell in a face-centred cubic lattice, only *alternate* cubelets of fluoride ions contain a central caclium ion. The 'missing' calcium cation is shown by a star in one of the two cubelet arrays of F^- ions in (d).

(c) The fluoride ions are tetrahedrally surrounded by four cations giving the fluorite structures 8:4 coordination. The structure may be described as a cubic close-packed array of calcium ions with fluoride ions occupying all the tetrahedral holes.

(d) Several compounds having the M_2X stoichiometry, e.g. the alkali metal monoxides M_2O, adopt the 'antifluorite' structure, in which the doubly charged anions replace the calcium cation and the singly charged cations replace the fluoride anions in the normal fluorite lattice.

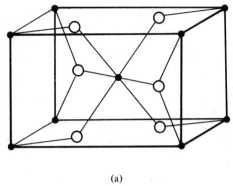

○ Ti

● O

(a)

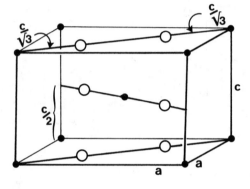

(b)

Fig. 152—Rutile (TiO_2) structure.

(a) Two sides of the unit cell are of length a with the third one, c, being about 30% shorter. The titanium atoms have octahedral and the oxygens trigonal planar coordination, giving a 6:3 structure in keeping with the TiO_2 stoichiometry. It is not possible for the ions to have perfectly regular coordination in this arrangement and hence slight distortions occur.

(b) Assuming regular coordination polyhedra around the ions, two oxygens lie at $c/2$ on the body diagonal and $c/\sqrt{3}$ (the length of the Ti—O 'bond') from the central titanium atom. The other four oxygens are on face diagonals $c/\sqrt{3}$ from their respective corners (i.e. from one of their coordinated titanium atoms).

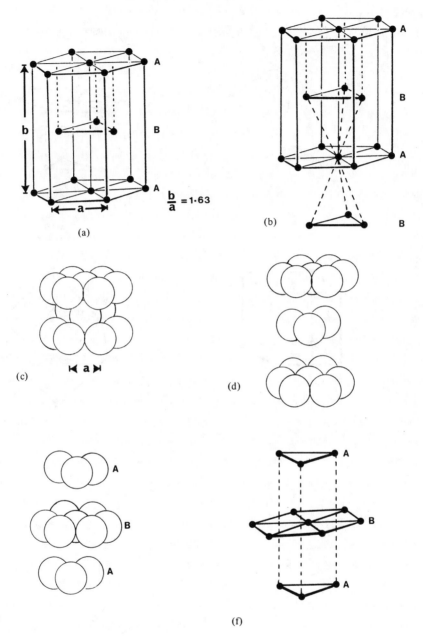

Fig. 153—Hexagonal close packing.

(a) View of a hexagonal close-packed structure showing the ABABAB ... layer sequence and the relative positions of the atomic nucleii; dimension a is the diameter of the spherical atoms making up the structure shown here as the sum of two radii.

(b) Demonstration of 12-coordination for the atoms in an HCP structure.

(c) Close packing of spheres in an HCP structure.

(d) Exploded view of (c).

(e), (f) Demonstration of 3:6:3 coordination in an HCP structure; compare the orientations of the A layer spheres in (e) with those of the A and C layers in a cubic close-packed structure (Fig. 154).

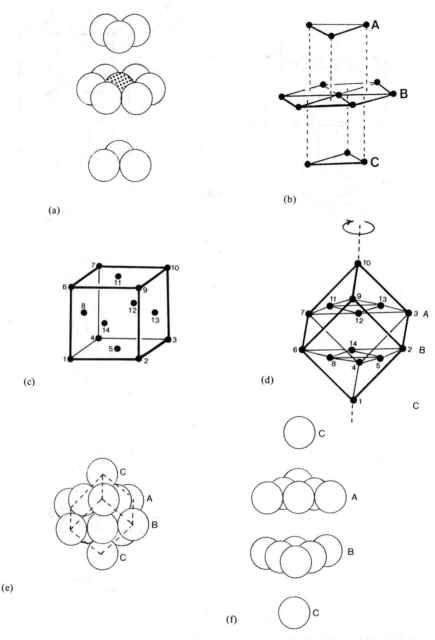

Fig. 154—Cubic close packing. (a), (b) The ABC arrangement of layers is shown for the cubic close-packed structure. Note the 3:6:3 12-coordination of the shaded sphere and compare the relative orientations of the three spheres in layers A and C to those in the hexagonal close-packed structure. (c) The face-centred cubic arrangement is another way of drawing the cubic close-packed structure. (d) The cube in (c) suspended by corner number 10 to demonstrate the ABC layers of cubic close packing. (e)Cubic close packing of spheres; note the outline of the face-centred cube superimposed on the diagram. (f) Exploded view of (e); compare to (d). Hexagonal and cubic close-packing arrangements represent the most space-economizing way of packing rigid spheres and both fill 74% of the available space. For n spheres in close-packed array there are $2n$ tetrahedral and n octahedral holes (see Fig. 155).

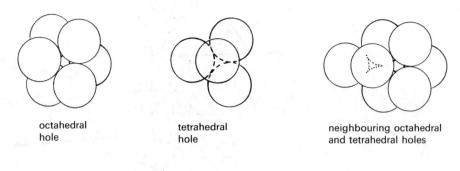

octahedral
hole

tetrahedral
hole

neighbouring octahedral
and tetrahedral holes

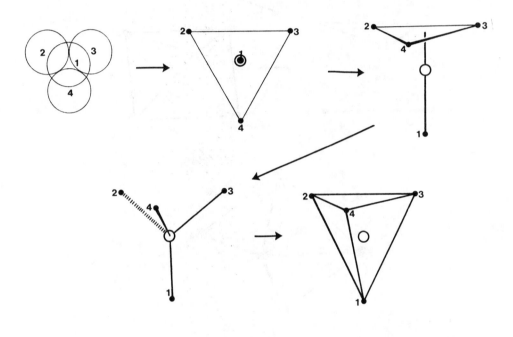

(a)

Fig. 155—Tetrahedral and octahedral holes in close-packed structures. (a) A tetrahedral hole is formed in a close-packed structure when one sphere (1) fits into the hollow formed by three neighbouring spheres (2, 3, 4). As shown in the diagrams, the hole gets its name from the arrangement of the four nucleii (sphere centres) which define a tetrahedron. The hole, which itself is not tetrahedral in shape since it has curved faces, can just accommodate a sphere of radius $0.225a$ (where a is the radius of the close-packed spheres) without distortion of the structure. For every n spheres in a close-packed structure there are $2n$ tetrahedral holes.

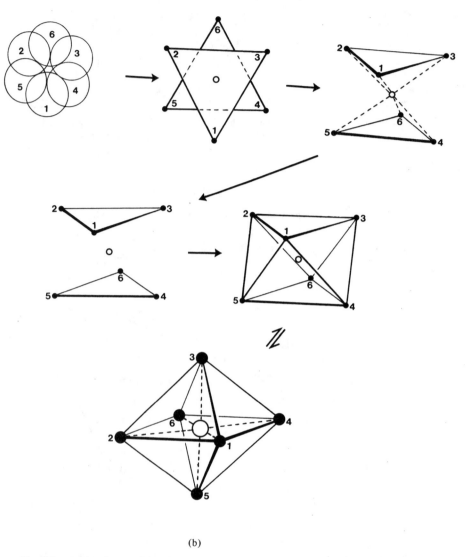

(b)

Fig. 155—*continued*

(b) Neighbouring hollows in a close-packed layer cannot, for steric reasons, each accommodate a sphere from the layer above. The hollow next to a tetrahedral hole is surrounded by a total of six spheres and, as shown in the diagrams, the centres (nuclei) of these six spheres describe an octahedron. The octahedral hole so formed can enclose a sphere of radius 0.414a whilst the larger spheres just remain in contact. There are n octahedral holes in a close-packed array of n spheres, which is only half the number of tetrahedral holes.

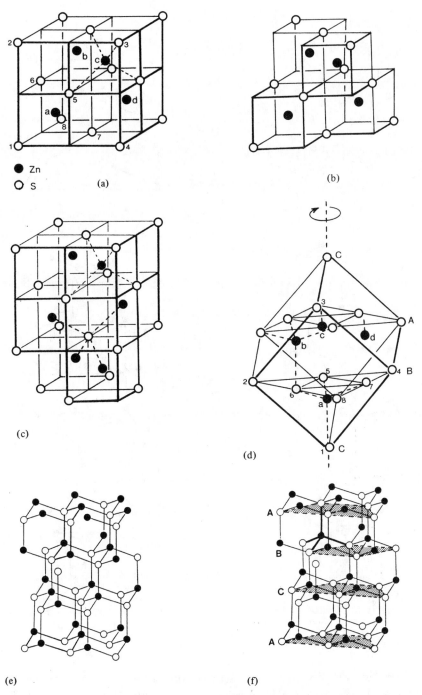

Zn ●
S ○

(a)

(b)

(c)

(d)

(e)

(f)

Fig. 156—Zinc blende (ZnS) structure. (a), (b) The sulphur atoms in zinc blende are in a cubic close-packed array (equivalent to face-centred cubic as shown in Fig. 154) with the zinc atoms occupying alternate tetrahedral holes. Since for n spheres in a cubic close-packed structure there are $2n$ tetrahedral holes, the desired 1:1 stoichiometry is achieved by having zinc atoms in half of them. (c) Demonstration of tetrahedral coordination of both Zn and S; i.e. the coordination in zinc blende is 4:4. (d) The cube in (a) is suspended by a corner to show the ABC layers of cubic close packing. (e) Another representation of the zinc blende structure; note the similarity to a diamond lattice. (f) Layers of sulphur atoms in (e) are shaded to emphasize the ABC sequence.

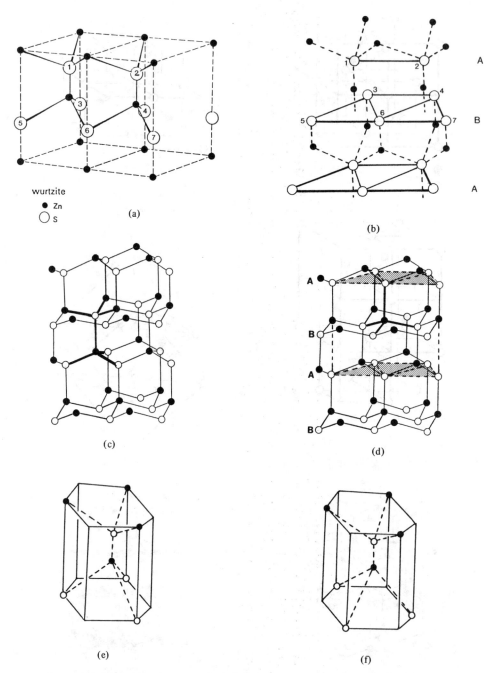

wurtzite

● Zn
○ S

(a)

(b)

(c)

(d)

(e)

(f)

Fig. 157—Wurtzite (ZnS) structure. (a) Simplified view of the 4:4 atomic coordination in wurtzite. (b) An extension of (a) to emphasize the ABAB ... layer sequence of sulphur atoms. In wurtzite the sulphurs are in a hexagonal close-packed array with zinc atoms occupying half of the tetrahedral holes to give the required 1:1 stoichiometry. (c) A different view of the wurtzite structure. (d) The layers in (c) shaded to demonstrate the ABAB ... sequence. (e), (f) The essential difference in relative orientation of the atoms in wurtzite (e) and zince blende (f).

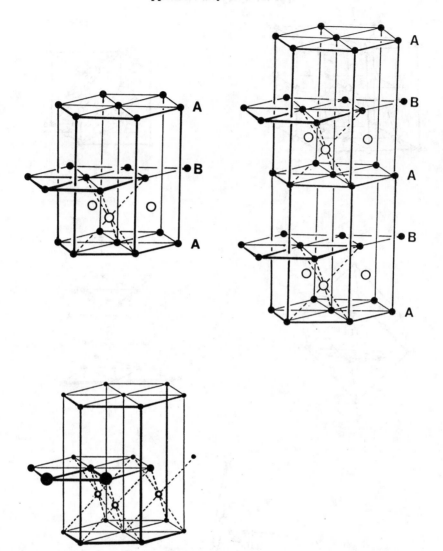

Fig. 158(a)—Cadmium iodide structure. This structure has a hexagonal close-packed array of iodines with the cadmium atoms occupying *half* the octahedral holes thus giving the required MX$_2$ stoichiometry. Only *alternate* layers of octahedral holes (i.e. those between every second two layers of iodine atoms) have cadmiums present. The cadmium atoms are octahedrally surrounded by six iodine neighbours; the iodine atoms have all three cadmium neighbours on the same side, giving them pyramidal coordination. Since adjacent atoms in the neighbouring unoccupied layers are identical (i.e. iodine), the forces between the layers are weak, being of the van der Waals type. Not unexpectedly, it is found that the crystals show a preference to cleavage *parallel* to the layers. In the very similar cadmium chloride structure, the chlorine atoms are in a cubic close-packed array with half the octahedral holes again occupied by cadmium atoms. In cadmium iodide the *metal layers* are arranged so that the Cd atoms of one layer lie directly above those in the next layer; in the cadmium chloride structure each metal atom is above (and below) a chlorine atom in the *next but one* layer of chlorines (note the relationship of the cadmium labelled a in Fig. 158(b), which is at the cube centre, and the chlorine atoms 1 and 10).

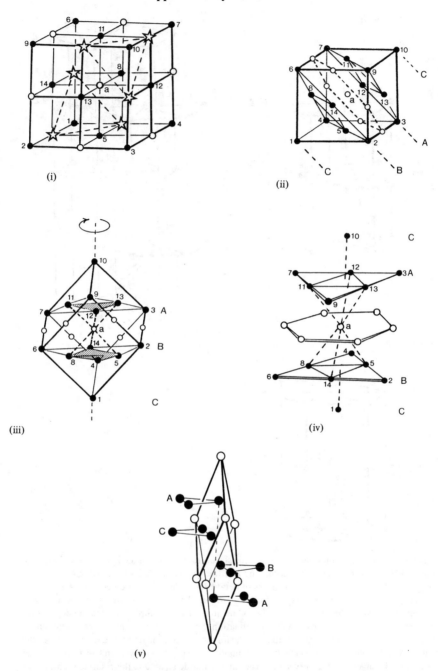

(i)

(ii)

(iii)

(iv)

(v)

Fig. 158(b)—Cadmium chloride structure. (i) Sodium-chloride-like cube with cadmium atoms in each alternate layer removed; the missing atom positions are shown by stars. (ii), (iii), (iv) Three views of the face-centred cubic (i.e. CCP) array of chloride atoms with one set of octahedral holes between layers A and B filled by Cd atoms; note the six-coordination of cadmium atom a in (i), (ii) and (iii). (v) Rhombohedral cell of cadmium chloride.

CRYSTAL STRUCTURE KIT

The apparatus used to build the crystal lattices consists of a wooden block drilled on both sides with a specific pattern of holes, plastic beads of diameter 10 mm and aluminium alloy rods on which the beads are supported. A structure is constructed by placing rods in the appropriate holes and adding the beads to them in sequence. For some structures, 'spacers' are required; these are made from pieces of plastic tubing which will slide comfortably over the rods. Instead of aluminium welding rod it is possible to use either cocktail sticks or 'boiling sticks' (widely used to prevent delayed boiling in organic solvents). Table 63 and the full-scale diagrams below show the position and numbering of the holes to be drilled in a hardwood block. The slight disadvantage of this kit, which should be recognized, is that all the 'ions' and 'atoms' are of identical size; however, with the given hole patterns *all* the simple structures can be built using a single block.

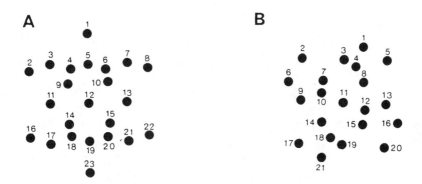

Table 63—Coordinates (mm) of the holes to be drilled for the structure kit

Side A		Side B	
x	y	x	y
+5.5	+5.5	−11.5	0
+16	+9	−6	+2
+5	+9	−15	+5
0	+18.5	−5.5	+5.5
+10.5	+10.5	−11.5	+11.5
−10.5	+10.5	0	+11.5
0	+10.5	0	0
−16	+9	+3	+9.5
−5	+9	+5	+15
−5.5	+5.5	+5.5	+5.5
−10.5	0	+11.5	0
0	0	+11.5	+11.5
+10.5	0	+6	−2
+5.5	−5.5	+5.5	−5.5
+5	−9	+15	−5
+16	−9	+11.5	−11.5
+10.5	−10.5	0	−11.5
0	−10.5	−3	−9.5
−10.5	−10.5	−5	−15
0	−18.5	−5.5	−5.5
−5	−9	−11.5	−11.5
−16	−9		
−5.5	−5.5		
(23 holes)		(21 holes)	

Sodium chloride

Place nine pegs in holes 3, 5, 7, 11, 12, 13, 17, 19, 21 of side A to give the pattern:

○ ○ ○
○ ○ ○   W = white bead (Na or Cl)
○ ○ ○   B = black bead (Cl or Na)

Row 1   W B W
               B W B
               W B W

Row 2   B W B
               W B W
               B W B

Repeat rows 1, 2, 1, 2, etc. to build up the structure.

(a)   (b)   (c)

(a) Unit cell based on the simple cube.
(b) Stacking of several unit cells together showing the face-centred cubic array of both the anions and the cations.
(c) Each sphere is surrounded octahedrally by six neighbours of opposite charge; this is illustrated by the top black bead which has its six white neighbours clearly visible.

Calcium carbide, CaC_2

Place nine pegs in holes 3, 5, 7, 11, 12, 13, 17, 19, 21 of side A to give the pattern:

○	○	○
○	○	○
○	○	○

W = white bead (Ca^{2+})

BB = two black beads ($C\equiv C^{2-}$)

S = spacer (5 mm long)

Row 1

BB SW BB

SW BB SW

BB SW BB

Row 2

W   BB   W

BB   W   BB

W   BB   W

Row 3

BB BB W

W   W   BB

BB   W   BB

Note: (1) The similarity of the structure to that of NaCl. (2) The parallel allignment of the C_2^{2-} ions causes a lengthening of the unit cell along the vertical axis, thus destroying the cubic symmetry of the NaCl structure. (3) That the stoichiometry is different to that in NaCl due to the C atoms being coupled in pairs; the same structure is adopted by the alkali metal superoxides MO_2.

Caesium chloride

Place thirteen pegs in holes 2, 3, 5, 7, 8, 9, 11, 13, 14, 15, 17, 19, 20 of side **B** to give the pattern:

```
   O     O     O
      O     O              W = white bead (Cs or Cl)
   O     O     O           B = black bead (Cl or Cs)
      O     O              E = position empty
   O     O     O
```

Row 1
```
        W     W     W
           E     E
        W     W     W
           E     E
        W     W     W
```

Row 2
```
        E     E     E
           B     B
        E     E     E
           B     B
        E     E     E
```

Repeat rows 1, 2, 1, 2, 1, 2 etc. to build up the structure.

Unit cell showing the eight-coordination of the black sphere by white spheres at the corners of a cube.

Stacking of several unit cells together.

Some of the white spheres have been removed and the structure extended to show that the coordination of the white and black spheres is identical. The lattice is made up of two interlocking cubic systems.

Note: (1) The array of eight white beads, at the corners of a cube, around each black bead. (2) The eight-coordination of white beads. This can be seen most clearly by building rows 1 and 2 normally but adding the central white bead on peg 10 *only* when putting in the third row; the fourth row is a normal row 2. Thus the structure can be seen to be two interlocking arrays of simple cubes, one of white beads and one of black.

Fluorite (CaF₂) structure

Place thirteen pegs in holes 2, 3, 5, 7, 8, 9, 11, 13, 14, 15, 17, 19, 20 of side B to give the pattern:

W = white bead

B = black bead

E = position empty

Row 1   W   E   W

                   E   E

          E   W   E

                   E   E

          W   E   W

Row 2   E   E   E

                   B   B

          E   E   E

                   B   B

          E   E   E

Row 3   E   W   E

                   E   E

          W   E   W

                   E   E

          E   W   E

Row 4 is identical to row 2.
Row 5 is identical to row 1.

Unit cell showing the face-centred cubic arrangement of the white beads (Ca^{2+} ions) and the tetrahedral coordination of the black beads (F^- ions).

Structure extended to show the coordination of the top white bead (Ca^{2+}) by eight black beads (F^-) arranged at the corners of a cube. Only alternate black cubes contain a white sphere.

Note how the coordination numbers of the ions are in agreement with the stoichiometry of the compound.

Rutile, TiO_2

Place nine pegs in holes 1, 4, 6, 10, 11, 12, 16, 18, 21 of side B to give the pattern:

```
            O 6
            O                   W = white bead (Ti)
                                B = black bead (O)
    OO   O   OO 1
            O                   E = position empty
                                S = spacer (6 mm long)
            O
```

Row 1
```
                    W
                    B
            WE    E    EW
                    B
                    W
```

Row 2
```
                    S
                    E
            EB   W   BE
                    E
                    S
```

To complete the unit cell, construct rows 1, 2, 1; to build up the structure further the row sequence is 1, 2, 1, 2, 1, 2, etc.

Note: (1) The six-coordination of the central Ti atom in the unit cell. (2) The planar three-coordination of the O atoms. (3) That the unit cell is not a cube; the *c* axis is about 30% shorter than the *a* and *b* axes.

Zinc blende (ZnS)

Place thirteen pegs in holes 2, 3, 5, 7, 8, 9, 11, 13, 14, 15, 17, 19, 20 of side B to give the pattern:

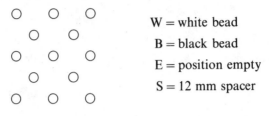

W = white bead
B = black bead
E = position empty
S = 12 mm spacer

Row 1

        W   E   W
            E   E
        E   W   E
            E   E
        W   E   W

Row 2

        E   E   E
            B   E
        E   E   E
            E   B
        E   E   E

Row 3

        E   W   E
            E   E
        W   E   W
            E   E
        E   W   E

Row 4

        S   E   E
            E   B
        E   E   E
            B   E
        E   E   S

Row 5 is identical to row 1.

Unit cell showing the face-centred cubic arrangement of the white spheres (S). The black spheres (Zn) occupy alternate tetrahedral sites. Structure is similar to that of fluorite, CaF_2, except that the Zn atoms occupy only half of the fluoride ion positions.

One of the corner beads removed to show the close-packed arrangement of the white spheres in the diagonal plane (structure is cubic close-packed with respect to white spheres, see page 527). Note the puckered six-membered (3 white, 3 black spheres) ring in the centre of the structure. The structure is closely related to that of diamond.

Body-centred cubic structure

This is the atomic arrangement adopted by about twenty of the metallic elements. It is closely related to the caesium chloride structure, the positions previously occupied by both caesium and chlorine in CsCl now being taken up by atoms of the metal. Place thirteen pegs in holes 2, 3, 5, 7, 8, 9, 11, 13, 14, 15, 17, 19, 20 of side B to give the pattern:

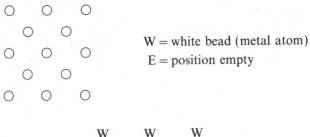

W = white bead (metal atom)
E = position empty

Row 1

	W	W	W
		E		E	
	W	W	W
		E		E	
	W	W	W

Row 2   E   E   E

                                   W   W

                          E   E   E

                                   W   W

                          E   E   E

Repeat rows 1, 2, 1, 2, etc. to build up the structure.

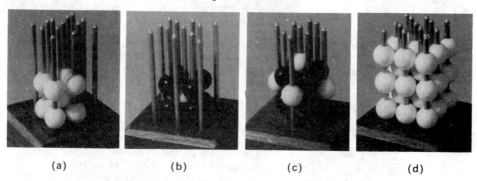

(a)   (b)   (c)   (d)

(a) Eight nearest neighbours of the central white bead arranged at the corners of a cube. (b), (c) The central white sphere (barely visible) is surrounded by eight black spheres as nearest neighbours at the corners of a cube. Six white spheres at the midpoint of the cube faces are the next-nearest neighbours and are very *slightly* more distant than the black spheres. The coordination number of the central sphere is thus 14 (8:6). (d) Several unit cells built up together; notice the similarity of this structure to that of caesium chloride.

Hexagonal and cubic close-packed structures

Place seven pegs in holes 4, 6, 11, 12, 13, 18, 20 of side A to give a hexagonal pattern:

                    O   O   W = white sphere
                 O   O   O   B = black sphere
                      12
                    O   O

Each sphere is coordinated in one plane by six others having their centres at the corners of a regular hexagon:

                    W   W

                  W   B   W

                    W   W

Remove peg 12 and its black sphere. Drop a white bead into the centre of the hexagon to complete the first layer and put three white beads into the space *between* the six pegs so that they rest in hollows in the first layer; these constitute part of the second layer which is the same in both hexagonal and cubic close-packing arrangements. (The layers can be extended laterally by using more pegs in holes 1, 2, 8, 16, 22 and 23.)

Hexagonal close packing: the third layer is identical to the first so that beads are added to each of the six pegs, with one bead again being dropped into the centre of the hexagon to complete the layer. The structure is built up by repeating the layer sequence 1, 2, 1, 2, 1, 2, etc.

Cubic close packing: layers 1 and 2 are built up as described; three white beads are placed *inside* the six pegs to lie in the hollows of the second layer:

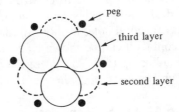

Closer inspection of the structure will show that the third layer of beads lies directly above the *unused* hollows of the first layer. The layers are repeated in the sequence 1, 2, 3, 1, 2, 3, etc. to build up the structure.

Coordination number in HCP and CCP structures

The coordination number of a sphere is 12 in a close-packed structure; six spheres lie in one plane with three more in the plane above and three in the plane below, i.e. a 3:6:3 arrangement. The hexagonal and cubic close-packed structures differ in the orientation of the two sets of three neighbours in the upper and lower layers. In HCP the 1, 2, 1, 2, 1, 2 layer sequence requires that the two sets of three neighbours are superimposed when viewed in plan, whereas in CCP the triangle containing the three neighbours in the top layer is rotated by 60° relative to the three lower neighbours:

hexagonal close packing   cubic close packing

Tetrahedral holes

Where a sphere of one layer lies in the hollow formed by three spheres of the next layer, a cavity is formed. Because the centres of the four spheres making up this cavity lie at the corners of a tetrahedron, the cavity is called a 'tetrahedral hole'. There are *two* tetrahedral holes associated with each sphere.

Octahedral holes

The second layer sits in hollows of the first layer; there is an equivalent set of hollows on which the spheres of the second layer cannot rest because of steric reasons (demonstrate this using layer 1 and two spheres of layer 2).

The first set of hollows result in the tetrahedral holes; the second set of hollows are surrounded by six spheres having their centres at the corners of an octahedron and hence are called 'octahedral holes'. Each sphere is associated with *one* octahedral hole—hence the number of octahedral holes is only half that of the tetrahedral holes.

When the spheres are in contact it can be shown by geometry that the tetrahedral holes can hold a sphere of radius up to 0.225 that of the close-packed spheres; the octahedral holes are larger and can house a sphere of radius 0.414 times that of the close-packed spheres.

Place six pegs in holes 4, 6, 11, 13, 18, 20 and demonstrate:

(a) a tetrahedral hole by placing three beads *inside* the six pegs and putting one bead in the hollow formed by these three

(b) an octahedral hole by placing three beads *inside* the six pegs and placing three more on top of them to form the next layer

(c) that a ball-bearing (diameter approximately 4 mm; diameter of beads 10 mm) will comfortably fit in an octahedral hole whereas the same ball-bearing is far too large to fit in a tetrahedral hole;

(d) that neighbouring hollows in a layer give rise to one tetrahedral and one octahedral hole. A further peg in hole 1 will allow four spheres of layer 1 to be used instead of three as previously (slide the fourth bead *between* pegs 1, 4 and 6). For the second layer, place one bead on peg 4 and one on peg 6; a third rests freely in the remaining hollow (which is directly above hole 12).

Close-packed structures

First layer identical in
HCP and CCP structures.

Second layer identical in HCP
and CCP structures; spheres rest
in half the hollows of first layer.

Third layer in HCP structure
lies directly above the spheres
in first layer.

Third layer in CCP structure lies
above the unused hollows of
first layer.

Hexagonal close packing.
Alternate layers are repeated, giving
rise to the layer sequence ABABAB.

Cubic close-packing. Here the layer sequence is
ABCABCABC; in the right-hand model the layers A
and B are constructed from white beads and
layer C from black.

Twelve-coordination of atoms in close-packed structures

The black sphere is surrounded by six white spheres
in one plane (identical coordination in both
HCP and CCP structures.

In hexagonal close packing the black sphere is also in contact with three white spheres in the plane above and three white spheres in *equivalent* positions in the plane below.

In cubic close packing the coordination is again 3:6:3 but the two equivalent triangles containing the centres of the spheres in the planes above and below the black bead are staggered by 60°.

Holes in close-packed structures

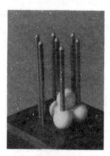

Tetrahedral hole.

Tetrahedral hole and a neighbouring octahedral hole, shown from above.

Octahedral hole containing a sphere of 4 mm diameter; side view.

Octahedral hole containing a sphere of 4 mm diameter; viewed from above.

Face-centred cubic

Face-centred cubic arrangement of spheres, as found for one of the ions in a sodium chloride lattice.

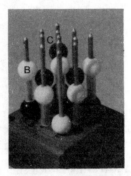
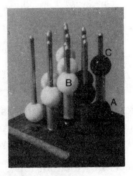

One corner bead removed from the above structure to reveal the 'close-packing' arrangement of the spheres (the spheres do not touch each other in the NaCl-type lattice).

In a face-centred cubic structure the atoms are actually cubic close-packed. Another way of describing NaCl is to assume the chloride ions are in a cubic close-packed array with sodium ions occupying all the octahedral holes.

Li (b)	Be (h)																
Na (b)	Mg (h)	Al															
K (b)	Ca (b,h,c)	Sc (h,c)	Ti (h,b)	V (b)	Cr (b)	Mn	Fe (b,c)	Co (h,c)	Ni (h,c,c)	Cu (c)	Zn (h)	Ga					
Rb (b)	Sr (c,b,h)	Y (h)	Zr (h,b,b)	Nb (b)	Mo (h,b,h)	Tc (h)	Ru (h,c,c)	Rh (c)	Pd (c)	Ag (c)	Cd (h)	In	Sn	Sb			
Cs (b)	Ba (b)		Hf (h,b,b)	Ta (b)	W (b)	Re (h)	Os (h,c,c)	Ir (c)	Pt (c)	Au (c)	Hg	Tl	Pb	Bi	Po		
Fr	Ra		104	105	106	107		109									

Lanthanides: La (h,c), Ce, Pr, Nd, Pm, Sm (b), Eu (h), Gd (h), Tb (h), Dy (h), Ho (h), Er (h), Tm (h), Yb (c), Lu (h)

Actinides: Ac, Th (c), Pa (b), U (b), Np (b,c), Pu (b,c), Am, Cm, Bk, Cf, Es, Fm, Md, No, Lw

Fig. 159—The crystal structures of metals: b, body-centred cubic; c, cubic close-packed; h, hexagonal close-packed.

DEFECTS IN CRYSTALS

Although we draw crystal structures as perfect assemblies of ions or atoms, this only occurs in ideal crystals at absolute zero; in real crystals entropy dictates that defects must occur at all other temperatures. Two types of simple defect have been recognized;

(a) Frenkel defects in which an ion or atom leaves its lattice site and moves to an empty interstitial site, often changing its coordination number as it does so. An example might be a zinc atom in zinc blende moving from a tetrahedral hole to a nearby octahedral hole in the close-packed sulphur array. The defect thus consists of an interstitial atom *and* a vacant site.

(b) Schottky defects arise when a pair of oppositely charged ions move to the surface of the crystal to leave *two* vacancies, one in the cation array and one in the anion array.

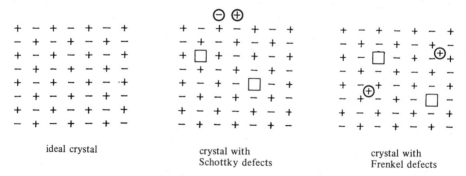

ideal crystal   crystal with Schottky defects   crystal with Frenkel defects

It might appear surprising that atoms and ions can move so freely within a solid crystal, but it has been known since the last century that gold, for example, will diffuse into a lead crystal at 300°C faster than sodium and chloride ions move about in water at room temperature! To illustrate the extent of Schottky defect formation, it has been calculated that in sodium chloride at 500°C about one ion site of each charge per million is vacant, and near the melting point this rises to about one site in ten thousand; at room temperature the lattice may be considered to be quite regular since only one site in 10^{15} is affected.

Index